International Conference on ICT, Society, and Human Beings (ICT2023), International Conference on e-Health (EH2023), International Conference on Connected Smart Cities (CSC2023), and International Conference on Big Data Analytics, Data Mining and Computational Intelligence (BIGDACI2023)

Held at the 17th Multi-Conference on Computer Science and Information Systems (MCCSIS 2023)

Porto, Portugal
15-18 July 2023

Editors:

Piet Kommers
Mário Macedo
Guo Chao Peng
Ajith Abraham

ISBN: 978-1-7138-8468-2

Printed from e-media with permission by:

Curran Associates, Inc.
57 Morehouse Lane
Red Hook, NY 12571

CURRAN ASSOCIATES INC. proceedings.com

Some format issues inherent in the e-media version may also appear in this print version.

International Association for Development of the Information Society (IADIS)
Rua Sao Sebastiao Da Pedreira
No. 100, 30, 1050-209
Lisbon, Portugal

Phone: 351 21 3151373
Fax: 351 21 3151244

secretariat@iadis.org

Additional copies of this publication are available from:

Curran Associates, Inc.
57 Morehouse Lane
Red Hook, NY 12571 USA
Phone: 845-758-0400
Fax: 845-758-2633
Email: curran@proceedings.com
Web: www.proceedings.com

TABLE OF CONTENTS

FULL PAPERS

SHORT PAPERS

REFLECTION

POSTERS

DOCTORAL

FULL PAPERS

CORĒHEALTH: OPERATING CENTER OF TELEMEDICINE FOR CHRONIC DISEASES AND CLINICAL NETWORKS OF APULIA REGION

Giovanni Gorgoni, Pasquale Notarangelo and Vito Petrarolo
AReSS Puglia – Regional Agency for Health and Social Care of Apulia,
Lungomare Nazario Sauro, 33 – 70121 Bari, Italy

ABSTRACT

The COVID-19 emergency has given a great acceleration to the transformation processes of remote health pathways (telemedicine), which Apulia had already begun to pursue with the POR Puglia 2014-2020 from which COReHealth, established at AReSS, derives directly. COReHealth is characterized by: Web-cloud platform with HL7-FHIR interoperability; Televisit, Teleconsultation, Telemonitoring, Telecooperation and Teleconferencing; Device traceability system; Clinical repository; Patient Synoptic; Central Alarms; Mobile App for the Citizens (Android and IoS); Integration with the Regional Health Information System, Electronic health record, etc. The APP promotes the patient's involvement in the care process: keeping in touch with his/her specialist doctor and caregiver (video calls and chat); consult the agenda of the televised programs with patient's care team; view his/her treatment plan; enter the vital parameters that are communicated to the doctor in real time; manage their clinical diary and possibly share it with the care team; facilitate the measurement of compliance with the treatment path (drug intake, lifestyle). Each patient is assigned a treatment plan based on the specific Integrated Care Pathway(s) for their pathologies; the medical team, based on the specific needs, can provide patients with a kit of medical devices (e.g. oximeter, multiparameter, scale, sphygmomanometer, ecg, etc.), interoperable with the control unit, suitable for real-time detection and monitoring of the salient vital parameters that allow constant detection and monitoring, and possible intervention by virtue of the automatic alarm system the control unit is equipped with. From January 2022 the transformation of the Apulian Oncological Network into a digital key was started and therefore the training of the operators of the Oncological Guidance Centers and the Breast Units and the enrollment of about 23,000 cancer patients (approx. Breast). Another fundamental element of the COReHealth is the systematic approach in Continuing Training and Consolidation of Digital Skills to enable informed access of all the actors involved to new digital technologies. And this is why AReSS has launched a virtuous path of training, change management and digital literacy for all users (medical and administrative teams) and patients.

KEYWORDS

Telemedicine, Value Based Care, Change Management, Custom Treatment Plan, Patients Empowerment, Digital Health Innovation

1. INTRODUCTION

When we started planning and designing the application model of digital healthcare in the Apulia Region, expectations were high, but the result that is there for all to see today was absolutely unexpected. The vision and design skills are certainly of enormous importance, but the emergency due to the pandemic has further accelerated, allowing even the most reluctant to perceive how much the use of digital technology in the healthcare world is a primary and now unavoidable objective.

In fact, in recent years, the Apulia Region has undertaken to meet the new and ever-increasing health needs of the Apulian citizens and to adapt the local health services, in compliance with the criteria of quality and sustainability. In this context, one of the fundamental drivers of innovation lies in the propulsion to make healthcare processes more efficient and in the improvement of the diagnostic-therapeutic outcome associated with the use of innovative ICT technologies.

The COVID-19 emergency has given a great boost to the transformation processes of remote health pathways (telemedicine), which Puglia had already begun to pursue with the POR Puglia 2014-2020; the launch of the experimental experiences of using telemedicine platforms in Puglia, therefore, starts from afar, and is based on the initiatives already undertaken previously in the context of the HLCM projects (HLCM Program Agreement - Health Life-Cycle Management - TITLE II) and TALIsMAn (Personalized assistance technologies for improving the quality of life - PON Research and Innovation 2014/2020 and FSC - winner of the Digital Innovation Award in Healthcare 2021), both based on the now tested process of taking charge of the patient of the historic Care Puglia project, and culminating in the H-casa COVID-19 platform and in the Operations Center of Chronicities and Clinical Networks of the Apulia Region (CORēHealth).

2. THE ESTABLISHMENT OF THE CORĒHEALTH

The Apulia Region in a pioneering way, by DGR n. 1088 of 2020/07/16, established the Regional Operating Center of Telemedicine of Chronic Diseases and Clinical Networks within AReSS (Regional Agency for Health and Social Care of Apulia), also providing the operational guidelines for the promotion and dissemination of the telemedicine in the health service and authorizing the financing of the information-technological infrastructure for the first start-up phase of the Operation Center.

The Agency followed up on the mandate conferred by the Region and, through acceptance of the Consip SPC Cloud Lot 1 Framework Contract, developed the infrastructure, the platform and the Mobile APP (available on the Android and iOS stores) of the Operating Center, called CORēHealth. Subsequently, with Resolution of the CEO n.65 of 07/04/2021, AReSS Puglia published the Guidelines for the Management of Outpatient Services in Telemedicine, transposing the "National indications for the provision of services in Telemedicine" approved by the Conference standing for relations between the State, the Regions and the Autonomous Provinces in the session of 2020/11/17. The document defines and regulates the procedures (recipients of the service, services, delivery methods, etc.) aimed at the creation of some specific telemedicine services for people who need remote health checks, such as, for example, televisit/health video call, remote control of devices in specific areas e.g. cardiology, diabetes, specialist teleconsultation, territorial health and social teleservices.

CORēHealth is part of the Regional Health Services Network by dialoguing and cooperating with the various institutional and operational stakeholders of the area and is the institutional and operational reference for all e-Health initiatives implemented by Apulia. The Regional Operating Center, in its full operational mode, provides for central coordination of the various ehealth solutions implemented region wide, guaranteeing technological interoperability and plurality of services; so it works as a Service Provider for regional e-Health services. To this end, periodic working tables have been organized to involve, compare and implement operational plans for new services to be developed. From an institutional point of view, the DGR 1088/2020 has identified the regional coordination for telemedicine and the main stakeholders belonging to the network of regional health services. Furthermore, these stakeholders are flanked, depending on the topic of discussion at the board and on the basis of specific interests, also by: Patients' associations, health trade unions, private healthcare associations (e.g. clinics, studios multi-specialist, etc.), manufacturers of medical devices and healthcare information systems and other stakeholders. The involvement is extended to almost all subjects belonging to the regional social-health sector.

The trial phase of CORēHealth, which ended in December 2021, focused on 4 regional pilot projects: Oncology Department of the San Paolo Hospital in Bari (about Breast cancer), Thalassemia Center in Brindisi (about Thalassemia), Putignano District (about Diabetes and Hypertension) and Local Health Authority of Taranto (rural medicine). As of January 2022, training of "COrO" and Breast Unit providers, enrollment of approximately 23,000 cancer patients (about breast), and initiation of a change management and digital literacy process involving both the care team and patients has begun. Another key element of CORēHealth is the systematic approach in Continuing Training and Digital Competence Consolidation to enable informed access of all the actors involved to new digital technologies. For this, AReSS has initiated a virtuous path of training, change management and digital literacy for all users (medical and administrative teams) and patients. Among the results to be achieved in the near future:

- Extension of the Telemedicine service to all chronic diseases in the Apulia Region;
- Apulian Oncological Network: training and enrollment started (about 200,000 oncological patients in Apulia);
- Rare Diseases: pilot project launched with the Rare Disease Center - Thalassemia of ASL Brindisi (around 23,000 rare disease patients in Puglia);

• Chronic: the pilot project of Putignano district of Local Health Authority of Bari, about Diabetes/Hypertension District started (about 1.7 ml of chronic patients in Apulia).

2.1 The Objectives of The Corēhealth

Through the COREHealth we want, first of all, to promote the centrality of the citizen-patient and this requires that the entire social assistance and protection network, the various actors who are involved in it, cooperate effectively and assiduously, with the aim of protecting their well-being and health, guaranteeing the satisfaction of their main needs and their active inclusion as a conjunction.

The CORēHealth, therefore, must operate as an enabling factor for the realization of this basic prerequisite, guaranteeing: an extensive digitization of processes, consistent with the reference organizational models (regional and corporate, district and clinical networks) in the local area, aimed at maximizing the adoption of the new system by operators by focusing on usability aspects and optimizing the feeding activities of the new system (operators will only have to enter the data necessary for the specific process they are carrying out and not those that are already available in other systems, which can be retrieved via, for example, integration mechanisms); the interaction between all the actors involved (e.g. GPs, pathology networks, accredited care providers, third-sector entities, etc.) and therefore, a "unified" taking care of the patient's needs, with particular reference to complex needs; a monitoring of processes, of the clinical-epidemiological and economic aspects to support both the continuous improvement of services offered and regional social-health planning. It is also necessary to always be aware that e-Health is not the end, but the tool to achieve the ultimate goal: to meet the health and well-being needs of citizens.

Critical success factors need to be addressed for the realization of this transformation: citizen-patient centricity, data centricity, overall system governance, digital skills enablement, and finally, innovative financing methods. Another key element is the systematic approach in continuous training and digital skills building to enable informed access of all the actors involved to the new digital technologies: among the enabling factors of digital health, the digital skills spread to all professionals are undoubtedly an essential strategic element. And it is for this reason that AReSS, by joining the Consip Framework Contract «DIGITAL HEALTHCARE - Clinical-Assistance Information Systems» - Lot 6 has designed a path of change management, training and digital literacy of all users of the system (medical, administrative teams) and patients, so that CORēHealth does not remain a mere tool, albeit technologically advanced, but represents a real organizational shift toward a new way of caring for chronic patients with benefits for both patients and caregivers.

The necessary skills that should be strengthened are above all digital managerial skills, to plan, govern and optimize the digital transformation process. The objectives of the Operating Center, therefore, are:

▪ To refine, through the digitization of processes, the regional organizational model of prevention, care and treatment of its frail citizens and/or those with chronic diseases in accordance with the provisions of the "National Plan of Chronicity" which identified the adoption of telemedicine models, techniques and tools as opportunities for improvement and benefits;

▪ Centralize citizens' epidemiological data on a single platform, as well as the entire process of taking charge and managing the defined prevention, assistance and treatment plans;

▪ Promote the cooperation and collaboration of all the structures and professional figures involved in the process of prevention, assistance and care of citizens according to the various levels of responsibility and action;

▪ Integrate the entire network of regional services with particular reference to the "clinical networks of rare diseases" and the "Oncological Orientation Centers (COrO)";

▪ Make the telemedicine network more efficient (definition of the operational processes of the Service Center and the Delivery Centres) for the management of data, reports and reports acquired through Telemonitoring, Televisit, Teleconsultation and Telecooperation healthcare tools;

▪ Have full health, organizational and economic governance of the interventions and plans activated;

▪ Strengthen the relationship with the citizen and improve the citizen's perception of services.

2.2 The Management of Patient Care

The platform will enable each Supply Center (District, COrO, Rare Disease Centre, etc.) to manage its clients according to defined and suitably modeled processes. The platform manages various methods of "Request/Access" to the services offered by the Control Unit:

- By integrating the medical records of GPs /PCPs (Primary Care Pediatricians): the proposed system allows GPs and PCPs to use the platform made available or, alternatively, their own medical records to fill out requests for taking charge, or the request to activate services to be provided to their patients, and transmit this request to the district services through bidirectional specifications. The GP/PCP has a constant update on their medical office software of the status of their requests and their patient's entire course of care;
- Through the platform by the Supplying Centres: the proposed system allows authorized and suitably profiled operators from COrOs, Rare Disease Centers and Districts to register an acceptance request in the system and start the process according to regional regulations;
- Through APP or WEB PORTAL: the system provides specific services to enable the publication of a citizen-facing service for reporting a need and, as was the case of the COVID-19 emergency, to enable the citizen to fill out a pre-triage questionnaire and be contacted by competent health services;
- Through the platform or through specific services with the hospital structures: the system provides hospital facilities with specific communication tools with the Delivery Centers for the management of "Hospital-Territory Continuity of Care". Requests will be visible from the Delivery Centers on the platform through a dashboard that allows for total operational governance of the requests received, the different phases of intake, the different types of Diagnostic Therapeutic Assistance Plans ("PDTA") activated, their distribution throughout the territory, and other information that allows Coordination Centers to govern and manage the patients in their care in a simplified way.

In the logical model of CORēHealth, the various stakeholders, can access the platform (back office and/or mobile APP) for patient enrollment, clinical assessment, protocol and treatment plan definition, and subsequent follow-up plan. The platform makes all the information available in the clinical repository and through interoperability tools it interacts with the services of: alarm center, logistics platform, patient synoptic, televisit/telemonitoring as well as with third-party systems (e.g. diagnostic systems, reporting systems, prescription systems, electronic booking systems, etc...)

3. THE CORĒHEALTH TECHNOLOGY

CORēHealth can be summarized in three macro-blocks: 1 - AReSS Service Center: the facility responsible for the management and maintenance of the entire information system; 2 - Dispensing Centers / Clinical Networks: the structures of the Health Service (authorized or accredited, public or private, GPs / PCPs, specialist clinics, etc.) who manage the relationship with citizens through the management of all phases of the process of taking charge and providing the services provided by the Diagnostic Therapeutic Care Plans; 3 - The telematic platform.

From a more strictly technical point of view, the platform implements an application solution that enables the model at a regional level, implementing the technological transformation towards a cloud environment, as well as the re-engineering and migration of its application systems on the Cloud infrastructure of the Service Center of Innovapuglia (in-house provider of the Region). The project includes, among other technological components, the following [Figure 1]:

- Platform for the management of chronicities and clinical networks;
- Televisit, Teleconsultation, Telemonitoring, Telecooperation and Teleconference Platform;
- Televisit and teleconsultation system;
- Device traceability system;
- Clinical repository;
- Patient Synoptic: certified medical device for the collection of parameters and clinical data;
- AReSS Alarm Center for centralized monitoring by the Agency;
- HL7 FHIR Interoperability Platform for integration management;
- Mobile App for Citizens, with guaranteed access via SPiD/eIDAS, which favors their involvement in the care process: keeping in touch with their specialist doctor and caregiver (video calls and chat); consult the

agenda of scheduled televisits with patient's care team; view the treatment plan; enter the vital parameters that are communicated to the doctor in real time; promote the measurement of adherence to the treatment pathway (drug intake, lifestyle), etc.;

- Integration with the Regional Health Information System ("Edotto");
- Integration with the EHR (Electronic Health Record) and with the LIS, Oncological Screening, RIS/PACS systems (in the planning phase).

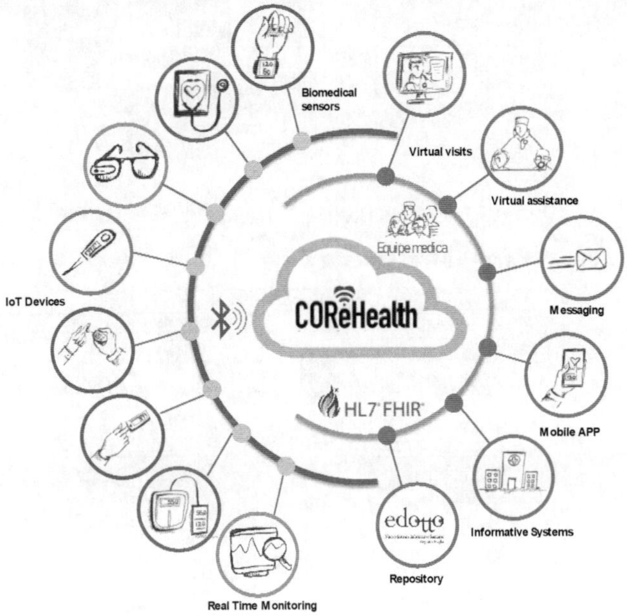

Figure 1. Interoperability of the COReHealth

Each patient is assigned an Integrated Care Pathway ("PAI") based on the specific treatment plans of their pathologies (Diagnostic Therapeutic Care Plan) and, if necessary, a medical device kit, interoperable with the Unit, for the detection and monitoring of key vital parameters. In fact, The medical team, based on specific needs, can provide patients with a medical device kit (tablet, pulse oximeter, multi-parameter, scale, etc.), interoperable with the Unit, suitable for real-time detection and monitoring of vital parameters, salient features that enable constant detection and monitoring and possible intervention by virtue of the automatic alarm system with which the control unit is equipped.

3.1 Corēhealth It Architecture

CORēHealth thus provides the care team with a web-cloud (backoffice) platform available for the telematic management of their patients by offering, among the main services: personalized patient monitoring paths (Telemonitoring), Teleassistance, Televisit, Teleconsultation and Remote Health Collaboration, digitized services for taking charge, personalization and management of patient care plans, logistics/warehouse management of medical device kits [Figure 2].

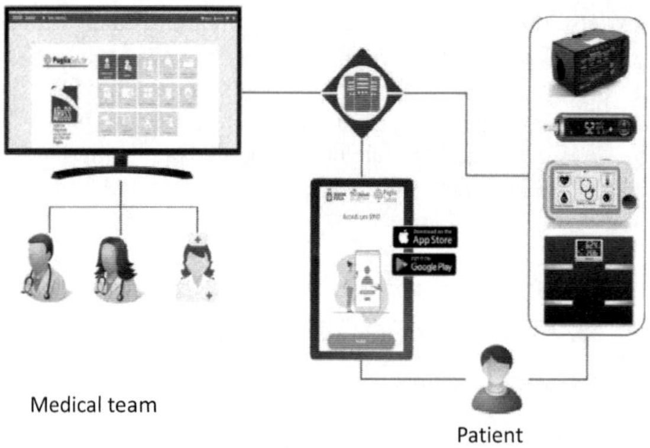

Figure 2. CORēHealth IT Architecture logical model

3.1.1 Corēhealth Server Farm Infrastructure

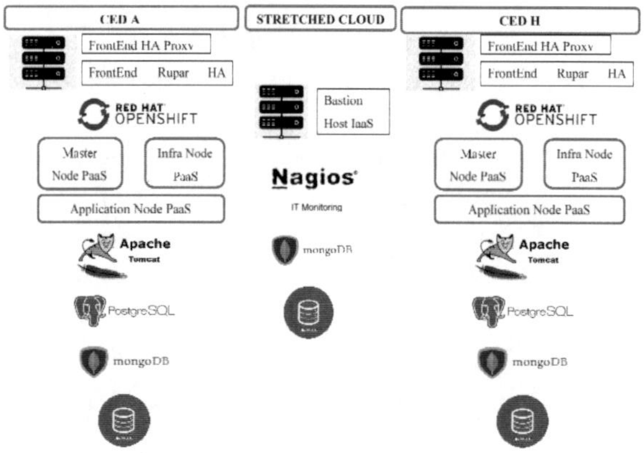

Figure 3. CORēHealth server farm architecture

Figure 3 illustrates the "*physical*" architecture of the CORēHealth central unit that resides on Innova Puglia's cloud. The cloud, dedicated to the plant, features 3 distributed nodes: CED A, CED H and the Stretched cloud configured together in High Availability. The PaaS platform is based on the Red-Hat Openshift system capable of easily integrating both open source and off-the-shelf products and projects. For each application component or service, within the solution, an "*architectural template*" is instantiated, which is the set of all the technological and operational components necessary for the functioning of the given component or service. The two nodes CED A and CED H are "*orchestrated*" by the Stretched Cloud characterized by the monitoring application, based on Nagios, and by the Bastion Host for the protection of the entire infrastructure. The CED A and the CED H are configured in mirror replication and in particular contain the following VMs: Oracle DB; Mongo DB; PostgreSqL DB; Apache Tomcat Web Server; OpenShift nodes (Master Nodes, Infra Node, Application Nodes, Front-End proxy in High reliability on local network and RUPAR (Unitary Network of the Apulia Regional Public Administration)).

In order to guarantee high security and availability of application services (not only of systems), easy scalability and adaptability to dynamic needs in terms of performance and a projection towards Business Continuity and Disaster Recovery solutions, even cross-site with high automation, the proposed technological architecture provides for the virtualization of systems capable of ensuring the provisioning of IaaS infrastructural elements, and the virtualization of applications using containers, through the adoption of the PaaS platform for container management (based on open standards such as Docker and Kubernetes) Enterprise level: Red Hat OpenShift. The solution is proposed with a high availability configuration through node redundancy (Master, Infra, Application). The Control Unit thus contemplates all the non-functional features that the new Platform must have (e.g. usability, security, compliance with the Privacy legislation, etc.), maximizes usability by users, ensures full satisfaction of all dimensions of ICT system security, complies with privacy regulations - GDPR: General Data Protection Regulation - and is consistent with the "privacy by design and by default" principle. The Center ensures the operational continuity of the system, also in line with the relevant AgID (Agency for Digital Italy) Guidelines, to guarantee the availability of all the elements that contribute to the functioning of the information systems (e.g. ICT resources, logistical conditions, etc.) and it operates on an adequate and robust technological infrastructure at Innova Puglia's Regional Data Centers. The plan defines the integration standards with the Regional Health Information Systems and through the interoperability platform can be integrated with the reference regional information systems, such as: Regional Single Registry System; Electronic Prescription System; Single Booking Systems ("CUP")/ Diaries/ Specialist booking systems; Reporting systems; Medical Devices for Telemonitoring; Regional Single Sign On system; Regional Repository/DHR; External systems of Accredited Structures and Third Sector for the assignment of Diagnostic Therapeutic Care Plan; Medical office SW (GP/PCP). Among the platform's special features is also the acquisition of compliance certification in terms of web accessibility according to AGiD guidelines and relevant legislation ("Legge Stanca" and D.lgs 82/2005, Code for Digital Administration). The Veeam Backup suite is used for backup, which allows to backup and recover instances directly in the cloud.

3.1.2 Corēhealth App Mobile

The Citizen Mobile App (with SPiD/eIDAS access) is also available on the Android and iOS stores [Figure 4]. The mobile app allows, for the first time, the patient to have a series of functions in the palm of his hands: televisit diary, his/her treatment plan, sending and historical archive of his/her vital parameters in real time, diary, therapeutic adherence questionnaires, help-desk service, etc... Further developments are planned with the release of a series of additional features such as the integration of laboratory tests, electronic booking, electronic payments, etc ...

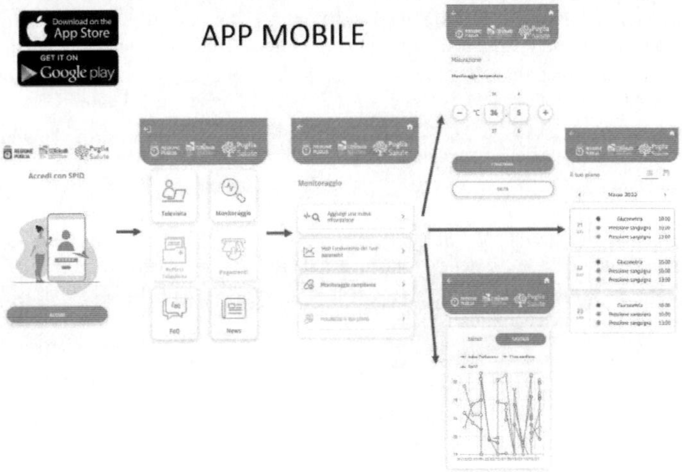

Figure 4. CORēHealth APP mobile interfaces

4. CONCLUSIONS

Telemedicine, now considered one of the key components for improving citizens' health, is a historic opportunity to be seized thanks to the NRRP which intends to create a real ecosystem in this regard, through a national enabling platform (implemented by the National Agency for Regional Health Services -AGENAS) and vertical regional implementations. The Apulia Region (during the Presentation at the Permanent Conference for relations between the State, the regions and the autonomous provinces, last 2 March) was identified by the Minister for Technological Innovation and Digital Transition to develop, together with the Lombardy region (Flag Regions), the applications for the provision of Telemedicine services nationwide under M6C1 of the italian National Recovery and Resilience Plan, (NRRP).

COReHealth overturns the traditional patient care pathways, bringing, for the first time, digital innovation directly into their hands (App Mobile) enabling them to have their own medical history, their care plan and maintain a constant, regulated dialogue with their doctor, medical team and care giver, as well as being able to take the most appropriate measurements for the management of pathology/ies. Doctors can manage the clinical history of their patients in an integrated environment, optimizing care time, reducing inappropriate access to hospital facilities, developing their professional skills, including digital skills, according to a multidisciplinary approach to care taking with a single common point of access, reducing the distance between hospital and territory, and moving the care activities of chronically ill patients to an environment as close as possible to the patient's home.

In November 2022 the Apulia Region received the "Innovation in Digital Health" award, organized by AGENAS, the National Agency for Regional Health Services, and SICS, publisher of the journals Quotidiano Sanità and Popular Science, for the regional telemedicine platform COReHealth. COReHealth was also included among the best practices of the AGENAS's PON GOV chronicity (2014-2020) as a simultaneous investment to improve the use of and access to ICT Technologies and strengthen the institutional capacity of public authorities.

REFERENCES

Gorgoni G., Maglio G., Notarangelo P., Petrarolo V. (2022). Article. Agenda Digitale.eu, COReHealth: com'è fatta la centrale di telemedicina pugliese

Gorgoni G., Notarangelo P., Petrarolo V. (2022). Award. Forum PA Sanità 2022. COReHealth - La centrale Operativa Regionale di Telemedicina della Regione Puglia

Gorgoni G., Notarangelo P., Petrarolo V. (2022). Abstract. XLVI Congress of the Italian Association of Epidemiology 2022. "Telemedicina made in Puglia: la Centrale Operativa delle Cronicità e delle Reti Cliniche (COReHealth)"

Gorgoni G., Notarangelo P., Petrarolo V. (2022). Poster. XLVI Congress of the Italian Association of Epidemiology 2022. Telemedicina made in Puglia: la Centrale Operativa delle Cronicità e delle Reti Cliniche (COReHealth)

Gorgoni G., Notarangelo P., Petrarolo V. (2022). Award. Digital360Awards. Non-winning finalist. COReHealth: Centrale Operativa Regionale delle Cronicità e delle reti Cliniche della Regione Puglia

Gorgoni G., Notarangelo P., Petrarolo V. (2022). Award. Innovation in Digital Health, Building Biotech Bridge. COReHealth

Gorgoni G., Notarangelo P., Petrarolo V. (2022). Poster. 17 th Forum Risk Management in Sanità 2022. Arezzo (Italy) COReHealth: La Centrale Operativa delle Cronicità e delle Reti Cliniche della Regione Puglia

Gorgoni G., Notarangelo P., Petrarolo V. (2022). Poster. 10 th Anniversary - Euregha. COReHealth — Telemedicine operations center for chronic conditions and clinical networks

THE ROLE OF BUILT HERITAGE WITHIN EDUCATION FOR SUSTAINABLE DEVELOPMENT THROUGH MOBILE AUGMENTED REALITY GAMES

João Ferreira-Santos[1] and Lúcia Pombo[2]

[1] *Research Centre on Didactics and Technology in the Education of Trainers (CIDTFF), Department of Education and Psychology (DEP), University of Aveiro (UA)Campus Universitário de Santiago, 3810-193 Aveiro, Portugal*

[2]*CIDTFF, DEP, UA, Campus Universitário de Santiago, 3810-193 Aveiro, Portugal*

ABSTRACT

Looking at heritage and considering it as a relevant piece for education is currently part of different narratives, whether of educational, cultural, political, or even legal nature. The aim of this paper is to explore the concept of heritage and its implications for education for sustainability, using technology as a facilitator. Through a narrative literature review, an analysis is made of the concept of heritage, with special focus on the concept of built heritage and its connection to education, as a learning space, and of its role in the development of competences related to education for sustainability, with technology assuming an aggregating role. It was found that although the important role that built heritage has for the development of the human being, there is still limited literature and research about this. We also saw the important role that technology plays in exploring these areas, particularly through a Mobile Augmented Reality Learning in an App. Since the Augmented Reality contents, the mobile educational game and the EduCITY App itself, are still under development, this document is an exercise of theoretical development. As final remarks it is possible to assume the effective connection of heritage and its role to the development of education for sustainability, taking advantage of the potentialities of technology as Mobile Augmented Reality Learning. Despite this recognition, research on the articulation of these concepts is still underdeveloped, and this paper seeks to contribute to the deepening these fields of human development and their connections.

KEYWORDS

Heritage, Education for Sustainable Development, Mobile learning, Augmented Reality, EduCITY

1. INTRODUCTION

Everyone recognizes the valuation of the cultural heritage, rather being tangible or intangible, as one of the pillars of the evolution of society, especially in a conception of society as post-materialist, whose values have moved from effective relations to material aspects, such as economic development, housing needs or food security, to values linked to immateriality, i.e., related to environmental concerns, social and cultural development, among others (Inglehart, 1977). Of course, this eminently political and sociological positioning emerges as a possible trigger for an awakening towards the valuation of heritage and the need for its preservation; however, it should be noted that the concern with heritage issues is much older, with records on Europe dating back to the year AD 44 on the need for heritage protection (Correia, 2014).

Alongside the progressive concern with heritage and as already mentioned above, there is also, in the mid-20th century, a growing concern with environmental issues. These issues, which emerged within the concept of Sustainable Development, have been deepened from 1987, the year of publication of the United Nations "Our Common Future" report, where concerns are introduced at the level of the relationship between human beings, their development and the environment (Brundtland, 1987). It was then defined that sustainable development ensures that the development of future generations is not harmed by the previous ones. However, despite this concern, at the end of the 20th century, several fragilities were highlighted, which at the present time, are experienced with greater impact, such as climate change or the growing scarcity of resources or even the extinction of animal species (UNDP, 2022), among many others.

We live in a time of uncertainty, constant reconfigurations and eminent complexity, making the 21st century, despite all the progress achieved by Man, a time of great uncertainty about the future. However, as Morin (2021) stated, human being, in the face of the unknown and the complexity, shows enormous versatility and capacity for action, reacting, adapting, creating and recreating new tools. Two of the most effective and powerful tools developed by Man, which allow us to respond to the unknown, are education and technology (UNESCO, 2021). In a world woven of multiple and growing complexities, education is a mean of transmitting and empowering knowledge through an effective resilience that everyone recognizes. This knowledge is more effective and efficient when produced and conveyed in a transdisciplinary way, grounded on the environment, starting from the known to the unknown, searching for bridges between the past and the future, in a clear commitment to a narrative and systemic thinking articulated with ecological awareness (Morin, 2002), which today is called as education for sustainability.

Needing to approach an active citizenship, education had to reformulate itself seeking for new allies, in order to approach the daily realities of its various agents. Currently, education coexists with the specificities and potentialities of technology, especially the one that allows us to see beyond the physical and immediate reality, towards other new contents, such as Augmented Reality or Virtual Reality. There are several potentialities of mixed realities, especially Augmented Reality that is recognized as a useful tool in education, considering its potentiality on mobile learning contexts (Crompton & Burke, 2018). School is no longer closed in the classroom, coming closer to experiences of students and their families, and teachers, taking place in new spaces, thus enabling new narratives and learning. This allows establishing bridges between different types of knowledge, searching for a global and transdisciplinary knowledge (UNESCO, 2021). These bridges are a concern in the PhD project of the first author, entitled "Built heritage in Mobile Game for development of key skills towards Education for Sustainability – from academia to society", where it is intended to develop a game to be explored while touring the city, with specific points of interest related to build heritage in order to promote people's environmental awareness, in a citizen science logic, promoting cities as safe, healthy and accessible for all, making aware of their built heritage, assuming it as a resource education for sustainable development.

This research focuses on the role that the built heritage assumes, as well as its valorisation in the development of key competences for Sustainability (Wiek et al., 2011), through technology, in particular the use of Augmented Reality in mobile learning games contexts.

This paper is organized into six sections. After an introduction, the second one is about the methods used in this paper, whose aim is to explore theoretically Heritage, Education for Sustainability and Technology and its connections. The third section concerns an analysis and theoretical development on heritage, aiming a deep understanding about the concept itself, and connecting it to the remaining fields of human action under analysis, education for sustainability and technology. On the fourth section an analysis is taken concerning the role that heritage plays in education for sustainability. The fifth section concerns about the role of technology on the development of education de-plotted and based on built heritage, with special attention to the use of contents in Augmented Reality through mobile learning. The last section stands as Final Remarks, constituting a brief presentation of the research developed, advancing also to future investigations.

2. METHODOLOGY

Despite the electronic search for academic literature through aggregators and databases such as Scopus and Web of Science, we did not carry out a systematic literature review, but rather a narrative review of the literature (Chaney, 2021) on the fields under analysis. Based on the aim of his paper, search string using Boolean operators was: KEY (education) AND KEY ("Augmented Reality"). Due the nature of this review, the above search string aimed only to the education field and the Augmented Reality, since previous searches were conducted. However, we would point out that the search should have been open to other search terms. We also point out that the Boolean AND was used in the sense of obtaining as much records as possible. This literature review was taken in order to develop theoretical understanding and aiming to search links between the concepts. Doing this type of research, a search was made to consolidate and summarize the concept of heritage and particularly built heritage, establishing connections of its role to the development of education for sustainability using technology as inductor. The use of other types of literature sources, as books, promoted a richer analysis of the first concept.

3. WHAT DO WE MEAN ABOUT BUILT HERITAGE?

Before we move on to an abbreviated analysis of the concept of heritage, it is essential to reinforce the idea that for our research, we are essentially concerned about built heritage.

The concept of heritage is complex, multidimensional and, for this very reason, it may be named and interpreted differently. The concept may also appear as Cultural Heritage or Historic Heritage, given its dynamic nature. It is a concept that has changed over time, being explored in different theoretical fields or fields of action of human being. Consequently, the legal concept of heritage differs from the aesthetic concept of heritage, among other contexts. Within these same different contexts that lies the important significance of this concept for society and particularly to its development (Marmion et al., 2010).

Despite the rarity of records prior to the 19th century regarding heritage protection processes, it is possible to realize that since Classical Antiquity there are some awareness about the vestiges of the past, however, it is important to emphasize that this concept cannot be understood as concerning the ancient or historical periods, but also as a contemporary concern (Choay, 2021). Moreover, if the temporal dimension tends to be a notion that no longer "ties" heritage to the past, other characteristics have also evolved considerably. It is possible to observe a clear transformation in its own understanding, moving from the tangibility of the monument to the intangibility of emotions (Choay, 2019), as far as heritage is not only considered as something material, but also something of expression, communication, transmission of knowledge and meaning (Smith, 2006). This new conception promoted an affirmation of oral traditions, and types of expression, such as music, customs and traditions, such as popular festivals. These types of heritage are called as "living heritage" (Martins, 2020, p. 23) or "practices of heritage" (Harrison, 2010, p. 9), which is nowadays translated as much more apparent representations of folklore, as it was called in the past. These types of heritage became understood as something complex, as it appears in the "Convention for the Safeguarding of the Intangible Cultural Heritage" (UNESCO, 2003).

Now, we will analyze another context of heritage, namely, the evolution of the concept building-object to object-involvement, which can easily be experienced in the case of built heritage. This evolution promoted an understanding of the building-object as part of its surrounding. This process became known as "historic ensemble", through the "Recommendation on the Safeguarding of Historic Ensembles and their Function in Contemporary Life" (UNESCO, 1976). There was then an integrated vision of the built heritage in relation to its surroundings, considering it no longer isolated, linking the built heritage to the community itself (Smith, 2006). To these previous contexts, others appear, as the one related to the transformation of the concept of heritage as a monument to something more subjective and connected with the values of a certain society, such as memory, identity and potentiator of dialogue between communities (Council of Europe, 2005). However, it is essential to understand that this subjectivity and relationships were developed through a process of "homogenization of the process values" (Choay, 2021, p. 106) that started taking place after the end of the Second World War.

This renewed vision on the role of heritage emerges after a moment of impact and transformation that occurred from the mid-twentieth century and, in particular from the 1970s/80s (Smith, 2006), especially in the affirmation about the need of protection of heritage, valuing it and integrating it into "contemporary life" (Choay, 2019, p. 233). At the same time, this approach to contemporary life, promoted one understating of heritage in an outward movement from the mystical aura of the museum concept, untouchable and not very accessible, giving it new functionalities, reinventing it, and using it in a sustainable way (Council of Europe, 2005).

Taking the example of the case of the "Grand Louvre" project (1981-1989), whose most visible structure are the metallic and glass pyramids designed by the architect I. M. Pei. If the physical transformation of the, now known as, "Cour Napoléon" has become evident, the identity and memory of the surrounding building did not disappear. However, we have come to witness what is called as a "heritage industry" (Choay, 2019, p. 239). This context appears associated with another one, massification of tourism or as "disneyfication of tourism" (Smith, 2006, p. 33). This happened to many historical sites and cities, as the case of the city of Venice, Italy.

Returning to the concept of heritage associated with memory and the construction of personal and collective identity (Harrison, 2010), we must understand the heritage-object as a "deliberated creation" (Choay, 2019, p. 19), as something that has been selected by someone as representative of a broader past than himself, being a representation that this someone had of a certain historical moment. As for memory, we can assume it in the relationship between heritage and the capital of knowledge itself, arising from the very development and human experience (Hosagrahar et al., 2016; López-Fernández et al., 2021). In this case, heritage appears as an entity that supports memory, recognizing its importance in the very evolution of the human being, even if there is no certainty about the function or meaning of that piece of heritage, or the reason of its construction, as it is stated in the "Krakow Charter" (ICOMOS, 2000).

Another context of built heritage is linked to its function, ceasing to be an immutable object, capable of being adapted to gain new functions, moving from object of worship to object of learning, assuming that the experience in art is global (Dewey, 2005). This assumption promotes a movement of seeing and enjoying built heritage itself, as a "verification of pleasure" (Choay, 2019, p. 12) to a place of learning and questioning. This connects to Rancière words, about the role of spectator as an active agent of its own narrative, making his own translation of the "story" and being capable of building new narratives (Rancière, 2010). In this sense, and focused on the built heritage, the change of position regarding the role of the monument itself allows it to be seen as a learning space. This vision about built heritage as a learning space is a concept that has been evolving since the 1950s, being considered since the "European Architectural Heritage Charter" (Council of Europe, 1975) as essential to the balance and development of society, also revealing a decisive educational value and tool. These characteristics, are clearly displayed in the "Krakow Charter", appearing within the importance of the relationships between heritage, training and education itself, both on training technical experts in the field of heritage(s) and on the actions in an educational context (ICOMOS, 2000). More recently, built heritage has become labelled as a non-renewable asset, reinforcing the need for its protection and enhancement, and understanding it, as an essential educational resource for the development of the human being itself (UNESCO, 2014).

4. FROM EDUCATION THROUGH HERITAGE TOWARDS EDUCATION FOR SUSTAINABILITY: BUILDING A FUTURE

We have seen how the evolution of the concept of heritage, especially the built heritage, promoting it to new realities or functions, and as a space for training and education (Council of Europe, 1975; ICOMOS, 2000). This process occurred within the change of vision towards built heritage, affirming its value and need of preservation, as well as the assumption of its role to education, and as a tool for transforming the world and enhancing a sustainable future (UN, 2015).

Currently, and as already mentioned, the importance and urgency of a "viable future" (Mayor, 2002) is a base to the presentation of the Sustainable Development Goals through the "2030 Agenda" (UN, 2015). In these goals, heritage appears in Goal 11, and it's about "Make cities and communities more inclusive, safe, resilient and sustainable", relating it to the need to protect and safeguard heritage. This goal stems from the presentation, in the report "Future We Want" (2012), of the section "Sustainable cities and human settlements", point 134. In this section emerged the need for recognition, conservation and revitalization and rehabilitation of the natural and cultural heritage of urban centers, in a perspective of promotion and development of urban sustainability (UN, 2012). As we have seen, there are several relationships between heritage, education and sustainability. education is being assumed as the connecting link or tool of union (Heng, 2010), however, in addition of being a tool, education is responsible for promoting an overview of reality in order to respond to the unknown by strengthening the unity and plurality of the human being and its productions. This occurs through processes that focus on the particularity and the unity of all the parts, supported by a clear vision antagonistic to the crystallization of knowledge (Morin, 2002), which can be called as integration of knowledge. It is supported in this vision of global unity that heritage can be understood as a learning space and arising from its multidimensional nature enabling various fields of thematic exploration, such as History, Citizenship Education or more specifically, Education for Sustainability (Van Doorsselaere, 2021).

It is understandable that a building can serve several functions, as it changes over time, however, in the great majority of the cases, when a building is constructed with a well-defined purpose, the initial function is maintained. That is the case of religious buildings, civil buildings, like libraries or palaces, and also buildings or constructions of military nature. However, we also see buildings that, by their identity and memory, are changed, most often for purposes related to education and culture. A good example of this, is the case of the Museum of Aveiro, which was built as a religious monastery, and after 1911, became a museum of this city, enriched by other religious institutions (DGPC, n. d.), or the "Museu Arte Nova", also in the same city, which was built as a private residence and today it is the interpretative centre of this artistic style, and considered one of the most emblematic buildings of this style in Portugal (MCA, n.d.).

It is through the process of safeguarding and valuing the built heritage that key competences for education for sustainability are sought to be developed. These competences were defined by Wiek et al. (2011) as: "Systems-thinking competence" - the ability to analyse or understand different systems; "Anticipatory competence" - the ability to analyse and understand "the big picture" and relate it to the resolution of future problems; "Normative competence" - the ability to map, connect or reconnect, negotiate and apply the values, principles and actions related to sustainability; "Strategic competence" - the ability to boost actions towards the implementation of solutions; and, "Interpersonal competence" - the personal ability to promote, facilitate and carry out actions (Wiek et al, 2011, pp. 207-211). According to these assumptions, the first author's research seeks to develop these competences, through the development of a mobile educational game "Art Nouveau Path" based on challenges that integrate educational resources in Augmented Reality, simulations, 3D animations and informative spots, which are triggered, in the app, in several strategic points of the city, where there is Art Nouveau built heritage, spread around the city of Aveiro. This mobile educational game is to be integrated in the mobile app EduCITY, developed under the Research & Development project "EduCITY - Smart and sustainable cities with Augmented Reality mobile educational games created by and for citizens", which intends to create better learning experiences, supported by the users' experience in situ and their performance while playing or creating pervasive and challenging games in order to enhance public understanding of built and natural heritage.

5. THE ROLE OF TECHNOLOGY IN DEVELOPING EDUCATION FOR SUSTAINABILITY THROUGH BUILT HERITAGE

Assuming, as already mentioned, the value of the heritage as a training and educational space, with special significance for the development of key competences for education for sustainability, it becomes essential to conceive educational actions that take place in the real context in which the heritage exists. This assumption arises from the vision of the transforming role that education has in society. However, this same role induces it, as we have seen previously, to new challenges. Those new challenges comprise the need of approach to students' daily experiences, in order to manage learning as an effective process (UNESCO, 2021) and assures that learning must be concrete, starting from the real to the abstract, towards the creation of new values and new learnings, promoting more solid and effective interdependencies of knowledge (OECD, 2018).

We have seen how the development of key competencies in sustainability can be supported by technological developments, since technology allows, when properly exploited, to enable new experiences, extending reality, using Virtual Reality, Mixed Reality or Augmented Reality (Alnagrat et al., 2022). These new realities complement the material world, enhancing and mobilizing new learning experiences (Chen & Duh, 2018). It also allows the exploration of learning in new contexts, whether formal, non-formal or informal, at school or in outdoors, in close contact to reality, near nature and in a perspective of lifelong learning (Pombo et al., 2019).

The potential of integrating the tangibility of real world with the intangibility of new realities, such as Augmented Reality, has proven to be significantly positive for students, both in terms of increasing their interest and performance, but also in terms of the effectiveness of their own learnings (Alnagrat et al., 2022). However, the use of Augmented Reality and its effective potential mobilization as an educational resource is enhanced through mobile learning (Marcel, 2019). That occurs attending the fact that built heritage exists in outdoor context, and mobile learning acts as a connector of the realities, and as a tool to the promotion of Augmented Reality and its use as education resource.

This situation can be facilitated through the massification of the use of mobile devices and its relevance to education, having proven to be prolific not only at individual learning, but also through cooperative strategies and dynamics learning activities (Chen & Duh, 2018). The use of Augmented Reality is still considered as recent, and there are few authors who study its potential to promote learning. This potential regards the fact that Augmented Reality enables contact with reality, through the process of overlaying new content (Barrado-Timón & Hidalgo-Giralt, 2019). By requiring the existence of real contexts, Augmented Reality induces learning in spaces outside classrooms or schools, bringing them closer to the real world (Pombo, 2022). This causes engagement and motivation in its users (Chen & Duh, 2018). However, as Scavarelli et al. (2021)

refer, it is necessary to continue developing work to make Augmented Reality more efficient, with especial regard to educational contents. This implies the development of Augmented Reality contents considering educational specificities, as proper exploration in terms of the context of the resources and its relations with school curricula, ensuring their educational or formative role.

5.1 The Mobile Educational Game with Augmented Reality "Percurso Arte Nova" on the Educity App

The mobile educational game using Augmented Reality content "Art Noveau Path" is being developed, based on the potential of Augmented Reality among with Mobile learning, the importance of situated learning (Barrado-Timón & Hidalgo-Giralt, 2019; Pombo, 2021) and assuming that built heritage is an educational and learning space (Council of Europe, 1975; ICOMOS, 2000). Summing to the set out above, built heritage can take an essential role on the development of key competences of education for sustainability (Van Doorsselaere, 2021). This educational mobile game is part of the EduCITY mobile app that emerges under the Research & Development project "EduCITY", which allows the creation of a smart learning environment in the city of Aveiro, constituting itself in the future, as a reference for other cities. The educational mobile game "Art Noveau Path" will be, like others, available for free on the EduCITY app for Android and iOS operating systems. The development of the game follows the dynamics of Design-based Research, considering several cycles of improvement. Regarding the research aims, the game will be applied, through a quasi-experimental research plan, using classes of sixth and ninth grades, with pre and post-test groups. With this study, it is intended to contribute to the development of key skills for Education for Sustainability, enhancing the local built heritage, through the creation of challenges that integrate educational resources in Augmented Reality, simulations, 3D animations and informative spots, which are triggered, in the app, in several strategic points of the city, namely near the built heritage in Art Nouveau style, taking place in the city of Aveiro.

This research ensures the active involvement of various stakeholders, including students, teachers, academic and general community, assuming the construction of science as a complex process and as an exercise of citizenship.

6. FINAL REMARKS

We can proceed to a definition concerning the concept of heritage, stating that it is a process of culture and education (Chaoy, 2019). As any process, it is dynamic, allowing change and transformation, and being linked to the development of human being. Nowadays it is considered as an integral part of the memory and identity of any society (Council of Europe, 2005), allowing the establishment of connections between Past, Present and Future. Although the research was not exhaustive, it is possible to mention that this concept presents different natures, such as material, as architecture, or immaterial, as music, connected to its origin as human production. Although the importance of human productions, we must also value and protect natural heritage, as a space of balance and essential to the effective development of human being. Is considered as a multidimensional concept, perceived in different ways, according to one's personal experiences (Smith, 2006). It is a deeply sociological concept, whether established in personal or collective values. International agencies and organizations, such as the United Nations and UNESCO, recognize the role of heritage as an educational enabler, revealing, through its multidimensionality, its potential as an agent for the development of key competences for education for sustainability (Van Doorsselaere, 2021).

Despite the ease with which relationships between the areas investigated can be perceived, the literature search process, although not exhaustive, allowed us to realize the small amount of research on the importance of heritage in the development of key competences for education for sustainability. This paper aimed to build bridges between these research areas, constituting a hinge point for further research. In this sense, a narrative literature research was carried out on the concept of heritage, with special emphasis on the built heritage. From this, it was possible to establish links with the area of education, particularly with education for sustainability. For this relationship, technology is involved, in the sense that it enables, through resources in Mobile Augmented Reality Learning, the exploration of the built heritage as a learning space, with implications to the development of education for sustainability (OECD, 2018; UN, 2015; UNESCO, 2014, 2021). These relationships, which appear explicitly in this paper, will enable the development of future research.

The future research will include the development of an educational mobile game using Augmented Reality content. From the implementation of this game to the school community, data will be collected through the development of a quasi-experimental, sequential explanatory research plan, evolving six classes (three test groups and three control groups) and their respective teachers. To validate the previous results, we will also use a survey and focus group. It is expected, therefore, to contribute to the development of key competences related to education for sustainability, through the valorisation of the Art Noveau built heritage of the city of Aveiro.

ACKNOWLEDGEMENT

This paper is part of a PhD project integrated in the Research & Development Project "EduCITY - Smart and sustainable cities with mobile educational games in Augmented Reality created by and for Citizens" (https://educity.web.ua.pt/), which is financed by Portuguese funds through FCT within the framework of the project PTDC/CED-EDG/0197/2021.
This work is financially supported by National Funds through FCT – Fundação para a Ciência e a Tecnologia, I.P., under the project UIDB/00194/2020.

REFERENCES

Alnagrat, A. J. A., Ismail, R. C., & Idrus, S. Z. S. (2022). Virtual Transformations in Human Learning Environment: An Extended Reality Approach. *Journal of Human Centered Technology*, *1*(2), 116–124. https://doi.org/10.11113/humentech.v1n2.26

Barrado-Timón, D. A., & Hidalgo-Giralt, C. (2019). The Historic City, Its Transmission and Perception via Augmented Reality and Virtual Reality and the Use of the Past as a Resource for the Present: A New Era for Urban Cultural Heritage and Tourism? *Sustainability*, *11*. https://doi.org/10.3390/su11102835

Brundtland, G. (1987). *Our Common Future: Report of the World Commission on Environment and Development*. http://www.un-documents.net/ocf-ov.htm

Chaney, M. A. (2021). So You Want to Write a Narrative Review Article? *Journal of Cardiothoracic and Vascular Anesthesia*, *35*, 3045–3049. https://doi.org/10.1053/j.jvca.2021.06.017

Chen, S. C., & Duh, H. (2018). Mixed reality in education: Recent developments and future trends. *Proceedings - IEEE 18th International Conference on Advanced Learning Technologies, ICALT 2018*, 367–371. https://doi.org/10.1109/ICALT.2018.00092

Choay, F. (2019)/(1982). *Alegoria do Património* (3rd ed.)/*L'Allégorie du Patrimoine*. Edições 70./Éditions du Seuil

Choay, F. (2021)/(2009). *As questões do Património. Antologia para um combate../ Le Patrimonie en questions. Anthologie pour un combat.*Edições 70. /Éditions du Seuil

Correia, M. (2014). Enquadramento histórico das normas internacionais sobre o Património Arquitetónico e Arqueológico. In *Património Cultural. Critérios e normas internacionais de proteção.* (pp. 13–23). Caleidoscópio.

Council of Europe. (1975). *European Charter of the Architectural Heritage*.

Council of Europe. (2005). *Council of Europe Framework Convention on the Value of Cultural Heritage for Society*. https://doi.org/10.1007/978-3-030-30018-0_1051

Crompton, H., & Burke, D. (2018). The use of mobile learning in higher education: A systematic review. *Computers and Education*, *123*(April), 53–64. https://doi.org/10.1016/j.compedu.2018.04.007

Dewey, J. (2005)/(1934). *Art as Experience*. Penguin Books Ltd.

DGPC. (n.d.). *Museu de Aveiro*. Retrieved January 8, 2023, from https://www.patrimoniocultural.gov.pt/pt/museus-e-monumentos/rede-portuguesa/m/museu-de-aveiro/

Harrison, R. (2010). What is Heritage? In R. Harrison (Ed.), *Understanding the politics of heritage* (pp. 5–42). Manchester University Press. https://doi.org/10.1515/9781400835423.25

Heng, C. L. (2010). Incorporating Education for Sustainable Development into World Heritage Education: Perspective, Principles and Values. In Asia and Pacific Regional Bureau for Education (Ed.), *Incorporating Education for Sustainable Development into World Heritage Education: a teacher's guide* (pp. 3–34). UNESCO.

Hosagrahar, J., Soule, J., Girard, L., & Potts, A. (2016). *Cultural Heritage, the UN Sustainable Development Goals, and the New Urban Agenda*.

ICOMOS. (2000). *The Charter of Krakow 2000: principles for conservation and restoration of built heritage.* http://hdl.handle.net/1854/LU-128776

Inglehart, R. (1977). *The Silent Revolution*. Princeton University Press.

López-Fernández, J. A., Medina, S., López, M. J., & García-Morís, R. (2021). Perceptions of heritage among students of early childhood and primary education. *Sustainability*, *13*, 1–16. https://doi.org/10.3390/su131910636

Marcel, F. (2019). Mobile augmented reality learning objects in higher education. *Research in Learning Technology*, *27*, 1–10. https://doi.org/10.25304/rlt.v27.2133

Marmion, M., Calver, S., & Wilkes, K. (2010). Heritage ? What do you mean by heritage? In E. Lira & R. Amoeda (Eds.), *Constructing Intangible Heritage* (pp. 33–44). Green Line Institute for Sustainable Develpment.

Martins, G. d'Oliveira. (2020). *Patrimínio Cultural - Realidade viva*. Fundação Francisco Manuel dos Santos.

Mayor, F. (2002). Prefácio do Director-Geral da UNESCO, 1987-1999. In *Os sete saberes para a Educação do Futuro* (pp. 11–13). Edições Piaget.

MCA. (n.d.). *Museu Arte Nova*. Retrieved January 8, 2023, from http://mca.cm-aveiro.pt/rede-de-museus/museu-arte-nova/

Morin, E. (2002). *Os sete saberes para a educação do futuro*. Edições Piaget.

Morin, E. (2021). *Leçons d'un siècle de vie*. Denoël.

OECD. (2018). *The future of education and skills Education 2030*. https://www.oecd.org/education/2030-project/contact/E2030 Position Paper (05.04.2018).pdf

Pombo, L. (2021). Reconhecer in loco o valor do património histórico e botânico – aprendizagens proporcionadas pelo projeto EduPARK. *Boletim Da AIA-CTS*, *14*, 34–37. https://aia-cts.web.ua.pt/wp-content/uploads/2021/03/Boletim_AIA_CTS_n14.pdf

Pombo, L. (2022). Exploring the role of mobile game-based apps towards a smart learning city environment – the innovation of EduCITY. *Education and Training*. https://doi.org/10.1108/ET-06-2022-0238

Pombo, L., Marques, M. M., Afonso, L., Dias, P., & Madeira, J. (2019). Evaluation of a mobile augmented reality game application as an outdoor learning tool. *International Journal of Mobile and Blended Learning*, *11*(4), 59–78. https://doi.org/10.4018/IJMBL.2019100105

Rancière, J. (2010). *O espectador emancipado*. Orfeu Negro.

Scavarelli, A., Arya, A., & Teather, R. J. (2021). Virtual reality and augmented reality in social learning spaces: a literature review. *Virtual Reality*, *25*, 257–277. https://doi.org/10.1007/s10055-020-00444-8

Smith, L. (2006). *Uses of Heritage*. Routledge.

UN. (2012). The future we want. In *A/RES/66/288 - The Future We Want*. https://doi.org/10.1017/S0020818300001806

UN. (2015). *Transforming our world: the 2030 Agenda for Sustainable Development (A/RES/70/1)*. UN General Assembly. https://doi.org/10.1163/15718093-12341375

UNDP. (2022). *Human Development Report 2021-22: Uncertain Times, Unsettled Lives: Shaping our Future in a Transforming World* (P. Conceição (ed.)). UNDP.

UNESCO. (1976). *Recomendação sobre a salvaguarda dos conjuntos históricos e da sua função na vida contemporânea.* (Obra original editada em 1976)

UNESCO. (2003). *Convenção para a Salvaguarda do Património Cultural Imaterial.* https://ich.unesco.org/doc/src/00009-PT-Portugal-PDF.pdf#page=2&zoom=auto,-106,772

UNESCO. (2014). *Culture for Sustainable Development: Sustainable Cities.* http://www.unesco.org/new/en/culture/themes/culture-and-development/the-future-we-want-the-role-of-culture/sustainable-cities/

UNESCO. (2021). *Reimagining Our Futures Together: A New Social Contract for Education.* https://en.unesco.org/futuresofeducation/

Van Doorsselaere, J. (2021). Connecting sustainable development and heritage education? An analysis of the curriculum reform in Flemish public secondary schools. *Sustainability*, *13*(4), 1–17. https://doi.org/10.3390/su13041857

Wiek, A., Withycombe, L., & Redman, C. (2011). Key competencies in sustainability: A reference framework for academic program development. *Sustainability Science*, *6*(2), 203–218. https://doi.org/10.1007/s11625-011-0132-6

EMPLOYABILITY OF ICT GRADUATES IN CROATIA

Ana Tecilazić
Algebra University College
Zagreb, Croatia

ABSTRACT

The employability of graduates has become an important topic of research interest. Generally, graduates with qualifications in driving sectors such as ICT are assumed to be more employable than others. Research on graduate employability based on data collected in the Eurograduate Pilot Survey and analysed using the operational employability model adapted to the Eurograduate Survey supports this assumption and shows that the field of study is one of the factors influencing graduate employability. The results of the analysis, conducted on a sample of graduates with university master's degrees obtained in Croatia in 2016/2017, show that graduates from the broad field of study "Information and Communication Technology - ICT" (ISCED F) are more employable than graduates with comparable qualifications in the reference educational field. The results indicate that for ICT graduates i) the odds ratio of finding employment in occupations vertically matching their qualifications is three times higher than the odds ratio of graduates in the reference field, ii) the probability of finding a job within a shorter period after graduation is 57.2% higher than the same probability for graduates in the reference education field and iii) the gross monthly income is on average EUR 701 higher than the gross monthly income of graduates with comparable qualifications in the reference educational field. These results support the main postulate of human capital theory that investment in high-quality and relevant education and training brings economic benefits to individuals and society.

KEYWORDS

Eurograduate, ICT, Employability, Human Capital Theory

1. INTRODUCTION

Recent developments in education and employment policies, as well as contemporary research findings, show that human capital is becoming increasingly important for the development of individuals as well as for economic growth and the development of inclusive societies. Possessing relevant skills and knowledge arguably increases an individual's chances of finding quality employment after graduation. Building on the basic principles of human capital theory, complemented by the stipulations of cultural capital and social capital theories, as well as constructivist education theory, this paper examines the relationship between the field of study and the employability of graduates and shows what advantages ICT graduates have in the Croatian labour market.

The operational employability model used in this analysis is embedded in the broad theoretical and conceptual framework. It builds on previous research in this area and on existing analytical frameworks that explore the concept of employability. It has been adapted to the Eurograduate instrument and the data collected in the pilot survey. The operational employability model was tested and piloted as part of a wider research of the Eurograduate data and is used in this paper to examine the hypothesis that stipulates that ICT graduates are more employable than graduates from other disciplines.

Employability is measured here by three indicators: i) finding employment in occupations requiring a higher education qualification, ii) in a shorter period of time after graduation, and iii) having a gross monthly income that is higher than the income of persons with comparable qualifications in the reference educational field. Focusing on the relationship between field of study and employability, the assumption that there is a causal relationship between field of study and the three dimensions of employability was tested and the hypothesis was confirmed.

The paper begins with the theoretical and conceptual framework that provides the background for the development of an operational model of employability, presented in Section 2. Section 3 describes the methodology which includes the presentation of the sample, research instrument and research design. Section 4 presents the results and the final Section 5 provides conclusions that highlight the benefits, limitations and potential applications of this study.

2. THEORETICAL AND CONCEPTUAL FRAMEWORK

The theoretical framework for the analysis presented in this paper is grounded in the human capital theory, complemented by the theories of cultural and social capital and constructivist education theory. The conceptual framework is based on the analysis of the chronological development of the construct of employability, the existing analytical frameworks and the models of employability that show the complexity of the different factors that influence the outcomes of individuals in the labour market. The theoretical and conceptual framework formed the basis for the development of the operational employability model applied in the analysis presented here.

2.1 Human, Social and Cultural Capital Theories

Human capital is generally and broadly defined as "a person's skills, knowledge and competencies that are developed through education and training" (Fuente and Ciccone, 2003). Human capital theory, developed in the 1960s by economists Gary Becker and Theodore Schultz (Schultz, 1961; Becker, 1962) has regained prominence with recent policy developments in the knowledge economy and society that depend on people's skills and knowledge. OECD data show that higher educational attainment increases the likelihood of being employed and that higher educational attainment also brings financial rewards. More specifically, the data show that "on average across OECD countries the employment rate is 61% for 25-34 year-olds without upper secondary education and 85% for those with tertiary education" and that "full-time workers with a tertiary degree earn 54% more than those with an upper secondary education" (OECD, 2020).

The focus on individuals' skills and knowledge as important determinants of employability places activating learning and teaching methods (O'Neill and McMahon, 2005; Hannafin and Hannafin, 2010) and engaging student practices or "high impact practices" (Hu and Wolniak, 2013; Wolniak and Engberg, 2019) at the centre of educational theories. According to constructivist theory, people construct their knowledge with their own different experiences and views of reality (Carlile and Jordan, 2005). This implies that students actively learn and engage in knowledge creation (Sfard, 1998) which requires active and deep learning and understanding (Lea, Stephenson and Troy, 2003) as well as high impact practices such as volunteering, studying abroad or work-based learning, that affect early career outcomes (Wolniak and Engberg, 2019).

However, possessing the right skills and knowledge is not enough to explain positive labour market outcomes for individuals. It is necessary to understand the role that a broader social context and access to social resources and networks play in creating educational and career opportunities. The theory of cultural and social capital advanced by the French sociologist Pierre Bourdieu emphasises the importance of family cultural and educational background (Bourdieu, 1977, 1986; Bourdieu and Passeron, 1990) and highlights the value of social networks and connections for advancement in society (Lin, 1999) which includes finding a suitable job after graduation.

2.2 Conceptual Framework

The conceptual framework for the analysis of employability of graduates is based on the chronological developments of employability as an exploratory construct, existing analytical frameworks and employability models that demonstrate the complexity of different factors affecting persons' outcomes in the labour market.

The conceptual framework for analysing graduate employability draws on the chronological development of employability as an exploratory construct, existing analytical frameworks and employability models that reveal the complexity of the various factors that influence individuals' labour market outcomes. A review of the literature shows that since the beginning of the last century, when the focus was exclusively on individual characteristics, i.e. the supply side (Gazier, 2001) the definition and concept of employability has evolved towards a more complex concept that additionally incorporates the demand side, i.e. characteristics of the labour market and local economy (Peck and Theodore, 2000; McQuaid and Lindsay, 2005). Examples of broad frameworks for employability analysis include individual characteristics, discipline-specific and transversal knowledge, skills and competences as well as personal circumstances on the one hand, and external factors such as labour market demand for certain skills and knowledge on the other hand (Hillage and Pollard, 1998; Dale and Krueger, 2002; Eby, Butts and Lockwood, 2003; de Grip, Loo and Sanders, 2004; McQuaid and Lindsay, 2005; Black and Smith, 2006; Van der Heijde and Van der Heijden, 2006).

The theoretical and conceptual framework provided the background for the development of an operational employability model, additionally adapted to the Eurograduate research instrument used to collect data on graduates. The operational employability model was tested on a set of data on individuals who graduated from a higher education institution in Croatia in 2016/2017 and some results are presented in this paper.

2.3 Operational Employability Model

The operational employability model used in this analysis (Figure 1) identifies cultural, social and human capitals as well as specific learning experiences during higher education as predictors of employability. The interaction between factors associated with cultural and educational family background, socio-economic status, social networking, personal characteristics, knowledge, skills and attitudes, and promoted by activating learning and teaching methods and specific high-impact practices during study, affects graduate employability as measured by three indicators, namely finding employment in occupations matching graduates' qualifications, income and the probability of finding a job in a shorter period of time after graduation. The operational employability model was developed as part of a broader research and doctoral thesis Employability of Higher Education Graduates: Transition from Higher Education to Labour Market (Tecilazić, 2023).

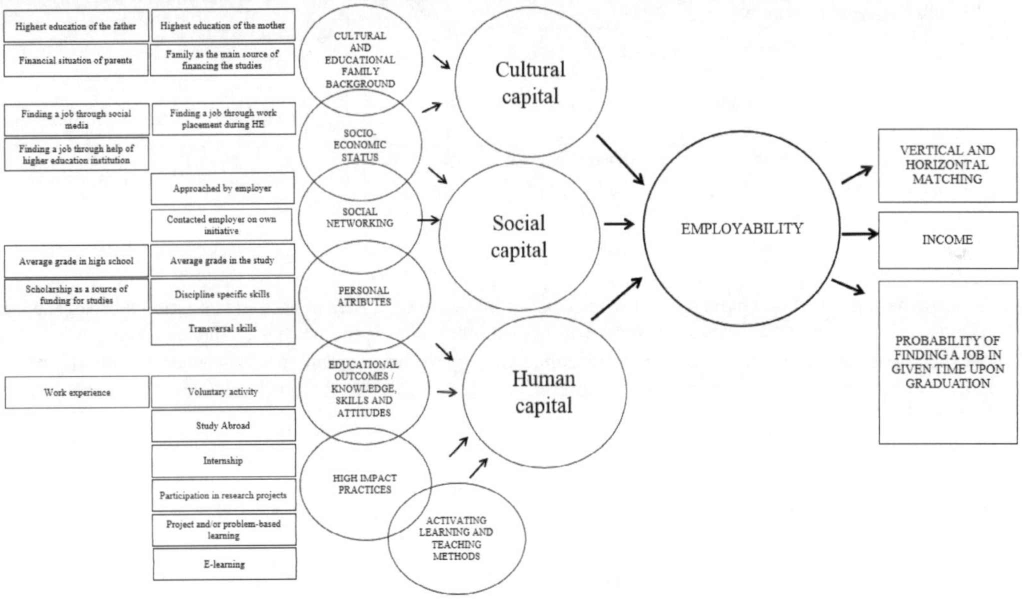

Figure 1. Operational employability model

The predictors of employability in the operational employability model include measured variables and factor scores, as summarised in Table 1. Factor analysis was applied to reduce the number of measured variables indicating the same constructs of human, social or cultural capital, activating learning and teaching methods and high impact practices. The measured variables and factor scores were used as predictors in the multiple regression analysis presented in the following section.

Table 1. Variables included in the operational employability model

MEASURED VARIABLE INCLUDED IN THE OPERATIONAL EMPLOYABILITY MODEL	CONSTRUCT
"Highest education of the mother"	CULTURAL CAPITAL (Cultural and educational family background; Socio-economic status)
"Highest education of the father"	
"Financial situation of parents"	
"Family as a source of financing the studies"	
Successful employment strategy - "Through social media"	SOCIAL CAPITAL (Social networking - Bridging social capital)
Successful employment strategy - "Through help of HEI"	
Successful employment strategy - "Through work placement during HE"	
Successful employment strategy - "Approached by employer"	
"Scholarship as a source of financing the studies"	HUMAN CAPITAL (Educational outcomes – knowledge, skills and attitudes)
"Average final examination grade in secondary education"	
"Average final grade in higher education"	
Successful employment strategy - "Contacted employer by own initiative"	
Discipline specific or different transversal skills	
"Participation in research projects"	Activating Learning and Teaching Methods
"Internships/work placement as formal part of the study programme"	
"Project and/or problem-based learning"	
"E-learning"	
"Study in another country"	High Impact Practice – Study Abroad
"Studying in another country: mobility"	
"Internship as part of or outside the curriculum (abroad)"	
"Voluntary activity related to higher education institution"	High Impact Practice - Volunteering
"Voluntary activity not related to higher education institution"	
"Work experience in the country of HEI not related to study"	High Impact Practice - Working experience
"Work experience abroad that not related to study programme"	
"Work experience before studying"	
"Internship as part of or outside curriculum (country of HEI)"	High Impact Practice - Internship

However, it has to be emphasised that the operational employability model, grounded in a broad theoretical framework, built upon the previous research and conceptual developments of the employability construct, does not include a variety of factors that may influence employability of graduates. The operational employability model was adapted and, therefore, limited to the set of data available from the questionnaire used in the pilot Eurograduate tracking survey.

3. METHODOLOGY

The research findings presented in this paper are part of a broader research and analysis of data collected as part of the Eurograduate Pilot Survey. The method used in this research is multiple regression analysis, using as predictors measured variables and factor scores resulting from factor analysis that indicate constructs of human, social or cultural capital.

3.1 Sample and Instrument

The data on graduates collected between October 2018 and February 2019 as part of Eurograduate, the first pilot study tracking graduates in Europe, includes responses from a total of 16,408 graduates from two cohorts graduating in 2012/2013 or 2016/2017 in eight European countries (Austria, Croatia, Czech Republic, Germany, Greece, Lithuania, Malta, Norway) (Meng *et al.*, 2020; Rimac, 2020). The final dataset was weighted using population data and then clustered based on the level of qualification (BA, MA) and the type of higher education institution (university, non-university). The analysis presented in this paper is conducted on a sample of 1,620 individuals who graduated from master's programmes at universities in Croatia in 2016/2017. The sample is distributed across nine broad fields of study according to the broad European classification ISCED-F, which includes: Education; Arts and Humanities; Social Sciences; Journalism and Information; Business, Administration and Law; Information and Communication Technologies (ICT); Engineering, Manufacturing and Construction; Agriculture, Forestry, Fisheries and Veterinary; Health and Welfare; and Services. Table 1 provides a description of the demographic characteristics of the sample distributed across the study areas.

Table 2. Description of the sample

Sample characteristics	%
Age	
<24	44.4
25 – 29	46.0
30 – 34	5.0
35 – 39	1.8
40 – 49	2.3
50<	0.4
Gender	
Male	36.6
Female	63.4
Study field	
Education	9.0
Arts and Humanities	11.9
Social Sciences, Journalism and Information	8.3
Business, Administration and Law	25.1
Natural sciences, Mathematics and Statistics	5.8
Information and Communication Technologies	4.1
Engineering, Manufacturing and Construction	14.9
Agriculture, Forestry, Fisheries and Veterinary	5.1
Health and Welfare	11.2
Services	4.6

The survey instrument used in this research was a standardised online questionnaire with data centrally collected and anonymised by the Deutsche Zentrum für Hochschul und Wissenschaftsforschung (DZHW) – German Centre for Higher Education Research and Science Studies.

3.2 Research Design

In analysing the data, multiple linear regression analysis was applied to determine the causal relationship between field of study and employability as measured by three indicators, controlling for other possible predictors identified in the operational employability model (Figure 1). First, logistic regression was used to examine the odds ratio of finding a job in an occupation vertically matched to the qualification. Second, survival analysis and Cox regression were used to examine the factors that influence the probability of finding a job within a short period of time after graduation. Finally, linear regression analysis was used to predict the relationship between field of study as the independent variable and gross monthly income as the dependent variable.

The independent variables were included as covariates in blocks in each analysis. The first block contained eight variables related to fields of study, with "services" being the reference field of study. The second block contained factor scores related to individual characteristics and family cultural and educational background, and the third block contained factor scores related to learning and teaching methods and high-impact practices (volunteering, internships, work experience, study abroad) during studies. The fourth block contained variables related to efficient strategies in finding the first significant job after graduation, as follows: "advertisements in electronic media"; "Croatian Employment Service"; "contacting the employer on one's own initiative"; "approached by an employer"; "family, friends and acquaintances". The research results are presented in the following section.

4. RESEARCH RESULTS

Research on the employability of graduates based on data collected in the Eurograduate Pilot Survey shows that graduates in the broad field of study "Information and Communication Technology - ICT" (ISCED F) are more employable than graduates with comparable qualifications in the reference educational field.

First, the logistic regression results (Table 3) show that, controlling for other variables, five predictors have a significant impact on success in obtaining a job requiring a master's level degree (vertical matching): field of study, cultural capital, human capital, social capital and work experience. The regression analysis shows that the odds ratio of finding employment in occupations that vertically match the qualifications of graduates with a master's degree from universities in Croatia is 309% higher for ICT graduates than the corresponding odds ratio for graduates with a master's degree in the reference educational field of "Services" (B = 1.407; Exp(B) = 4.085; p = 0.014).

Table 3. Predictors for vertical matching of qualifications and occupations

	B	S.E.	Wald	df	Sig.	Exp(B)
Status (full time)	0.458	0.259	3.125	1	0.077	1.582
ISCED field			25.157	9	0.003	
Education	1.225	0.457	7.197	1	0.007	3.404
Arts and Humanities	0.781	0.439	3.163	1	0.075	2.184
Social Sciences, Journalism and Information	0.520	0.433	1.441	1	0.230	1.682
Business, Administration and Law	0.666	0.368	3.288	1	0.070	1.947
Natural Sciences, Mathematics, Statistics	1.078	0.509	4.489	1	0.034	2.939
Information and Communication Technologies	1.407	0.570	6.093	1	0.014	4.085
Engineering, Manufacturing and Construction	0.276	0.391	0.499	1	0.480	1.318
Agriculture, Forestry. Fisheries and Veterinary	-0.181	0.571	0.101	1	0.751	0.834
Health and Welfare	1.393	0.433	10.350	1	0.001	4.026
Family cultural capital	0.312	0.084	13.691	1	0.000	1.366
Human capital (education outcomes)	0.355	0.092	14.789	1	0.000	1.426
Passive learning and teaching methods	-0.063	0.082	0.591	1	0.442	0.939
Active learning and teaching methods	0.038	0.076	0.252	1	0.616	1.039
Study abroad	0.001	0.076	0.000	1	0.990	1.001
Volunteering	-0.018	0.080	0.050	1	0.823	0.982
Work experience	0.200	0.081	6.112	1	0.013	1.221
Internship	-0.124	0.083	2.206	1	0.137	0.884
Finding a job through advertisement in (online) newspaper	-0.276	0.193	2.041	1	0.153	0.759
Finding a job through the Croatian Employment Service	-0.179	0.194	0.853	1	0.356	0.836
Finding a job through social media	0.062	0.338	0.033	1	0.855	1.064
Finding a job by contacting the employer - own initiative	0.329	0.191	2.972	1	0.085	1.390
Finding a job by being approached by an employer	0.920	0.271	11.551	1	0.001	2.510
Finding a job through work placement	0.369	0.367	1.011	1	0.315	1.446
Finding a job through family, friends or acquaintances	-0.787	0.196	16.079	1	0.000	0.455
Finding a job through higher education institution	0.290	0.362	0.640	1	0.424	1.336
Finding a job through internship	-0.108	0.448	0.058	1	0.810	0.898
Constant	-0.142	0.415	0.117	1	0.732	0.868

Secondly, the results of the Cox regression (Table 4) show that when controlling for other variables, the factors that affect obtaining a job in a shorter time after graduation for graduates with a master's degree from universities in Croatia are the field of study and the specific strategies used in obtaining the first job. The regression analysis shows that the probability of finding a job in a shorter time after graduation is 57.2% higher for graduates from the ICT sector than for graduates from the reference field of study "Services" (Exp(B) = 1.572; p = 0.031).

Table 4. Predictors for probability of finding a job in a shorter period of time after graduation

	B	**SE**	**Wald**	**df**	**Sig.**	**Exp(B)**	**95.0% CI for Exp(B)**	
Gender	-0.035	0.071	0.240	1	0.625	0.966	0.840	1.111
Status (full time)	0.595	0.108	30.189	1	0.000	1.813	1.466	2.242
ISCED field			18.841	9	0.027			
Education	0.092	0.192	0.230	1	0.631	1.097	0.752	1.599
Arts and Humanities	-0.258	0.199	1.682	1	0.195	0.773	0.524	1.141
Social Sciences, Journalism and Information	-0.005	0.199	0.001	1	0.981	0.995	0.673	1.471
Business, Administration and Law	0.044	0.179	0.061	1	0.804	1.045	0.737	1.483
Natural Sciences, Mathematics, Statistics	0.058	0.192	0.091	1	0.763	1.060	0.728	1.543
Information and Communication Technologies	0.452	0.210	4.653	1	0.031	1.572	1.042	2.371
Engineering, Manufacturing and Construction	0.067	0.189	0.124	1	0.725	1.069	0.738	1.548
Agriculture, Forestry, Fisheries and Veterinary	-0.015	0.399	0.001	1	0.970	0.985	0.450	2.155
Health and Welfare	0.001	0.190	0.000	1	0.997	1.001	0.690	1.451
Family cultural capital	-0.036	0.032	1.286	1	0.257	0.965	0.907	1.026
Human capital (education outcomes)	0.002	0.035	0.004	1	0.947	1.002	0.936	1.073
Passive learning and teaching methods	-0.033	0.030	1.179	1	0.278	0.968	0.913	1.027
Active learning and teaching methods	-0.040	0.030	1.776	1	0.183	0.960	0.905	1.019
Study abroad	0.019	0.027	0.481	1	0.488	1.019	0.967	1.073
Volunteering	0.061	0.031	3.803	1	0.051	1.063	1.000	1.130
Work experience	-0.040	0.033	1.448	1	0.229	0.961	0.901	1.025
Internship	-0.009	0.031	0.080	1	0.778	0.991	0.932	1.054
Finding a job through advertisement in (online) newspaper	-0.129	0.077	2.792	1	0.095	0.879	0.756	1.022
Finding a job through the Croatian Employment Service	-0.312	0.075	17.491	1	0.000	0.732	0.633	0.847
Finding a job through social media	0.142	0.126	1.280	1	0.258	1.153	0.901	1.475
Finding a job by contacting the employer - own initiative	-0.254	0.073	12.147	1	0.000	0.775	0.672	0.895
Finding a job by being approached by an employer	0.142	0.089	2.530	1	0.112	1.153	0.968	1.373
Finding a job through work placement	0.357	0.116	9.522	1	0.002	1.428	1.139	1.791
Finding a job through family. friends or acquaintances	-0.009	0.082	0.011	1	0.915	0.991	0.845	1.163
Finding a job through higher education institution	0.196	0.120	2.695	1	0.101	1.217	0.963	1.539
Finding a job through internship	0.105	0.155	0.459	1	0.498	1.111	0.820	1.504

Finally, the results of the linear regression (Table 5) show that when controlling for other variables, the following factors predict obtaining a job with higher income: the field of study and the use of social media when finding employment, while finding a job through the Croatian Employment Service has a statistically significant and negative effect on obtaining a job with higher monthly income. The regression analysis showed that among graduates with a master's degree obtained at universities in Croatia, those with qualifications in ICT are likely to find a job with an average gross monthly income that is EUR 701 higher than that of graduates with comparable qualifications in the educational field of "Services" ($B = 701.420$; $p = 0.004$).

Table 5. Predictors for finding employment with a higher gross monthly income

	Unstand. Coef.		Stand. Coef.	t	Sig.	95.0% Confidence Interval for B	
(Constant)	1217.462	185.739		6.555	0.000	852.909	1582.016
Education	-318.067	215.681	-0.084	-1.475	0.141	-741.389	105.254
Arts and Humanities	-350.737	217.003	-0.088	-1.616	0.106	-776.653	75.180
Social Sciences, Journalism and Information	-196.521	221.292	-0.047	-0.888	0.375	-630.856	237.814
Business, Administration and Law	-100.993	190.103	-0.040	-0.531	0.595	-474.111	272.126
Natural Sciences, Mathematics, Statistics	-174.432	235.777	-0.036	-0.740	0.460	-637.197	288.333
Information and Communication Technologies	701.420	241.484	0.137	2.905	0.004	227.455	1175.385
Engineering, Manufacturing and Construction	181.022	200.601	0.059	0.902	0.367	-212.702	574.746
Agriculture, Forestry, Fisheries and Veterinary	-303.795	279.884	-0.046	-1.085	0.278	-853.128	245.538
Health and Welfare	172.708	206.354	0.052	0.837	0.403	-232.306	577.723
Family cultural capital	-8.237	36.452	-0.008	-0.226	0.821	-79.782	63.308
Human capital (education outcomes)	46.029	40.247	0.041	1.144	0.253	-32.965	125.022
Passive learning and teaching methods	-24.708	36.874	-0.022	-0.670	0.503	-97.081	47.666
Active learning and teaching methods	-32.597	35.639	-0.030	-0.915	0.361	-102.547	37.352
Study abroad	-57.355	32.599	-0.059	-1.759	0.079	-121.338	6.628
Volunteering	53.456	35.922	0.051	1.488	0.137	-17.050	123.962
Work experience	-42.498	37.311	-0.038	-1.139	0.255	-115.729	30.732
Internship	-66.171	38.098	-0.060	-1.737	0.083	-140.946	8.605
Finding a job through advertisement in (online) newspaper	96.128	89.351	0.038	1.076	0.282	-79.243	271.499
Finding a job through the Croatian Employment Service	-307.094	89.477	-0.119	-3.432	0.001	-482.712	-131.476
Finding a job through social media	488.582	157.008	0.104	3.112	0.002	180.421	796.744
Finding a job by contacting the employer - own initiative	-156.351	84.469	-0.065	-1.851	0.065	-322.139	9.437
Finding a job by being approached by an employer	112.995	104.578	0.037	1.080	0.280	-92.263	318.253
Finding a job through work placement	-5.729	138.644	-0.001	-0.041	0.967	-277.848	266.390
Finding a job through family. friends or acquaintances	-56.982	92.593	-0.021	-0.615	0.538	-238.716	124.753
Finding a job through higher education institution	-213.325	143.295	-0.050	-1.489	0.137	-494.573	67.924
Finding a job through internship	154.125	197.230	0.026	0.781	0.435	-232.982	541.231

The results of the analysis show that, controlling for other variables, the field of study has a significant predictive effect on employability and that individuals with a degree in ICT have a higher probability of obtaining a better-paid job that matches their qualifications in a shorter time after graduation than their counterparts with qualifications from the reference field. It can be concluded that those who manage to find quality employment soon after graduation, where they can use the skills and knowledge they acquired during their education and where their qualifications pay off, are arguably more employable than those who have difficulty finding employment, or who take jobs that do not match their skill level and profile, or jobs with lower incomes.

5. CONCLUSION

Research on the employability of graduates based on data collected in the Eurograduate Pilot Survey shows that graduates from ICT programmes are more employable than graduates with comparable qualifications in the reference educational field. More specifically, ICT graduates are three times more likely to find employment in occupations vertically matching their qualifications than graduates in the reference field. The probability of finding a job within a shorter period of time after graduation is 57.2% higher for ICT graduates than for graduates of the reference educational field. The gross monthly income for ICT graduates is on average 701 EUR higher than the gross monthly income of graduates with comparable qualifications in the reference educational field.

It should be noted, however, that much of the data has been anonymised or categorised into broader groups in accordance with the General Data Protection Regulation. For example, study programmes have been grouped into broad fields of study, which limits the researcher's ability to conduct a more detailed analysis and provide evidence of comparative differences between qualifications from different educational fields and from different higher education institutions. Another limitation of this study is the limitation of the operational employability model itself. Although the theoretical and conceptual framework, elaborating on the complexity of the employability construct, acknowledge the labour market demand as a predictor of graduates' employability, the model was adapted to the Eurograduate data set which does not contain data on the employability factors from the side of the labour market demand.

Nevertheless, these research findings support the main postulate of human capital theory that investment in high-quality and relevant higher education brings economic and non-economic benefits to individuals and society. Based on the theories of human capital, cultural capital and social capital, as well as constructivist education theory, the operational employability model has proven to be effective in demonstrating a causal link between the field of study and graduates' employability. The model can be further applied in the analysis of a different data sets and cross-national comparisons can be made to further improve the understanding of graduate employability.

REFERENCES

Becker, G.S. (1962) 'Investment in Human Capital: A Theoretical Analysis', Journal of Political Economy, 70(5, Part 2), pp. 9–49. Available at: https://doi.org/10.1086/258724.

Black, D. and Smith, J. (2006) 'Estimating Returns to College Quality with Multiple Proxies for Quality', Journal of Labor Economics, 24, pp. 701–728. Available at: https://doi.org/10.1086/505067.

Bourdieu (1977) 'Cultural Reproduction and Social Reproduction'. In J. Karabel & A. H. Halsey (Eds.), Power and Ideology in Education (pp. 487-511). New York: Oxford University Press.

Bourdieu, P. (1986) 'The Forms of Capital'. In Richardson, J., Handbook of Theory and Research for the Sociology of Education', Westport, CT: Greenwood, pp. 241–58, p. 15.

Bourdieu, P. and Passeron, J.C. (1990) Reproduction in education, society, and culture. 1990 ed. London ; Newbury Park, Calif: Sage in association with Theory, Culture & Society, Dept. of Administrative and Social Studies, Teesside Polytechnic (Theory, culture & society).

Carlile, O. and Jordan, A. (2005) 'It works in practice but will it work in theory? The theoretical underpinnings of pedagogy'. In O'Neill, G., Moore, S. & McMullan, B. (Eds.) Emerging issues in the practice of teaching (pp. 15–26). Dublin: AISHE.

Dale, S.B. and Krueger, A.B. (2002) 'Estimating the Payoff to Attending a More Selective College: An Application of Selection on Observables and Unobservables, The Quarterly Journal of Economics, 117(4), pp. 1491–1527. Available at: https://doi.org/10.1162/003355302320935089.

de Grip, A., Loo, J. and Sanders, J. (2004) 'The Industry Employability Index: Taking account of supply and demand characteristics', International Labour Review, 143(3), pp. 211–233. Available at: https://doi.org/10.1111/j.1564-913X.2004.tb00269.x.

Eby, L., Butts, M. and Lockwood, A. (2003) 'Predictors of Success in the Era of the Boundaryless Career', Journal of Organizational Behavior, 24, pp. 689–708. Available at: https://doi.org/10.1002/job.214.

Fuente, A. de la and Ciccone, A. (2003) Human capital in a global and knowledge-based economy: final report. Luxembourg: Office for Official Publications of the European Communities (Employment & social affairs).

Gazier, B. (2001) ''Employability: the complexity of a policy notion'. In P. Weinert (Ed.), Employability: From Theory to Practice. New Brunswick, NJ: Transaction Publishers (International social security series, v. 7).

Hannafin, M. and Hannafin, K. (2010) 'Cognition and Student-Centered, Web-Based Learning: Issues and Implications for Research and Theory', in Learning and Instruction in the Digital Age, pp. 11–23. Available at: https://doi.org/10.1007/978-1-4419-1551-1_2.

Hillage, J. and Pollard, E. (1998) Employability: developing a framework for policy analysis. DfEE London.

Hu, S. and Wolniak, G.C. (2013) 'College Student Engagement and Early Career Earnings: Differences by Gender, Race/Ethnicity, and Academic Preparation', The Review of Higher Education, 36(2), pp. 211–233. Available at: https://doi.org/10.1353/rhe.2013.0002.

Lea, S.J., Stephenson, D. and Troy, J. (2003) 'Higher Education Students' Attitudes to Student-centred Learning: Beyond "educational bulimia"?', Studies in Higher Education, 28(3), pp. 321–334. Available at: https://doi.org/10.1080/03075070309293.

Lin, N. (1999) 'Social Networks and Status Attainment', Annual Review of Sociology, 25, pp. 467–487.

McQuaid, R.W. and Lindsay, C. (2005) 'The concept of employability', Urban studies, 42(2), pp. 197–219.

Meng, C. et al. (2020) Eurograduate pilot survey: design and implementation of a pilot European graduate survey. LU: Publications Office. Available at: https://data.europa.eu/doi/10.2766/149071 (Accessed: 30 May 2021).

O'Neill, G. and McMahon, T. (2005) 'Student-centred learning: What does it mean for students and lecturers?', Emerging Issues in the Practice of University Learning and Teaching, 1.

OECD (2020) Education at a Glance 2020: OECD Indicators. OECD (Education at a Glance). Available at: https://doi.org/10.1787/69096873-en.

Peck, J. and Theodore, N. (2000) 'Beyond "employability"', Cambridge Journal of Economics, 24(6), pp. 729–749.

Rimac, I. (2020) Eurograduate pilot survey: country report Croatia. LU: Publications Office. Available at: https://data.europa.eu/doi/10.2766/805719.

Schultz, T.W. (1961) 'Investment in Human Capital', p. 20.

Sfard, A. (1998) 'On Two Metaphors for Learning and the Danger of Choosing Just One', Educational researcher, 27, p. 4. Available at: https://doi.org/10.2307/1176193.

Tecilazić, A. (2023) 'Zapošljivost osoba s kvalifikacijama visokog obrazovanja: tranzicija iz visokog obrazovanja na tržište rada'. Available at: https://www.pravo.unizg.hr/_download/repository/Ana_Tecilazic_-_Disertacija.pdf.

Van der Heijde, C. and Van der Heijden, B. (2006) 'A competence-based and multidimensional operationalization and measurement of employability', Human Resource Management, 45, pp. 449–476. Available at: https://doi.org/10.1002/hrm.20119.

Wolniak, G.C. and Engberg, M.E. (2019) 'Do "High-Impact" College Experiences Affect Early Career Outcomes?', The Review of Higher Education, 42(3), pp. 825–858. Available at: https://doi.org/10.1353/rhe.2019.0021.

PROCEDURE DEVELOPMENT PRACTICES IN FINNISH NUCLEAR POWER PLANTS

Jari Laarni and Marja Liinasuo
VTT Technical Research Centre of Finland Ltd, Tekniikantie 21, Espoo, 02044 VTT, Finland

ABSTRACT

Since operating procedures play an important role in nuclear process control by guiding and standardizing how to cope with demanding operational situations, it is important to have a clear understanding of how operating procedures are developed in the nuclear domain. We have investigated procedure development practices in two Finnish nuclear power plants. In this paper, we present a method for the analysis of these practices and some examples of the application of the method. We have interviewed procedure writers and other experts who are involved with procedure development work and explored the results using methodological tools of Practice and Activity Theory. Procedure development activity can be seen as a social practice, loosely supported by methodological guidance. Especially, the early phases of the development project seem to escape from generic characterization. Based on the interview data, some potential tensions and contradictions were identified, and they are discussed through the lens of Activity Theory.

KEYWORDS

Procedure Development, Nuclear Power Plant, Practice Theory, Human Factors Engineering

1. INTRODUCTION

Procedures provide instructions and guidance on how to perform work responsibilities in due manner. Operating procedures play an important role in nuclear process control by guiding and standardizing how to cope with demanding operational situations. The vast majority of accidents in the nuclear domain are associated with use of operating procedures (Marsden, 1996). From this perspective, it is not insignificant how the procedures describe the actions needed to perform a task, which, in turn, is dependent on how the procedures are developed. The development of procedures is a challenging endeavor including several steps extending from planning and writing from validation (Stanton et al., 2010). In safety critical domains it has been argued that a formal procedure development process is required, and rigorous guidance or method must be developed to ensure that the formal process is followed (Stanton et al., 2010). On the other hand, it is well known that any method can never be rigorously adhered to, but it has to be adopted and adapted to the characteristics of each situation (Dittrich, 2016; Fitzgerald, 1996; Laarni et al., 2022). Since it is important to have a better understanding of how operating procedures are developed in the nuclear domain, our aim is to study the procedure development practices in Finnish nuclear power plants (NPPs).

This paper will present the methodology and the first results of the research. It is a continuation of our former paper, in which the methodology was presented for the first time (Laarni et al., 2022). The paper is structured as follows: Chapter 2 introduces a typical procedure life cycle, regulatory and other guidance on procedure development, and the idea of procedure development as social practice. Chapter 3 presents the methodological basis of our research, and Chapter 4 describes the preliminary results of expert interviews. The results are discussed in Chapter 5, and in Chapter 6 we present concluding remarks.

2. PROCEDURE DEVELOPMENT

2.1 Procedure Development Process

The procedure development process and the procedure life cycle in safety-critical domains include a particular set of phases that follow each other. Figure 1 depicts the procedure life cycle, which is based on System Development Life Cycle (Ahmed et al., 2020). There, procedure planning, design and development are subdivided into seven subsections. Procedure development is an iterative process, and the start of a new 'round' may be caused by various reasons, e.g., a new technical system requires extensive changes in process control, or an inventory of a group of procedures signals a need for harmonization of the procedures.

Terms 'design' and 'development' are used interchangeably in the paper, even though they refer to distinct phases in the procedure life cycle.

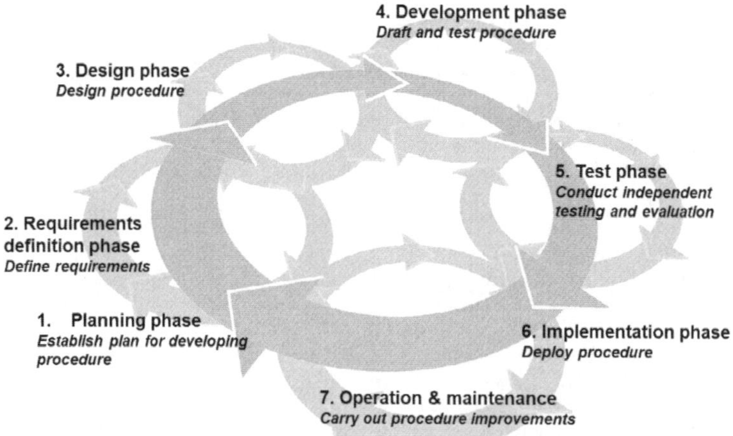

Figure 1. Procedure life cycle (cf., Ahmed et al., 2020). Arrow circles indicate iterations both within each phase and over phases

2.2 Regulatory and Other Guidance on Procedure Development

Methodological guidance used in procedure development in the nuclear domain is based on relevant standards and guides (Laarni et al., 2022). Guidance comes in many forms, and some of it is plant-specific, some of it is national or international. International Atomic Energy Agency's (IAEA) reports (e.g., IAEA-TECDOC-1058, Safety Reports Series No. 48), the U.S. Nuclear Regulatory Commission's (NRC) reports in NUREG- series (e.g., NUREG-0711) and the Electric Power Research Institute's (EPRI) reports (e.g., EPRI 3002004310) are examples of international nuclear guides, and the Finnish Radiation and Nuclear Safety Authority's (STUK) YVL-A.6 is an example of a national guide.

From the perspective of the procedure life cycle, any of the above-mentioned guides does not address all the procedure life-cycle phases. The Procedure Professionals Association's (PPA) AP-907-001 Procedure Process Description is the only document that presents comprehensive guidance for the activities required in procedure development in safety-critical domains. However, as a high-level document, the guide does not provide detailed guidance on any topic.

2.3 Procedure Development as Social Practice

Figure 2 shows our approach for the development of a plant-specific HFE framework for procedure development. National and international standards and guidelines (as described above) have an influence on plant-specific guidance on procedure development. Standards and guidelines also have an impact on the content of plant-specific programs (e.g., HFE program) and reports (e.g., safety analyses). All these plant-specific documents have an impact on procedure development practices, but they do not completely determine them.

Methodological guidance provides tools for developing a procedure. It specifies what is produced, and processes through which the endpoint is achieved. As proposed by Fitzgerald (1996) and Dittrich (2016), it is a normal state of affairs that methodological guidance is never applied as was originally intended by their developers. Experts in different domains do not typically follow the method closely but utilize only those elements that are relevant to their tasks. Different experts also interpret methods differently, and because situations differ, the same person does not apply the method in the same way in different situations.

According to Dittrich (2016), methods can be considered as practice patterns that have to adapted to the contingencies of each situation. These methods in action are social practices consisting of interrelated activities that are connected through shared teleo-affective structures and shared understandings of what it means to develop a procedure in the nuclear domain (Laarni et al., 2022).

To understand procedure development as a practice we should specify tools and equipment that are used, and the settings in which the practices are actualized (Dittrich, 2016). In addition, as said above, we have to take into account that procedure development practices are flexible so that they can change, develop, and adapt to the characteristics of an ongoing situation (Dittrich, 2016).

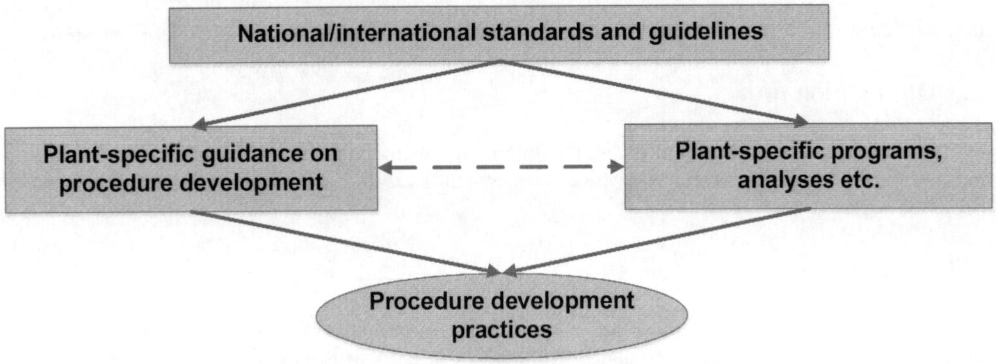

Figure 2. Our approach for the development of a plant-specific HFE framework for procedure development

3. METHODOLOGY

We will investigate procedure development practices by ethnographic methods providing an insider's perspective to understand designers' behaviors, perspectives and experiences (Laarni et al., 2022). The application of the ethnographic approach results in the comprehensive, detailed and contextual interpretation of research data. The ethnographic approach we apply has a special interest in procedure developers' point of view, and its focus is on everyday situations and settings. It is based on an iterative and flexible process of inquiry requiring constant redefinition of the research questions (e.g., Jorgensen, 1989).

3.1 Toolbox Approach

An approach to study practice at work is used (Nicolini, 2012), which is based on a set of stimulating research questions, the answers to which are sought by ethnographic methods of data collection (Laarni et al., 2022). First, we will zoom in on the details of the procedure development practice; after that we will zoom out to enlarge the scope of the study (Nicolini, 2012). In the latter phase, our aim is to focus on connections between practices and their results (Nicolini, 2012). The iterative zooming in and out will continue alternatively until we can provide a feasible account of the practice and its effects on other practices.

Zooming in –questions address topics such as philosophy and strategy on the use of procedures, contributing stakeholders, characteristic features of the design process, tools and artefacts used in the practice guidance used, main constraints of the procedure development practice and main concerns and worries

(Nicolini, 2012). Zooming out –questions address topics such as connections between the procedure development practice under consideration and other engineering practices in the NPP, history of the practice of procedure development, and tensions and contradictions in procedure development practices and their remedies (Nicolini, 2012).

3.2 Procedure Development as an Activity System

By using ethnographic methods and the toolbox approach, we are able to better understand the factors that may contribute to the development of a new procedure. We explore the procedure development project through the lens of Activity Theory and consider the endeavor of procedure development as an activity system. According to the Activity Theory framework, the main elements of procedure development activity are the following: procedure designer (Subject), process control activities requiring proceduralization (Object), new procedure (Outcome), design guides (Rules), NPP personnel (Community), engineering vs. operating personnel (Division of Labour) and modelling tools and simulators (Tools) (Engeström, 2005). In applying the Activity Theory framework, we aim to identify potential tensions and conflicts emergent in attributes of the activity system and between different activity systems (e.g., between different engineering domains or between suppliers and power companies). According to the Activity Theory, these contradictions are not necessarily harmful, but they may stimulate learning and the development of the whole activity system.

3.3 Data Collection

We have conducted a couple of semi-structured interviews by interviewing experts of two Finnish NPPs. Seven experts participated in the interviews, which were conducted at different moments of time during a period of about ten years.

3.4 Data Analysis

Audio recordings of semi-structured interviews were transcribed. The interview transcripts are coded to identify main themes and important concepts. Some main parts of the transcriptions of the discussions are analyzed in a more detailed fashion.

Data analysis is based on a dialectical strategy, according to which data is broken down into elements and components, and they are analyzed further to identify patterns and relationships (Nicolini, 2012). When an interesting topic has been identified, the data is again put together, providing an interpretation of a question of a particular problem. The synthesis is then assessed and critically examined. This zooming-in – zooming-out process can be repeated several times. Finally, the results of the interviews are carefully documented.

4. RESULTS

4.1 Procedure Development Process in Finnish NPPs

Next, we will present preliminary answers to the above-mentioned zooming-in and zooming-out topics. Our answers sketch a Finnish way to develop operating procedures in nuclear domain.

4.1.1 Procedure Development Process

The procedure development project varies much depending on the extent and size of the technical project to which it is associated. The project is much larger and longer, if it is related to an extensive automation renewal project than to an update of an individual procedure. The project is also more extensive if the aim is to harmonize all procedures or to change the format of a large amount of procedures. The process is also different depending on whether there is an existing procedure on the topic or not.

The interviewees agreed on the main steps in procedure development. The following steps were identified: 1) Identification of the initial event and reflecting the plant's response to it; 2) outline of the strategy and specifying the safety functions; 3) analyzing by modelling tools how the strategy actually works; 4) preparation of the first draft (shift supervisor's version); 5) performing additional analyses and testing the draft at the training/development simulator in order to identify the critical details; 6) involvement of operators and testing the procedure at the full-scope training simulator; 7) development of possible operator-role specific procedures (i.e., versions for reactor operator and turbine operator); 8) development of background material and additional information; 9) verification/staff reading review of the procedure; 10) validation planning, incl. determination of criteria for the technical acceptance of a procedure and determination of the safe limits for critical parameters; 11) preliminary simulator validation; 12) final validation; and 13) final check-up and approval.

From the perspective of the procedure life cycle (see Figure 1), steps 1-3 are included in Planning, Requirements development and Design phases of the life cycle, and the rest of the steps are included in Development and Test phases. The steps in the early stage of the development process partially overlap each other, and there are iterations both within steps and across several steps.

Typically, HFE issues play an increasing role at later developmental phases. The amount of modelling and testing in the early stage of the development process is highly dependent on the scope of the project: in new-build and large modernization projects more extensive testing and modelling is needed than when developing a single procedure for a well-analyzed task.

Next, in sections 4.1.2-4.1.8 we present the most important zooming-in topics; in sections 4.1.9-4.1.10 two zooming-out topics are described.

4.1.2 Perspectives Taken on Procedure Development

We first analyzed what positions and perspectives were taken on development work. Procedure philosophy is materialized in a procedure strategy. Regarding emerging operating procedures (EOPs), the procedure strategy can be defined as a description of the main steps to achieve the safe state of the nuclear power process. Typically, designers cannot define the strategy at their work desk. Instead, they have to conduct tests and monitor the plant response, and based on the test results, they can finally settle the strategy.

A Concept of Operations (ConOps) was another term that was used to describe the starting point for the procedure development process – especially for the development of EOPs. A ConOps can be defined as a high-level description of how the proceduralized activity is conducted in a particular situation in order to achieve the safety goals (Fairley and Thayer, 1997). The ConOps, thus, describes the general structure of the procedure, and it is further specified throughout the development process.

In one of the two NPPs, the interviewee had a very practical attitude towards the designers' work: the procedure designer sees him-/herself as a member of the operative unit, and by developing good procedures for operators his/her work contributes to plant operation. Procedure development is also seen as a part of configuration management process, in which functional and physical characteristics of plant systems are controlled based on plant knowledge.

4.1.3 Contributing Stakeholders

We were interested to know who are involved in procedure development. Typically, the total number of practitioners involved in the development work varies much depending on the size of the project. In one Finnish NPP, the typical number is 20-30. Often, there is one person in charge and another one acting as a deputy. The core team also includes, for example, the head of the procedure design department, the main simulator trainer, some people from the safety department, and some heads of divisions. Their role varies depending on the expertise needed in various phases of the development project. Typically, they provide background information, review drafts of the procedure and participate in the verification, validation and approval of the final versions.

In another NPP, there is the core group, which is responsible for the development work itself. In addition to that, a group of experts conducting modelling and testing is involved as well as a group of operating personnel (e.g., operators) from the plant who play an important role in procedure development, because they have an overall view of the state of the process in a particular situation. They present comments on the first drafts, based on which a revised version is drafted. Operators also participate in the validation of a new procedure.

As a whole, it was thought that procedure design is a collective activity. The procedure development group jointly considers alternative solutions in different phases of the development process. Review events at the NPP in which several process experts from relevant domains participate play an important role in the development work. In these events, procedure designers tell the operating personnel that a particular kind of procedure is on the way, what the strategy and the end state will be etc. The aim is to receive early feedback from process experts and get confirmation about whether the development work is proceeding on the right path. There is also collaboration with the safety-engineering group. Safety engineers are able to tell what process changes may occur, and the procedure designers' task is to determine how the effects of the changes should be taken into account in the procedure.

Operator feedback is received through different events: Feedback is based on operators' experiences in simulator training, when they read and familiarize themselves to the background material of the procedure, and when they prepare themselves for the license exam or participate in refresher training.

4.1.4 Characteristic Features of The Design Process

Regarding timing and schedule of the procedure development practice, one interviewee thought that the design process is often rather long: it may last for one year or even longer. Sometimes, there is a fly in the ointment, and the design work has to be started anew from the beginning; sometimes, several partial iterations are needed.

The share between independent work and teamwork varies from one phase to another. Some interviewees thought that the design work is a solitary grind in the early phases of the project: you prepare a draft version alone, and then you submit the draft to reviewers and wait for their comments. Interestingly, it was also considered as creative work: the procedure designer is like an artist, who designs the draft from all bits and pieces (raw material) he/she has. Especially, creativity is needed in the early phase of the design process and also in the development of a totally new procedure. The designer has to overcome the 'white paper fear' and make the first guess, because there is no earlier version from which he/she could copy the first ideas. Typically, the designer can prepare the first draft in different ways: there are several degrees of freedom, and there is no one right answer how to solve a particular problem and present the solution in a procedure.

In one of the two NPPs, meetings with other design team members are arranged about once a month. The progress of the work is discussed, and it is also checked if additional modelling is needed. Nowadays, remote electronic meetings are preferred. Participants are better prepared for them, it is easier to focus on the subject matter, and there is no additional small talk. For many team members working at the plant area, it is also easier to participate in electronic meetings. Remote work during the COVID 19 pandemic has become a part of a normal routine and changed the work practices. In general, the change was considered positive: The interviewee thought that working at home is more suitable for creative work than working in an open office.

One critical question is how the procedure development practice is kept on track. The proverb 'well planned is half done' holds good in procedure development. The procedures must be rigorous enough, and the administrative procedure plays an important role by providing basic guidance about the process. The correct order of progress is that you first familiarize yourself with the topic, after which you can recruit other people to participate in the design work. It is also important that meetings are arranged in a regular fashion.

4.1.5 Tools and Artefacts Used in The Practice

In order to take into account the process effects of plant changes, modelling and simulations play an important role. According to one interviewee, modelling plant processes by Apros (Advanced Process Simulation Software) or other modelling software is very important and nearly obligatory, because the designer cannot take into account everything when he/she is working at his/her desk. Their wish to Father Christmas was a part-scope simulator with which it would be possible to develop and test the procedure with the laptop. According to the interviewee, they do not use specific graphics programs in their work. MS Word and Visio are the most important tools, and they also have a graphical template for the design of flowchart procedures.

4.1.6 Guidance on Procedure Development

High-level administrative guidance on procedure development provides an overall structure for the development work in both Finnish NPPs. In one of the two plants, according to the interviewee, it is the only administrative document providing guidance on procedure development work – more detailed administrative guidance, e.g., in the form of a procedure writer's guide does not exist. In another NPP, there has been intensive collaboration with foreign companies, which have provided guidance on the development of EOPs over a period of about two decades.

As noted in Introduction, there is little detailed regulatory guidance on procedure development in Finland. The interviewees in one of the two NPPs said that international guidance is not utilized, and it is not known very well. Instead of reading documents, discussions with experts from other NPPs are preferred. They receive advice, backup and useful material from their colleagues in another plant. They had also participated in a Nordic project in which plant-specific best practices were shared and compared. Procedure formats were found quite similar in Nordic NPPs.

4.1.7 Main Constraints of The Procedure Development Practices

A constraint means anything that is preventing the designer to achieve the goal. How much the development process is constrained depends on the type of procedure: there are more constraints in the development of EOPs than normal operating procedures. Typically, there is detailed guidance on the development of flowchart-based procedures, incl. general principles, writing and drawing guidelines and guidance on crew behavior. Guidance comes in different levels of detail.

New procedures may be necessary if some equipment having significant process effects is renewed. In this case, the equipment supplier may have a large role in revision of the procedure, since only it knows how the new device functions. For example, it has to be decided with the supplier whether it is possible to test the new system at the factory and whether it can be tested before the launch of the procedure.

The designer has also to incline his/her ears to the opinions and preferences of safety engineers and of operating personnel. Safety analyses place constraints on what direction the development work should take, and as already mentioned, operators take a close look at the usability of new procedures, when they comment on the draft versions and participate in verification and validation activities.

Procedure changes are informed to the Finnish nuclear authority. Principally, the designer has to take into account the nuclear authority's guidance and its remarks but, in general, the authority only keeps a watch that plant changes have been considered in procedure development. As said, the guidance provided by the YVL guides is quite cursory. Overall, the nuclear authority is not much interested in the procedure design process or the contents of the procedures as such.

4.1.8 Main Concerns and Worries

Typically, the designer hopes that the whole process will proceed smoothly and as scheduled, all critical information has been gathered and all critical issues are taken into account so that there is no need to return to previous phases of the development process.

One of the designers' main concerns is whether they properly understand how the new device works and whether all critical process effects of plant changes have been taken into consideration in the design work. In the beginning, there is a discussion on what boundary conditions have to be taken into consideration, what consequences a particular event may have, and what human errors may occur. A special challenge may be that the coverage of a procedure may be difficult to determine beforehand complicating the identification of critical source information. Furthermore, it must be kept an eye on consistency with other procedures, which may turn out to be difficult, if the pool of procedures is big.

At the time the designer makes the first guess, he/she does not know all the details; so, his/her guess may be wrong. Something surprising may happen that the designer has not been able to anticipate, for example, related to a particular process parameter or the functioning of the automation system. It is also very difficult to ask others' opinions, when you do not have drafted anything – you only have the white paper. When the first guesses have been made, discussions with other stakeholders can be started.

Sometimes it is found in testing that the new procedure does not seem to work. The simulation freezes, it has to be interrupted, and it must be analysed what went wrong and why. It is also possible that some experts raise an issue that has not been properly addressed, and therefore the whole process must be started again from the beginning. Experts from different domains notice different things, and the procedure designer can become blind to the details. Sometimes the procedure may be so complicated that the reader is not able to understand it correctly. However, it is not only a matter of facts, but also opinions may differ. In that case, experts have to find a compromise, which all can agree on.

Resistance to change is sometimes a problem - especially in a case of upgrading existing procedures, which operators have used for years. Active operator participation throughout the process is the best remedy for that.

4.1.9 Connections Between the Procedure Development Practice Under Consideration and Other Engineering Practices

Procedure development is linked to several other processes. As was already said, often procedure changes are linked to plant changes: a new device is installed, which has implications to the nuclear process, to monitoring and control activities, which in turn requires changes to the procedures. Therefore, several groups of experts must be consulted in the design phase, such as HFE and training personnel as well as members of the safety analysis team and the operation manager. It is challenging to take into account all the possible implications of other engineering activities on the procedure development practice.

One issue that was raised in the interviews was the relationship between the HFE program and procedure design activities. Both plants have recently developed their HFE practices which also has implications on procedure development activities. Some new requirements have been laid down, stating, for example, that operator training must be arranged before new systems and procedures are taken into use.

Especially, in new-build and extensive modernization projects, the orchestration of interactions between several engineering practices involving practitioners from different organizations is a challenging task.

4.1.10 History of The Practice of Procedure Development

Procedure design is also considered from the perspective of how it has changed and developed over years and decades. The designer sees him-/herself as one link in a chain of historical development. In one NPP, much improvement has been seen: for example, previously, there was no administrative guidance on procedure development. The lack was justified by saying that the development practices were built-in to the courses of action.

In another NPP, in the early 2000's, a decision was made what direction and approach will be taken in the procedure development. International collaboration has played a major role in the development of the procedure development practices. Philosophy of use and ergonomics were 'bought' from a foreign organization and adapted to the Finnish context. There is still intensive ongoing collaboration with foreign stakeholders in procedure design.

4.1.11 Summary

Procedure design activity is still a mystery: after several interviews, we still know quite little about the development process – it seems that the designers are not able to verbally describe the process in a detailed fashion. This is understandable if procedure development is considered as creative work – then, the details of which are understandably difficult to verbalize. Another, not exclusive reason can be that the procedure design can proceed in various ways, depending on the scope of the procedure and perhaps also on the person who leads the process.

An additional problem is that the design process is not addressed in existing administrative guidance in detail. Due to this, it is not possible to compare the design practices to guidance; such guidance just does not exist, and the existing Finnish guidance documents are in such a general level that it does not support procedure development in practice.

Rigorous guidance is a necessary but not a sufficient prerequisite for a good product. On the other hand, high requirements for the product shape the process to a certain mould. Even without strict guidance we can, thus, end up with rather uniform practices.

4.2 Description of The Example Case

Next, we present briefly an example case of a procedure development activity in a Finnish NPP. The aim is to demonstrate how a fluent development process evolves. It was found that there are different opinions on what good communication is in accident situations. Therefore, it was decided to make changes to some procedures regarding communication and change the administrative procedure addressing operative personnel's tasks and duties. There are potential benefits of more detailed guidance on communication practices. First of all, it can promote team situational awareness in more complicated situations (simulator runs) so that the operator has a better understanding of what other operators are doing.

A group of experts was established to conduct the work, including outage coordinators, heads of divisions and shift supervisors (SSs). Their task was to define what good communication is in accident situations. The interviewee was asked to update some procedures from this perspective.

First, an annex of an administrative procedure addressing expectations on communication in accident/incident situations was designed. Topics included are, among others, what has to be verified and who should do that. Two-way communication is typically required, and exceptions must be communicated immediately. Guidance was provided on how the information should be presented in a procedure.

Simulator testing of the new procedures with communication related instructions was conducted in spring 2022. Simulator trainer and head of divisions discussed the progression of the run and the impact of the change of the procedures. They were quite satisfied with the new procedures and thought that the procedures have been improved: the amount of communication was increased, and the crews' performance was more consistent.

Shift supervisors have to make several announcements in accident situations, and previously they typically preferred these announcements over other activities (such as asking information about the process situation from other operators). It was found in the simulator tests that they now inquire other operators before they send the announcements.

There still was some need for improvements, since all the details of the new procedures were not adopted by operators. Some crews delivered too big chunks of information in one utterance, which was not the original idea. It was also found that even though operators should notify that they aim to turn the page, they did not do it in a regular fashion. According to the interviewee, the run proceeded a bit slower than normally, which is understandable, since adoption of anything new takes time and effort.

In the beginning, there was quite much resistance, but it quieted down after the simulator test. The new administrative procedure was submitted to the shift personnel, and also some example procedures were delivered to them. These were discussed in training days in spring.

Regarding SS's briefings, a special form has been developed; it is expected to be filled in at a suitable moment of time during the simulator run. SS determines when there is a suitable time for a briefing.

The work is in progress, and some issues are still open: For example, it must be decided whether there will be communication boxes in all accident/incident procedures. If yes, it will be a very laborous task to realize. The next step is to update the new procedures based on comments. Then, heads of division will once again check them, and after that, it is monitored how the crews apply the new guidance. Human performance simulator runs will be conducted in the following year, in which special attention will be paid to crew communication.

This is an example of a rather simple case, in which existing procedures were improved. Instead of computer modelling, a couple of simulator runs were conducted in a training simulator. Overall, the development work seemed to proceed quite fluently, and the project was kept well on track.

5. DISCUSSION

5.1 Tensions and Contradictions in Procedure Development

In Methodology chapter we described the procedure development process capitalizing on Activity Theory. Figure 3 shows an activity triangle for procedure development demonstrating the interactions between the main attributes of the procedure development activity. According to Activity Theory, if we think backwards, the Outcome may change if one of the corner elements changes (e.g., new rules or tools are introduced). For example, new administrative guidance may be launched, new tools such as part-scope simulators and virtual meeting tools may be taken into use and division of labour may also change. Let's take as an example the effects of COVID 19 pandemic on procedure development work. One of the main dilemmas in the era of COVID 19 was, on one hand, the requirement to continue normal work arrangements and, on the other hand, the prevention of infections by reducing face-to-face contact. One solution to this contradiction was to introduce new work tools and new ways of working based on virtual remote work, which, in turn, challenged one of the fundamental principles of the work in nuclear industry, that is, the principle that all critical work activities are performed inside the fences of a NPP. Apparently, new guidance is needed to orchestrate both individual and collaborative work in the new situation.

It is also possible that new tools (e.g., virtual conferencing tools and part-scope simulators) will challenge the existing division of work between the procedure developer and those practitioners conducting modelling work, which, in turn, may put pressure on skills requirements. Furthermore, these changes may have implications for guidance, and possibly new rules are needed to address the new ways of conducting development work.

Contradictions may also emerge between activity systems. For example, in a new-build project or in a large modernization project typically a group of suppliers is involved, of which some may also be active in procedure development. This may have implications to all attributes of the activity system, for example, through the fact that Community includes not only practitioners from the power company but also from supplier organizations. Some tensions and contradictions between the attributes were identified in expert interviews, and they are illustrated by lightning bolts in Figure 3.

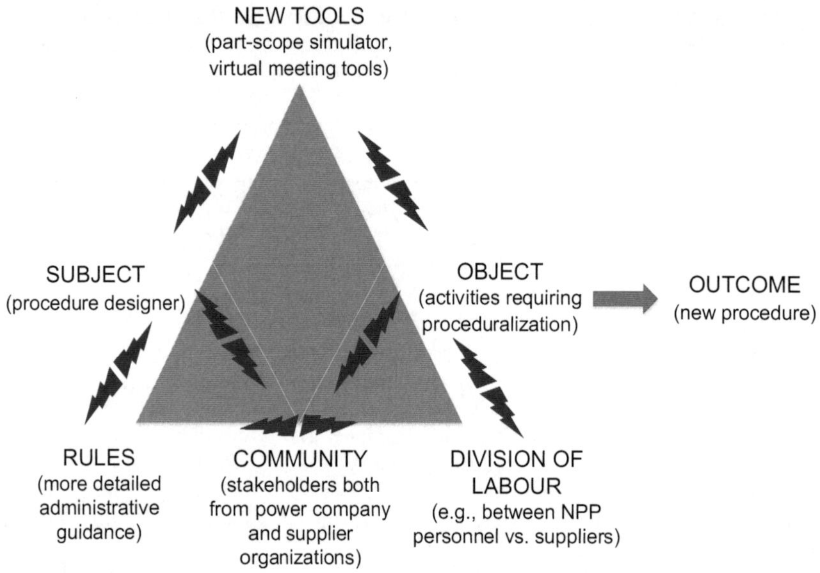

Figure 3. Activity triangle for procedure development (Engeström, 2005). Some tensions and contradictions between the attributes are shown with lightning bolts

5.2 Procedure Development Practices

Is there a need for methodological guidance on procedure development, and if yes, what information should the guidance include? In a general level, the procedure development process follows a particular path, which is well described in several handbooks and guidance documents. In a more specific level, it depends on what kind of procedure we are talking about. There are quite detailed instructions for the development of EOPs, providing guidance both on the process and the format of the procedure.

A plant-specific administrative guidance on procedure development can provide a general framework for procedure design work. Our analysis of one plant-specific guidance showed that the some topics are well addressed in detail in the Finnish administrative guidance such as review of the procedure draft, and approval of the new procedure. On the other hand, some topics are not addressed in the Finnish administrative guidance such as formal procedures for the management of the procedure change request (e.g., evaluation of the request, determination of the request's priority) and formal procedures for planning the development of a new procedure (e.g., determining the workflow for the alteration, detailed description of the development process and finalization of the development process and implementation of the change management process).

Let's assume that we have practical guidance on procedure development. The next question is how the guidance is followed. As was suggested by Dittrich (2016), procedure development practices are flexible so that they can change and develop and adapt to characteristics of an ongoing situation. This means that the activity of designing a new procedure is itself an object of development that can evolve.

What we have found is that it is quite difficult to proceduralize the early phases of the design process, which are an indication of collective creativity and collaborative learning. This kind of collective creativity is especially needed in the planning, requirements specification and design phases of the procedure lifecycle, when the work is still searching for direction and firm basis.

Even though the procedure development is a collective enterprise, there is always the designer in charge, who writes down the first guess and who has to overcome the white paper's fear. As one of the interviewees said, after drafting the first guess, it is easier for others to contribute to the design work.

As was suggested earlier, procedure development can be considered as a social practice: There is the shared intention to design and develop a new procedure, shared understandings what it means to develop a new procedure, explicit rules and principles providing guidance on the work and tools and artefacts that are important in accomplishment of the practice. Implicit in the conceptualization of procedure development as a social practice is the idea that there is always a possibility to do things differently: methods have to be adopted and adapted, and in some way always constructed on the fly in actual implementation.

5.3 Further Work

This paper only scratches the surface in understanding the procedure development practices in nuclear domain due to the fact that our results are based on interviews with a small group of practitioners. To paint a richer and more nuanced picture of the development process, a more versatile selection of ethnographic data collection methods should be included into the methodological toolbox. For example, we should participate in virtual meetings in which the procedure developers take part. Ethnographic observation of these meetings provides insight on how a procedure is drafted, how decisions are made, what drives and guides decision making and how it is implemented.

6. CONCLUSION

In order to create a better understanding of the procedure development process, a new methodological approach is needed, aggregating elements from both rigorous HFE analyses and an ethnographic case study approach. Since the type and scope of projects varies much, there is no universally applicable method for procedure development, but only quite general guidance can be presented. This is especially true in the early phases of the procedure development process, which seems to hide from detailed characterization. On the one hand, it is an individual endeavor, on the other hand it is a collective enterprise of several teams; on the one hand, it is a strictly proceduralized, analytical activity, on the other hand, both individual and collective creativity is needed. According to our interviewees, the periods of individual and collaborative and analytical and creative work typically follow each other in a quite regular fashion.

We have explored the procedure development through the lens of Practice Theory. But in order to be able to specify in more detail the understandings, rules, intentions and usage patterns of tools and technologies we must dig deeper into the implementation of procedure development practices. Only this way we are able to see the full potential of the ethnographic approach.

ACKNOWLEDGEMENT

Research was conducted under The Finnish Research Programme on Nuclear Power Plant Safety 2019 – 2022 and funded by The National Nuclear Waste Management Fund. The authors would like to thank the procedure designers, the simulator trainers and other personnel involved in this study.

REFERENCES

Ahmed, L. et al., (2020). Development of a Procedure Writers' Guide Framework: Integrating the Procedure Life Cycle and Reflecting on Current Industry Practices. *International Journal of Industrial Ergonomics*, Vol. 76, 102930.

Dittrich, Y., (2016). What Does it Mean to Use a Method? Towards a Practice Theory for Software Engineering. *Information and Software Technology*, Vol. 70, pp. 220-231.

Engeström, Y., (2005). *Developmental Work Research: Expanding Activity Theory into Practice*. Lehmans Media, Berlin.

Electric Power Research Institute (EPRI), 2015. *Human Factors Guidance for Control Room and Digital Human-System Interface Design and Modification (EPRI Report 3002004310)*. EPRI, Palo Alto, CA.

Fairley, R.E. and Thayer, R.H., (1997). The Concept of Operations: The Bridge from Operational Requirements to Technical Specifications. *Annual Software Engineering*, Vol. 3, pp. 417-432.

Finnish Radiation and Nuclear Safety Authority (STUK), (2019). *Conduct of Operations at a Nuclear Power Plant,* 15.6.2019. *YVL A.6.* https://www.stuklex.fi/en/ohje/YVLA-6.

Fitzgerald, B., (1996). Formalized Systems Development Methodologies: A Critical Perspective. *The Information Systems Journal*, Vol. 6, No. 1, pp. 3-23

International Atomic Energy Agency (IAEA), (1998). *Good Practices with Respect to the Development and Use of Nuclear Power Plant Procedures (IAEA-TECDOC-1058).* IAEA, Vienna.

International Atomic Energy Agency (IAEA), (2006). *Development and Review of Plant Specific Emergency Operating Procedures (Safety Reports Series No. 48).* IAEA, Vienna.

Jorgensen, D.L., (1989). *Participant Observation – A Methodology for Human Studies*. SAGE, Newbury Park.

Laarni, J. et al., 2022. Understanding Procedure Development in Nuclear Domain with Practice Theory. *Proceedings of the 13th International Conference on Applied Human Factors and Ergonomics (AHFE2022).* July 24-28, 2022, New York, USA.

Marsden, P., (1996). Procedures in the Nuclear Industry. In Stanton, N. (ed.) *Human factors in nuclear safety*, pp. 99-116. Taylor & Francis, London.

Nicolini, D., (2012). *Practice Theory, Work, & Organization. An Introduction*. Oxford University Press, Oxford.

Novatsis, E. and Skilling, E.J., (2016). Human Factors in the Design of Procedures. In Edmonds, J. (eds.) *Human Factors in the Chemical and Process Industries: Making it Work in Practice*, pp. 291—307, Elsevier, Amsterdam.

Procedure Professionals Association (PPA), (2016). *Procedure Process Description (AP-907-001)*, Rev. 2. https://www.ppaweb.org/.

Stanton, N.A. et al., (2010). *Human Factors in the Design and Evaluation of Central Control Room Operations*, Chapter 4: Procedures. CRC Press, Boca Raton.

U.S. Nuclear Regulatory Commission (NRC), (2004). *Human Factors Engineering Program Review Model (NUREG-0711)*. NRC, Washington, DC.

ONLINE TEACHING IN A UNIVERSITY ENVIRONMENT WITH LIMITED INTERNET ACCESS: CASE OF CONGOLESE UNIVERSITIES

Efrem Mbaki Luzayisu
Professor at University of Kinshasa and at Catholic University of Congo, Congo

ABSTRACT

For the past couple of decades, online learning and working remotely has become a daily reality in western countries. Governments, companies and universities continuously invest in equipment and software that will help promote and better facilitate working and learning remotely. With COVID-19 pandemic, working and learning remotely became indispensable as traveling was restricted to stop the spread of the virus. In developing countries such as Democratic Republic of the Congo, (DRC), realities are unfortunately different. There is a strong mistrust of online learning. In addition, authorities tarry to invest in educational infrastructures regardless of the fast-increasing number of students. However, the global health crisis of 2019 as well as other factors has shifted the mindsets of many Congolese authority figures in regards to online learning. The purpose of this reflection is to examine these developments before proposing some avenues for, on one hand, safeguarding the achievements and, on the other hand, working to improve these avenues. Ultimately, the main idea would be to examine the experiment conducted in four DRC's universities in order to build a solution, the cheapest possible, that could be applied to sub-Saharan Africa nations' universities where the education's condition is similar to DRC's. Internationally, in a context where the Internet is almost everywhere continuously, such a study might seem of little use. However, in a poorly connected environment, it is a challenge to implement a web campus.

KEYWORDS

Online Teaching, Educational Scenario, COVID-19, Virtual Classroom, Strategies, Web Campus, Limited Internet

1. INTRODUCTION

Distance learning, also known as distance training, online education, or e-learning, is a paradigm of learning delivered electronically (Anderson, 2001). There are also many different elements that can constitute an e-learning program, such as live or pre-recorded lecture, video, quizzes, simulations, games, activities and other interactive elements. Indeed, online learning encompasses a range of technologies such as the worldwide web, email, chat, new groups and texts, audio and video conferencing delivered through computer networks to impact education. It helps the learner to learn at their own pace, according to their own convenience. Online Education requires a great deal of resources and careful planning. Therefore, teachers act as facilitators rather than transmitters of content, knowledge, and Information and Communications Technology (ICT) is regarded as resource that enhances the learning experience of students (Katsaris, 2021). In this context, students learn through e-learning tools which are available to all (Lavidas, 2022). e-Learning has brought back the joy in the learning process through its innovative and interactive contents delivery and has been proved to be more appealing among students (Indira, 2016). Nevertheless, one relevent advantage of public schooling is that most schools are local, and all children can therefore attend. Thus, for online learning to be acceptable as an alternative to in-class instruction, it should be available to all children. This was clearly not the case during Covid-19, in even the most economically advanced countries (Tony, 2021). Therefore, facing the challenges of the underdevelopment of education in Africa, particularly in the Democratic Republic of Congo (DRC), this article seeks to analyze how and why e-learning could be one of the answers. The ultimate goal is to put forward proposals that would allow the introduction of e-Learning without an explosion of installation and operating costs for the country's higher education and academic institutions.

1.1 Problematic

Despite the valuable efforts of both academic and political authorities, the real observation is that the buildings of the public universities of the Democratic Republic of Congo continue to deteriorate. At the same time, the student population continues to increase each year with a relatively lenient policy of hiring personnel. As if this was not enough, the authorities granted authorizations for the creation of new academic orientations without having the classrooms to receive these new students. Thus, in order to accommodate the exponential growth of students per classroom space, some faculties were forced to hold lectures in disorganized segments. For example, in the Mathematics, Statistics and Computer Science Department of the Faculty of Science and Technology of the University of Kinshasa twelve classes shared six auditoriums. The Table 1 below illustrates the growth between 2015-2016 and 2020-2021 regardless of the category of the five estimated populations.

Table 1. Recorded variation in the intervals 2015-2016 and 2019-2020(*)

N°	Variables	Academic Year		Growth rate %
		2015-2016	2019-2020	
1	Establissements	843	971	15
2	Students	462604	564421	22
3	Academic personnel	3137	3692	18
4	Scientific personnel	36589	47866	31
5	Administrative, technical and labor staff	37575	50092	33

(*) Source = Statistical directory of higher and university education for the 2019-2020 academic year, produced by the technical unit for education statistics with the technical and financial support of Unesco, Kinshasa, February

Table 2. Sharing of classrooms between classes of the Faculty of Science and Technology of the University of Kinshasa, Mention Mathematics, Statistics and Computer Science (**)

ROOM	CLASSES	
B1	2nd Year of Computer Science	3rd Year of Computer Science (LMD)
B16	1st Year of Computer Science (Focus on software engineering)	1st Year of Computer Science (Focus on Management)
B27	2nd Year of Computer Science (Focus on software engineering)	2nd Year of Computer Science (Focus on Management)
D5	1st year of Mathematics (LMD) and 1st Year of Computer Science (LMD). Both classes take all courses together.	
D12	3rd Year of Mathematics (LMD)	2nd Year of Mathematics (LMD)
D20	1st year of Mathematics	1st year of Biology

(**) Source = Office of the Department of Mathematics and Computer Science

This being the case, the purpose of this article is to make a plea for Congolese universities to resort to online teaching not only to guarantee the continuity of teaching, despite insufficiency of classrooms, but also to improve the quality of university pedagogy with the exploitation of new information and communication technologies via online teaching. Ideally, this should be done properly without exploding the operating costs of universities. Therefore, the main challenge of this article is to answer the following question: *What is the least expensive strategy for optimizing education in DR Congo?*

1.2 Methodological Approach

With the size of the DR Congo being 2, 345, 408 square kilometers and a population of 89.56 million, the answer to the above question is not obvious. To achieve this, we opt for an experimental approach by implementing a web campus ourselves and making it available to students who follow our courses in four Congolese universities, three being private and one public. Before expanding on the different classes' schedule, we must mention that we have examined the possibility of exploiting, based on the context of this work, the four scenarios of online teaching: (1) *In person - Synchronous*: the teacher and the student are in relatively the same place and work live (at the same time); (2) *In person - Asynchronous*: the student and the teacher are in relatively the same place but work off-line; (3) *Distanced - Synchronous*: the teacher and the student are physically distant from each other but work live (at the same time). Students attend real-time videoconferences according to a defined schedule and; (4) *Distanced - Asynchronous*: the student and the teacher are physically distant from each other and work offline. Students have access to their courses in different formats at any time.

Generally, a fifth scenario called *mixed*, or *hybrid* learning refers to the joint use of e-learning and the traditional learning method, often called face-to-face. The learner will alternate between remote online sessions and face-to-face sessions with the trainer.

1.3 State of Play

First of all, it is worth recalling before moving forward, that e-Learning is the acquisition of knowledge through electronic technologies and media. In simple terms, e-Learning is defined as learning where knowledge transmission and activities are done electronically (Brandon Hall, 2008). It has been shown that online learning gives students greater flexibility in what courses they can take, when they can take them, and where they need to be to take them. Today, the world is witnessing a rapid increase of online courses known as Massive Open Online Courses (MOOCs) that are accessible to a wide audience. Most MOOCs found on the Internet are free (Gerardus, 2022). Some, however, charge a fee in order to award a certificate of completion. In terms of material, the learner will only need a computer and a good internet connection to ensure their studies (Pablo, 2017). Additionally, the year 2020 has served as a laboratory for large-scale online education. Some universities have even taken the steps towards committing completely to distance learning. Indeed, with the COVID-19 pandemic and health restrictions, many schools and universities have closed their doors and distance learning has become a necessity (Benigno, 2020). The effects of the COVID-19 pandemic are present in nearly every aspect of our lives, and the resulting disruption is expected for months and even years to come. With the closing of schools and social distancing measures comes a unique set of challenges for the education sector. (Ambigapathy, 2021). In a very short period of time, teachers have been forced to move from textbooks to online platforms and social networks in order to ensure pedagogical continuity for their learners. They have been forced to find, adapt and develop educational resources for use in these online and offline environments (Alessia, 2021).

However, the Democratic Republic of Congo (DRC) lacks infrastructure in almost all areas. Unfortunately, education is not immune to this reality. Both private and public universities face serious development challenges due to the limited availability of funding. Ideally, CLOMs, the new trend in training and education, will be one of the answers to integrating e-learning into the DRC's education system. Few years ago, education looked like a classroom filled with students and a teacher at the front leading the process. Physical presence was a given, and any other type of learning was questionable at best. Then the Internet came along, radically changing pedagogical paradigms through the exploitation of digital online teaching platforms (Bates, 1996). Without seeking to cast stones at anyone, we notice that some professors are succeeding brilliantly in this transition by transforming their "paper-based" courses into effective teaching materials that can be used online. In the same way, while we are witnessing many dropouts of students disgusted and disoriented by distance learning, there are some who are very studious and very concerned with their course of study despite the difficulties, such a poor quality of the Internet connection and/or the financial overhead (Bernard Michel, 1999). Also, after the lockdown, some universities have evolved their mode of operation by changing their internal regulations and integrated distance learning (Roy, 2021). To achieve this, they have doubled or tripled classroom sizes to accommodate the large number of students. Similarly, they have set up rooms with good network connections to allow students to take courses online. The objective of this paper is to ideally move the future of the DRC education system in three directions: (1) Identify and highlight best practices in online education; (2) Make some proposals to enhance the knowledge and skills of teachers and learners concerning the use of online platforms and educational technologies and (3) Encourage political authorities to implement mechanisms to encourage potential investors to provide capital for hardware and software infrastructure for online education. We hope that this reflection will serve as a basis to initiate constructive exchanges that will allow everyone to progress thanks to the experiences of others in this time of health crisis.

2. TODAY'S ONLINE EDUCATION

Moore et al. (2021) argue that there can now be no return to 'normal', in the sense of what existed before Covid-19: *"for those institutions that have already invested in online and hybrid learning, "normal" is not an idealized past but is a continuation, a process of leaning into multimodal learning ecosystems to further expand access and opportunities*."

Prior to moving on, it is important to remember that the term "Online Education" refers to the use of new multimedia technologies and the Internet to improve the quality of learning by facilitating access to resources and services as well as exchanges and collaboration remotely. In other words, online education is a type of educational instruction that is delivered through internet to students using their home computers. During the last decade, online degrees and courses have become popular alternative for a wide range of nontraditional students, include those who want to continue working full-time or raising families. Most of the time, online degree programs and courses are offered via the host school's online learning platform, although some are delivered using alternative technologies (Online Education, 2020). Accordingly, an online campus, a virtual campus or a virtual university would be a multimedia computer environment dedicated to online teaching and generally accessible via a website. In continuity, a virtual classroom is an online course space that can complement ex-cathedra sessions. Today, almost every major university in the world is involved in e-learning and each has at least one virtual campus. In other words, almost every institution of higher education and university runs one or more multimedia environments dedicated to virtual classrooms. Now that affordable e-learning solutions exist for both computers and the Internet, all it takes is a good e-learning tool to make education easier and virtual from anywhere. Technology has advanced so much that the geographic gap is bridged by using tools that make one feels like being in the classroom (Morrison, 2003). In other words, e-learning is an affordable solution (often free when internet connection is available) that offers learners the opportunity to fit learning into their lifestyle, allowing even the busiest person to pursue a career and gain new skills. However, in Africa the bell is timidly ringing for universities to follow the same trend by accompanying traditional teaching with online training. Certainly, we note here and there a few e-Learning projects, most of which are animated and financed by international organizations, but to date the African university community has shown little interest in these rare e-learning investments.

3. COVID-19 AND EDUCATION

In 2020, the world faced an unprecedented crisis. The International Monetary Fund (IMF) and its members from various countries took action. At the same time, national authorities took bold steps to save lives and prevent the global economy from collapsing. Even reluctant businesses have switched to remote working. And, to prevent the spread of the virus through pupils and students, policymakers in many countries have taken the difficult step of temporarily closing schools and universities. In other words, because of the dangers of a fast-spreading pendemic, most schools, colleges and universities across the world were forced to close to protect students and staff from infection (Oecd, 2021). But education did not stop (Tzavara, 2023). Certainly, in North America, all post-secondary instructors, and many k-12 teachers, pivoted within two weeks to emergency remote learning, using mainly Internet-based video-conferencing technology such as Zoom, Microsoft Teams or Google Meet. This enabled teaching to continue and students to complete their courses and programs (Bccampus, 2021). However, in this section, we will first briefly review the crisis of 2020 and then analyze its impact on education.

3.1 Covid-19, An Unprecedented Crisis

By the fall of 2021, France had acknowledged that it had officially entered the fifth wave with an accelerating Covid-19 epidemic with nearly 20,000 new cases daily, or 7,000 more cases than seven days earlier (Ministere, 2021). At this rate, the country could quickly reach the levels of the United Kingdom and Germany, which recorded nearly 40,000 new infections each day, while the French average over a week remained above the 10,000 new cases mark, which was more than in Italy and Spain. After "Delta", the new variant was now called "Omicron". In Geneva that same year, experts mandated by the World Health Organization (WHO) decided that it should be taken as "a concern". Experts called for increased surveillance and sequencing (Figaro, 2021). With nearly 100 million more people pushed into extreme poverty by 2020, 1.4 billion children affected by school closures, and record unemployment, COVID-19 represents a monumental obstacle to human progress. Before moving on, it is worth recalling that coronavirus (COVID-19) is a disease caused by a new coronavirus now called severe acute respiratory syndrome coronavirus 2 (SARS-CoV-2; formerly known as 2019-nCoV), which was first identified amid an outbreak of respiratory disease cases in Wuhan City, Hubei Province, China. It was initially reported to WHO on December 31, 2019. On January 30, 2020, WHO declared the COVID-19 outbreak a global health emergency. On March 11, 2020, WHO declared COVID-19 a global pandemic, its first designation since the 2009 H1N1 influenza pandemic declaration (Philip, 2021).

3.2 Education During The COVID-19 Crisis

Global data suggests that the risk of transmission of COVID-19 between children in a school setting is minimal when protective measures (social distancing) are adequate (mask, distance, hygiene). However, many governments around the world have instituted temporary school closures. In fact, as early as March 2020, as part of their efforts to contain the COVID-19 pandemic, several governments including the DRC's decided to shut down schools nationwide, denying more than 27 million children access to education. DRC schools first reopened in October 2020, only to close again after two months in response to a second wave of the COVID-19 epidemic. Following advocacy efforts by national and international organizations operating in DRC, the government reopened schools and universities on February 22, 2021 (Unicef, 2021). Overall, more than one billion children are at risk of falling behind due to school closures aimed at containing the spread of COVID-19. To keep the children around the world learning, countries have implemented distance learning programs. Yet many children in the world, especially those in the poorest households, do not have Internet access, personal computers, televisions, or even radios at home. This amplifies the effects of existing learning inequalities. Students without access to the technologies needed for home-based learning have limited means to pursue their education. As a result, many run the risk of never returning to school, undoing years of progress in education around the world. In countries or regions where community transmission of COVID-19 was high, and in settings where social distancing was not feasible, the World Health Organization (WHO) recommended that certain criteria be applied to avoid prolonged closure of schools and/or universities. For example, the WHO encouraged children aged 5 years or younger not to wear masks. For children aged 6-11 years, it advised them to take a risk-based approach to deciding whether or not to wear a mask. Children and teenagers aged 12 years or older should follow national guidelines for adults regarding mask use.

Lockdown in response to COVID-19 have interrupted conventional schooling with nationwide school closures in most OECD and partner countries, the majority lasting at least ten (10) weeks. While the educational community have made concerted efforts to maintain learning continuity during this period, children and students have had to rely more on their own resources to continue learning remotely through the Internet, television or radio. Teachers also had to adapt to new pedagogical concepts and modes of delivery of teaching, for which they may not have been trained. especially students in the most marginalized groups, who don't have access to digital learning resources or lack the resilience and comittement to learn on their own, are at risk of falling behind (Andreas Schleicher 2020). Indeed, Children and adolescents tend to have more mild disease compared to adults if they catch the coronavirus that causes Covid-19, but they are still struggling to cope with the pandemic. Curfews, closures, and lockdowns are taking their toll on their emotional wellbeing (Elizabeth 2021, Fernando 2020, Wayne 2022). In other terms, the outcome of school closures and declining school attendance not only have serious consequences for education, but also carry significant health and safety risks for children and teenagers.

4. EXPERIMENT AND RESULTS

The purpose of this section is to present both the environment in which I have worked and the results I have achieved. Then we introduce a discussion about this experience.

4.1 Research Context

We have opted for a simple solution by taking advantage of our teaching to directly test our hypotheses. Indeed, we currently have an hourly load in four institutions in the DRC as indicated in the Table 3 below where the first column gives the name of the institution. The second gives the nature or type of the university. The third indicates the cities and the number of universities in those areas. The last three columns give the number of faculties, courses and students concerned by our teachings in each of these four universities.

Table 3. Institutions concerned by the experience

Institution	Type	City and #Implantations	#Faculties	#Courses	#Students
Institut Supérieur d'Informatique Programmation et d'Analyse (ISIPA)	Private	Three location in Kinshasa and 1 in Matadi	1 in 2 orientations	2	300
Catholic University of Congo (CUC)	Private	Two locations in Kinshasa	4	8	2000
University of Kinshasa (UNIKIN)	Public	One location in Kinshasa	2 in 5 departments	8	1500
New Horizons University (NHU)	Private	One location in Lubumbashi	4	3	22

4.2 Teaching Tools

For the past twenty years, we have been constantly learning and following the pedagogical and technological evolutions concerning online teaching in order to disseminate them through our courses. Therefore, each new academic year, we spare no effort to make available to our students a web campus developed with Moodle software and presented in figure 1 below to support the various courses in our charge (https://efremmbaki.com/Moodle2022/). by exploiting the list of students, the access parameters to the web campus for each of them were dynamically generated. In this way, each student receives a username and a password to connect to the web campus. We counted 3,355 users as shown in Figure 2 below. Students can find a variety of multimedia learning resources such as simple or commented *.docx, *.pptx, *.PDF or images.

Figure 1. Efrem's Web Campus

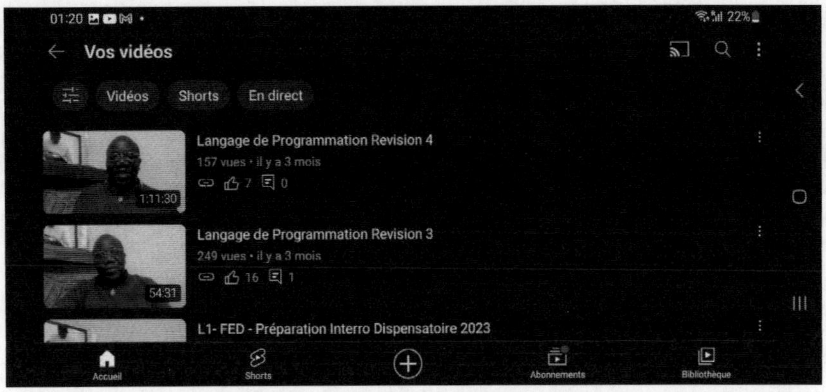

Figure 2. Efrem's YouTube Chanel

Where possible and appropriate, we have also posted explanatory videos on our YouTube channel, see Figure 2 above, before sharing the links in the Web Campus. We opt for "Unlisted" visibility to prevent the videos from being exploited in a context not compatible with our teachings. In addition, we also used activity-type resources to provoke student interactivity and thus follow their respective progress. In fact, after each classroom or distance learning session (with the Zoom Meeting application), we propose interactive tests (Scorm Quizzes) to the learners. Generally, we impose a delay of three days before the exercises expire. It is possible to use Moodle to design and implement quizzes but we have exploited the iSpring Suite to carry out our exercises (Bornouane, 2011). In the same order of idea, we opted for the Wooclap platform to design digital quizzes and/or exercises that can be used in synchronized mode. Moreover, regardless of the telecommunication operators in the DRC, Internet connections are unstable and relatively very expensive. Often it is a miracle to download a video and/or a small document of about ten pages. Under these conditions, to minimize the overall cost of student connections, we have designed and published a nine (9) Gigabyte multimedia DVD (Figure 4) with our course notes, a recent computer dictionary, software setups to be installed as well as explanatory and illustrative videos of the concepts covered in the different courses.

Figure 3. Class notes and videos' DVD

4.3 Discussion

In this section, we will examine the way we have applied the pedagogical scenarios of online teaching in the four universities mentioned above. Indeed, at ISIPA, we teach two courses in the first year of a computer science degree in Software Engineering and Database. On one hand, the Computer Marketplace course was entirely taught online except for the first two sessions which were held in person. As soon as ISIPA does not have the infrastructure to help the students to follow the course in videoconference, we had to find a way around. Therefore, we share three files weekly via WhatsApp: an explanatory video published via YouTube, lecture notes in PDF or PPTx format with comments, and a WooClap link of the formative test. In contrast, the second course on advanced Java programming language was organized as a face-to-face session with substantial support from online tools. Indeed, at the beginning of each session, we offered the students a live WooClap quiz. And at the end of the session, students were invited to solve the Moodle quiz that we posted on our web campus. In the end, we used two of four online pedagogical scenarios: in person-Synchronous and Distanced-Asynchronous.

At the Catholic University of Congo, we taught ten courses during the 2021-2022 academic year as mentioned in the Table 3 in section 4.1 relative to research context. Generally, the lecture sessions always started with an online quiz with the WooClap platform. And after each session, we would give the students a week to go into the web campus to solve the test associated with the material seen. In this way, we were leveraging two of the four online teaching scenarios mentioned above, namely In person-Synchronous and Distanced-Asynchronous. I applied this pedagogy at the University of Kinshasa. However, at New Horizons University, things went differently. In the first semester, we taught the System Administration course totally online with the Zoom application on the university's account. We must emphasize that the university's IT department provides students with well-equipped facilities to connect and follow the course remotely. Since the students were in Lubumbashi and the teacher in Belgium, the only two possible scenarios were: (1) Distanced-Synchronous for the presentations and the Wooclap questions and, (2) Distanced-Asynchronous for the web campus tests. In the second semester, we had the privilege of joining the students on site in Lubumbashi where some sessions were online with Zoom for the Master 2 Project Management course.

5. WINNING STRATEGY

The health restrictions imposed by COVID-19 justify the increased use of online education. As a kind of training, schools and universities are preparing themselves pedagogically and materially in the use of NICTs in order to overcome any closure. Also, in case that computer equipment is no longer a barrier, the establishment of virtual campuses is undoubtedly a positive response to the shortage of documentation so decried in the DRC. This being the case, we will propose to any university that does not yet have a virtual campus to start the process. To begin with, it will reserve a pedagogical space on its web server, which space will be accessible at first by professors who will upload electronic notes, even if it'd be draft only. In a second phase, this space will be opened to students to consult the courses and/or the notes uploaded by the teachers. Furthermore, we will emphasize that ICT and e-learning skills are a key indicator of the ability to engage in e-learning. Certain skills in the use of information and learning technologies, such as Internet-related tools, but also the ability for self-directed learning processes, can be described as necessary preconditions for a successful approach to e-learning. Ideally, therefore, computer courses should give all teachers and learners the opportunity to experience good practice in the use of the Internet and communication technologies. Of course, there is a clear difference between a paper-based course and an electronic course, but in order to move things forward, we sincerely believe that it is unnecessary to set the bar very high in terms of computer quality for the first notes to be produced. Little by little, and according to a strategy to be set up, teachers will learn to exploit existing techniques and/or tools to enrich their courses in order to make them more interactive, and then, more attractive for the learners. For example, it is now very easy to transform an MS Word document into a pedagogical HTML document thanks to one of the tools produced in the e-Learning platforms.

"Win-Win" collaborations can be considered at different levels (governmental, academic or even personal) in order to allow exchanges of pedagogical data from one university campus to another. For example, we could negotiate the possibility of moving all or part of the courses of the virtual campus of the "*Facultés Notre-Dame de la Paix*" in Namur to one or more African universities. Similarly, students from a European university could benefit from video sequences of an erupting volcano, with pedagogical comments by a professor from the South. Only, to avoid any ambiguity, each university will set the rules and manage registrations to their virtual campus(es). The success of the virtual campus will also depend on the active participation of the learners. It would therefore be recommended that students take part in the improvement of course content (Clark 2007).

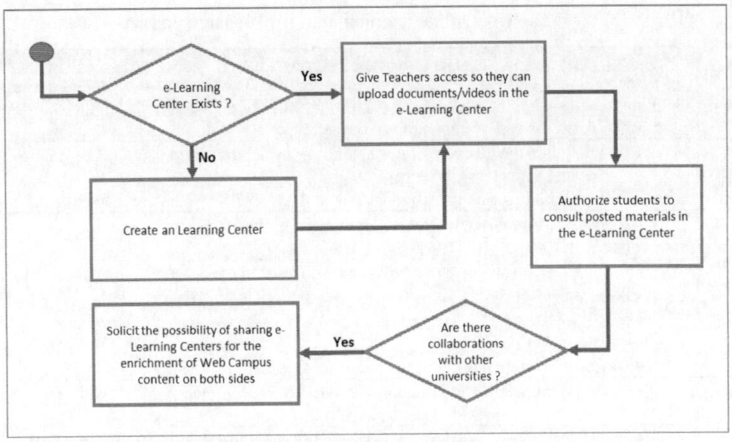

Figure 4. Initialization phase

For example, taking advantage of seminars or programming assignments, they can create PowerPoint documents, video sequences, flash objects or java applets for a given course. In this sense, the competent authorities would be asked to promote and disseminate information about projects or courses to be implemented which are essential factors for the future. This also means raising awareness of the potential of e-learning methods and techniques. In terms of the sustainability of thc course and of e-learning in general, promotion, information dissemination and awareness raising will produce a cascading effect that will reach a wider audience than that of a single course, also helping to increase public interest in e-learning (Leplatre, 1998). To summarize, the ideal would be to organize the implementation of e-Learning in two phases: a first phase of initialization (figure 4 above) before moving on to a second phase of consolidation (Figure 5 below). Indeed, the first phase would concern institutions that do not have a digital environment shared by their academic community. In these conditions, it would be best to start slowly by encouraging each teacher to upload files and/or videos without worrying about their quality. Then, proceed to a second stage of consolidation by insisting on the continuous training of teachers, on free Internet connection and on a pedagogical and technical evaluation of contents.

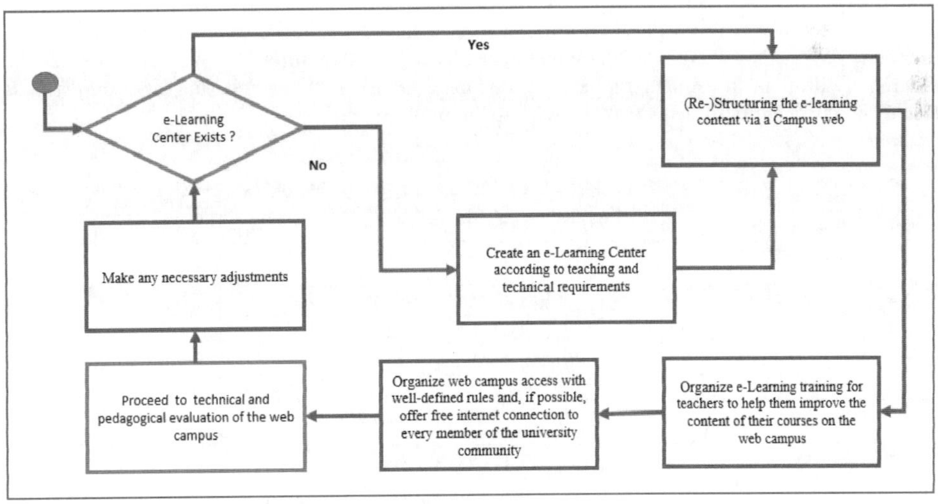

Figure 5. Consolidation phase

Specifically, we propose the implementation of web campuses with the characteristics described in table 4 below:

Table 4. Online learning platform's characteristics

Property	Details
Map	Implementation of a digital platform in order to amplify the online teaching in all the sites of the university by defining, on one hand, the components of the platform and on the other hand, the Business Process (Protocols) of access and/or Publication of contents
Goals	•To strengthen the capacity of our institution, which has been using e-Learning to support teaching for several years (Ex-Cathedra) •To cope with repeated interruptions due to the COVID-19 pandemic by ensuring continuity of teaching
Components	•The Digital Portal (www.isipa.eu or www.isipa.com) •The Web Campus (Moodle and/or Claroline) •A YouTube Channel (of the University dedicated to the cause) •Two Laptops (1 Server and 1 Client) with a good internet connection •A Zoom license •A Quiz Maker license •The rules of internal order of the platform
Enrollment process	•Plan a Zoom meeting in collaboration with the teacher and possibly with the students involved in the course
	•The teacher gives a synchronous lesson with Zoom installed in the Laptop Client
	•If necessary, he/she shares his/her course notes (PPT, PDF, Quiz or Other) with the Administrator
	•At the same time, the Administrator records the course with Zoom installed in the Laptop called Server
Publication process	•Administrator posts recording to YouTube channel (Unlisted and not for children)
	•Administrator shares the YouTube link on the Web Campus
	•Eventually, the Administrator publishes the course notes in the Web Campus
	•Automatically, the rights holders view the new item in their Web Campus accounts
Access protocol	•Each Teacher receives an individual Login and Password with the Courses Creator' profile (Read/Write for his courses)
	•Each student receives an individual Login and Password with the Student profile (ReadOnly)
	•Each member of the community signs the internal rules of the digital platform
Global Approach	•Analyze the Acceptability and the Technical and Financial Feasibility
	•Analyze the conditions to obtain the Adhesion of the teachers
	•Analyze the financial modalities by the students
	•Analyze the possibilities of exploitation of the products archives (Quizzes, notes, videos...)
	•Write the internal order rules

Finally, the table 5 below provides a rough idea of the cost of the essential components of the solution described above. PCs are a one-time cost that generally pay for themselves in four years. On the other hand, it is extremely difficult to estimate the cost of producing DVDs and/or purchasing USB drives. Indeed, it all depends on the quantity of items that you wish to put in place.

Table 5. Components prices

N°	Objects	Source	Cost/Month	Cost/Year
1	Zoom subscription	https://www.zoom.us/	12 x 20€	240 €
2	License IsPring Suite	https://www.ispring.fr/		800 $
3	Website .eu GoGeek	https://www.siteground.com	12X40$	480 $
4	PC Administration Server	https://www.dell.com/		1200 $
5	Multimedia PC Client	https://www.dell.com/		1200 $
5	YouTube	https://www.youtube.com/		Free
6	Moodle	https://moodle.com/		Free
7	Web Campus administration	Price compilation based on congolese market	200 $	2400$
8	Two 360° cameras	https://www.amazon.fr/		2200$
8	DVDs or USB drives	https://www.amazon.fr/	5€ /item	

Clearly, the solution built is the cheapest in the sense that it minimizes costs in the four axes that make up the implementation of a web campus: (1) For the hardware Infrastructure a desktop or laptop computer is enough to serve as a web and application server. The best option would be the subcontracting of the hosting of the web server by paying for a space. And thus, minimize and relieve administrative and / or security tasks related to the website. Since it is a web solution, students use their smartphones, tablets or computers to connect. Under these conditions, we do not think of investing for a large laboratory with connected computers; (2) For the software, we chose Moodle which is free, and we buy a relatively cheaper application to create interactive exercises. In addition, we also subscribe to Zoom with the cheapest formula that can allow us to organize synchronous sessions in small groups; (3) For the administration, one person is enough for the begining. Indeed, the Administrator must ensure access to the web campus by managing the creation of user accounts (students and teachers), the creation of courses and other equivalent tasks and; (4) For the content, we worked alone. Ideally, we would hire a person whose role would be to accompany the professors in the design of their online course.

6. CONCLUSION

Although COVID-19 has caused many human losses around the world, this study has shown that it is a great opportunity for the establishment of online education. Under these conditions, it would be a real shame if the DRC, or any other sub-Saharan country, were to lag behind. Convinced of the usefulness of the web, this article encourages actors in higher education and university to work hand in hand so that each institution can have at least one virtual campus. Of course, it would be naïve to believe that virtual campuses, by themselves, would overcome the digital divide and the cruel shortage of documentation that Africa suffers. Nevertheless, it is certain that electronic data, electronic pedagogy and online learning will mitigate some of the damage. At least students, researchers and professors will enjoy the benefits of online education as they will be optimally browsing for well-formatted and good quality literature. The ideal would be that the cheaper solution proposed in this study be implemented to allow sub-Saharan universities to enjoy the benefits of online education. An idea to continue this study would be to analyze and compare the real cost of implementing e-Learning in a university in a developed country.

REFERENCES

Alessia Plutino (2021), *Languages at work, competent multilinguals and the pedagogical challenges of COVID-19*, Research-publishing.net,

Andreas Schleicher (2020), *The impact of covid-19 on education insights from education at a glance*, OECD, https://www.oecd.org/education/the-impact-of-covid-19-on-education-insights-education-at-a-glance-2020.pdf, Accessed 2022/09/20

Ambigapathy Pandian (2021), *COVID-19, Education, and Literacy in Malaysia: Social Contexts of Teaching and Learning (COVID-19 in Asia)*, Routledge

Anderson, L.W., Krathwohl, D.R. (2001), *A Taxonomy for Learning, Teaching and Assessing*. A Revision of Bloom's Taxonomy of Educational Objectives, Addison Wesley, 2001

Bates Peter J. (1996), *Telematics for flexible and distance learning: final report*, Commission européenne DG XIII/Bruxelles/Belgique

BCCampus (2022), The impact of covid-19 on teaching and learning, https://pressbooks.bccampus.ca/teachinginadigitalagev3/chapter/1-9-the-impact-of-covid-19-on-teaching-and-learning/, 2021, Accessed 2022/09/20

Benigno Núñez Novo (2020), *L'enseignement en ligne à l'époque des coronavirus (COVID-19): Leçons à distance*, Editions Notre Savoir

Benraouane S.A (2011), *Guide pratique du e-learning : stratégies, pédagogie et conception avec le logiciel Moodle.* Malakoff : Dunod : fonctions de l'entreprise

Bernard Michel (1999), *Penser la mise à distance en formation*, L'Harmattan/Paris, Éducation et formation

Brandon Hall (2008), *e-learning, le guide de référence.* Former vos salariés par l'Internet, https://cursus.edu/formations/19709/e-learning-le-guide-de-reference-former-vos-salaries-par-linternet, Consulted on 2021/08/05

Clark R.C. (2007), *The New Virtual Classroom: Evidence-based Guidelines for Synchronous e-Learning*, Pfeiffer

Elizabeth A. K. Jones, Amal K. Mitra,* and Azad R. Bhuiyan (2021), *Impact of COVID-19 on Mental Health in Adolescents: A Systematic Review,* Int J Environ Res Public Health. 2021 Mar, Published online 2021 Mar 3

Fernando Jose Guedes da Silva Junior (2020), *Impact of COVID-19 pandemic on mental health of young people and adults: a systematic review protocol of observational studies*, National Library of Medecine, 2020/07/01, https://pubmed.ncbi.nlm.nih.gov/32611746/, Accessed 2022/09/20

Figaro (2021), *Covid-19 : le variant B.1.1.529 classé «préoccupant» par l'OMS et baptisé Omicron*, Published on 2021/11/26, https://www.lefigaro.fr/sciences/covid-19-le-variant-b-1-1-529-classe-preoccupant-par-l-oms-et-baptise-omicron-20211126, Consulted on 2022/05/08

Gerardus Blokdyk (2022), *Massive Open Online Course MOOC Standard Requirements*, 5STARCooks, 2022

Indira Dhull and MS. Sakshi (2016), *Online learning*, International Education & Research Journal

Katsaris, I., & Vidakis, N. (2021). Adaptive e-learning systems through learning styles: A review of the literature. Advances in Mobile Learning Educational Research, 1(2), 124-145. https://doi.org/10.25082/AMLER.2021.02.007

Lavidas, K., Apostolou, Z., & Papadakis, S. (2022). Challenges and opportunities of mathematics in digital times: Preschool teachers' views. Education Sciences, 12(7), 459.

Leplatre Françoise (1998), *Formations ouvertes et à distance: enjeux et perspectives, pratiques et expériences*, Actualité de la formation permanente, n° 156 et 157

Ministère des Solidarités et de la Santé (2021), *Vaccination contre la Covid en France - au 17 novembre 2021*, plus de 101 960 500 injections ont été réalisées, Published on 2021/11/17, https://solidarites-sante.gouv.fr/actualites/presse/communiques-de-presse/article/ vaccination-contre-la-covid-en-france-au-17-novembre-2021-plus-de-101-960-500, Consulted on 2022/05/08

Moore and als. (2021), *One Year Later . . . and Counting: Reflections on Emergency Remote Teaching and Online Learning*, EDUCAUSE Review, November 10

Morrison D. (2003), *E-Learning Strategies - How to Get Implementation and Delivery Right First Time*, John Wiley

OECD (2021), *The state of school education: one year into the Covid pandemic Paris*, France: OECD

Online education (2020), *What is Online Education ?* https://www.online-education.net/articles/general/what-is-online-education.html, Accessed 2022/09/20

Pablo Achard (2017), *Les Moocs. Cours en ligne et transformations des universités*, PU Montreal

Pascal Picq et Denis Lafay (2020), *S'adapter ou périr: Covid-19 : faire front*, Editions de l'Aube

Philippe A. (2021), *Covid-19, enquête sur un virus : manipulations, vols, meurtres, influences et guerres médiatiques*, le Jardin des livres, 2021/03/16

Roy Chan (2021), *Online Teaching and Learning in Higher Education during COVID-19*, Routledge

Tzavara, A., Lavidas, K., Komis, V., Misirli, A., Karalis, T., & Papadakis, S. (2023). Using Personal Learning Environments before, during and after the Pandemic: The Case of "e-Me". Education Sciences, 13(1), 87.

Tony Bates (2021), *Online learning and (k-12) schools: 2. Technology and cost issues*, https://www.tonybates.ca/2021/01/10/online-learning-and-k-12-schools-2-technology-and-cost-issues/, Accessed 2022/09/20

Unicef (2021), *COVID-19 School closures in the DRC: Impact on the health, protection and education of children and youth*, Published on 2021/05/06, https://reliefweb.int/report/democratic-republic-congo/covid-19-school-closures-drc-impact-health-protection-and-education, Consulted on 2022/05/08

Wayne Deeker (2022), *The Covid generation: the effects of the pandemic on youth mental health*, Horizon (The EU Research & Innovation Magazine), https://ec.europa.eu/research-and-innovation/en/horizon-magazine/covid-generation-effects-pandemic-youth-mental-health, Accessed 2022/09/20

"UNLOCKING MEMORIES, REIGNITING JOY": A FEASIBILITY STUDY ON PERSONALIZED INTERACTIVE GAMES TO ENHANCE REMINISCENCE FOR PEOPLE WITH ALZHEIMER'S DISEASE

Anis Hasliza Abu Hashim[1], Emma Nuraihan Mior Ibrahim[2], Marina Ismail[2] and Juhaida Ismail[3]
[1]*Faculty of Electrical Engineering, Mathematics and Computer Science*
University of Twente, Enschede, The Netherlands
[2]*College of Computing, Informatics and Media*
Universiti Teknologi MARA (UiTM), Shah Alam, Malaysia
[3]*College of Computing, Informatics and Media*
Universiti Teknologi MARA (UiTM) Raub Branch, Raub, Malaysia

ABSTRACT

This paper discusses the significance of research focused on progressive dementia due to the pressing needs of an aging population. Technological solutions have been explored as useful tools to diagnose, manage and treat elderly patients with Alzheimer disease (AD). Interactive games have proven to be an effective therapeutic approach for individuals with AD, improving cognitive function and quality of life. An interactive game was incorporated in my-MOBAL app, a personalized digital memory book, was developed to enhance reminiscence and stimulate cognitive function in individuals with AD. A feasibility study was conducted with clinical experts, an elderly user, and caretakers to collect data via interview and direct observation. Qualitative data collected over 8 weeks showed that personalized interactive games tailored to specific user can promote a sense of familiarity and connection to identity, providing comfort and reducing disorientation in individuals with dementia. It also encouraged communication between the user and the caretakers. Incorporating personalized elements in the interactive games could help to promote autonomy and independence, which are crucial for maintaining self-identity in individuals with dementia.

KEYWORDS

Interactive Games, Alzheimer Disease (AD), Non-Pharmacological Therapy, Reminiscence Therapy (RT), Cognitive Stimulation Therapy (CST)

1. INTRODUCTION

The elderly population is rapidly increasing worldwide, with the number of people over 65 years old growing and expected to continue rising for the next 20 years (United Nations, Department of Economic and Social Affairs, Population Division 2019). Dementia is a syndrome that is associated with ageing factor (World Health Organisation 2021). Alzheimer's disease (AD) is the most common type of dementia, accounting for 60-80% of cases globally. By 2050, the number of people with dementia is projected to reach 135 million worldwide, with 23 million in the Asia-Pacific region alone (Alzheimer's Disease International 2019). According to the "National Dementia Registry Report 2020" by the Ministry of Health Malaysia (2020), there were 41,251 registered cases of dementia in Malaysia, with an average age of onset of 73.4 years. It also mentioned that Alzheimer's disease (AD) was the most common type, affecting 69.8% of cases.

AD has no cure or treatment to reverse its progress, but pharmacological and non-pharmacological interventions are available to slow its development. Nevertheless, pharmacological treatments have limitations such as side effects and limited effectiveness. This leads to the increased of interest in non-pharmacological interventions such as reminiscence therapy (RT) and cognitive stimulation therapy (CST) (The Lancet Neurology Commission 2020; Imtiaz, Khan & Seelye 2018; Lee et al. 2019; Cammisuli et al. 2022). There is a growing body of literature supporting the use of computer technology, including interactive games, as effective interventions to stimulate cognitive function (Nguyen et al. 2020; Kammrath et al. 2021; Li et al.

2022), social interaction and communication (Groenewoud et al. 2017) for people with AD and their caregivers. Interactive games can support AD patients by providing cognitive stimulation and engaging their attention, memory, and problem-solving skills which can improve their mood and quality of life. Besides pleasure, games also could be used as a communication tool among youngsters and adults (Yusof, Mohd Rias & Yusoff 2014).

Over the last few years, several video games were developed, and they focus on multiple perspectives as well as stages of dementia. McCallum and Boletsis reported that dementia games do have an effect on cognitive impaired patients. However, this field still needs to be explored and need more research (McCallum & Boletsis 2013). Further research and development in this area are necessary to improve the effectiveness of AD non-pharmacological therapies in diverse populations. Hence, the aim of this research is to conduct a feasibility study on the design features of our interactive game prototype in my-MOBAL app, a personalized digital memory book with some interactive activities to assess its potential as a culturally tailored intervention for AD patients from diverse backgrounds.

2. RELATED WORK

2.1 Dementia and Alzheimer's Disease

Dementia is a complex and multifaceted condition that affects various aspects of daily life. It is a general term that refers to a group of symptoms, including memory impairment, language difficulties, and changes in behavior and personality, that can interfere with daily functioning (Dementia of the ageing population in Malaysia, 2021). AD is the most common cause of dementia leading to the deterioration of cognitive and functional abilities. It is important to note that while AD is a specific type of dementia, not all cases of dementia are due to AD. Early detection and management of symptoms can help improve quality of life for people with dementia and AD. Caretakers and family members should aim to provide stimulating activities and social interactions to maintain their cognitive and social abilities and prevent isolation, which can negatively impact their self-confidence and overall wellbeing (Al-Khafaji et al. 2013).

In Malaysian society, dementia is often surrounded by stigma and misconceptions. The prevailing belief is that dementia is a part of normal aging or a type of mental illness. This can lead to a delay in diagnosis and inadequate treatment further causing isolation and discrimination for those with dementia and their caregivers. Research has shown that dementia carries a significant cultural burden in Malaysia, and there are cultural taboos around discussing death and end-of-life care, which can lead to underutilization of available resources and services (Kua 2017; Nadarajah See & Narayanan 2018; Ong et al. 2019; Zainal, Ismail & Ahmad 2020). Greater awareness and education about dementia are needed to address these cultural and social barriers and provide appropriate care and support to people with dementia and their families.

2.2 Non-Pharmacological Therapies

Non-pharmacological therapy, also known as psychosocial therapy, is becoming increasingly popular for treating people with Alzheimer's disease, aiming to enhance their quality of life, reduce symptoms, and avoid the limitations of pharmacological interventions. There are several types of non-pharmacological therapies available, including cognitive-based, psychosocial, movement, and sensorial therapies. For this study, reminiscence therapy and cognitive stimulation therapy were selected as practical and effective. RT involves discussing past experiences using artifacts such as photographs and videos to evoke memories, increase personal identity, and reduce stress for caregivers. In addition, CST promotes cognitive processing and enhances cognitive function through group activities centered on specific themes such as physical games, childhood, and categorizing objects. RT and CST have been the subject of numerous studies in the field of dementia care. For example, researchers found that reminiscence therapy reduced depressive symptoms and improved overall well-being (Damianakis et al. 2009) also increased communication, and personal identity and reduced caregiver stress (Spector, Woods & Orrell, 2008).

RT has been studied in various countries, including Malaysia. Recent study conducted in Malaysia explored the effects of RT on the cognitive function and depression levels of elderly individuals with mild cognitive impairment (MCI) (Chong et al. 2021). RT consists of discussions on various topics such as childhood memories, family, and cultural events. Another study conducted in Malaysia investigated the effects of group reminiscence therapy on the quality of life and depression levels of elderly individuals with dementia (Roslan et al. 2021). The results showed that the experimental group had significant improvements in the quality of life and a reduction in depression levels compared to the control group. In addition, recent researchers showed the efficacy of CST that has also been studied extensively in the field of dementia care (Ranasinghe, Kane & Gillman 2020; Jøranson et al. 2021; van Alphen et al. 2021; Jeong et al. 2021; Vagnarelli et al. 2021). Hence, it can be summarized that CST was effective in improving cognitive function, behavioral and psychological symptoms of dementia, as well as reducing depressive and anxiety symptoms.

2.3 Interactive Game as a Therapy

There is growing evidence to suggest that interactive digital games can be a valuable tool in therapy for a range of health and mental health issues. Recent studies have provided further evidence of the potential of interactive digital games in dementia treatment. For example, a randomized controlled trial by Liao et al (2020) found that playing a virtual reality game improved cognitive function, mood, and quality of life in older adults with mild cognitive impairment. Similarly, a study by Sánchez-Cubillo et al (2021) found that a virtual reality game-based intervention improved cognitive and functional outcomes in patients with AD. Furthermore, a systematic review by Sezgin and Özdemir (2021) found that RT using digital games could improve cognition and psychosocial outcomes in patients with dementia. In addition, a study by Groenewoud et al (2017) suggests that video games can have positive effects on psychosocial aspects such as motivation, emotion, and social interactions, which are crucial in the context of mental health therapy. This is also supported by a study by Cipresso et al (2021), which found that playing a virtual reality game improved mood, attention, and executive function in patients with Alzheimer's disease. Similarly, the study by Yamagata et al (2013) found that using brain, memory and problem-solving games could help to stimulate the brain and lower the symptoms of AD. These findings support the use of interactive digital games as a valuable tool in dementia treatment and highlight their potential to improve cognitive, psychosocial, and functional outcomes in patients with cognitive impairments. As such, the integration of digital games into clinical practice could lead to improved patient outcomes and quality of life.

Ning et al (2020) described that in a study by Chang et al (2013), it is reported that by combining RT in games, it could help to restore the memories and cognitive ability of people with dementia. In Chang et al studies, they simulated the scene of buying food with Food Stamps that was set in the 1960s setting. While there is growing evidence to suggest that digital games can be an effective tool for dementia treatment, there is less research addressing the development of personalized digital memory books that embed cultural factors for reminiscence therapy in dementia patients. Personalized reminiscence therapy that incorporates cultural factors has the potential to improve the effectiveness of treatment for individuals with dementia, as cultural backgrounds and experiences can shape personal memories and influence the effectiveness of RT.

However, as noted by a recent scoping review by Cheng, Huang & Chang (2021), there is a lack of studies that have developed and tested personalized digital memory books that incorporate cultural factors in the context of reminiscence therapy for dementia patients. In addition, study by Sezgin & Özdemir (2021) found that a culturally sensitive reminiscence therapy program that incorporated music, food, and religious symbols improved cognitive function and quality of life in Turkish individuals with dementia. The study highlights the importance of incorporating cultural factors into reminiscence therapy for individuals with dementia. Additionally, a review by Kim et al (2021) found that using cultural-specific stimuli in RT could enhance the effectiveness of the therapy for individuals with dementia. The authors suggest that incorporating cultural factors into RT can improve the individual's emotional, psychological, and social well-being. This highlights the need for further research in this area, which could lead to the development of more effective and culturally sensitive therapies for individuals with dementia. Therefore, this project is aimed to further research and develop a personalized digital memory book that incorporates cultural factors in the context of RT for dementia patients.

3. METHODOLOGY

3.1 Study Design

A single study design was used in this research. Single subject study has been used by many researchers to assist people with Alzheimer's in their daily activities (Henley and Lusty, 1984; McPherson et al. 2001; Foloppe et al. 2018). Massimi et al (2008) also conducted on a single participant in their exploratory case study to improve the sense of identity for this target group. Using single case study could use the benefit of saving time, less expensive, help in creating high-quality theory and having a deep understanding of the exploring subject (Gustafsson 2017).

Though, it is not uncommon for focus group studies to have a small sample size of one participant, especially when the focus is on exploring in-depth experiences or perspectives of that individual. In this case, the study is designed to explore the effectiveness of a personalized digital interactive game in the context of the specific participant. By focusing on one individual, the study could gain a more comprehensive understanding of the participant's experiences and how the application impacted her cognitive and psychosocial functioning. Razak (2008) recommended that single case study could be used when we want to study a person or a group of people. Sometimes it is also possible to generalize from a single person. The findings of this small-scale study could also serve as a starting point for future research with larger sample sizes.

3.2 Design Process for the Personalised Digital Interactive Game

The personalized digital interactive games are part of my-MOBAL application. In this app, we developed two types of games with the purpose of stimulating cognitive function and enhancing the memory of the participant. We used the Mobile Application Development Lifecycle (MADLC) to design and develop the personalized digital mobile application (Vithani and Kumar 2014). After performing literature review for the design features, we began to design and develop the application. A few diagrams were sketched to assist with visualizing and understanding the process.

3.3 Screen Designs

Storyboards were drawn to visualize the layout of the game (Figure 1-2). The storyboards contained sketches of graphics, text, animation, audio, and other functional components such as buttons and production notes. Based on the literature study on designing application for people with Alzheimer's, it is essential that the games need to be simple and straightforward so that the user can easily view all items and understand all functions. Big buttons with vibrant colors were selected so that users could easily see the functions. In contrast, dark green was selected as a background color. The same background color was used throughout the whole application so that it gave a look of consistency.

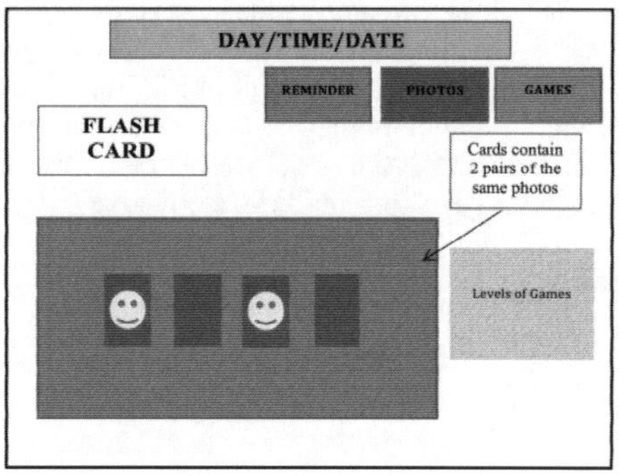

Figure 1. Memory card game for Level 1 screen design

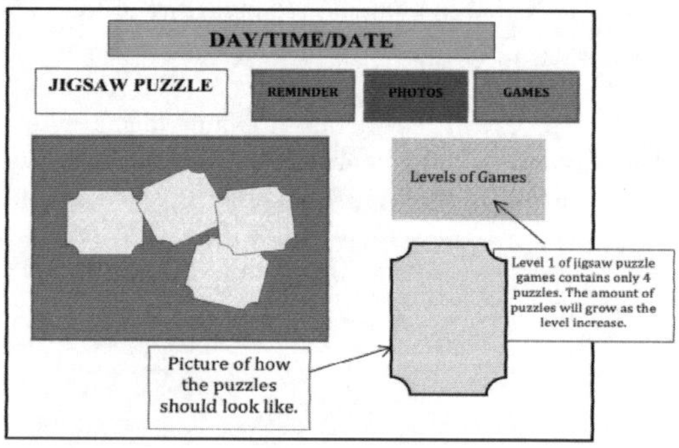

Figure 2. Jigsaw puzzle game for Level 1 screen design

For the media content in this study, photos related to the participant were used. These photos included pictures of family members, including the participant herself. All photos were carefully selected by family members to avoid evoking sad or bad memories of the participant. Mizen (2003) recommended selecting photos of the patient smiling. All photos were colored, but the quality was not very good. We edited the color in the photos to make them clearer and more recognizable. Photos of family members and familiar artifacts were used to assist in enhancing the participant's memory. As the participant did not keep many personal artifacts, only a few old photos were available for this research.

3.4 Personalised Digital Interactive Game Prototype

The my-MOBAL app application was developed using several tools such as Adobe Flash Air and Photoshop. It consists of daily activities and reminder page, reminiscence page that shows family members and events photos and games activities page (Figure 3), for cognitive training. The application for the personalized memory book introduces two games, namely the memory card games (Figure 4) and jigsaw puzzle (Figure 5). The games were designed in the participant's native language, Bahasa Malaysia/Malay Language, so that the participants can understand the game's requirements. These games offer a few levels. However, the purpose of these levels is not to challenge the user but to provide variety so that the games will be more fun and enjoyable. Memory card game comes with 2 levels of difficulty. The first level contains three pairs of photos (six cards), and second level contains four pairs of photos (8 cards). The user needs to find the matched pair by flipping the card one by one. If the photos do not match, the application will display messages that indicate the pair was not correct.

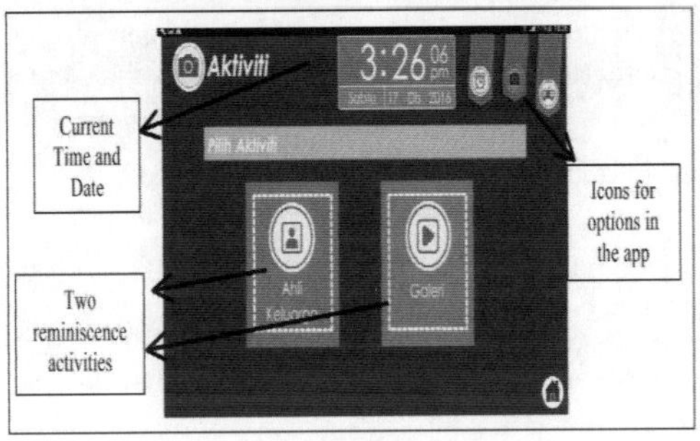

Figure 3. The games activities screen

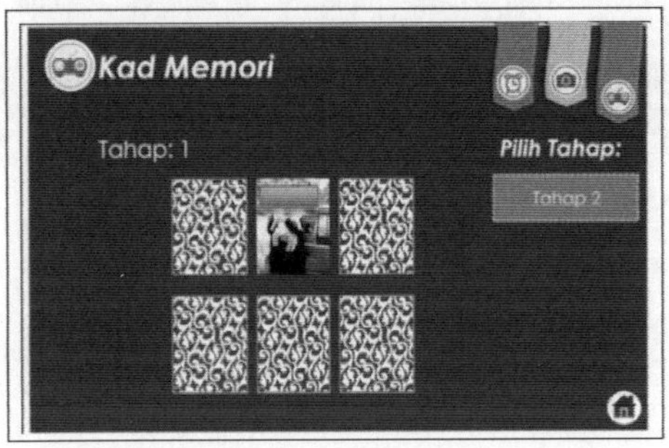

Figure 4. Memory card game for Level 1 screen

Another interactive game included in this app is the jigsaw puzzle. To play this game, the participant needs to drag the puzzle pieces into the correct frame. The pictures used events that related to the participant. The first level has four puzzle pieces, and it was the simplest puzzle, which was a wedding picture - one of the participant's favorite events. The second puzzle was a picture of the graduation day, which has six puzzle pieces. The third puzzle was a bit different because it was not a picture of her family members, but a picture of a bouquet of flowers, specifically a tulip bouquet. For this part, we did not use any family pictures to observe how the participant would react to non-familiar content. These two games were designed in a way that the participant could easily navigate through them and be error-free. We also included audio feedback to compliment the user when they succeeded, as well as for motivational purposes. The user was free to choose which game she wanted to play and to stop playing. Although the games provided levels of difficulty, the user could jump to any level she would like to play. The user did not need to follow the order of the levels. my-MOBAL application has also included an event-logging tool that recorded patient's activities during therapy sessions (Figure 6). These logs could assist the researcher to follow the patient's activities even without the researcher's presence by looking at the dates for every activity. Figure 6 displayed the date when the participant played the games and time spent to complete the activity. This information is valuable especially when the researcher was not present during the session.

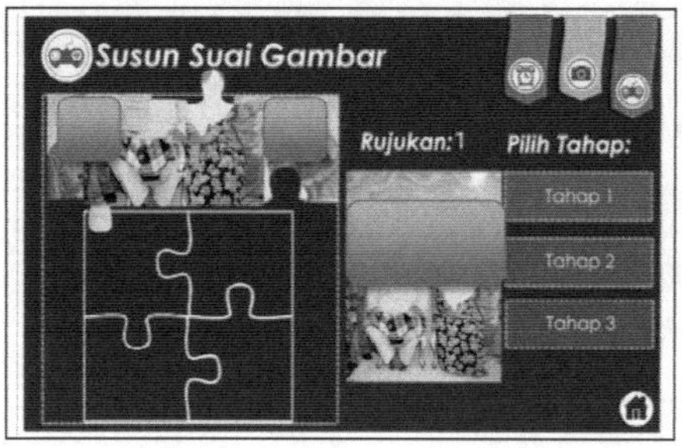

Figure 5. Jigsaw puzzle game for Level 1

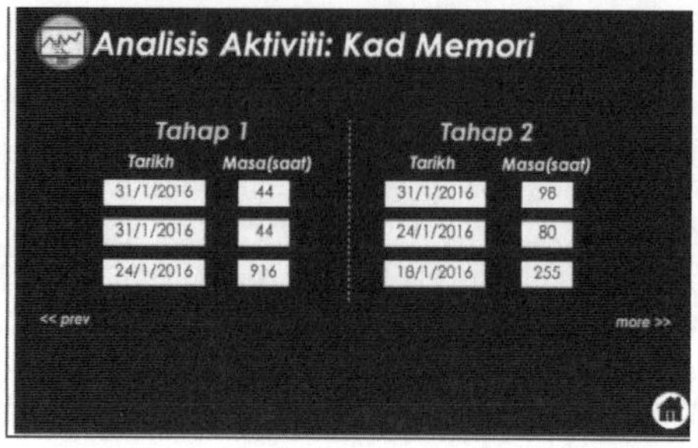

Figure 6. Analysis page on frequencies viewing the photos

3.5 Expert Review Assessment

An expert review technique was used to evaluate the application. Expert review is a form of evaluation where a group of domain experts evaluate a product or intervention. In this case of digital games for reminiscence therapy, domain experts would be individuals with knowledge and expertise in dementia, gerontology, psychology, and game design. We include expert review in this study because to ensure that digital games are designed to meet the specific needs and challenges of dementia patients and to ensure digital games are evidence-based and effective in improving reminiscence and cognitive function in dementia patients. Four (4) clinical experts participated in the interview to assess the usability and capability of the application, including the games elements. All the experts have more than 10 years of experience in the field of gerontology and psychology. To assess the concept and usability of the interactive games, six (6) related questions were asked. Those are as in table 1.

Table 1. Related questions on assessment of concept and usability of the interactive games

Numbers	Questions
Q1	Can computer-based past game activities enhance reminiscence for Alzheimer Disease (AD) patients?
Q2	Does computer-based games stimulate the cognitive function of AD patients?
Q3	Does using personal/family member photos in game activities help patients to reminisce?
Q4	Does repetition use of past or traditional games using application/computer enhance reminiscence for AD patients?
Q5	Can repetition use of the games stimulate cognitive function of AD patients?
Q6	Can different levels in game activities assist in reminiscence therapy and cognitive stimulation therapy?

Besides the expert review, for this feasibility study, we also recruited a person with mild Alzheimer's disease.

3.6 Participant's Selection

Prior to participant selection, this research has undergone the university's ethical approval process and has been granted approval from Universiti Kebangsaan Malaysia (UKM) as a research collaborator (UKM 1.5.3.5/244/NN-102-2014) to carry out data collection with elderly individuals with dementia. The recruitment of the participant was through a friend's referral. A few meetings were carried out with the caretakers and the participant to establish a good relationship with the participant. The participant was 74 years old, female and was from Malay race. She has no major health issues, and her level of education was until primary school. The participant has two caretakers that will take turns to look after her. Prior to the recruitment of the study, the patient visited Memory Clinic and underwent two assessments, MMSE (Mini Mental State Exam) and CDR (Clinical Dementia Rating Scale). Her score for MMSE was 19/20 and for CDR was 1.0. For this study, the assessments were carried out before and after using the application. MMSE for this study was in Bahasa Malaysia (Malay language) because of the participant's mother-tongue language and suitable for the culture of Malaysian.

During this study, the interactive games were tested by the participant during her therapy sessions. The testing was carried out at the caretakers' house. According to O'Conner et al (2009), the presence of family members could give a positive result in therapy sessions. For this study, the sessions took place twice weekly for eight weeks which lasted between 30 to 45 minutes per session. During the sessions, the researcher would take notes and observe the participant. All the interviews with the expert reviews, AD participants and their caretakers were analyzed using content analysis.

4. RESULT AND DISCUSSION

4.1 Feeback From the Experts

Based on the expert review interviews, all of them agreed that the interactive games in my-MOBAL application could help in enhancing reminiscence and stimulate the cognitive function of people with AD. They also agreed that using personal artifacts related to the patients could assist in reminiscing. Using personal/family member photos in game activities can help patients to reminisce. It provides a more personal and emotional connection to the memories being recalled, which can enhance the effectiveness of the therapy. Whereas the repetition use of past or traditional games using application/computer can enhance reminiscence for people with AD. Repetition has been found to be an effective way to reinforce memories and improve cognitive function. They further commented, the brain responds well to repetition and reinforcement, and this can lead to improved cognitive function. In addition, according to the experts, different levels in game activities can assist in reminiscence therapy and cognitive stimulation therapy. The experts also agreed that different levels of difficulties in the games should not be used to challenge them, but to provide more options to solve the games and make the games more attractive.

In addition, the expert reviewers provided us with some suggestions on how to enhance the interactive games of the my-MOBAL app. Three experts recommended that we use cultural language for the mobile application's overall content. By doing so, we can ensure that the games are culturally appropriate and sensitive to the needs of the population. Currently, the content language is focused on Bahasa Malaysia (Malay language), which may only be relevant to a single race, even though it is the national language. Therefore, it is important to consider using other languages such as English, Tamil, and Mandarin, as these languages represent the races in the Malaysian community. Furthermore, two experts suggested using symbols and imagery in the computer game activities. This can help create a sense of familiarity and connection with the targeted population. Additionally, they recommended that the app should provide options for customization of the games to accommodate individual cultural preferences and differences. This can enhance the user's experience and make the games more appealing to a wider audience.

Lastly, the experts recommended that we conduct a feasibility study of the prototype with various races. This can be achieved through feedback from different races, caregivers, and healthcare professionals to gauge the effectiveness of the games and identify possible cultural elements for the content creation. This will help us to understand the specific needs and preferences of our target population and make necessary adjustments to the app. In summary, the expert reviewers provided valuable suggestions to enhance the interactive games in my-MOBAL apps. By incorporating cultural language, symbols, imagery, and customization options, we can create a more culturally sensitive and appealing application. Conducting a feasibility study will enable us to evaluate the effectiveness of the games and gather feedback from different stakeholders, which can help us to improve the application and better meet the needs of our target population.

4.2 Feedback from Participant and Caretakers/Family Members

Based on the researcher's observation, the 74 years old female participant with mild AD had positive experience with the personalized digital memory book application. The observation revealed that the participant's mood varied and was often affected by her environment and surroundings. Additionally, the participant's physical condition influenced how the therapy sessions went. For instance, if the participant tired, she would prefer shorter sessions.

The participant usually accompanied by her grandchildren who were also curious about the app in those sessions. The grandchildren will ask the grandmother if she knew how to play the games. Example of the conversation be like:

Grandchild 1: *Grandma, do you know how to play this game?* (Since the game also showed pictures of past events, the grandmother would tell the story of the event while playing the games).

Grandchild 2: *Grandma, who is this?* (Grandmother replied, "*this is your mother*")

From the observation, when she surrounded by her grandchildren, she could go for more than 45 minutes and thoroughly enjoy the sessions. It is worth noting that a participant's mood and environment can significantly affect the outcome of therapy sessions. Therefore, it is essential to take these factors into consideration when planning therapy sessions. In addition, the bonding between the grandmother and the grandchildren indicates that the app has the potential to bridge the generational gap and promote intergenerational interaction.

The exchange between the participant and grandchild 2, where the participant identified a picture of their grandchild's mother, demonstrates the potential for the app to facilitate communication and strengthen family bonds. It also highlights the importance of family members' involvement in reminiscence therapy and the potential benefits it can have for both the participant and their family members.

Participant (Feedback 1): *How am I going to play this game? What should I do?*

Another interesting observation, the app not only enhances the participant's memory but also stimulates their cognitive function. In participant's feedback 1, she showed her readiness to learn new skill. In the first few sessions, she was not sure what she supposed to do. Since the participant also not familiar with computer technology, she was quite unsure what she should do with the games. She was curious and kept asking what she should do. However, after a few sessions, she stopped asking, and just played with the games once it appeared on the screen. The caretakers were quite shocked and very pleased with the improvement. This is because the participant has never used a computer or any mobile devices previously.

During the session, the participant also expressed her enjoyment in using the app especially when the grandchildren or other family members involved in the sessions. The used of pictures of past events seems to have triggered the participant's memories, prompting her to share stories of these events while playing the interactive games. According to the caretaker, her mother is usually a very quiet person. However, during the sessions, her mother became more talkative. Their mother loves to talk about the event in the photographs especially when she is in the picture. She also showed more motivation to play the games with the presence of her grandchildren. There are quite a few times when her grandchildren were able to join the sessions, and usually those sessions would take more than 40 minutes to end. In a few sessions, the caretakers were impressed and said that they had never seen their mother talked so much and still remember some of the events. This suggested that the app could be an effective tool for reminiscence therapy, a technique that uses the recollection of past events and experiences to improve psychological wellbeing. It also could be used for cognitive stimulation therapy as the games could be used to encourage cognitive processing in order to improve cognitive function of the participant. This positive feedback is a good sign that the app is engaging and useful for the participant.

Furthermore, it is important to note that cultural elements should also be considered in the design of non-pharmacological therapy apps. In the case of the participant in this study, she accompanied by her caretakers and grandchildren who were likely to be from the same cultural background as her. In fact, my-MOBAL app was developed in a native language which is Bahasa Malaysia where both parties could understand it well also the symbols and imagery used in the content are relevant and familiar to the participant's cultural background. While the study provides valuable insights into the potential benefits of using a personalized digital memory book app for psychosocial therapy in individuals with mild AD, it is important to note that the findings cannot be generalized to other races or cultural backgrounds. The participant in this study and her grandchildren were from a specific cultural background and spoke a specific language, which may have influenced their experience and interaction with the app. The use of language, symbols, and imagery in the app's design that are relevant and familiar to the participant's cultural background may not be as effective or engaging for individuals from different cultural backgrounds. Therefore, it is important to consider cultural elements in the design of reminiscence therapy apps for future enhancement to ensure their effectiveness and relevance for individuals from diverse cultural backgrounds. Conducting similar studies with participants from different cultural backgrounds can help to determine the generalizability and effectiveness of reminiscence therapy apps for diverse populations.

5. CONCLUSION

This study aimed to test the feasibility of a digital interactive game that is embedded in my-MOBAL, a personalized digital memory book, that is a tool for non-pharmacological therapy. We wanted to explore further the app design features whether it is viable and practical for use in a larger study or clinical setting in the future. As this is the first version of the prototype, we wanted to gather preliminary data on the potential, challenges and limitations of the app from the design perspectives.

The preliminary data gathered from this study revealed the need for the app to be culturally sensitive and appropriate. The current version of the app lacks cultural elements such as varied languages, symbols, and imagery relevant to the participant's cultural background, which may limit its effectiveness for individuals from different cultural backgrounds. The limitation in this study is that we only manage to test and interview one participant from a single race and the design of the mobile application content is prone towards one single race influence including the language usage that resulted a positive response from the participants. We believed that it might not be the case for other races, which probably have culturally appropriate and sensitive content preference and tolerance in which we must investigate further. Therefore, the findings from this research cannot be generalized and represent the entire view of elderly with dementia as our conclusions are based on one person study. However, this finding will serve as the future development of my-MOBAL prototype enhancement which will incorporate designs that resemble multi-cultural metaphors. Furthermore, we plan to test the prototype with different races.

In addition to the limitations regarding cultural generalizability, we noted the importance of considering the potential impact of user impairment on the use and effectiveness of personalized digital memory book apps for individuals with AD. Alzheimer's disease is a progressive neurodegenerative disorder that can lead to cognitive and functional impairment, making it difficult for individuals to learn and use new technologies. While the participant in this study had mild AD, individuals with more severe forms of the disease may struggle to use digital memory book apps effectively or may not be able to use them at all. It is also important to consider the potential impact of user impairment on the efficacy of reminiscence therapy using digital memory book apps. Individuals with advanced AD may have difficulty recalling past events or experiences, making reminiscence therapy less effective. Furthermore, individuals with AD may experience agitation, confusion, or frustration when trying to recall past events, which could negatively impact their experience with the app and their engagement in reminiscence therapy.

Therefore, future research should also consider the potential impact of user impairment on the use and effectiveness of digital memory apps for reminiscence therapy in individuals with Alzheimer's disease. This could involve testing the app with individuals at different stages of Alzheimer's disease or with different levels of cognitive impairment to determine the app's effectiveness and usability across a wider range of individuals with AD. Additionally, it may be necessary to develop specific adaptations or modifications to the app to ensure its effectiveness for individuals with more advanced Alzheimer's disease. Therefore, it is important for app developers and therapists to consider cultural elements in the design and implementation of reminiscence therapy apps to maximize their potential benefits as well as test it with a different user background and capabilities.

REFERENCES

Al-Khafaji, H. A. H., Al-Shaher, A. A. A., Al-Khafaji, S. S. H., & Ahmed Asmail, I. A. 2013. The impact of social isolation on the self-confidence of Alzheimer's patients. *Journal of Medicine and Medical Sciences*, 4(4), 148-155.

Alzheimer's Disease International (ADI). *World Alzheimer Report 2019: Attitudes to dementia.* 2019. https://www.alz.co.uk/research/WorldAlzheimerReport2019.pdf

Cammisuli, D. M., Cipriani, G., Giusti, E. M., & Castelnuovo, G. 2022. Efficacy and safety of non-pharmacological interventions for cognitive impairment in older adults: A systematic review and meta-analysis of randomized controlled trials. *Aging Clinical and Experimental Research*, 34(1), 1-14.

Chang, K et al. 2013.Food Stams: A reminiscence therapy tablet game for Chinese seniors. In *Proc. ICME Int. Conf. Complex Med. Eng.*, May 2013, pp. 556-561.

Cheng, S.-T., Huang, C.-M., & Chang, Y.-C. 2021. Reminiscence Therapy for Dementia: A Scoping Review of Digital Interventions that Incorporate Cultural Factors. *Journal of Geriatric Psychiatry and Neurology*, 34(2), 94–104. https://doi.org/10.1177/0891988720969613.

Chong, C. K., Tan, K. H., Chong, S. P., & Ting, K. S. 2021. Reminiscence therapy on cognitive function and depression in mild cognitive impairment: A randomized controlled trial. *International Journal of Geriatric Psychiatry*, 36(2), 279-285. doi: 10.1002/gps.5464.

Cipresso, P., Riva, G., Giglioli, I. A. C., Pallavicini, F., Triberti, S., & Monaro, M. 2021. Playing digital games improves mood, attention, and executive functions in people with Alzheimer's disease: A randomized controlled trial. *Journal of Alzheimer's Disease*, 79(3), 1231-1243. https://doi.org/10.3233/JAD-201171.

Damianakis, T., Crete-Nishihata, M., Smith, B., Baecker, R., & Marziali, E. 2009. A computer-based platform to support reminiscence therapy in dementia care. *ACM Transactions on Computer-Human Interaction (TOCHI),* 16(4), 1-24.

Dementia of the ageing population in Malaysia: A scoping review of published research. 2021. *Dementia & Neuropsychologia*, 15(1), 23-31.

Foloppe, D., Richard, P., Yamaguchi, T., Etcharry-Bouyx, F., & Allain, P. 2018. The potential of virtual reality-based training to enhance the functional autonomy of Alzheimer's disease patients in cooking activities: A single case study. *Neuropsychological Rehabilitation*, 28, 709 - 733.

Groenewoud, H., Apps, M. A., & Westerhof, G. J. 2017. Computer-based social activation in older adults with dementia: A pilot study. *Aging & Mental Health,* 21(2), 171-179. doi: 10.1080/13607863.2015.1132826

Groenewoud, J. G., de Lange, J., Schikhof, Y., Astell, A. J., Joddrell, P., & Goumans, M. 2017. A social robot and electronic game system for people with dementia: an exploratory study. *Journal of Alzheimer's Disease,* 55(2), 541-548.

Groenewoud, J. H., et al. 2017. Video games and social competence: A review. *Social Science & Medicine,* 194: 85-92.

Gustafsson, J. 2017. Single case studies vs. multiple case studies: A comparative study.

Hussain, S. 2017. Dementia in Malaysia: Time to address the elephant in the room. *Journal of the ASEAN Federation of Endocrine Societies*, 32(2), 138-140.

Imtiaz, B., Khan, A., & Seelye, A. 2018. Pharmacological management of Alzheimer's disease: A systematic review. *Journal of Clinical Neurology*, 14(2), 147-153.

Jeong, H., Na, H. R., Choi, S. H., Kang, J., Moon, S. Y., & Seo, S. W. 2021. Individualized cognitive stimulation therapy for dementia in community-dwelling older adults: A randomized controlled trial. *International Journal of Geriatric Psychiatry*, 36(1), 36-45. https://doi.org/10.1002/gps.5494.

Jøranson, N., Pedersen, I., Rokstad, A. M. M., Ihlebaek, C., & Askvik, L. W. 2021. Cognitive Stimulation Therapy for persons with dementia in a Norwegian community setting: a randomized controlled trial. *BMC Geriatrics*, 21(1), 317. https://doi.org/10.1186/s12877-021-02213-5.

Kammrath, L. K., Bohnen, N. I., Müller, M. L., Rizk-Jackson, A., & Lockhart, S. N. 2021. Effects of cognitive training interventions for people with Alzheimer's disease: A systematic review and meta-analysis. *Alzheimer's & Dementia: Diagnosis, Assessment & Disease Monitoring*, 13(1), e12129. doi: 10.1002/dad2.12129.

Kim, J. H., Kim, J., & Kim, K. 2021. Effects of culture-specific stimuli on reminiscence therapy for individuals with dementia: A systematic review and meta-analysis. *Archives of Psychiatric Nursing,* 35(4), 337-346. doi: 10.1016/j.apnu.2021.04.008.

Kua, E. H. 2017. Dementia in the Asia Pacific region. Dementia and Geriatric Cognitive Disorders Extra, 7(2), 253-259.

Lee, Y., Lee, K. E., Kim, H. R., & Choi, M. 2019. Effects of a computer-based cognitive rehabilitation program on cognitive function and activities of daily living in patients with Alzheimer's disease: A single-blind, randomized controlled trial. American *Journal of Alzheimer's Disease & Other Dementias*, 34(2), 90-99.

Li, J., Chen, S., Yu, L., Hu, J., Chen, G., & Wang, M. 2022. Virtual reality-based cognitive training for patients with Alzheimer's disease: A randomized controlled trial. *Aging & Mental Health*, 26(1), 116-124. doi: 10.1080/13607863.2019.1689139.

Liao, Y.-Y., Tseng, H.-Y., Chen, Y.-H., Lin, Y.-J., & Lin, Y. 2020. Virtual reality-based exercise improves cognitive function and quality of life in older adults with mild cognitive impairment: A randomized controlled trial. *Frontiers in Aging Neuroscience*, 12, 575740. https://doi.org/10.3389/fnagi.2020.575740.

Massimi, M., Berry, E., Browne, G., Smyth, G., Watson, P., & Baecker, R.M. 2008. An exploratory case study of the impact of ambient biographical displays on identity in a patient with Alzheimer's disease. *Neuropsychological Rehabilitation* 18(5/6), 742-765.

McCallum, S., Boletsis, C. (2013). Dementia Games: A Literature Review of Dementia-Related Serious Games. In: Ma, M., Oliveira, M.F., Petersen, S., Hauge, J.B. (eds) Serious Games Development and Applications. SGDA 2013. *Lecture Notes in Computer Science*, vol 8101. Springer, Berlin, Heidelberg. https://doi.org/10.1007/978-3-642-40790-1_2.

McPherson, A., Furniss, F. G., Sdogati, C., Cesaroni, F., Tartaglini, B., & Lindesay, J. 2001. Effects of individualized memory aids on the conversation of persons with severe dementia: a pilot study. *Aging & mental health*, 5(3), 289–294. https://doi.org/10.1080/13607860120064970.

Ministry of Health Malaysia. National Dementia Registry Report 2020. https://www.moh.gov.my/moh/resources/Penerbitan/Laporan/Umum/National_Dementia_Registry_Report_2020.pdf.

Mizen, R. 2003. Therapeutic uses of life story work: A practice guide. Jessica Kingsley Publishers.

Nadarajah, S., See, T. A., & Narayanan, V. 2018. Dementia in Asia: A review of the literature. *Malaysian Journal of Medical Sciences*, 25(5), 5-14.

Nguyen, T. T., Ta, T., Pham, T., Nguyen, T., & Tran, T. 2020. Computerized cognitive training for persons with mild cognitive impairment and early Alzheimer's disease: A pilot study. *Archives of Gerontology & Geriatrics*, 87, 104007. doi: 10.1016/j.archger.2019.104007

Ning, H., Li, R., Ye, X., Zhang, Y., & Liu, L. 2020. A Review on Serious Games for Dementia Care in Ageing Societies. *IEEE journal of translational engineering in health and medicine*, 8, 1400411, https://doi.org/10.1109/JHETM.2020.2998055.

O'Connor, D. W., Ames, D., Gardner, B., & King, M. 2009. Psychosocial treatments of behaviour symptoms in dementia: A systematic review of reports meeting quality standards. *International Psychogeriatrics*, 21, 225-240.

Ong, H. C., Ibrahim, N., Din, N. C., Tengku Ismail, T. H., Kaur, G., & Chan, K. W. 2019. Dementia caregivers in Malaysia: prevalence and predictors of burden and depression. *Asia Pacific Journal of Public Health*, 31(1), 47-56.

Ranasinghe, D. C., Kane, R. T., & Gillman, A. 2020. Cognitive Stimulation Therapy for dementia: A systematic review and meta-analysis. *SAGE Open Medicine*, 8, 2050312120949653. https://doi.org/10.1177/2050312120949653.

Razak, F.H. 2008. Single Person Study: Methodological Issues (Doctoral thesis). Retrieved from https://www.hcibook.net/docs/Hanis-Thesis.pdf.

Roslan, N. S., Mohamad, N., Md Zain, A., & Ismail, H. 2021. The effects of group reminiscence therapy on quality of life and depression levels among elderly with dementia. *Malaysian Journal of Medical Sciences*, 28(1), 1-13. doi: 10.21315/mjms2021.28.1.1.

Sánchez-Cubillo, I., Anguera, J. A., Rodríguez-Molinero, A., & Burón, A. 2021. Effectiveness of a virtual reality game-based intervention on cognitive and functional outcomes in patients with Alzheimer's disease: A randomized controlled trial. *Journal of Alzheimer's Disease*, 80(3), 1223-1237. https://doi.org/10.3233/JAD-201164.

Sezgin, E., & Özdemir, E. 2021. The effectiveness of reminiscence therapy using digital games on cognitive and psychosocial outcomes in patients with dementia: A systematic review. *Dementia,* 20(5), 1795-1808. https://doi.org/10.1177/1471301221997821.

Spector, A., Woods, B., & Orrell, M. 2008. Cognitive stimulation therapy (CST): effects on different areas of cognitive function for people with dementia. *International Journal of Geriatric Psychiatry*, 23(3), 274-281.

The Lancet Neurology Commission. Dementia prevention, intervention, and care: 2020 report of the Lancet Commission. *The Lancet Neurology*. 2019. 18(3), 225-266.

United Nations, Department of Economic and Social Affairs, Population Division. *World Population Ageing*. 2019 (ST/ESA/SER.A/444). New York: United Nations 2019. https://www.un.org/en/development/desa/population/publications/pdf/ageing/WorldPopulationAgeing2019-Report.pdf.

Vagnarelli, F., Berardi, V., Gabola, P., Bonacchi, R., Moretti, R., & Caffarra, P. 2021. The effects of cognitive stimulation therapy on cognitive and behavioural symptoms in patients with dementia: a randomized controlled trial. *Aging Clinical and Experimental Research*, 33(10), 2727-2736. https://doi.org/10.1007/s40520-021-01863-1.

van Alphen, H. J., Volkers, K. M., Blankevoort, C. G., Scheltens, P., & van der Flier, W. M. 2021. A randomized controlled trial of group cognitive stimulation therapy for people with mild cognitive impairment or mild dementia: a pilot study. *International Psychogeriatrics,* 33(3), 275-285. https://doi.org/10.1017/S104161022000133X.

Vithani, K., & Kumar, A. 2014. Analysis of mobile application development life cycle. *In 2014 5th International Conference- Confluence The Next Generation Information Technology Summit* (pp. 177-183). IEEE (2014).

World Health Organization (WHO). *Dementia.* 2021. https://www.who.int/news-room/fact-sheets/detail/dementia

Yamagata, C., Kowtko, M., Coppola, J.F., & Joyce, S. 2013. Mobile app development and usability research to help dementia and Alzheimer patients. *In Systems, Applications and Technology Conference (LISAT),* 2013 IEEE Long Island, pp. 1-6

Yusof, N., Mohd Rias, R., & Yusoff, E.H. 2014. Serious Games in Mental Health Treatment: *Review of Literature", Knowledge Management International Conference* (KMICe) 2014, Malaysia, 12-15 August, 2014.

Zainal, N. Z. A., Ismail, Z., & Ahmad, N. H. 2020. Dementia and the Malay-Muslim community in Malaysia: a scoping review. *BMC Public Health*, 20(1), 1-12.

A COMPARATIVE ANALYSIS OF URBAN AND RURAL BROADBAND PENETRATION AND ACCESS TRENDS IN SOUTH AFRICA

Sicelokuhle Oscar Ngwenya[1], Reolyn Heymann[2], Theo G. Swart[2] and Albert Lysko[3]
[1]*Postgraduate School of Engineering Management*
[2]*Department of Electrical and Electronic Engineering Science*
University of Johannesburg, South Africa
[3]*Department of Electrical Engineering*
University of Cape Town, South Africa

ABSTRACT

Various studies have suggested that broadband connectivity is an enabler of rapid development in the socio-economic landscape, in countries where it is readily accessible. Consequently, the demand for Information and Communication Technology (ICT) has risen sharply, as countries continually recognize the benefits and potential of ICT as a tool for social and economic development. However, while this adaptation is seen globally, the observed trend is that it is mainly readily accessible in urban areas, while in rural communities the ICT services, including internet access, is not widely seen. This paper verifies these assertions and findings through non-intrusive research methodologies that are deemed effective in obtaining the relevant information required to that end, in an effort to ascertain the cause of this disparity. It is further asserted that the slow pace of internet penetration and access in rural areas is attributed primarily to the fact that rural areas are commonly characterized by low income, poverty, underdeveloped infrastructure, illiteracy, and unemployment. These are demonstrated in this paper using results obtained from a broadband survey conducted by the authors. This paper further aims to highlight the extent of the digital divide, in South Africa (SA) through the same survey, which has been conducted in both rural and urban areas. This comparative study was done to highlight the disparity in broadband access and penetration statistics between the rural and urban communities in South Africa observed in recent years, with the aim to better understand the extent of the digital divide, as it persists today. The parameters that are used to achieve this is employment trends, availability of enabling infrastructure and broadband services, and technology availability trends, including the broadband data rates thereof. These results will clearly demonstrate the status of the ICT sector, particularly highlighting the discrepancy between rural and urban communities in SA, and these will be unpacked in this paper.

KEYWORDS

Broadband, Digital Divide, Wireless, Rural, South Africa, Urban, Wi-Fi, Cellular

1. INTRODUCTION

According to statistics from the world bank (2021) ('World Bank', 2021), it is estimated that, on average, 39.4% of the global population is residing in rural areas. In South Africa (SA), the rural population was estimated at 31.4% at the beginning of 2023, according to Datareportal (*Datareportal*, 2023). It has also been reported that 31.8% of the South African population remained offline at the beginning of the same year. With the large dependence of advanced economies on modern ICTs, broadband internet access has become an essential tool for establishing the modern knowledge-based economy and information society (Stork, Calandro and Gamage, 2014). Inadequate, or a complete lack of access to ICT services, is therefore viewed as hugely regressive in today's economic landscape.

The International Telecommunications Union (ITU) in collaboration with the United Nations (UN), with their report titled "Partnership on Measuring ICT for Development" (ITU, 2012), have called on countries to develop their National Broadband Policies, with the purpose of outlining targets on their respective country's broadband access and penetration. In South Africa, this National Broadband Policy is known as the South Africa Connect (Communications, 2013), and this policy aligns to goals set in the National Development

Plan (NDP) Vision 2030 (National Planning Commission, 2011). This policy also emphasizes the aspiration of "a seamless information infrastructure by 2030 that will underpin a dynamic and connected, vibrant information society and knowledge economy that is more inclusive, equitable and prosperous." These two policy documents, and the objectives outlined therein, serves as guidelines with respect to targets and timelines set, and progress in achieving them. The main aim of this paper is to attempt to understand the pertinant contributing factors of the so-called digital divide, in order to formulate frameworks that will assist in averting it or minimizing its impact. The formal definition of the digital divide given in (Souter *et al.*, 2009) is "the gap which has arisen between those that make use of ICTs and those that do not".

In the second section of this paper, we discuss the problem statement in relation to the current global landscape and position on broadband, with respect to access and penetration. This includes the research conducted by the ITU and Independent Communications Authority of South Africa (ICASA), where the findings from the resulting reports are summarized. This section then briefly discusses the research that was conducted by the authors. In the third section we discuss the research methodology that was employed for this study, which entails the objective as well as the research design. The fourth section discusses the results, and the analysis thereof. Here, the data is presented together with the results and a qualitative comparative analysis is provided. The fifth section then discusses the findings drawn from the results obtained, followed by a section to finally conclude on the findings of the research study.

2. URBAN VS RURAL BROADBAND

2.1 Problem Statement

ICT has become a great enabler of advancement in modern economies, with broadband access being an essential tool. However, broadband is not readily available in rural areas, which impedes the economic growth of a country. Although rural areas are not considered as economic hubs, a simple example of how broadband connectivity can assist in rural areas is by providing access to information to informal businesses, which are prevalent in rural, remote, and isolated areas. This information can include, weather reports, related market prices and provision the ability to do research on their products, as well as other better income-earning opportunities, which they currently lack. However, without access to ICT, these services are a pipeline dream for the rural population, and the social ills experienced by the inhabitants of those communities persists unabated, while their urban counterparts improve. Should this observation persist, it would be a misalignment with the targets stipulated in the NDP Vision 2030 and SA Connect Policy and may potentially further exacerbate the gross inequality experienced in a developing country such as South Africa.

This paper attempts to highlight this inconsistency and ascertain its severity. To achieve this, it is prudent to highlight the targets set in the SA Connect Policy, shown in Figure 1. These will be used as a point of reference to findings from an online and physical survey performed by the authors, discussed in a later section. Reference will also be made to findings made by telecommunications bodies like the ITU as well as the South African Telecommunications Regulator, the ICASA, for comparison purposes.

Broadband Policy Target	Penetration/ Measure	Baseline (2013)	By 2016	By 2020	By 2030
Broadband access in Mbps user experience	% of population	33.7% Internet access	50% at 5Mbps	90% at 5Mbps 50% at 100Mbps	100% at 10Mbps 80% at 100Mbps
Schools	% of schools	25% connected	50% at 10Mbps	100% at 10Mbps 80% at 100Mbps	100% at 1Gbps
Health Facilities	% of health facilities	13% connected	50% at 10Mbps	100% at 10Mbps 80% at 100Mbps	100% at 1Gbps
Public sector facilities	% of government offices		50% at 5Mbps	100% at 100Mbps	100% at 100Mbps

Figure 1. Key SA Connect Policy targets (Communications, 2013)

2.2 Current State of ICT

Entities such as the ITU and ICASA conduct research to ascertain the current state of the ICT sector, both globally and locally. Internationally, the ITU has released statistics in their report titled "Measuring the Information Society Report (International Telecommunication Union (ITU), 2018)". Locally, ICASA has undertaken to issue an annual report that provides information on the state of the ICT sector in South Africa. The first annual report was released in 2016 and was made available to the public. They have since released follow up reports annually from 2017 with the latest one released in March of 2023 ('The State of the ICT Sector Report in South Africa', 2023).

The authors have, in addition, undertaken a survey to obtain more insight on the demographics of those affected, and the extent to which they are affected. The results, as well as the methodology used to conduct the survey, are discussed in Section 4.

2.2.1 ITU and ICASA

From their recent reports (International Telecommunication Union (ITU), 2018)('The State of the ICT Sector Report in South Africa', 2023), both the ITU and ICASA have emphasized notable upward trends, and key findings, in the global ICT sector such as the availability of communication services, growth in broadband, growing predominance of mobile over fixed services, increase in revenue in the telecommunications sector, the increase in usage of mobile and fixed data over mobile and fixed voice respectively, the significant annual increase in mobile subscriptions over fixed subscriptions, and cellular population coverage for 3G is recorded at 100%, while 4G/LTE stood at over 98% in SA, as of 2022.

2.2.2 Public Survey

The public survey is essentially a questionnaire that was formulated by the authors, for the purposes of distribution to ordinary South African citizens. The survey aims to elicit qualitative data that is indicative of the state of the ICT in urban and rural areas in SA, and draw a distinction, if any, between the two.

3. RESEARCH METHODOLOGY

3.1 Purpose of the Survey

The survey aims to ascertain the extent of the digital divide between urban and rural dwellers in South Africa. The survey targets both these areas in order to provide a comparison between the two "communities".

3.2 Research Design

The technique that has been employed by the researcher to collect relevant data for this research is public surveys. The authors conducted the survey in the form of a questionnaire. These questionnaires have been made available to participants using various methods. The questionnaire is driven at collecting information about the respondent that pertains to demographics, location, internet access and penetration trends. The questionnaire consists of non-intrusive, concise questions, presented in multiple-choice format. The same questionnaire was made available both in electronic (online survey) and physical copies. The online copy was envisioned to be available nationwide (all provinces), and reach mainly the urban population, who were the target group for this part (online) of the survey, while the physical copies were intended to reach mainly the rural dwellers (restricted to the communities visited) as they are expected to have challenges accessing the online survey. This assumption was informed by the very objective of this research study. The physical copies were therefore distributed physically in the rural areas of Vryheid and Mbazwana, two rural communities in the province of KwaZulu-Natal (KZN), South Africa. The samples that were targeted were randomly selected, and not limited by race, gender or even sample size, however, it only excluded minors (below ages of 18), in compliance to the University ethical clearance approvals. As such, the results and analysis thereof would be qualitative, due to limitations relating to pre-organization of the participants. All participants were required to complete consent forms prior to completing the questionnaire.

4. RESULTS AND COMPARATIVE ANALYSIS

The online survey closed as of the 30th of May 2019, and received 295 responses, with one individual opting not to give consent to participate. The online data captured, therefore, includes that of 294 consenting respondents.

The physical walkdown survey was concluded as of the 07th of June 2019, and received 103 responses from two separate rural areas, where it was conducted. The first survey was conducted on the 20th of April 2019 in the small town of Vryheid, where 21 individuals gave consent to participate and completed the survey. The second survey was conducted on the 13th of May 2019 in the village of Mbazwana, where 44 individuals gave consent to participate and completed the survey. The third and final survey was conducted again in the village of Mbazwana, in the weeks from the 27th of May 2019 to the 06th of June 2019, where a further 38 individuals completed the questionnaire. In total, 82 respondents from Mbazwana provided completed surveys.

The results for both the online and physical walkdown surveys were as follows:

4.1 Employment Status

From the online survey, the majority of the respondents are employed (71.1%) compared to the physical walkdown survey, where the majority of the respondents are students (43.7%) and unemployed (36.9%), respectively, as depicted in Figure 2. In addition, those who reported to be employed in rural areas, are in the lowest salary bracket, as depicted in Figure 3.

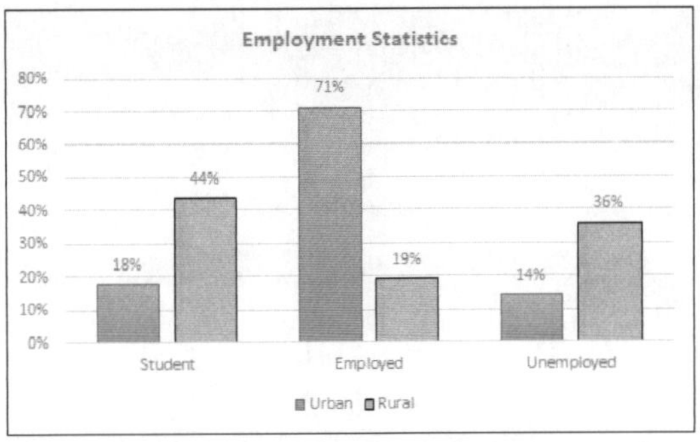

Figure 2. Employment Statistics

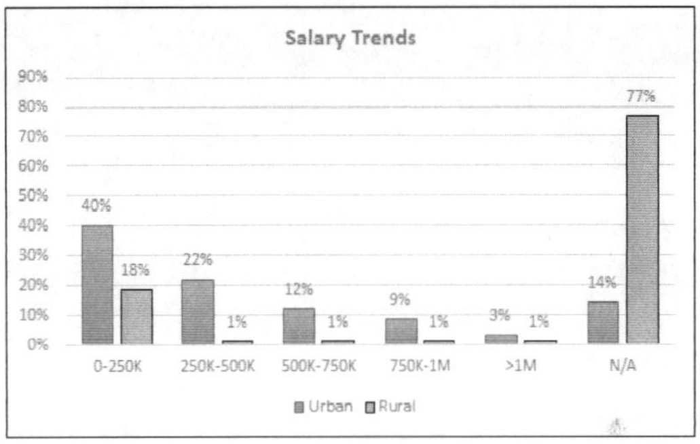

Figure 3. Salary Trends

4.2 Energy/Power Infrastructure

From the online survey, almost all the urban respondents (99.3%) have electricity at their place of residence, compared to 81.5% in the physical walkdown survey. From the online survey, the majority of respondents (75%) do not have a telephone line installed at their residence, compared to 93.2% in the physical walkdown survey. This is depicted in Figure 4, and can be viewed as a positive statistic, as it shows that South Africa generally has good availability in terms of electrical power infrastructure nationally and is also in line with global technology advancements and trends, with the move from fixed telephony to mobile.

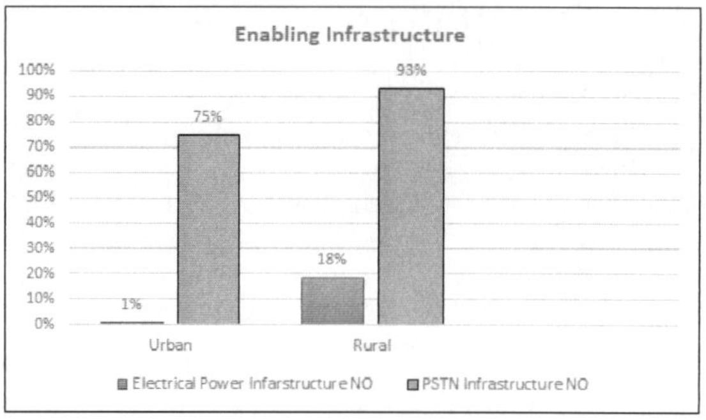

Figure 4. Broadband Enabling Infrastructure

4.3 Broadband Access Terminals

From the online survey, the overwhelming majority of the respondents' own smartphones (96.2%), as is the case in the physical walkdown survey (85.4%). The ownership of laptops and tablets follows as the most popular access terminals, in both surveys, respectively. This is shown in Figure 5. Additionally, in the online survey, 95.5% participants stated that they have access to the internet compared to 85.4% in the physical walkdown survey.

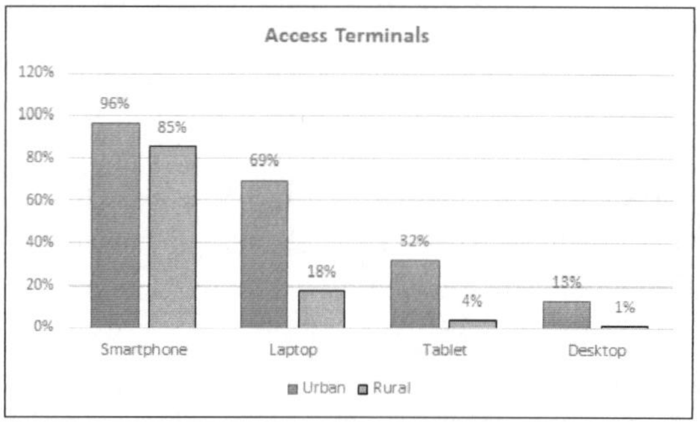

Figure 5. Internet Access Terminal

4.4 Broadband Access Location

From the online survey, 79.5% of the respondents stated that they could access the internet at home compared to 35.9% from the physical walkdown survey. The next popular point of access is the workplace and schools in both surveys, respectively, as depicted in Figure 6. The key difference between these two results is the significant number of respondents from the physical walkdown survey who claimed that they access the internet from the local community center and library, and those who do not have access at all, represented by those who picked the "Other" and "Not Applicable" options respectively.

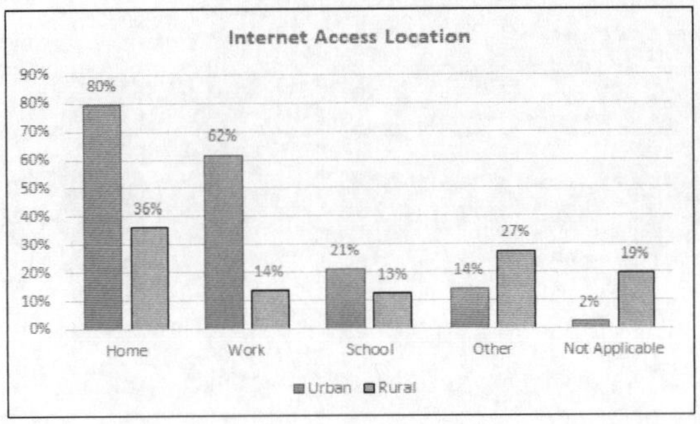

Figure 6. Broadband Internet Access Points

4.5 Broadband Access Technology

From the online survey, the majority of the respondents stated that they use Wi-Fi (63.9%) as the broadband access technology, compared to 44.7% from the physical walkdown survey. This is followed by cellular technologies in both surveys, as shown in Figure 7. This shows that wireless access is more readily available in both urban and rural communities.

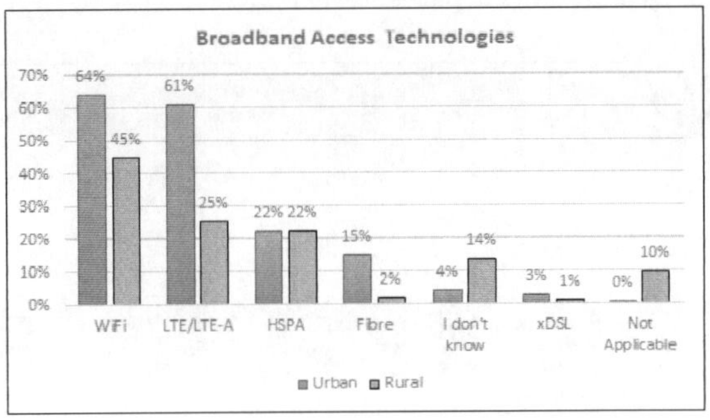

Figure 7. Broadband Access Technologies

4.6 Data Usage

From the online survey, over half of the respondents (51.4%) stated that they use 1-10GB of data per month compared to a quarter of the respondents (25.2%) in the physical walkdown survey. In the physical walkdown survey, the majority stated that they use 100MB-1GB (32%) of data per month, as shown in Figure 8. This translates to the majority of the respondents (52.1%) in the online survey spending between R100-500 per month on data costs, in comparison to 33% of the physical walkdown survey respondents, who spend less than R100. This data is depicted in Figure 9 and collaborates with the one found in Figure 2 and 3 above, where it is shown that many urban respondents are employed and, as a result, can afford to spend more money on data, and hence access more ICT services.

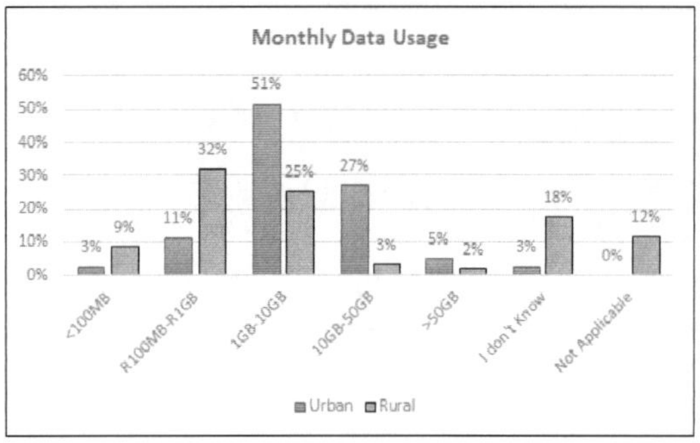

Figure 8. Monthly Data Usage

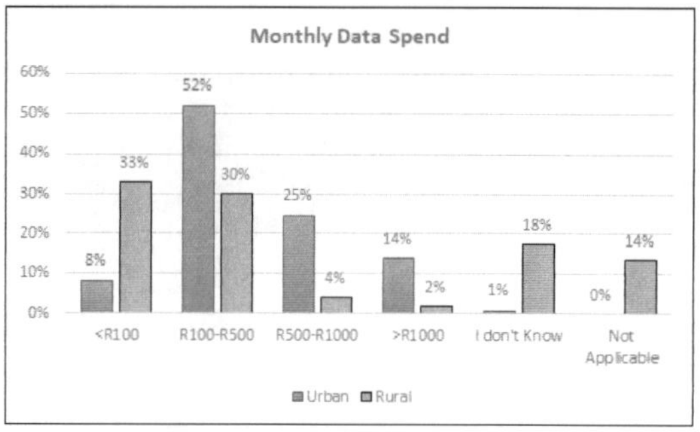

Figure 9. Monthly Data Spend

4.7 Broadband Speeds

When asked what their current broadband speeds were, the response for both the online survey and physical survey showed similar trends. That is, the majority of the respondents in both samples stated that they are not aware of the speeds that they currently have access to. However, for those who claimed to know this information, the majority asserted to have access to 10 Mbps or more, followed by between 5-10 Mbps, in both surveys, respectively. This result is depicted in Figure 10.

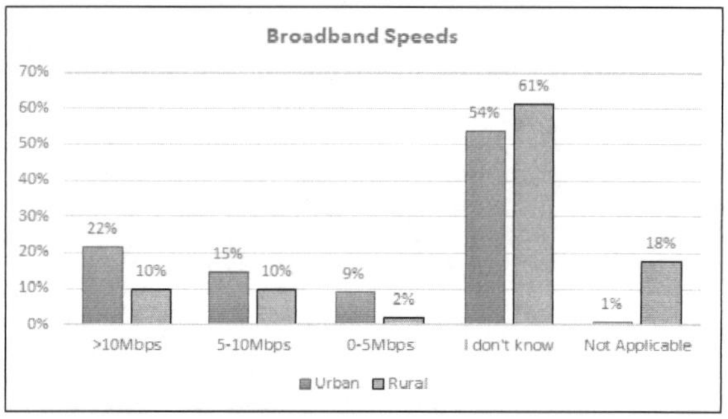

Figure 10. Broadband Data Rates/Speeds

5. DISCUSSION OF RESULTS

From both the surveys, the pattern that emerges is that the online respondents comprise of mostly employed individuals, some of which are also still enrolled in academic institutions, possibly pursuing postgraduate qualifications. An assumption was made that the majority of this sample resides in the urban and suburban areas. From the physical walkdown survey, it is clear that most of the respondents are in fact unemployed. A positive statistic in both surveys is the availability of power infrastructure, and the obvious apparent shift from fixed telephony to mobile telephony, in line with the ITU and ICASA findings. Another positive statistic is the overwhelming majority of respondents who have reported that they own a smartphone, which is also in line with global trends. As expected, a larger number of respondents from the physical walkdown survey reported to have no access to the internet at all, compared to the online survey. This implies that internet access, particularly in the samples and areas considered in this paper, is indeed often readily available in urban areas compared to rural areas. In addition, the internet access in urban areas is, for the most part, available in the convenience of the respondents' homes, while some of the rural respondents depend solely on community facilities, like for example the local library in this survey. Wireless broadband technology has emerged as the most abundantly available technology for connectivity, which is also in line with the findings from the telecommunications bodies.

The majority of the urban respondents have also reported to utilizing and spending more money on data, as compared to their rural counterparts. This collaborates with the data collected, which reflects that most of the rural respondents are unemployed, and many other factors, including availability of broadband infrastructure, and low-income levels. Global statistics show that SA had access to speeds of up to 55 Mbps in 2021, according to studies done by Speed Test Global Index. The respondents from both surveys have reported that they indeed do have access to speeds in excess of 10 Mbps. This response cannot be confirmed, especially in light of the number of respondents who have responded that they are not aware of their broadband speeds. However, 55 Mbps is still not in line with the stipulated target of 50% population having access to speeds of 100 Mbps, as per the SA Connect Policy objectives stipulated in Figure 1.

6. CONCLUSION

The surveys done by the authors were issued in two formats, i.e. the online version as well as one conducted physically, using physical copies of the same questionnaire. The respondents that completed the online survey came from all parts of South Africa, with all nine provinces represented. The respondents from the rural areas came from two small communities in KZN, which had representation from various surrounding villages. The assumption made is that rural areas nationwide have similar characteristics, more or less, with the exception of minor and negligible deviations, for the purposes of this research study. The samples from both surveys were from respondents between the ages of 18 and 65 years.

From the results obtained in both the online and physical surveys, the discrepancy between the urban and rural respondents is observed. Although in both surveys, the majority own smart devices, the access patterns are not the same. It is seen that the urban respondents can readily access the internet from the comfort of their homes, workplace and schools, whereas that is not the case with the rural respondents. This can be attributed to a barrier that will be labelled as "Accessibility". Additionally, the consumption pattern for those in the rural areas who do have access to internet connectivity is not the same as the respondents from the rural areas i.e. the amount of data consumed and the amount of money spent. This can be attributed to a second barrier, which will be labeled as "Affordability". Furthermore, the reasons cited for this are visible from the results. That is, the high poverty and unemployment levels in the rural areas in South Africa, which are highlighted by respondents who have indicated that they obtain little to no income. This despite the fact that most of the survey respondents reported that they own smartphones, more than any other communication device and are able to receive 3G cellular broadband signal, at minimum. Notably, this statistic collaborates with both ITU and ICASA's finding of the growing predominance of mobile over fixed telecommunications services as well as ICASA's estimate of in excess of 90% 3G and LTE coverage, nationwide.

With indicative evidence of the status quo, it is the authors' view that broadband access is indeed one of the key drivers of economic participation and development in a society and country as indicated by the state of and discrepancy in socioeconomic situations between rural and urban areas in SA. The analysis of the data from the SA Connect, ITU and Global Index, backed up by the surveys conducted by the authors, indicate that the goals set in the SA Connect for 2020 has not been fulfilled at the time of this publication, and would need to be prioritized and expedited if economic inclusion is to be achieved. To this end, it is proposed that community-based broadband systems are implemented in rural areas, which would be bespoke to the community in question, and particularly address the key barriers, which will be posed as research questions, i.e. "Accessibility" and "Affordability". In addition, this system must be sustainable, in order to facilitate preservation and long-term profitability thereof, which should impact the community in a positive way, and facilitate growth and prosperity in rural areas. These three barriers (Accessibility, Affordability and Sustainability) are the key research questions for future work.

REFERENCES

Communications, M. of (2013) 'South Africa's Broadband Policy', *Government Gazette* [Preprint], (November).

Datareportal (2023). Available at: https://datareportal.com/reports/digital-2023-south-africa#:~:text=51.3 percent of South Africa%27s, percent lived in rural areas

International Telecommunication Union (ITU) (2018) *Measuring the Information Society Report 2018 - Volume 1. Geneva, Switzerland., ITU Publications.*

ITU (2012) 'Partnership on Measuring ICT For Development'.

National Planning Commission (2011) *Our future - make it work.* Available at: https://doi.org/ISBN: 978-0-621-41180-5.

Souter, D. *et al.* (2009) *The apc icT policy Handbook Editor David Souter With contributions from.* Available at: www.apc.org.

Stork, C., Calandro, E. and Gamage, R. (2014) 'The future of broadband in Africa', *info*, 16(1). Available at: https://doi.org/10.1108/info-10-2013-0055.

'The State of the ICT Sector Report in South Africa' (2023), (March).

'World Bank' (2021). Available at: https://data.worldbank.org/indicator/SP.RUR.TOTL.ZS.

AN EXPLORATORY STUDY ABOUT THE ROLE OF GENDER IN CYBERSECURITY AWARENESS FOR CHILDREN

Farzana Quayyum and Letizia Jaccheri
Norwegian University of Science and Technology (NTNU)
Trondheim, Norway

ABSTRACT

Gender disparities in technology use and cybersecurity are a persistent and growing concern among researchers and society. This paper presents the results from a qualitative study conducted with 25 parents in Norway, exploring the role of gender (of either the child or the parent) in parents' perceptions of their children's cybersecurity awareness and the mediation strategies of children's online activities. Our research reveals certain patterns regarding parents' technology use, perceptions of their children's technology preferences, and exposure to online risk depending on gender. By investigating parents' views from a gender perspective, our study offers insights for practitioners and researchers engaged in gender research, designing and implementing cybersecurity awareness solutions. In addition, we provide suggestions for future research and application.

KEYWORDS

Cybersecurity Awareness, Parents, Children, Gender

1. INTRODUCTION

Managing technology use at home is an evolving endeavor that has recently developed rapidly, posing new issues and concerns for parents (Quayyum et al., 2021). Parents play a significant role in mediating children's online activities and are essential stakeholders in raising cybersecurity awareness among children. Therefore, exploring and understanding any gender influence among the parents is crucial. Moreover, demographic changes, culture, customs, and parental socioeconomic status can also impact socially structured parenting patterns in the face of children's cybersecurity needs (Elstad & Stefansen, 2014).

Many researchers have acknowledged the gender gap in the cybersecurity domain. Researchers have explored gender differences in many areas, such as in computer self-efficacy (e.g., Aesaert & Van Braak, 2015), enjoyment of STEM fields (e.g., Miller et al., 2006), programming and coding activities (e.g., Papavlasopoulou et al., 2020), cybersecurity engagement and self-efficacy among teens (e.g., Amo, 2016) and so on. However, little research has been done on how and whether the gender of adults (parents in our case) has any influence when it comes to children's cybersecurity awareness and behavior. In this study, we have investigated the role of gender in parents' perceptions of their children's cybersecurity awareness as well as among the children themselves in terms of cybersecurity risks and practices. We believe understanding parental perceptions and any possible gender influence can support the development of fairer and more inclusive cybersecurity awareness solutions and technologies for users. Thus, in this study, we pose the following research questions:

RQ1: *What role does gender play in parents' perceptions of their children's cybersecurity and concerns?*
RQ2: *What role does gender play in how parents mediate their children's online activities and internet use?*

In Section 2, we describe the background and context of this study. We then present our research methods in Section 3 and the results in Section 4. In Section 5, we discuss the findings and study limitations. Finally, we conclude by presenting the implications of the findings and suggesting future research directions in Section 6.

2. GENDER DISPARITIES AND CYBERSECURITY

Gender disparities in attitudes toward technology use and education have long been a concern for researchers and society in general (Cai et al., 2017). Studies have revealed evidence of gender differences surrounding beliefs and behavioral intentions regarding cybersecurity (e.g., Anwar et al., 2017). Gender has been associated with exposure to risky or uncomfortable situations (Savoia et al., 2021). Amo (2016) found gender differences in cybersecurity engagement and self-efficacy among teenagers. While the use of the internet and social networking sites are universal for all genders, research also shows gender-related differences in the scope and depth of its use (Iwilade, 2015). Perceptions of and exposure to risks can also vary with gender. Gustafsod (1998) investigated how gender can contribute to understanding risk and revealed differences in how males and females perceive risks in general and regarding cybersecurity risks in particular. Many other studies, including Dempsey et al. (2018), Desimpelaere et al. (2020), and Hoy & Milne (2010), show that gender is related to the degree of online privacy concerns, with females showing greater privacy concerns than males. Women place greater importance on perceived control and privacy risks when sharing information on social networking sites than men (Hajli & Lin, 2016). Exposure to some types of harmful content may also depend on gender. A recent study (Kvardova et al., 2021) shows that girls can be exposed more often to harmful content on the internet, at least in some contexts. Gender differences are also relevant in terms of cyberbullying. Notar et al. (2013) conducted a literature review on cyberbullying and presented a stark gender comparison: girls are more likely to experience cyberbullying than boys.

In addition, Sezer et al. (2015) investigated the awareness levels of teachers concerning cyberbullying, finding that male teachers have a higher level of awareness than female teachers. Mıhcı Türker & Kılıç Çakmak (2019) found that female students had significantly higher awareness than male students in terms of internet addiction, cyberbullying, netiquette, online privacy, inappropriate content, copyright, and cybersecurity. Though these two studies had different target groups (students and teachers), it is clear that gender differences can exist among children and adults. Therefore, it is crucial to investigate whether and how gender influences adults' perceptions of and actions toward children's cybersecurity.

Research suggests that parental mediation strategies and parents' security concerns influence children's online behavior and internet use (Wisniewski et al., 2015). However, findings regarding the relationship between gender and the amount or type of parental mediation are mixed. While some researchers have found a relationship between a child's gender and the amount or style of parents' mediation (Nikken & Jansz, 2014; Sonck et al., 2013), others found no such connection (Livingstone & Helsper, 2008). Considering all the literature mentioned above and their findings, we extend the gender studies in the cybersecurity domain by exploring whether parents' cybersecurity concerns and mediation strategies have any relationship with gender (of either the child or the parent) in our study context.

3. RESEARCH METHODOLOGY

We conducted semi-structured interviews with 25 parents living in Norway. Since this study aimed to concentrate on parents' perceptions and actions, we performed the interviews with only the parents and did not include the children. The interviews were exploratory in nature, and we asked the participants open-ended questions. We asked questions about parents' perceptions of children's cybersecurity awareness, concerns about online activities, the experience of risk exposure, and how they mediate their children's internet use at home. We asked open-ended questions so that they could respond in detail to our questions and introduce any issues that they considered relevant. We did not ask the parents any direct questions related to gender as we wanted to determine whether the parents' behavior or perceptions differed depending on gender in a natural setting rather than what they said when questioned directly on the issue.

Moreover, as our sample was not gender-balanced (eight fathers and 17 mothers), we attempted to identify if there were any gender-related patterns in the parents' answers rather than making direct comparisons. By "pattern," we mean repeated selective elements found in parents' responses to the interview questions. We conducted a thematic analysis of the interview data following the steps recommended by Cruzes and Dyba (2011) to identify any gender-specific patterns.

We recruited participants with at least one child in our target age group (10–15 years old). Interviews were conducted face-to-face or online, lasting from 10 to 34 minutes (depending on how much the parents shared and elaborated on their answers). The sample comprised eight fathers and 17 mothers (including five couples), and their ages ranged from 38 to 56. The interviewees had 53 children, whereas 26 children (six girls and 20 boys) were in our targeted age range. All the participating parents had higher education, varying from a bachelor's degree or higher. Before starting the data collection process, the researchers received approval from the Norwegian Centre for Research Data (NSD). Following ethical approval from the NSD, we obtained informed consent from all the participants.

4. RESULTS AND FINDINGS

4.1 Parents' Risk Perceptions

We have identified multiple privacy- and security-related risks about which parents are concerned regarding their children (Figure 1). We have categorized these concerns by their nature and divided them into five major categories of risks: (i) privacy risks, (ii) online content, (iii) online contact, (iv) online conduct, and (v) consumer risks.

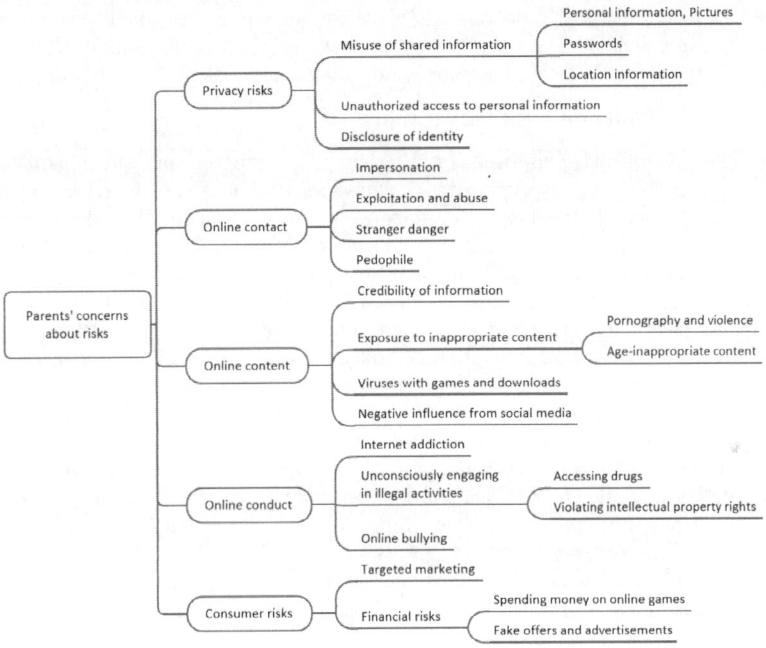

Figure 1. Parents' risk concerns for children

Privacy risks. Twelve of 25 parents expressed concern about their children's privacy. Parents worried that children could share sensitive personal information (location, pictures, identity, etc.) on social media and game forums. Sharing personal information makes it easy for people (both known and unknown) to track or misuse it. One parent also expressed concern about freedom of speech and the opinion that today's children share information on social media that others can exploit in the long term; this parent cited Facebook and Cambridge Analytica as examples. Parents also expressed concerns about how sharing one's location online could result in physical harm. Moreover, children's poor password practices can also result in privacy breaches. Risky practices like sharing passwords with online friends can result in disclosing one's identity and sensitive information, losing control of devices and information, exploitation, and abuse by friends or others connected—however remotely—to someone with whom a child shares a password.

Online contact. Another major cybersecurity concern is online contact, which 14 of 25 parents mentioned. It is related to the fact that children can easily contact strangers through social media or online games. Parents are worried that their children can get into danger by contacting strangers. Impersonation, pedophiles, exploitation, and abuse are some of the common forms of danger that parents mentioned.

Online content. Parents' online content-related concerns include exposure to inappropriate content, the credibility of online information, the negative influence of online content, and the risk of viruses that can result from downloading online content. Nine parents noted concerns related to content that children can access over the internet.

Online conduct. Seven parents mentioned issues related to online conduct. The most common form of online conduct-related risk is cyberbullying. The parents' interviews showed that most online bullying cases occur on social media platforms or online video game forums. Parents are concerned that such online bullying experiences can affect their children mentally and that children might feel uncomfortable sharing such experiences, even with their parents. Becoming addicted to the internet or online games, unconsciously engaging in illegal activities like accessing illegal products, and violating intellectual property rights were the other forms of conduct-related risks that parents mentioned.

Consumer risks. The fifth category of parental concerns is related to the risks that arise when children are treated as potential consumers. Three parents reported concerns about how third-party organizations access personal data and children's usage patterns from their online activities to employ in targeted marketing. One parent also mentioned that a child's internet behavior and activities might not significantly affect them today; however, when their data are collected and monitored by third parties, children could be affected over the long term. Such risks can also come from online games and advertisements. In addition to the parents who expressed data and usage-related privacy concerns, seven other parents expressed concerns about children being tricked into spending money on online games, fake offers, or online advertisements.

4.1.1 Influence of Gender on Risk Perceptions

The parents in the present study expressed concerns about their children similarly, regardless of the gender of the parents. As presented above, parents have expressed a wide range of security concerns about their children's online presence. While some of these concerns are relevant for all children regardless of gender, we have seen some concerns that do relate to the gender of the child. For example, some cybersecurity risks that the parents mentioned referred only to their sons. Risks like spending money on online games and exposure to pornography were specifically mentioned for male children; none of the parents expressed these same concerns for their daughters. Similarly, the concern regarding the negative impact of online games and the risk of becoming addicted to them were noted only for sons. Seven parents from our sample referred to worries about their children sharing personal information, such as photos, on social media. Though the parents had this concern for both boys and girls, four of the seven parents emphasized sharing personal pictures more in relation to girls, as stated by this parent,

"It is perhaps the consequence that can occur in ten years if you have shared something that should not have been shared then. Typically, perhaps more in relation to girls, but it applies to boys as well. You are sharing photos where you're barely dressed or things like that, and it can be abused by someone. That may be what I think of first and foremost as the biggest concern."

Another parent who has an 11-year-old daughter stated that *"[I am concerned] mostly in terms of being tricked by someone, so they're tempted to suddenly send some sensitive or personal photos or meet some people other than what they pretend to be."*

4.2 Parental Mediation Strategies

The parents in our study reported using a variety of strategies to mediate children's internet use and related cybersecurity risks. Following the literature (Livingstone & Helsper, 2008; Nathanson, 2001), we have divided the strategies into three broad categories, as presented in Figure 2.

Active mediation. The most common strategy in active mediation is talking with children. Parents felt that raising awareness and disseminating knowledge are important, rather than simply being strict and controlling the use of devices or the internet in general. A majority of parents (16 of 25) explicitly mentioned talking about online security issues with their children, using either their own cybersecurity experiences or a cybersecurity case in the media as examples.

Restrictive mediation. Many parents also mentioned setting rules for and restrictions on internet use. Using device- or site-driven parental control tools is also a common strategy to set specific rules and restrictions. Thirteen of 25 parents reported using parental controls, mostly a device's default features. The parents used parental controls to limit screen time, restrict access to different websites, set age limits for accessing websites, and limit download opportunities. Other restrictive strategies include limiting children's access to specific activities (such as downloading apps or buying anything on their own).

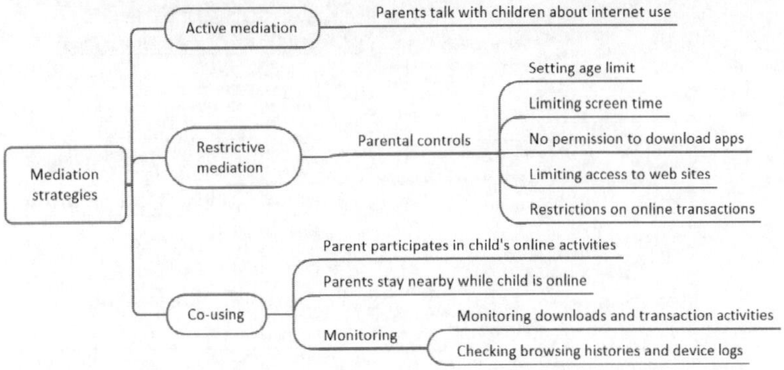

Figure 2. Parents' mediation strategies

Co-using. The co-using strategies mentioned by our study parents include strategies in which a parent tries to participate in the child's activities or simply stays nearby and watches those activities. Two parents said that they try to engage in online activities with their children (such as playing online games together or creating accounts on social media to follow each other). Nine parents reported staying nearby and observing the child using the internet and devices, but they do not necessarily have any direct interaction with the child at those times. Ten parents also mentioned that they occasionally or routinely check their children's devices and device logs to monitor the activities that their children are participating in, even if they don't restrict their activities. Because the parents reported doing so with an agreement with the children, we have classified this device log-checking as co-using. These parents said that their children are aware of the fact that they periodically check their phones, even though the children may not like it.

4.2.1 Gender Patterns in Mediation Strategies

The parents use active and restrictive mediation strategies similarly regardless of their gender and their children's gender. However, we have observed that when it comes to monitoring children's personal devices and activity logs, mothers from our sample do this more than fathers. Just over a third (six of 17) of mothers reported checking their children's mobile phones and devices to monitor their online activities, whereas a quarter (two of eight) of fathers reported checking their children's phones or devices.

Even though more mothers reported checking children's devices, some mothers (four of 17) reported that fathers take care of technical issues in the home, such as setting parental control tools, controlling downloads, and assessing specific software or games and related permissions. For example, when asked about using parental control tools, one mother stated, *"I guess we have one of those apps that limits [the access on phones], but I'm not sure. It's my husband who fixes it"*. Another mother responded that *"I didn't do it; it was my husband—but I think we set an upper age limit. It was also for not accidentally getting into inappropriate websites."*

4.3 Parents' Views on Children's Online Activities and Gender Influence

In terms of children's online activities and behaviors, we observed some patterns that relate to the gender of the children (Figure 3). Before presenting them, we note that they are based on the parents' descriptions, as our interviews were conducted only with parents. Thus, we acknowledge that the children may have different perspectives on their choices and preferences.

Participation in programming activities. Seven parents reported that their children were taking or had taken elective courses on programming (either at school or as an extracurricular activity): notably, all seven children are boys. None of the parents' daughters were taking or had taken any elective programming courses in schools or outside.

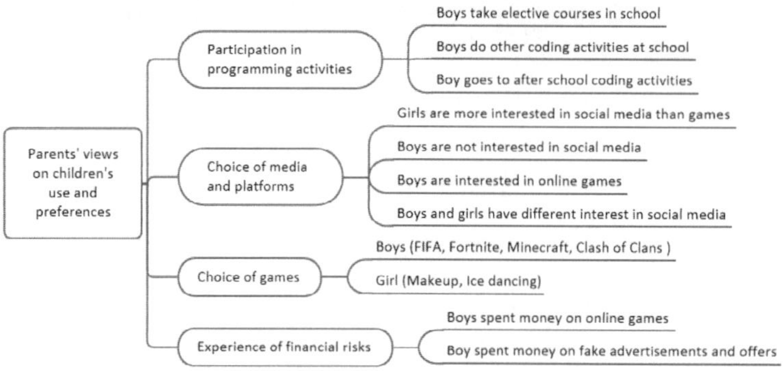

Figure 3. Parents' views on children's online activities, media use, and preferences

Choice of media and platforms. Following the parents' descriptions, we have identified a few patterns in the children's media and platform choices. Only one parent noted her daughter playing online games, while some parents (seven of 25) acknowledged their sons frequently playing online games. There can also be variations in the games available. The 10-year-old daughter enjoys playing games like dress-up, ice dance, and makeup, while the sons apparently enjoy playing FIFA, Fortnite, and Minecraft (as mentioned by the parents). Additionally, seven parents claimed that boys are not interested in social media; rather, they believe that boys are more into games and that girls are more interested in social media.

Risk experiences. The parents mentioned multiple episodes in which their children were exposed to different risky situations that resulted from online activities, such as meeting strangers they contacted online, sharing passwords with friends, being bullied on social media platforms, exposure to inappropriate content, falling victim to online frauds and advertisements, and so on. Interestingly, we have noticed that all the experiences related to spending money online were reported for boys. Nine parents said that their sons had spent money on online games and fake advertisements without fully understanding what they were doing. The most commonly cited example was related to the game "Fortnite". However, none of the parents mentioned having their girls experience the same occurrences.

5. DISCUSSION

5.1 Parents' Cybersecurity Perspectives and the Role of Gender

As presented in Section 2, previous studies have shown the effect of gender on various aspects of cybersecurity, including cybersecurity engagement, behavior, and users' risk experiences. Our study also revealed some patterns among the parents, as presented above. One example is taking care of technical issues. As Cai et al. (2017) show, gender-based differences in attitudes toward technology have long been a concern. These differences still may exist to date. Our study results showed that some mothers prefer fathers to take care of technical issues at home, even though both parents are equally responsible for ensuring their children's online security. This finding can indicate the existence of the stereotypical view of technology use and gender. However, we must acknowledge that only four mothers from our study (out of 17) revealed this behavior; the majority did not show similar behavior regarding technology usage.

Another gender pattern we observed is the gender influence in terms of risk exposure, which is supported by the findings in other studies (Kvardova et al., 2021; Savoia et al., 2021). Being tricked into spending money on games or fake advertisements is a risk we have found parents mentioned only for their sons. This finding also relates to another pattern we found; some parents reported that their boys like to play online games (such as FIFA and Fortnite), whereas girls like to spend time on social media. We also noticed that parents worry about some risks more for their sons but not at the same level for their daughters, such as exposure to pornography, internet addiction, and online shopping. A few parents also emphasized the risk of exposing sensitive personal photos more for girls. These results once more raise the possibility of gender stereotypes and beliefs. However, as was previously indicated, our sample size was small and unbalanced in terms of gender. As a result, we were unable to directly compare these findings based on gender.

5.2 Parental Mediation and the Role of Gender

In terms of mediation, we have seen that all the parents expressed about using similar strategies for the children's internet use and activities regardless of the gender of the children. This supports the findings of Livingstone & Helsper (2008). We did see, however, that mothers were more likely than fathers to check their children's device logs. Nevertheless, it is important to note that given that we included considerably fewer fathers than mothers in our study, this pattern could change. In our upcoming research, we can delve more into this topic. Apart from that, we have seen that mothers and fathers use active, restrictive, and co-using mediation techniques similarly.

5.3 Limitations

While we note the importance of the findings in this study, we do acknowledge certain limitations. The main limitation is the lack of a gender-balanced sample. Our study had more mothers than fathers, making direct comparisons of gender-based findings impossible. Our goal in this study, however, was to identify patterns that could serve as a valuable foundation for our future work and future researchers. Future gender-focused studies can explore the perceptions of parents using a more gender-balanced sample and identify any possible differences from the results of this study. Another limitation is that the result of this study reflects only parents' perceptions. As we have already noted, the children may have different opinions about their online activities and choices than what their parents reported. Thus, we would like to acknowledge that understanding and investigating both children's and parents' perceptions could complement our findings.

6. CONCLUSION

This study presents the results of a qualitative study exploring the role of gender in parents' perceptions and mediation strategies regarding children's cybersecurity awareness and attitudes. Our study provides some insights about potential gender influences in terms of parents' attitudes toward their children's cybersecurity, risk exposure, and parental mediation. As gender influence in cybersecurity engagement, behavior, and risk experiences have been demonstrated in both previous research and our study, it is essential to explore this factor more in future endeavors.

It is natural that children learn and develop the behavioral attributes they see in their parents and at home. Thus, the presence of any gender stereotypes in parents could lead to the development of similar stereotypes and attitudes among their children. Such stereotypes could also influence the children's future lives and choice of career paths, prolonging the gender inequality problem of today's society over the long term and resulting in inequalities for all genders. Understanding the role of gender regarding cybersecurity and technology use, in general, could be crucial for designing and developing inclusive and effective cybersecurity educational and awareness opportunities for both children and parents. Thus, we suggest that cybersecurity awareness solutions should be more inclusive, and designing inclusive technologies and solutions that fully consider the gender perspective is essential. Our current research findings will serve as inspiration for our future work, which will focus on designing an inclusive cybersecurity awareness application for children that considers parental expectations and influences.

REFERENCES

Aesaert, K., & Van Braak, J. (2015). Gender and socioeconomic related differences in performance-based ICT competences. *Computers & Education*, 84, 8–25.

Amo, L. (2016). Addressing gender gaps in teens' cybersecurity engagement and self-efficacy. *IEEE Security & Privacy*, 14(1), 72–75.

Anwar, M., He, W., Ash, I., Yuan, X., Li, L., & Xu, L. (2017). Gender difference and employees' cybersecurity behaviors. *Computers in Human Behavior*, 69, 437–443.

Cai, Z., Fan, X., & Du, J. (2017). Gender and attitudes toward technology use: A meta-analysis. Computers & Education, 105, 1–13.

Cruzes, D. S., & Dyba, T. (2011). Recommended steps for thematic synthesis in software engineering. *In 2011 International Symposium on Empirical Software Engineering and Measurement* (pp. 275–284). IEEE.

Dempsey, J., Sim, G., & Cassidy, B. (2018). Designing for GDPR-investigating children's understanding of privacy: A survey approach. *In HCI '18: 32nd Human Computer Interaction Conference*. Association for Computing Machinery.

Desimpelaere, L., Hudders, L., & Van de Sompel, D. (2020). Knowledge as a strategy for privacy protection: How a privacy literacy training affects children's online disclosure behavior. *Computers in Human Behavior*, 110, 106382.

Elstad, J. I., & Stefansen, K. (2014). Social variations in perceived parenting styles among Norwegian adolescents. *Child Indicators Research*, 7(3), 649–670.

Gustafsod, P. E. (1998). Gender differences in risk perception: Theoretical and methodological perspectives. *Risk Analysis*, 18(6), 805–811.

Hajli, N., & Lin, X. (2016). Exploring the security of information sharing on social networking sites: The role of perceived control of information. *Journal of Business Ethics*, 133(1), 111–123.

Hoy, M. G., & Milne, G. (2010). Gender differences in privacy-related measures for young adult Facebook users. *Journal of Interactive Advertising*, 10(2), 28–45.

Iwilade, A. (2015). New spaces, New interactions? Young people's online social networks and gender relations in Africa. *Gender, Technology and Development*, 19(1), 70–90.

Kvardova, N., Smahel, D., Machackova, H., & Subrahmanyam, K. (2021). Who is exposed to harmful online content? The role of risk and protective factors among Czech, Finnish, and Spanish adolescents. *Journal of Youth and Adolescence*, 50(12), 2294–2310.

Livingstone, S., & Helsper, E. J. (2008). Parental mediation of children's internet use. *Journal of Broadcasting & Electronic Media*, 52(4), 581–599.

Miller, P. H., Slawinski Blessing, J., & Schwartz, S. (2006). Gender differences in high-school students' views about science. *International Journal of Science Education*, 28(4), 363–381.

Mıhcı Türker, P., & Kılıç Çakmak, E. (2019). An investigation of cyber wellness awareness: Turkey secondary school students, teachers, and parents. *Computers in the Schools*, 36(4), 293–318.

Nathanson, A. I. (2001). Parent and child perspectives on the presence and meaning of parental television mediation. *Journal of Broadcasting & Electronic Media*, 45(2), 201–220.

Nikken, P., & Jansz, J. (2014). Developing scales to measure parental mediation of young children's internet use. *Learning, Media and Technology*, 39(2), 250–266.

Notar, C. E., Padgett, S., & Roden, J. (2013). Cyberbullying: A review of the literature. *Universal Journal of Educational Research*, 1(1), 1–9.

Papavlasopoulou, S., Sharma, K., & Giannakos, M. N. (2020). Coding activities for children: Coupling eye-tracking with qualitative data to investigate gender differences. *Computers in Human Behavior*, 105, Article 105939.

Quayyum, F., Bueie, J., Cruzes, D. S., Jaccheri, L., & Vidal, J. C. T. (2021). Understanding parents' perceptions of children's cybersecurity awareness in Norway. *In Proceedings of the Conference on Information Technology for Social Good* (pp. 236–241). Association for Computing Machinery.

Savoia, E., Harriman, N. W., Su, M., Cote, T., & Shortland, N. (2021). Adolescents' exposure to online risks: Gender disparities and vulnerabilities related to online behaviors. *International Journal of Environmental Research and Public Health*, 18(11), 5786.

Sezer, B., Yilmaz, R. and Karaoglan Yilmaz, F.G. (2015). Cyber bullying and teachers' awareness. *Internet Research*, 25(4), 674–687.

Sonck, N., Nikken, P., & de Haan, J. (2013). Determinants of internet mediation: A comparison of the reports by Dutch parents and children. *Journal of Children and Media*, 7(1), 96–113.

Wisniewski, P., Jia, H., Xu, H., Rosson, M.B. and Carroll, J.M., (2015). "Preventative" vs. "Reactive" How Parental Mediation Influences Teens' Social Media Privacy Behaviors. *In Proceedings of the 18th ACM conference on computer supported cooperative work & social computing* (pp. 302-316).

DIGITAL EXPERIENCES FOR THE PROMOTION OF CULTURAL HERITAGE

Dalia Gallico
Università San Raffaele Roma
Italy

ABSTRACT

The incorporation of digital technology in museums can offer a more engaging and interactive experience for visitors. It can also allow you to reach a wider audience through resources and online platforms and to provide additional context and information about their collection. However, it is important for museums to carefully consider how to integrate digital technology into their exhibits and ensure that it enriches rather than detracts from the overall visitor experience. By carefully incorporating digital experiences into their museums, they can provide visitors with a more immersive and engaging experience. Studying the intersection between digital technology and cultural heritage, this thesis aims to explore the potential of digital experiences in the cultural heritage sector and to identify best practices for creating immersive and educational digital experiences that engage and educate users. We will first examine the new technologies at the service of the cultural heritage sector and then move on to market numbers and target users. Through a literature review and successful case studies, we will identify the key factors contributing to the effectiveness of digital experiences applied to cultural heritage and then develop a concrete implementation strategy of these technologies by sector actors.

KEYWORDS

Digital Experience, Cultural Heritage, Virtual Reality (VR), Augmented Reality (AR), Social, Promotion

1. INTRODUCTION

The digital age has brought about unprecedented changes in the access, experience and interaction with cultural heritage. From virtual reality exhibits to online collections and interactive educational resources, the use of digital technologies in the cultural heritage sector has the potential to democratize access by increasing enjoyment to a wider audience. However, simply digitizing cultural heritage does not necessarily result in a meaningful and engaging user experience. Therefore, it is crucial to carefully consider the design and implementation of digital experiences in this area in order to effectively convey the meaning and value of cultural heritage.

Before proceeding further, it is also necessary to consider the challenges that this digital change has brought to the cultural heritage sector and how it has instead affected the use of content during the COVID-19 pandemic.

In 2020, in Italy, following the legal provisions which ordered the mandatory closure of museum structures to contain the spread of Covid-19 (between the end of February and May and between November and December), almost all museums (92 %) managed to guarantee services and activities, reopening the exhibition spaces to the public of visitors when permitted. Of these, over half (56.7%) remained open to the public for more than five months in the year, 28.1% for three to five months while 15.2% ensured physical access to visitors only for two months. The museums of the central regions (61.1%) and the museums of large urban centers (63.2%) and metropolitan cities (59%) have managed more than others to guarantee prolonged opening of the structures (more than five months). On average, during 2020, the facilities were open to the public for around 116 days. In the months in which the public had access to the exhibition spaces, 82.4% of the museums adopted at least one of the health measures required to reduce the risk of contagion. In addition to the use of disinfectant products for hand hygiene (85.8%) and devices for detecting body temperature at the entrance to the structure (63.1%), many museums have organized visits in such a way as to limit the presences (82%) or to allow only programmed entrances by telephone and online appointment (58.1%). Just over half of the museums (54.2%) also set up visit itineraries with horizontal signs and 44.9% created the delimitation of the contact areas with adequate physical barriers.

In the face of the measures that have limited the possibility of attending public places indoors due to the health emergency, seven out of 10 museums (73%) have used alternative tools and methods to stay in touch with the public. Of these, the majority (63.6%) carried out remote communication and information activities through the main social media (Facebook, Instagram, Twitter), 46.1% increased or launched information initiatives through dedicated web platforms, 39.1% made streaming presentations of their collections or proposed video interviews with industry experts. Three museums out of 10 (30%) have chosen to make virtual tours of the museum facility available to users, to allow remote guided tours of the exhibition spaces.

To organize and provide online services, museums for the most part (85.4%) made use of internal staff who already had the professional skills necessary to manage digital technologies. Relatively few museums have resorted to external professionals (24.3%) or provided specific training or professional retraining interventions to internal staff (11.7%). The digital information and online communication activities with the public involved a larger number of structures than in previous years: in 2019, in fact, 63.4% of the institutes were present on the web with their own dedicated website, 57.4% had opened an account on major social media and 27% of museums offered tours and virtual visits on their website.

On the other hand, as a negative aspect to be attributed to digital change, there is certainly the proliferation of fake news and digital reproductions to be taken into consideration, which have cast doubt on the authenticity and value of the entire heritage. Furthermore, the growing demand for digital experiences has led to an oversaturation of online content, making it difficult for institutions stand out and effectively engage their audience.

Against this backdrop, it is more important than ever to carefully consider the design and implementation of digital experiences in order to effectively convey the meaning and value of cultural heritage.

The cultural heritage sector is vast and diverse, encompassing a wide range of museums, sites and traditions valued for their historical, artistic or cultural significance. Therefore, it is important to consider the specific needs and context of each heritage organization and its public.

Table 1. Percentage of people who have visited museums and/or exhibitions at least once in the last 12 months (2019 and 2020). Elaboration by Openpolis - With the Children on Istat data

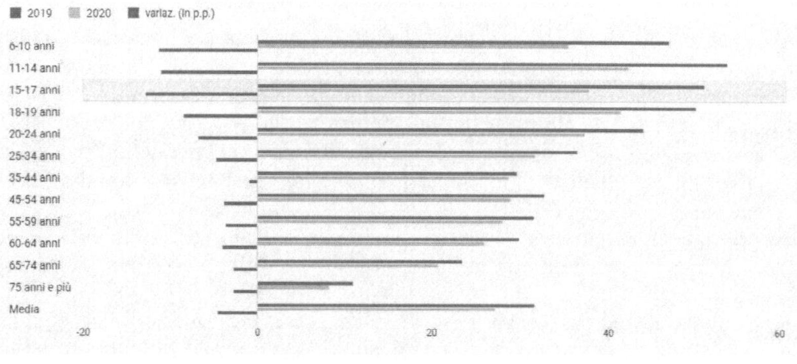

Table 2. Percentage of minors who have visited museums and/or exhibitions at least once in the last 12 months (2010-20). Elaboration by Openpolis - With the Children on Istat data

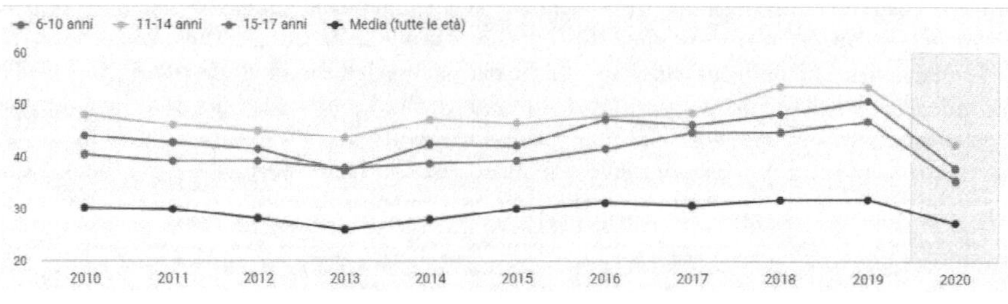

Table 3. Percentage composition of Italian museums based on reopening in 2021.
Elaboration Openpolis with Children on Istat data

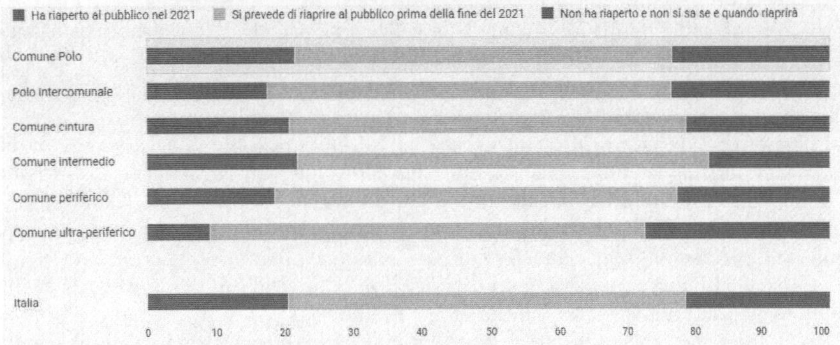

2. DIGITAL EXPERIENCE

Digital experiences can take many forms, including:

• Virtual Reality (VR) exhibitions: VR technology allows visitors to immerse themselves in a digital environment, often linked to the museum's collection. For example, a history museum might use VR to transport visitors back in time to experience an event that happened or delve into a historical location. VR can also be used to create interactive exhibits that allow visitors to explore and interact with a digital environment in a more active way. This can be especially effective for science museums, where visitors can learn complex concepts in a more immersive and interactive way.

We can find some Italian case studies such as: the exhibition “A Space Adventure”. In Fiumefreddo Bruzio (Cosenza) the Space Adventure exhibition offers "astronaut" experiences thanks to simulators, installations and virtual reality. Made by Extramuseum, by the RADAR Association in partnership with the Municipality of Fiumefreddo Bruzio, the Province of Cosenza and the Calabria Region. It is a production of the US Space & Rocket Center, the space museum in Huntsville in Alabama, USA.

The second Italian case study is the exhibition "The R-Evolution Park" created by Way Experience and Milanoguida with the scientific support of WWF and open to the public from March 2022. "The R-Evolution Park" is a revolutionary experience that addresses the theme of the evolution of Planet Earth. A journey through time, from the dawn to the present day, to find out how save the world. An adventure, with a scientific and environmental sustainability theme, to be experienced at 360° thanks to the Virtual Reality viewers.

• Augmented Reality (AR) exhibits: AR technology overlays digital information onto the real world, often through a smartphone or tablet app. This can be used to enrich traditional exhibits, for example by adding additional context or historical information to an exhibit on display. For example, a museum might use AR to provide additional information or context about a specific painting or sculpture, or to allow visitors to virtually explore a 3D model of an exhibit. The most virtuous case studies were conceived by: AGO Art Gallery of Ontario, with the exhibition "ReBlink". The brainchild of artist Alex Mayhew, the Toronto museum has created ReBlink, an innovative Augmented Reality experience that draws on innovations in technology to offer visitors the ability to see AGO's works in a whole new way through their phone's camera and a specially developed app.

The National Museum of Natural History Smithsonian in Washington DC allows visitors to reconstruct the skeletons featured in the "Bone Hall" exhibit via a specially developed mobile app.

• Interactive kiosks and touchscreens: Many museums now have interactive kiosks or touchscreens that allow visitors to learn more about specific exhibits or the museum's collection as a whole. These may include video content, quizzes and other interactive elements. Interactive kiosks can be especially effective for museums with large collections, allowing visitors easy access to information about specific items in the collection.

• Online resources and digital collections: Many museums now offer online resources and digital collections that visitors can access before or after their visit. These can include insights, virtual tours, online exhibits and digitized versions of the museum's collection. Online resources can be a great way for museums to provide additional context and information about their collection, and they can also be a valuable resource for researchers and educators. The reference Italian case studies are: Museion of Bolzano, which organizes online events, among which the one dedicated to art therapy stands out, with conferences, talks by experts and workshops that can be accessed directly from the museum's Facebook page.

From the official website of the Pinacoteca di Brera, it is possible to access the over 600 works in the collection, with very high definition photographs that allow you to discover details that may be indecipherable live. Furthermore, the Brera On air channel of the art gallery offers music, readings, documentaries and exhibitions, the first of which is dedicated to Raphael. The contemporary dialogues with the Peggy Guggenheim Collection in Venice are also online. The virtual programming of the Brescia Musei Foundation allows visitors to visit current exhibitions, educational activities and features as well as new titles in streaming directly from the latest edition of the Trieste Film Festival.

• Social Media and other web platforms: Museums can use social media and other online platforms to interact with visitors and provide additional content related to their exhibitions. This may include behind the scenes at museums, live broadcasts of events and interactive elements such as polls and quizzes. These media can therefore be a great way for museums to reach a wider audience and build a sense of community among their visitors.

This last point deserves a dutiful study as the dynamics of social networks can be very complex, especially if intersected with the cultural heritage sector. Looking at the current trends, important clues emerge about the change of direction in progress. If we take into consideration the top hundred most visited museums in the world, last year the three major social networks had all recorded growth in terms of followers, and Instagram had surpassed Twitter in fan preferences. In 2021, however, Twitter and Facebook lost 4.2 million and 2.7 million followers respectively, and Instagram, while growing, had a 30% lower increase. The scenario suggests that museums will be forced to use platforms like TikTok to find new and younger audiences. The main obstacle is of a cultural nature, because there are so many old-fashioned institutions that take themselves very seriously, while TikTok is synonymous with energy, speed, constantly evolving news.

For its part, the Chinese social network is moving to enter the world of culture, as demonstrated by the partnerships with the Cannes Film Festival and with the Royal Shakespeare Company. Some museums, however, have taken action and obtained the first results. The Prado in Madrid has the largest following on the platform, with more than 360,000 followers, mainly thanks to a series of "behind the scenes" videos of the museum's activities. Among the most interesting contents, we should also mention those of the National Gallery in London, the Rijksmuseum in Amsterdam and, in Italy, the Uffizi Galleries in Florence, where the followers on TikTok exceeded those of Twitter (96,400 followers and 10% total) compared to all the social networks on which the museum is present. However, very far from Instagram which collects 681,556 users of the total 962,987 of the Uffizi (71%), followed by Facebook, with 126,674 followers and 13% total.

After the boom of the past years, growth in 2021 has slowed, perhaps as a reflection of the physical reopening of many museums. The Museum of Modern Art in New York has been the undisputed number one for years and, even in the top twenty, the names are mostly the same as they were a year ago.

Table 4. Top 10 museums on social

	Sede o complesso museale	Città	Follower	Incremento dal 2020
1	Museum of Modern Art	New York	13.191.000	1,4%
2	Metropolitan Museum of Art	New York	10.439.000	1,9%
3	Tate Galleries	Londra, Liverpool, St Ives	10.116.000	1,6%
4	Louvre	Parigi	9.080.000	5,2%
5	Guggenheim	New York	6.927.000	0,2%
6	Van Gogh Museum	Amsterdam	6.452.000	4,6%
7	British Museum	Londra	5.769.000	0,1%
8	National Gallery	Londra	3.921.000	3,7%
9	Museo Nacional del Prado	Madrid	3.811.000	20,3%
10	Victoria and Albert Museum	Londra	3.789.000	3,1%

3. THE RESEARCH

In carrying out this work we have tried to offer a sufficiently complete overview of the evolution of new technologies and we have seen how they have acquired a predominant role in everyday life and how they have changed the way we absorb information, communicate and socialize. At the basis of all this is the birth and development of the Internet on a global level which, in parallel with technological evolution, has drastically changed the social, cultural and economic landscape. Starting from this, we have outlined how technological progress in a general sense has impacted the cultural sector, the main subject of this text. The path towards the digitization of art and cultural heritage has already led to remarkable results, highlighting numerous examples of excellence among museums, creative companies and Italian organizations. We have therefore observed, through well-known case studies, how the process of digitization of artistic and cultural works has led to excellent examples of how organizations have managed to implement and exploit the enormous potential of new technologies, such as AR, VR, digital experiences, to offer innovative and high quality products, services and experiences. However, this process has often been difficult and complex from the start and looks set to have a long way to go. Before the expansion of the pandemic, the cultural and creative sector had seen strong growth, especially in terms of creating new jobs and positively impacting other key sectors for economic development. However, the sector has encountered numerous difficulties, such as the lack of support and lack of trust in digital planning, both from governments and cultural institutions, and the low levels of cultural consumption in our country. Through the qualitative and quantitative research presented, we were able to outline the cultural sector and observe the changes and transformations that have taken place in the information and cultural consumption sector; data that reflects the difficult situation of recent years caused by the pandemic and the slow recovery we are still facing. The "Io Sono Cultura 2022" research by the Symbola Foundation and Unioncamere has allowed us to understand what the great powers are producers of artistic and cultural content on a global level and how the Cultural and Creative Productive System is the engine of our country's economy, representing 5.6% in terms of value and 5.8% in terms of employment, thus proving to be fundamental in the development capital and territory.

The investigations of ICOM Italia, on the other hand, have allowed us to better understand the impacts of the pandemic on Italian museums and archaeological areas and how these institutions have approached digital with the aim of staying in touch with their audiences; in fact, we have seen how 89.24% of museums created ad hoc digital content after the closure and how 64.31% created new social channels or reactivated channels they already had. The pandemic and related restrictions have significantly accelerated digitization processes, especially in the world of culture and museums. Numerous cultural institutions have reacted to the situation by moving at least half of their normal cultural activities and offerings online. However, this phenomenon also seems to have influenced the way the public perceives culture, increasing the risk of making its meanings and nuances more ephemeral. After analyzing the cultural heritage sector in its breadth and depth, we focused on the reference target for the launch of the digital strategy, i.e. young people belonging to Generation Z, the so-called digital natives. People, therefore, who grew up during the full development of Information and Communication Technologies, and the first to have known a world already pervaded by new digital tools that have influenced how we live everyday life and the processes of acquiring information, purchasing behavior and ways to communicate. In fact, these prove to be perfect early adopters, users therefore familiar with the digital environment capable of educating previous generations and guiding future ones.

Table 5. Gen Z media diet. BUZZOLE survey

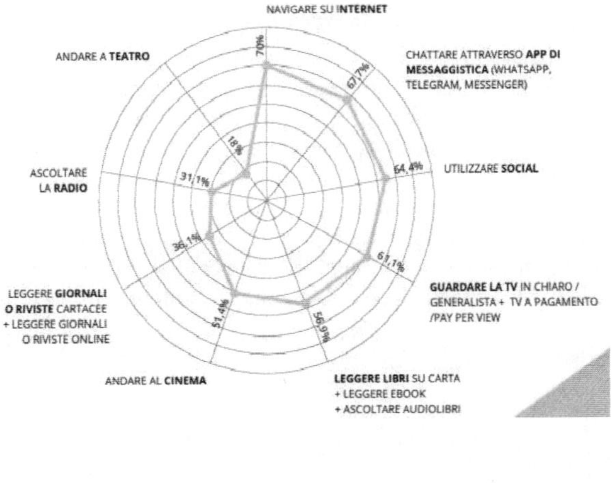

Through the surveys conducted by BUZZOOLE we have understood how Gen Z's media diet is made up, still finding a strong presence of traditional media such as TV, books and cinema, albeit in a reduced way compared to more recent media, probably also due to the cultural heritage of family of origin. This analysis then allowed us to cluster our target into four groups, taking into consideration what the likes and needs are and how much individuals want to know the brand in depth or how much instead they allow themselves to be influenced by a by external actors. This division allows us to further refine the digital campaign launch strategy, demonstrating how followers and effortless users, the most active online, represent the user group on which the brand should concentrate its efforts. After this first quantitative analysis, we focused on understanding the values pursued by Gen Z to date, noting how we find ourselves interacting with individuals who are very attentive to both political and environmental issues. In fact, brands have had to deal with a new dogma: being activists and taking a position to increase sales. This has led to the birth of Brand Activism, of which we have seen the definition given by P. Kotler, and how this must become a guide for companies and can no longer be limited to just being an element of communication strategies. To better understand this phenomenon, some of the most virtuous case studies have been reported. In recent years, the continuous increase of the Internet has become evident, which has confirmed its position as a privileged means of information for Italians, especially among the youngest. Only television has a greater impact than the Internet, but the latter is mainly used through social networks and by consulting specialized websites on various cultural topics. It has also been observed how web communication strategies, in particular those that make use of influencers, are able to involve a vast audience in the world of culture, especially among young people. However, this has often led to extreme and conflicting considerations, as in the well-known case of the " Museum Uffizi- influencerFerragni". It is therefore important to continue monitoring the long-term effects of these strategies, to evaluate whether the large number of "likes" and followers on the social pages of museum institutions will actually translate into a real increase in visitors.

Table 6. The new geographies of cultural and creative production in the world. Unioncamere - Symbola Foundation - I am culture - Report 2022

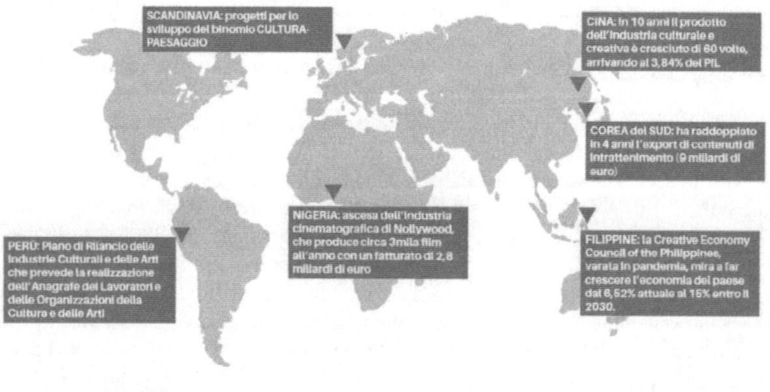

4. CONCLUSION

In conclusion, we can state that, as also revealed by Digital Report 2023, in Italy there is a growing maturity in the use of the Internet and social media. These tools are used in increasingly diversified ways, above all for the purpose of entertainment, information, sharing and socializing. They offer the possibility of providing cultural products and services adapted to different types of users and users, simplifying their narration, involvement and experientiality, generating consumption practices and proactive learning. Furthermore, they constitute a potentially useful tool for individual growth and intellectual capacity of people, thanks to the vast range of training contents that can be shared and used through the various sharing platforms. This part of the analysis was the basis for the development of a digital strategy defined in two parts: on-site and off-site. For everything that concerns the off-site, therefore the part relating to good practices to be implemented in the digital world to increase brand awareness and interact with the public, we started with the presentation of a new brand, muvi, whose main purpose is to support cultural institutions in the creation and promotion of digital content using a platform accessible from mobile or desktop devices. After presenting the Brand Guidelines, we went on to define the aspect and main functions of the application through which all the information on current events and exhibitions is made available together with the digital content created by the institutions. These contents will therefore be available for consultation and above all shareable by the entire community, offering greater visibility to the organizers. Thanks to a fully customizable notification system we will be able to create a totally personalized environment based on the preferences of the individual user, improving the user experience. To complete this first strategic part, we then deepened communication on the web. Starting from the key concepts, such as accessibility and usability and, once understood how indispensable they are in order to be able to create an inclusive virtual environment that responds to the needs of all users, all the tools and best practices to be implemented to respond to these were presented requirements and also to comply with the regulatory framework, see the Stanca law. We therefore took advantage of the guidelines provided by WCAG to define the basic principles to follow for the creation of a fully accessible website, also defining the right practices to adopt. Finally, the tools available to verify the accessibility of the institutional website were presented, which can also help to evaluate its usability. As regards the part relating to social networks, we have underlined the importance of how collaboration and co-creation can offer multiple opportunities for improvement and development, if properly studied. In fact, the not carefully considered choice of a possible collaborator, especially if one aims primarily to generate economic profit and whose work deviates too much from cultural purposes, could prove to be highly risky and counterproductive, leading to damage to the credibility, reputation or image of the cultural institution. To avoid this, it is advisable to conduct careful research on social media, with the aim of identifying the influencers who are closest to the principles of your brand and who reflect your personality. Turning now to the on-site strategy, we have defined the positive and negative aspects of phygital contents such as AR and VR, emphasizing how the size of the structure and the economic capacity can drastically influence the design choices; the case presented concerning museum houses was useful for us to better understand this aspect. In any case, given the high competence and specialization necessary for the creation of contents like these with a high technological content, we have reiterated the importance of co-creation. Therefore, the comparison with partners in the sector is the right path to follow to bring significant added value to the experience of visiting cultural sites. All this highlights how the experience of the lockdown experienced in recent years has led to greater awareness and new perspectives for the future, which institutions should consider to facilitate a correct digital transformation of cultural heritage and generate more revenue for museums. The latter should once again become places of cultural production, as well as consumption, creating new art forms and experiences through the support of digital technology. For the last part of the strategy, detaching ourselves from the digital world, we explored the theme of merchandising and capsule collections to increase brand awareness. We have therefore observed the phenomenon on the basis of two well-known case studies, Lidl and Ikea, effectively verifying what benefits this marketing strategy has brought to the two brands in question. In both cases, the result was undoubtedly positive, leading to an increase in traffic in their stores and generating potential future sales as regards the core business of the brands. Ultimately, we can state that, despite the growing presence of the Internet and digitization in society, many areas of the Italian cultural and artistic industry are struggling to adapt to these changes, highlighting a persistent gap. It happens that the initiatives promoted by holders of cultural contents do not take into account scientific or commercial innovations. For this reason, cultural institutions need guidance in understanding the importance of developments in digital research and applications in order to meet their innovation needs. The challenges facing every cultural institution include adopting innovative strategies to attract new visitors and make their collections and initiatives more accessible, thus promoting the democratization of knowledge. Digital and the web can be useful development, enhancement and communication tools at various levels, but require specific skills, adequate resources and appropriate investments. However, these actions are particularly complex for small and medium-sized enterprises and emerging realities, which make up a large part of the Italian cultural sector, and represent the subjects most at risk and to be valorised. One of the main challenges of the moment is therefore to ensure that these subjects too have access to the opportunities offered by digital.

REFERENCES

Accessible Resources for Cultural Heritage EcoSystems. (2020) *ARCHES project funded by the European call for proposals of the Horizon 2020 programme.* https://cordis.europa.eu/article/id/413505-technical-innovations-help-overcome-access-barriers-to-cultural-spaces/it

Bone Hall (2022). *National Museum of Natural History Smithsonian.* https://naturalhistory.si.edu/exhibits/bone-hall

Boracchi, C. (2020, December 17). *The rediscovery of the podcast and its potential* ICOM Italia. https://www.icom-italia.org/wp- content/uploads/2020/12/ICOM Italia. ICOM_RL. Comunicazione.17dicembre.2020_ Boracchi.pdf

BUZZOOLE (2021). *Influencers and Gen Z: the role of influencers in purchasing processes, surveys* https://buzzoole.com/resources/influencer-e-genz/it

Carraro, R. (2019). *The immersive digital transformation of Cultural Heritage 2019: the example of M9, Agenda Digitale.* https://www.agendadigitale.eu/cultura-digitale/la-digital-transformation-immersiva-dei-beni-culturali-lesempio-di-m9/#Linnovazione_costante_Sic_transit_tecnica_mundi

Clarelli, M.V. M. (2011). *The museum in the contemporary world.* Carocci editore

Criconia, A. (2011). The architecture of museums.Carocci editore

Dara,V. (2017). *Who are the digital natives? A portrait and some myths to debunk.* Inside Marketing, https://www.insidemarketing.it/chi-sono-i-nativi-digitali-caratteristiche/

Dawson, A. (2022). *Museums & Social Media.* https://www.ilgiornaledellarte.com/articoli/musei-e- social-media/138963.html

Digital Mosaik. (2018). *The difference between augmented reality (AR) and virtual reality (VR)* https://www.digitalmosaik.com/blog/differenza-ar-vr

Europa Creativa (2021-2027). *Creative europe* https://www.europacreativa- media.it/europa-creativa

FOCUS. (2021). *Space Adventure simulators and virtual reality to experience life as astronauts.* https://www.focus.it/scienza/spazio/space-adventure-simulatori-realta-virtuale- mostra

Foschini, C. (2021). *Ferragni effect on the Uffizi, boom in visits from the youngest. Director Schmidt: Stable growth for the 19-25 age group.* https://firenze.repubblica.it/cronaca/2021/04/02/news/firenze_ferragni_uffizi_influencer_boom_di_visite-294813857/

Haupt, R. (2021). *Netflix politics: who dictates the agenda in the age of social media.* Willmedia, Milano,

ICOM Italia (2020, October). "*Museums and COVID-19*". https://www.icom-italia.org/musei-e-covid-19- questionario-di-icom-italia/

ICOM Italia. (2020). *Digital Comunication and COVID-19.* https://www.icom-italia.org/comunicazione-digitale-dei-musei-ai-tempi-del-covid-19/

Il Museo del '900. (2018). https://www.m9museum.it/il- museo/ https://www.carraro-lab.com/2018/11/29/inaugurazione-di- m9-il-piu-grande-museo-multimediale-italiano/

IPSOS, Digital Audio Survey (2022). P*odcast 2022, quality as a way to grow.* https://www.ipsos.com/it-it/podcast-indagini-ipsos-digital-audio-survey-podcast- qualita-strada-crescere

ISTAT. (2022). *Report Museums and similar institutions in Italy.* https://www.istat.it/it/files/2022/02/REPORT_MUSEI-E- ISTITUZIONI-SIMILARI-IN-ITALIA.pdf

Kotler, P., Sarkar, C. (2020). *Brand Activism,* Hoepli, Milano

Maurizio, C. (2021). *Museum at the time of covid, the Italian solutions: virtual, immersive, interactive tours.* https://www.agendadigitale.eu/cultura-digitale/tour-virtuali

Meschini, A. (2011). *Digital technologies and communication of cultural heritage. State of the art and development prospects: Technologies for the communication of cultural heritage.* https://disegnarecon.unibo.it/article/view/2565/1943

New European Bauhaus (2022) https://new-european- bauhaus.europa.eu/index_en

Openpolis. (2022). *Access to culture and museums for minors, after two years of the pandemic.* https://www.openpolis.it/laccesso-alla-cultura-e-ai-musei-per-i-minori-dopo-due- anni-di-pandemia/

Symbola (2022). I'm Culture 2022. https://www.symbola.net/ricerca/io-sono-cultura-2022/

We Are Social & Hootsuite Digital (2023). *Report Italia 2023.* https://wearesocial.com/it/blog/2023/02/digital-2023- i-dati-italiani/

We Are Social & Hootsuite. (2022). *Digital 2022 Global Overview Report.* https://wearesocial.com/it/blog/2022/01/ digital-2022-i-dati-globali/

UNIVERSITY STUDENTS' ATTITUDES TOWARDS ONLINE LEARNING DURING THE COVID-19 PANDEMIC

Hesham Alomyan
University of Petra, Jordan

ABSTRACT

This study assessed undergraduate students' attitudes towards online learning during the Covid-19 Pandemic, specifically three months after the initiation of the March 14, 2020 lockdown in Jordan. Six hundred and twelve (*N=612*) undergraduate students from the University of Petra and University of Jordan across majors and academic years were surveyed using a user-designed online survey. The surveyed students exhibited negative attitudes towards online learning during the Covid-19 pandemic. More specifically results showed that the majority of the students responded that communicating with teachers and students through online learning was annoying. Also, over half of the students indicated that online learning did not increase their motivation to learning and 60% of them expressed that online learning was uncomfortable, indicating that they did not enjoy the online learning experience and would not take online courses if were used again. Overall, students displayed a wide range of responses showing mostly negative attitudes towards online learning. This suggests that students still prefer learning in the classroom to online learning. Finally, the paper highlighted various recommendations and implications for better online learning.

KEYWORDS

Online Learning, Distance Learning, Covid-19, Students' Attitudes

1. INTRODUCTION

The Covid-19 pandemic in 2020 forced most educational institutions globally to halt on-campus academic activities and quickly transition to online learning to guarantee continuity in the educational process. While online learning has several advantages, including simplicity of use, flexibility, and greater control over the learning environment (Abbasi, Ayoob, Malik, & Memon, 2020), the abrupt shift to this method of learning posed numerous challenges for students. The most significant of these challenges was the need for students to change their lifestyle to adapt to this new learning environment, which included spending extended period of time in front of electronic devices, losing opportunities for face-to-face interaction, and dealing with various technical issues. These and other obstacles may have affected students' encounters with online learning. As a result, it is critical to evaluate students' experiences and comprehend their perspectives on online learning during the Covid-19 pandemic.

2. LITERATURE REVIEW

Learning from distance has been an integral part of education since the 1940s, adapting and evolving along with the rapid advancement of technology. However, it was the advent of the Internet that truly revolutionized the way education was delivered and gave birth to the concept of online learning. This innovative approach to distance learning has proven to be extremely effective, prompting researchers to conduct extensive studies to examine its effectiveness and efficiency. Most of these studies were carried out in conventional learning environments (e.g., Oliveira, Cunha, & Nakayama, 2015; Yildiz, Tezer, and Uzunboylu, 2018; Zainuddin, Idrus, & Jamal, 2016), with online learning used as a supplement to traditional teaching and learning. These studies have found a number of challenges that students experience in this method of learning, such as increased homework and staff reluctance to incorporating online learning into their teaching practices.

Previous research examined stand-alone online learning systems (e.g., Dodo, 2013; Musingafi et al., 2015; Sharath, 2014; Roth et al., 2020), and revealed that despite the convenient and affordable accessibility of online learning, students still faced challenges that affect their learning. These challenges included insufficient study time, inadequate ICT skills, delayed feedback on students assessments, and absence of technical support, resulting in low levels of satisfaction with the course and academic achievement.

When the Covid-19 pandemic emerged, universities worldwide were forced to swiftly transition to online learning, presenting students with unexpected obstacles and concerns. In response to these circumstances, several researchers endeavored to recognize and examine these challenges (e.g., Li et al., 2020; Mailizar et al., 2020; Moawad, 2020; Prokopenko & Berezhna, 2020). Most of their studies found that students faced challenges such as restricted interaction, a dearth of hands-on learning opportunities, more tasks with less time to complete them, and difficulties using online learning management systems.

Despite the aforementioned challenges, online learning has its advantages. For example, students have the flexibility to study at their own pace and convenience. Furthermore, online learning offers access to a broader range of tools that may not be available in their physical location. (Holmes et al., 2020). Additionally, by eliminating the need for bodily interaction and travel to educational institutions, online learning has the potential to reduce the spread of infectious illnesses. (Li & Lalani, 2020). Nonetheless, to make the most of the advantages of online learning, it important to identify and address challenges that may hinder the achievement of learning in online learning. This will improve online learning as a reliable platform, especially in times of unanticipated crises, and enhance the quality and efficacy of online learning. To contribute to the ongoing research in this area, the present study aims to assess students' attitudes towards online learning during the COVID-19 pandemic.

3. STUDY QUESTIONS

The current study aimed to address the following questions:

1. What are the students' attitudes towards online learning amid the Covid-19 Pandemic?
2. Are there significant differences in students' attitudes towards online learning based on gender, education stream, academic year, type of university, level of computer skills, and online learning devices?

4. METHOD

4.1 Participants

The current study included a total of six hundred and twelve (N = 612) undergraduate students (384 females and 228 males) who were enrolled in various academic programs at the University of Petra and the University of Jordan during the second semester of the 2019/2020 academic year. The participants were selected using purposive sampling technique, and they completed an online questionnaire while under lockdown measures. A total of 612 responses were collected. Table 1 shows the demographic information of the participants.

Table 1. Participants' demographic information

Variable	Category	Frequency	Percentage
Gender	Male	228	37.3
	Female	384	62.7
Education stream	Humanities	222	.363
	Sciences	390	.637
Academic year	First	100	16.3
	Second	168	27.5
	Third	188	30.7
	Fourth	156	25.5
Online learning device	Computer	234	38.2
	Mobile Phone	369	60.3
	Tablet/iPad	9	.15
Computer skill levels	High	134	21.9
	Average	362	59.2
	Low	116	18.8

4.2 Study Instrument

The study used a self-administered online questionnaire to assess undergraduate students' attitudes towards online learning during the Covid-19 pandemic. The 14-item questionnaire was made available through the universities' Facebook account and other Facebook groups specifically created for university students, where students could access and respond to it. A five-point Likert scale was used, with responses rated as strongly agree, agree, neutral, disagree, and strongly disagree receiving weights of 5, 4, 3, 2, and 1, respectively. The degree of agreement for each questionnaire item was determined by calculating the means of the participants' responses using the following formula:

Scale range = (Highest score – lowest score) ÷ Number of levels
Scale range = (5 - 1) ÷ 3 = 1.33
Based on the above calculation the judgment on the degree of agreement (DA) is as shown in Table 2.

Table 2. Classification of degree of agreement

Class	Degree of agreement
1.00 -2.33	Low
2.34 - 3.66	Average
3.67 -5.00	High

4.2.1 Validity and Stability of the Instrument

To ensure the validity of the study instrument, it was evaluated by eight experts in the field of educational technology. They were asked to evaluate the instrument's language, content, and its ability to fulfill the study's objectives. By considering the experts' feedback and the obtained values, the Content Validity Index (CVI) was calculated and yielded a value of 0.84, indicating a high level of content validity and confirming the instrument's suitability.

Additionally, the stability of the study instrument was assessed using Cronbach's Alpha. The overall stability coefficient of the instrument was calculated to be 0.950, indicating strong internal consistency and reliability. This suggests that the instrument is well-suited for the specific purpose of the study and can reliably measure the intended constructs or variables.

5. STATISTICAL ANALYSIS

To answer the study questions, the following analyses were used within SPSS:

- Frequencies, percentages, means and standard deviations were used to describe the characteristics of the study participants.
- The analysis of five-way covariance (ANCOVA) was used to test for differences according to personal variables,
- T-test for one sample was used to test the differences between the mean of the answers as a whole and the fixed (hypothetical) mean.
- Cronbach Alpha was carried out to test the stability of the study instrument.

6. RESULTS

The study results are presented and discussed in relation to the study questions as follows:

Question one: What are the students' attitudes towards online learning during the Covid-19 Pandemic?

To answer this question, means were calculated for the participants' answers to the questionnaire items that measure students' attitudes towards online learning from their point of view. Findings are presented in Table 3.

Table 3. Means, standard deviations, ranks and degree of agreement of the items of the attitudes questionnaire

Item No.	Item	Mean	SD	Rank	RII	DA
1	I like the idea of online learning	2.04	1.30	7	40.8%	Low
2	Online learning is an innovative concept that should be encouraged	2.42	1.39	2	48.4%	Average
3	I enjoy learning from distance	2.01	1.23	11	40.3%	Low
4	Online learning has increased my motivation to learn	1.92	1.20	14	38.4%	Low
5	I encourage others to use online learning	1.97	1.25	13	39.4%	Low
6	Online learning has increased my academic performance	2.05	1.22	5	41.0%	Low
7	The content of the educational material in online learning is more organized	2.10	1.23	3	41.9%	Low
8	I would like to take online courses if online learning is used again	2.02	1.33	9	40.5%	Low
9	Online learning is a comfortable mode of learning	1.98	1.23	12	39.6%	Low
10	Online learning achieves the objectives of the curriculum	2.02	1.19	10	40.4%	Low
11	Online learning encourages creativity	2.06	1.20	4	41.1%	Low
12	Communication between teachers and students through online learning is annoying	3.63	1.40	1	72.6%	Average
13	I am generally satisfied with the learning environment of online learning	2.05	1.23	6	41.0%	Low
14	I am satisfied with the course management of online learning systems	2.04	1.25	8	40.8%	Low
	Total degree	2.17	0.98	-	43.3%	Average

As noticed in Table 3, students' attitudes towards online learning were of an average degree of agreement, where the total mean score of the scale as a whole was 2.17 with a standard deviation of 0.98. At the level of the scale items, the mean scores ranged from 1.92 to 3.63, with item 12 ranking the highest of all, which states that "communication with teachers and students through online learning is annoying"; this followed by item 2 with a mean score of 2.42 which states that "online learning is an innovative concept that should be encouraged". Finally, the lowest mean score was ranked by item 4, which states that, "Online learning increased my motivation to learn"

Further, to test for positivity in students' attitudes towards online learning, a one-sample t-test was used to test for significant differences between the mean of the participants' responses and the hypothetical mean of the students' attitudes scale which is 3, indicating neutrality.

Table 4. Results of One Sample T Test to test positivity in students' attitudes towards online learning

Variable	Mean	SD	T-value	DF	Sig.
Students' attitudes towards online learning	2.17	0.98	21.037-	611	*0.000

* The differences are statistically significant at ($\alpha \leq 0.05$).

As shown in Table 4, the one-sample t-test revealed significant differences ($p \leq 0.05$) between the mean of the students' responses (mean of 2.17 and the hypothetical scale's mean (mean of 3), where the calculated T-value was -21.037 with a significance level of 0.000. These findings indicate that, overall, students held negative attitudes towards online learning, as their mean responses were lower than the hypothetical scale's mean

Question two: Are there significant differences in students' attitudes towards online learning based on gender, education stream, academic year, type of university, level of computer skills and devices used in online learning?

Table 5. Means and standard deviations of students' attitudes towards online learning based to gender, education stream, academic year, type of university, online learning device, and level of computer skills

Variable	Variable classification	No	Mean	SD
Gender	Male	228	2.22	1.02
	Female	384	2.13	0.96
Education stream	Humanities	222	2.23	0.95
	Sciences	390	2.13	1.00
Academic year	First	100	2.12	0.97
	Second	168	2.15	1.01
	Third	188	2.14	0.99
	Fourth	156	2.25	0.95
Type of university	Public (University of Jordan)	168	2.11	0.93
	Private (University of Petra)	444	2.19	1.00
Online learning device	Computer	234	2.32	1.02
	Mobile phone	369	2.05	0.94
	iPad, Tablet	9	2.81	0.95
Computer skill level	High	134	2.44	1.12
	Medium	362	2.20	0.96
	Low	116	1.74	0.71

As shown in Table 5, there are apparent differences between the means of the levels of the participants' demographic variables. These differences are in favor of males, humanities, fourth-year students, private university students, tablet users, and finally in favor of those with high levels of computer skills. However, to determine if the differences between the means were significant, hexagonal covariance analysis was used. Results are presented in Table 6.

Table 6. Result of hexagonal covariance analysis

Source of difference (variable)	Sum of squares	DF	Mean of squares	F-value	Sig.
Gender	0.572	1	0.572	0.628	0.428
Education stream	4.625	1	4.625	5.080	0.025*
Academic year	0.641	3	0.214	0.235	0.872
Type of university	0.186	1	0.186	0.204	0.651
Online learning device	4.605	2	2.302	2.529	0.081
Computer skill level	22.960	2	11.480	12.611	0.000*
error	547.120	601	0.910		
Total	3457.852	612			

Table 6 illustrates the results of the hexagonal covariance analysis. The results showed significant differences at the level of $(0.05 \geq \alpha)$ in students' attitudes towards online learning based on education stream (in favor of the humanities stream) and level of computer skills, as their calculated F-values were 5.080, 12.611 respectively and the significance level was less than (0.05). The results of the analysis did not show any significant differences based on gender, academic year, university type, and online learning device, where the calculated F-values were 0.628, 0.235, 0.204, 2.529 respectively with significance levels higher than (0.05).

To test the differences between the computer skill levels, the Tukey's test was used for the post-tests and results presented in Table 7.

Table 7. Results of Tukey's test

Computer skill level (A)	Computer skill level (B)	Level of difference	Sig.
High	Average	0.2392	*0.036
	Low	0.6997	*0.000
Medium	High	0.2392-	*0.036
	Low	0.4605	*0.000
Low	High	0.6997-	*0.000
	Average	0.4605-	*0.000

As shown in Table 7, Tukey's test revealed significant differences at the level of (α 0.05) in the attitudes of students towards online learning between. The test showed that students with high computer skills had more positive attitudes towards online learning than those with average and low computer skills. Furthermore, the results showed that students with high computer skills had significantly more favorable attitudes than those with average computer skills. Finally, significant differences were observed between students with average and low computer skill levels, with the former having more positive attitudes towards online learning.

7. DISCUSSION

The main goal of the present study was to evaluate how undergraduate students perceived online learning during the Covid-19 pandemic. Additionally, the study aimed to identify any variations in attitudes among students based on factors such as gender, education stream, academic year, type of university, level of computer skills, and online learning devices. Findings of the study revealed negativity in students' attitudes towards online learning. That is, the majority of the students considered online learning annoying and over half of them did not feel motivated to learn, and 60% said that they would not take online courses again. Furthermore, the students did not enjoy the learning experience and that they did not want to continue using it. These results suggest that students were not comfortable with this mode of learning. These findings lend support to other recent studies (e.g., Abbasi, Ayoob, Malik, Memon, 2020; Neziri, Ahmeti, Memeti, 2021; Unger & Meiran, 2020) which overall reported negative perceptions towards online learning during the Covid-19 pandemic.

It appears from the above findings that the unexpected move to online and consequently the loss of lectures and in-class activities, along with the lack of prior experience with online learning, prior to the spread of the Covid-19 pandemic has dampened student's motivation to learn and resulted in negative perceptions towards online learning. Furthermore, students' concern about their academic performance and the lack of effective interaction with peers and teachers probably lead to decreased students' ability to learn entirely online. Dzakiria, Idrus, and Atan (2005) argue that engagement and interaction play a crucial role in online learning, as a high level of interaction positively affects students' psychology (Bryant, 2017) and thus their learning. According to Akther (2013), online learning necessitates a significant degree of independence and self-regulation from students, which may have been lacking among the participants in the current study.

Furthermore, the study results showed significant differences in students' attitudes towards online learning due to education stream with humanities students significantly showing less negative attitudes towards online learning compared to science students. This result may be attributed to the fact that humanities programs are typically theoretical and do not require students to be physically on campus, making online learning more acceptable. On the other hand, science programs are typically more hands-on, requiring in-person instruction, lab work, and collaboration with peers and instructors. This may be why science students had a more negative attitude towards online learning.

Additionally, the results of the study indicated significant differences in students' attitudes on online learning based on their computer skill levels. Specifically, students with advanced computer skills exhibited more favorable attitudes compared to those with average or lower computer skills. This result is to be expected, as students proficient in technology are better equipped to manage the online learning environment, troubleshoot technical challenges, and effectively engage in their studies. Consequently, their experiences with online learning tend to be more positive, leading to a reduced likelihood of developing negative attitudes.

8. CONCLUSION AND RECOMMENDATIONS

In summary, the present study assessed undergraduate students' attitudes towards online learning during the Covid-19 Pandemic. The surveyed students displayed a wide range of responses showing negative attitudes towards online learning indicating that students still prefer face-to-face learning. However, it must be noted that this study was conducted in the first three months of the onset of the pandemic. Therefore, the results of this study are limited to the initial outbreak of the disease and reflect students' opinions during that period when they had to adjust to online learning for the rest of the semester, and are likely biased. Hence, further research is needed to reassess students' opinions and attitudes after the lockdown was left, and students returned to conventional learning and had further experience with online learning.

Based on the study results, several recommendations have been devised for universities and educators to enhance the overall quality of online learning experiences in the future.

- *Improve Communication*: Since the majority of students found communicating with teachers and classmates through online learning annoying, universities and educators should strive to improve the lines of communication. This can be achieved by providing more online consultation opportunities, setting clear expectations for communication, and providing timely feedback.
- *Increase Motivation*: With more than half of the students reporting in the present study that online learning did not increase their motivation to learn, it's important to identify ways to increase student engagement. This can be achieved by incorporating engaging content, interactive activities and real-world examples into the curriculum.
- *Address discomfort*: With 60% of the students, reporting that online learning is not comfortable, universities and educators should make efforts to tackle this discomfort. One approach is to design shorter and more focused online sessions, build in frequent breaks, and offer multiple formats for content delivery, e.g. videos, podcasts and interactive quizzes.
- *Offer support*: Universities and educators should offer additional support to help students adapt to the online learning environment. This may include offering orientation sessions, providing technical support, and designing resources for self-paced learning.
- *Continuous Improvement*: Universities and educators should continuously monitor and improve the online learning experience based on student feedback. Regular surveys and assessments can help identify areas for improvement and guide future efforts to improve online learning. Finally, faculty professional development should be considered to ensure students have the best online learning experience.

REFERENCES

Abbasi, S., Ayoob, T., Malik, A., & Memon, S.I. (2020). Perceptions of students regarding E-learning during Covid-19 at a private medical college. *Pak J Med Sci*, S57-S61. https://doi.org/10.12669/pjms.36.COVID19-S4.2766

Akther, U. M. (2013). Psychological impact of distance education technologies in students' personal development. *International Journal of Developmental and Educational Psychology*, *2*(1), 203-211.

Antona, M., Ntoa, S. (eds) HCI International 2021-Posters. HCII 2021. *Communications in Computer and Information Science*, vol 1421. Springer, Cham. https://doi.org/10.1007/978-3-030-78645-8_15

Dodo, O. (2013). An analysis of challenges faced by students learning in virtual and open online learning system: A case of Bindura University of Science Education (BUSE). *Journal of Global Peace and Conflict*, *1*(1), 28-40.

Dzakiria, H., Idrus, R. M., & Atan, H. (2005). Interaction in open online learning: Research issues in Malaysia. *Malaysian Journal of Distance Education, 7*(2), 63-77.

Li, H. Y., Cao, H., Leung, D. Y.P., & Mak, Y. W. (2020). The psychological impacts of a COVID- 19 outbreak on college students in China: A longitudinal study. *International Journal of Environmental Research and Public Health*, *17*(11), 3933. DOI: 10.3390/ijerph17113933.

Mailizar, M., Almanthari, A., Maulina, S., & Bruce, S. (2020). Secondary school mathematics teachers views on eLearning implementation barriers during the covid-19 pandemic the case of Indonesia. *Eurasia Journal of Mathematics, Science and Technology Education, 16*(7). em1860.

Moawad, R. A. (2020). Online learning during the COVID-19 pandemic academic stress in university students. *Revista Romaneasca pentru Educatie Multidimensionala*, *12*(1Sup2), 100-107.

Musingafi, M. C.C., Mapuranga, B., Chiwanza, K., &., Zebron. S. (2015). Challenges for open and online learning (ODL) students: Experiences from students of the Zimbabwe Open University. *Journal of Education and Practice, 6*(18), 59-66.

Neziri, I., Ahmeti, K., & Memeti, A. (2021). The Relationship between Student Attitudes towards Online Learning and Environmental Factors during Covid-19 Pandemic: The Case of the University of Tetova. In: Stephanidis, C.,

Oliveira, P., Cunha, C., & Nakayama, M. (2015). Learning management systems (LMS) and e-learning management: an integrative review and research agenda. *Journal of Information Systems and Technology Management*, *13*(2), 27-44.

Prokopenko, I., & Berezhna, S. (2020). Higher education institutions in Ukraine during the Coronavirus, or COVID-19, Outbreak: New challenges vs new opportunities. *Revista Romaneasca pentru Educatie Multidimensionala, 12*(1Sup2), 130-135. Retrieved April 5, 2023, from: https://doi.org/10.18662/rrem/12.1sup1/256

Sharath, A. M. (2014, December, 29-30). The challenges of online learning education system in karnataka state for higher level education. Nitte University, *Fourth International Conference on Higher Education: Special Emphasis on Management Education* [Paper presentation]. Retrieved: May 12, 2023, from: https://papers.ssrn.com/sol3/papers.cfm?abstract_id=2600943

Unger, S., & Meiran, W. R. (2020). Student attitudes towards online education during the COVID-19 viral outbreak of 2020: Online learning in a time of social distance. *International Journal of Technology in Education and Science*, *4*(4), 256-266.

Yıldız, E., Tezer, M., & Uzunboylu, H. (2018). Student opinion scale related to Moodle LMS in an online learning environment: Validity and reliability study. *International Journal of Interactive Mobile Technologies*, *12*(4), 79-107. DOI: 10.3991/ijim.v12i4.9205.

Zainuddin, N., Idrus, R., & Jamal, A. F. (2016). Moodle as an ODL teaching tool: A perspective of students and academics. Electronic Journal of e-Learning, *14*(4), 2016-282.

CYBERAGGRESSION AMONG YOUNG STUDENTS IN THE DEMOCRATIC REPUBLIC OF THE CONGO (DRC): PREVALENCE AND RISK FACTORS

Efrem Mbaki Luzayisu[1] and Jean-Pierre Zamwangana Tungu[2]
[1]*Professor at University of Kinshasa and at Catholic University of Congo, Congo*
[2]*PhD in demography and independent researcher, Congo*

ABSTRACT

Intentionally aggressive acts perpetrated on the Internet are steadily increasing. In the digital lexicon, these acts are known as cyberaggression. Contrary to what some may think, these behaviors are also very present in Africa, particularly in the Democratic Republic of the Congo (DRC) where young Africans are becoming more and more addicted to their smartphones connected to the Internet almost continuously. This study aimed to examine the prevalence of and factors affecting online aggression in The Democratic Republic of the Congo. We undertook a field work among 1,500 university students about cyberaggression and we found that students who undertake highly risky behavior online such as sharing the internet connection with others, publishing contents online or having dates with virtual friends, are among those who experience higher probability of being victim and/or perpetrator of online attacks. Parental control of students Internet use is also consistent with the literature, as students living within families where there is family communication around the internet have lower probability of being victim and/or author of online attacks. However, psychological based attributes such as student self-esteem seem not to be significantly associated with online aggression. Based on the findings we formulated some recommendations that can help reduce the prevalence of cyberaggression.

KEYWORDS

Cyberaggression, Digital Behavior Risk, Hyperconnectivity, Parental Control on Student Internet Use, Cyberbullying, Self-Esteem, University Students, The Democratic Republic of The Congo (DRC)

1. INTRODUCTION

Nowadays, cyberaggression or online aggression is gaining more and more ground, particularly among children, teenagers and young adults (Bauman and Baldasare, 2015; UNICEF,2019). This is a vulnerable and Internet-avid population, “digital natives” or “digital children”, as they are sometimes called. Born with a mouse in hand, young people find it difficult to separate themselves from the Internet, which is an integral part of their lives. This is not just about children in Western countries; those of the countries of the South are also involved. But browsing the Internet also exposes them to attacks of all kinds. Why are these young people victims and/or perpetrators of online aggression? Is it because of their hyperconnectivity? Is this a consequence of a lack of awareness of the dangers that the Internet can represent? Or is this the expression of new forms of violence that find favorable ground among young people? The theories on the subject are juxtaposed and overlapped to try to understand this social phenomenon.

Two types of studies are currently of interest to researchers working on attacks on the Internet. On the one hand, methodological approaches are developing to better define and better assess online aggression among young people (Baldwin et al., 2015; Bauman and Baldasare ; Blaya, 2011; Menesini et al., 2012; Yusuf et al. 2021); and on the other hand, more and more studies are moving towards the identification, if not the causes, of the risk factors at the origin of this online phenomenon (Blaya, 2013; Etudes du Center Jean Gol, 2017; Macilotti, 2019).

The present study aimed to evaluate the prevalence of cyberaggression among university/college students in the Democratic Republic of the Congo (DRC) and examine risk factors associated with cyberaggression among these students. The paper is organized around six sections. In the first section, we have summarized the problem statement of the study by specifying why we decided to conduct it. In the second section we have summarized our conceptual framework based on a review of literature around online aggression towards the World. We have described our materials and methods used to analyze data and the study participants in the third section before presenting the main findings in the fourth section. Those results are briefly discussed in the fifth and last section, where study limitations and recommendations are also presented.

2. PROBLEM STATEMENT

Attacks on the Internet, mobile phones and other digital platforms, such as Facebook, SnapChat, Instagram, Pinterest, WhatsApp, Twitter and Tiktok, are becoming a worrying societal phenomenon, especially when they affect children, adolescents and young adults. They often take the form of insults, threats and other forms of harassment towards their victims and they can have various mental, educational and psychosocial repercussions (European Parliament, 2016; Macilotti, 2019; UNICEF, nd; Yusuf et al., 2021). On the school level, for example, some scholars (Even, 2019; Yusuf et al., 2021) mention, among other things, school and psychosomatic disorders which considerably affect the development of adolescents and young adults, not to mention situations such as the ingestion of drugs and other substances to escape the threat of these assaults of shame. Indeed, and as clearly mentioned by Yusuf et al. (2021, p240), youths who were attacked online "reported eating disorders, alcohol, drugs and substance abuse".

Several studies (Bauman and Baldasare, 2015; Li et al., 2020; Lee, C., & Shin, 2017), resulting from theoretical reflections or field surveys, have tried to understand the prevalence, the etiology, the characteristics as well as the numerous consequences of this new form of crime or violence on young victims, but these remain still limited in the face of the constantly growing scale of the phenomenon. Although we did not perform a systematic literature review, that was beyond the scope of our study, we know from similar research (Chen et al., 2016; Even, 2019; Li et al., 2020; Maron-Cortés et al., 2019; Menesini et al., 2012) that most of studies on cyberaggression have focused more on children and in particular schoolchildren and adolescents, especially in developed countries, where the frequency of use of the Internet is particularly high (Petrosyan, 2022). However, these studies have paid less attention to what happens among young adults, especially among those who attend higher education institutions or universities; one of exceptions might be the study by Bauman and Baldasare (2015). As everyone knows well, these students are very keen on the Internet, video games and social networks, in particular because of (i) their level of education, (ii) their easier access to new information and communication and (iii) their relative autonomy. Furthermore, although they highlight various factors as being at the origin of these behaviors, studies on cyberaggression have not focused much on the role of the digital profile of young people (more or less use of Internet and social media sites for example) in the online aggression (Yusuf et al., 2021).

This study, based on an exploratory survey performed on a sample of Congolese students, has attempted to fill some of these gaps. The objectives of the study were threefold. First, it tries to determine the extent of the phenomenon of online aggression among university students before indicating how the prevalence of this phenomenon varies according to a certain number of socio-demographic variables. Second, it tries to identify the factors influencing participation as a victim or perpetrator of these online attacks before examining the respective impact of the hyperconnectivity of young people themselves. It provides some recommendations that can help reduce this phenomenon among college students in the Democratic Republic of the Congo (DRC). The study is be based on a conceptual framework presented in the next section and developed from our previous knowledge about aggression online and its effects (Chen et al., 2016; Even, 2019; Li et al., 2020; Maron-Cortés et al., 2019; Menesini et al., 2012).

3. CONCEPTUAL FRAMEWORK

Following are a couple of ideas around our study; they will be our logical line of reasoning in this investigation. First, studies on online attacks distinguish the victims from the perpetrators of these criminal acts, thus making it possible to highlight the profile of the aggressors and that of the victims. Our study combines the two groups and talks instead about young people involved in cyberaggression as victims and/or aggressors (or better perpetrators) with the aim of assessing the real extent of this phenomenon in our society. Second, to designate these acts of aggression on and via the Internet, several synonymous concepts are used in studies and research on the phenomenon. Thus, we sometimes speak of online aggression (cyberaggression), online violence (or cyberviolence), cyberharrassment, and especially cyberbullying, to distinguish them from traditional forms of violence between young people (Menesini et al., 2012; Yusuf et al. 2021). For our part, we prefer the concept of online aggression or cyberaggression, which seems to encompass various acts falling within this register, whether insults, dissemination of rumors, humiliating images, acts of harassment or identity theft on and via the Internet.

In this sense, despite the abundance of definitions encountered in the literature, we retain the following definition of online aggression: "*an aggressive, intentional act perpetrated by an individual or a group of individuals by means of electronic forms of communication in a whether or not repeated against a victim who cannot easily defend himself*" (Smith et al., 2008, p376). Such a definition highlights three important aspects of the phenomenon: (i) the existence of humiliating and nauseating acts on the net (Internet, social networks and other digital platforms) and recognized as such, (ii) the presence of a perpetrator (or group of perpetrators) and a victim, (iii) the intention to harm others, although this aspect is difficult to operationalize in studies. However, the repetitive nature will not be retained in this study, in the face of all forms of aggression, occasional or repeated.

To understand the development of these aggressive acts in cyberspace and grasp the main determinants among young Internet users, there are various analytical perspectives put forward in the scientific community. The first perspective considers that aggression online or on the Internet characterizes the new way of experiencing conflict between young people, given that they spend more and more time in the digital world and have the majority of their friends and acquaintances there (Mendez-Baldwin et al., 2015; Vale et al., 2018). All the conflicts and problems they encounter in the real world are mainly resolved in cyberspace. It is in this context that Vale et al. (2018) as well as Macilotti (2019) mentioned that “cyberaggression is the new form of interpersonal violence among adolescents”. It appears that online attacks against young people are often “the result of arguments, teasing and face-to-face actions which are then continued on the internet”. According to Macilotti (2019), these attacks are characterized by “a continuity between online and offline experiences”.

Another perspective, no less attractive, tends to underline that the increasingly frequent online attacks reflect the hyperconnectivity of young people (Athanasiades et al., 2016; Alvarez-Garcia et al., 2015 ; Beyazit et al., 2017; Çakir et al., 2016; Cho et al., 2019; Gozlan, 2018; Park et al., 2014; Peker, 2015; Stahel and Weingartner, 2019; You and Ah Lin, 2016). According to the proponents of this thesis, cyberspace is now considered an extension of oneself. Social networks have changed communications, relationships and the very notion of intimacy (Gozlan, 2018). Authors like Tisseron (2003) even argues that these young people have “a desire for extimacy”, which he defines by a “desire to communicate about his inner world to be validated in his existence, in his originality. This hyperconnectivity thus pushes young people to vent their anger and resentment there, as they would do offline, in the real world”. There are other representations of cyberaggression among young people, ideas that deviate from these two mainstreams.

One could also mention the idea that online aggression characterizes a certain specific category/profile of young people, although the phenomenon does not spare anyone. As such, young people with a certain profile are more likely to commit or suffer attacks online (Lee and Shin, 2017; Merril and Hanson, 2016; Mishna et al., 2010). This perspective, which could be described as social determinants (Li et al., 2020 ; Stahel & Weingartner, 2019), underlines the existence of a certain number of factors at the origin of these attacks and which are more or less present among the perpetrators and/or victims of this phenomenon. This is the perspective we adopt in this study, the one that consists of examining the determinants of online aggression among young people.

In this very context, several studies (Alvarez-Garcia et al., 2015 ; Cho et al., 2019; Mendez-Baldwin et al., 2015; Val et al., 2018) conducted among children and adolescents have shown that those who use more social networks and other digital platforms are more likely to experience online violence, but also to commit it. Val et al. (2018), for example, showed that “a higher frequency of information and communication technology and cyber-practices/risks were associated with victim-perpetrators”. Other studies have not found a strong link between the frequency of Internet connection and online aggression, but researchers are not ready to abandon this hypothesis, given the ambivalence of the results of the available studies. The question, which is also at the heart of our study, is therefore whether the probability of being the victim/perpetrator of an online attack depends on the digital profile of those concerned.

Another aspect refers to the parental control or intrafamilial communication. The role of parental control on children’s digital practices has been pointed out in many studies (Mendez-Baldwin et al., 2015; Palermiti et al., 2017; Wright 2017; Val et al., 2021; Zang et al., 2016) considering, among others, that the existence of such control or of inter-family communication around this phenomenon reduces the risk of perpetrating or suffering aggression online. Although this awareness, little is known about the mechanisms (of this factor 's effect. Some studies also point to the role of parental control on children's digital practices and consider that the existence of such control or of inter-family communication around this phenomenon reduces the risk of committing or suffering aggression online, but mechanisms of the effect of this factor are not yet clearly elucidated. Although cyberaggression can affect all ages and populations, it has been reported in some countries and regions around the world that girls are more affected by the phenomenon than their male counterparts (Li et al., 2020). These studies do not make it possible to decide on this gender inequality in terms of online aggression, in particular because of the lack of control of other variables such as age, social background and family control.

Other authors mention, not without surprise, the overexposure of racial and ethnic minorities to violence on the Internet, particularly in terms of intimidation, but once again, this is not yet sufficiently supported by data from the field (Bauman and Baldasare, 2015; Merril and Hanson, 2016). While demographic factors such as age and sex, digital profile and parental control are cited as determinants of online aggression, other authors (Chang et al., 2015; Sasson and Mesh, 2017) also mention the risky behavior of some young people which overexposes them to this type of aggression or attacks. Among these behaviors we usually point out the exposure of young people to video games with violent content, the meeting with unknown virtual friends, the sharing of the connection with third parties, the publication of personal information. The more young people engage in these acts, the more likely they may commit and/or suffer attacks online. However, it is also necessary to take into account the personality traits of these young people which could contribute to the reinforcement or, on the contrary, to the reduction of the risks incurred (Antipina et al., 2020; Brewer and Kerslake, 2015 ; Festl and Quandt, 2016; Malinowska-Cies'lik et al., 2022). In this regard, no one is unaware of the impact that men's psychological attributes (openness, conscientiousness, extroversion, friendliness and narcissism to consider the big five personality traits, as defined by McCrae and Costa (1987), exert on the human actions and behaviors. Instead of these key traits of personality, some authors (Brewer and Kerslake, 2015; Palermiti et al., 2017; You and Ah Lim, 2016; Yusuf et al., 2021) point out the role of self-esteem as key psychological determinant of online aggression.

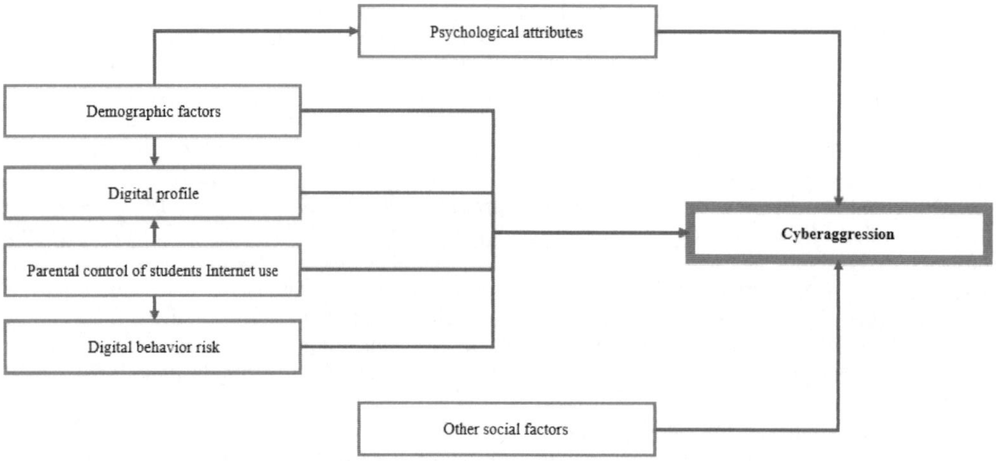

Source: developed by the authors.

Figure 1. Our proposed conceptual scheme for the study of cyberaggression factor risks

As we can notice it, online aggressions (or cyberaggression) can be the result of different factors of various kinds, and the most important of them would be, according to some literature, the digital profile, the degree of parental control of students Internet use, some demographic factors, digital behavior risk, personality type and other social factors, as we have schematized in the above diagram (Figure 1). To what extent do these factors influence online aggression among Congolese students? This is one of the two research questions addressed in this study and outlined above. After these theoretical considerations, we are going to describe how we have tried to gather relevant data and analyze them to try to answer our research questions.

4. PARTICIPANTS AND METHODS

4.1 Participants

Data used in this study have been collected from a sample of 1,500 participants who are currently university students from two Congolese universities (Catholic University of Congo, CUC, https://ucc.ac.cd/ and New Horizons University, NHU, https://www.unhorizons.org/) located respectively in Kinshasa and Lubumbashi, the two biggest cities of the DRC; more information around the survey can be found in (Mbaki Luzayisu and Zamwangana Tungu, 2023). As shown in the Table 1, there is a balanced number of female and male participants. Participants ages range between 18 and 30, with a higher share of those aged less than 20 years old. Students aged more than 24 years made up 14% of the total sample. Furthermore, participants mostly live with their (biological or non-biological) parents and are from large families (families with 5 children and more). Of the students surveyed, the majority (75%) is studying first grade of university, 10% are in second grade, 15% in higher grades (Table 1).

All the participants have been selected at random among those who have been enrolled in both Universities and have followed any computer science course since 2020 with one of the two paper's authors; he keeps a long list of email addresses made up of more than 8,000 students. The field work took place in November 2022.

Table 1. Breakdown of study participants by various socio-demographic attributes

Variable	Category	Frequency	Percentage*
Age group	18-20	705	48%
	20-22	400	27%
	22-24	170	11%
	24+	207	14%
Sex	Female	747	50%
	Male	748	50%
Family residence	Live with biological parents	1086	72%
	Live with other parents	131	9%
	Live with friends & acquaintances	26	2%
	Live with other persons	258	17%
Nb of siblings	1	57	4%
	2	88	6%
	3	184	12%
	4	229	15%
	5	287	19%
	6+	652	44%
Elder position or not	Elder	472	32%
	Other position	1025	68%
University grade	First grade	1128	75%
	Second grade	154	10%
	Third grade	91	6%
	Higher grades	128	9%

* Percentages calculated from the total number of participants (1500) with valid answers.

Source: developed by the authors using survey data.

4.2 Questionnaire

An ad hoc questionnaire with sixty-five questions/items has been prepared and submitted online to all the participants who filled in it without any human assistance (self-administered questionnaire). This questionnaire contains ten questions concerning students' attributes (date of birth, gender, number of siblings, mother language, school grade, family residence to name only a few) and use of electronic devices, Internet and social media sites as well as the usage frequency and duration of Internet. The questionnaire also contains students experience about posting/publishing contents on the Internet and friendship management online to assess their online behavior risk in terms of online aggressions. One questionnaire's module has been devoted to cyberaggression measurement using a set of items likely to occur online. Along with items regarding parental control and student knowledge and attitudes towards online aggression, a few questions have been also used to evaluate student self-esteem and their sociability level.

4.3 Data Analysis

Gathered data were analyzed using the SAS software package, especially its SAS/STAT module (version 15.2). Firstly, all the distribution of participants by study questions/items have been examined to ensure data quality and gather preliminary insights. That also allowed (i) to evaluate composite indicators such as online behavior risk, propensity to be victim and/or author of online aggression and connectivity level; (ii) and to better understand students digital usage patterns. Secondly, the prevalence of cyberaggression was analyzed in terms of frequency and percentages as well as associations between cyberaggression and various participants attributes. Finally, the factors associated with the cyberaggression were identified using logistic regression analysis.

We used logistic regression technique to examine the effect of different explanatory variables on the probability of being a victim and/or perpetrator of online aggression. This last variable is based on two values: if the student has already been a victim and/or perpetrator, the variable takes the value 1, otherwise 0. The binary nature of the variable makes it favorable to logistic modelling. We had six groups of explanatory variables: demographic factors (age group, sex, siblings, promotion and rank in the family, used separately in the model), digital profile (low user, medium and heavy Internet user), the digital risk characterizing the students in 3 groups (low risk, medium risk and high risk), the psychological factors grouped around self-esteem and the conflicting nature or not of the student in his living space, parental control of students Internet use as well as other variables such as knowledge of articles and laws on cyberaggression as well as the attitudes to adopt at the University in the face of an online attack. Results of logistic regression are interpreted in terms of odds ratios, obtained by the exponentials of regression coefficients (exp. (ß)), along with their associated p-values to evaluate their statistical significance. Only direct effects of different factors are considered. Hence, the main focus was to present the effect of each retained factors on cyberaggression; no interaction effects have been included in the final model, because the level of interrelationships (or associations) between risk factors was low. Data also showed that following logistic regression assumptions were met (independence of observations and no perfect multicollinearity of risk factors). To cope with the lack of linearity, each of the categorical variables has been transformed into buckets and then into dummy variables that were included in the model. Main results obtained using logistic regression are presented in the next section.

5. RESULTS

5.1 Digital Usage Patterns Among Participants

Before reporting our main findings on the prevalence of and factors affecting online aggression s, we first give some digital usage patterns characterizing participants (Table 2). The results show that most of the participants (94%) use smartphones to navigate on the Internet and only a few of them still go to cybercafés for that (Table 2). Participants also use the Internet almost daily (93% use it at least 5 days per week) for things other than university work. They reported also have used social media sites in the previous 12 months before the survey:

WhatsApp is the most used application with 99% reporting this usage and Twitter being the least used one. The majority of the participants (63%) reported spending more than 2 hours online and at least 10 dollars per week when they go on Internet. Most of them (77% in total) reported that they regularly publish various contents (photos, images, videos and personal info) on the net (Table 2). An important share of participants do share the internet connection with other when they are running out of credit. Our surveyed students also reported that they mainly use WhatsApp to communicate with their parents: the channels such as text messages (SMS), phone calls and emails are less and less used in the students ecosystem. We'll talk about the prevalence of cyberaggression among study participants in the next section.

Table 2. Key digital usage patterns among study participants

% of students accessing the web via their smartphones	99%
% of students who access the web via cybercafés	19%
% of students reported have used the following social media sites in the previous 12 months	
-WhatsApp	99%
- Facebook	78%
-Tiktok	73%
-Instagram	76%
-Snapshot	78%
-Twitter	38%
% of students communicating with parents via WhatsApp	90%
% of students spending at least 2 hours online per day	53%
% of students using the web between 5 and 7 days per week	93%
% of students spending at least 10$ per week for the internet	68%
% of students who have published contents on the internet in the previous 12 months	77%
% of students who have used Excel or Word applications in the previous 12 months	70%

Source: developed by the authors using the survey data.

5.2 Prevalence of Cyberaggression Among Participants

Based on the combination of all items related to cyberaggression submitted to the participants to evaluate and characterize the rate of participants being victim or authors of cyberaggression during the previous 12 months preceding the survey, the following results have been gathered:

- 61% of participants did not report any experience of cyberaggression during the reference period;
- 32% of participants have reported at least one of the aggression experiences as victim only;
- 5% of participants have reported any experience of the aggregation experiences as both victim and author during the period of reference;
- 2% of participants have reported any experience of cyberaggression as perpetrator only

Putting all together, the overall prevalence of cyberaggressions has been estimated to 39% of participants having involved in cyberaggression as victims and/or author during the study period. This relatively high prevalence of cyberaggressions among surveyed students could be linked to different factors such as their high connectivity and high risky of their digital behavior, as suggested in our conceptual framework. The results concerning these relationships are developed in the following section of the paper.

5.3 Factors Associated with Cyberaggression Among Participants

Table 3 provides information about the effect of the factors investigated in the analysis using logistic regression. Among others, the table highlights all the odds ratios (OR hereafter) and p-values associated with each of the variables. To evaluate if the effects are statistically significant, we have relied on the p-values less than 0.05. Such small p-values mean that there is a small probability that the effect observed is due to chance. In other words, it's the probability that the null hypothesis is true, as usual.

Considering digital behavioral risk, one of the facets of students connectivity, we found the following 3 groups: (i) students with low digital risk (19%), (ii) those with moderate digital risk (58%) and (iii) those belonging to the high digital risk group (23%). As we can see, nearly 80% of the students surveyed are in fact in a situation of moderate or high risk of cyberaggression. Results showed a significant correlation (OR=1.51; p-value=0.0001) between high digital risk and probability of being victim or author of cyberaggression. Furthermore, as already mentioned, the majority of the participants (80%) have high or medium digital connectivity that results from their usage frequency and duration of Internet and social media sites use in the previous 12 months. There is a good correlation between digital connectivity and cyberaggression, but it's not statistically significant (OR for high digital connectivity is 1.14; p-value=0.1488 and OR for medium connectivity is 1.05; p-value=0.5746). Results also showed that 43% of participants belong to families where parents or tutors discus with their children about the risk of online aggressions (Table 3). Significant correlation was found between parental control of internet usage and probability to be victim and/or author (OR =0.878; p-value=0.045). Speaking about parental control, it is also important to mention the effect of living together with biological or non-biological parents vs. other relatives or friends. Results revealed that participants who live in the former type of environments are significantly less likely to be victim and/or author of cyberaggression than their counterparts living without parents (Table 3).

Results also revealed the effects of psychological attributes used in the study (Table 3). There is a significant correlation between associability level and online aggression. Compared to those who are conflict averse, participants who tend to have hot discussion or arguments with their friends are more likely to be victim and/or author of cyberaggression (OR=1.873; p-value=0.001), but participants self-esteem measured through the Rosenberg scale (Rosenberg, 1965) and cyberaggression are not significantly linked. With regard to the demographic variables analyzed, gender obtained an unexpected result. According to prior available evidence, female exhibit higher cyberaggression than male students, but the data obtained in this study showed that every other things being equal there is no significant correlation between gender and the probability of being a victim and/or author of cyberaggression (OR=0.942; p-value=0.373). However, results revealed significant effect for age and university grade. Participants studying in first university grade do experience high probability of being victim and/or perpetrators than those enrolled in the highest grade (Table 3). Those key results will be discussed in the next section.

Table 3. Results of logistic regression of students' attributes on the propensity of being victim and/or author of online aggression

Variable	Parameter	Standard Error (SE)	Chi-Square (Wald stat.)	P-value	(Odds ratio)**
Intercept	-0.2588	0.316	0.6712	0.4126	
High internet user	0.127	0.088	2.08	0.1488	1.136
Medium internet user	0.049	0.089	0.31	0.5746	1.051
Low internet user (ref..)	****				1.000
High digital	0.4081	0.104	15.34	0.0001	1.504*
Moderate digital risk	0.0369	0.085	0.19	0.6636	1.038
Low digital risk (ref.)	****				1.000
Knows laws & instructions	0.051	0.073	0.486	0.486	1.052
Do not know laws & instr. (ref.)	****				1.000
Agree with academic sessions	-0.059	0.088	0.448	0.503	0.943
Do not agree (non) (ref.)	****				1.000
First grade	0.3904	0.163	5.737	0.017	1.478*
Second grade	0.1298	0.212	0.375	0.541	1.139
Third grade	-0.2366	0.252	0.882	0.348	0.789
Forth grade	-0.2672	0.397	0.452	0.501	0.766
Fifth grade (ref.)	****				1.000
18-20 years old	-0.2376	0.3257	0.5323	0.4656	0.788
20-22 years old	0.4948	0.1801	7.5447	0.0060	1.640*
22-24 years old	-0.1253	0.1906	0.4318	0.5111	0.882
24+ (ref.)	****				1.000
Female	-0.0594	0.066	0.794	0.373	0.942
Male (ref.)	****				1.000
Parental control (oui)	-0.013	0.065	4.017	0.045	0.878*
No parental control of internet use	****				1.000
Elder among siblings	0.0801	0.071	1.284	0.257	1.083

Other position among siblings (ref.)	****				1.000
Nb of siblings	0.0133	0.026	0.286	0.611	1.013
Strong self-esteem	0.0639	0.0914	0.488	0.4848	1.066
Medium self-esteem	0.0848	0.0911	0.867	0.352	1.089
Low self-esteem (ref.)	****				
Argue much with friends	0.6276	0.1342	21.88	0.0001	1.873*
Argue a little with friends	-0.1320	0.0935	1.993	0.1580	0.876
Do not argue with friends (ref.)	****				1.000
Live with biological parents	-0.2861	0.1515	3.5654	0.059	0.751*
Live with other parents	-0.2817	0.2052	1.8844	0.170	0.754
Live with friends or acquaintances	0.5979	0.3664	2.6622	0.103	1.818
Live with other persons (ref.)	****				1.000
Have many friends	-0.1244	0.1226	1.029	0.3104	0.883
Have few friends	-0.1919	0.1245	2.377	0.1231	0.825
Do not have friends (ref.)	****				
Sample size	1500				
Number of valid responses	1235				
Pearson's Chi-Square	91.3				
Pr>ChSq	<0.0001				

*Regression coefficient is statistically significant at 5% ; ref = reference category.
Source: developed by the authors using survey data.

6. DISCUSSION, STUDY'S LIMITATIONS AND RECOMMENDATIONS

The aim of this study was to examine the prevalence and risk factors of online aggression among university students in Congo. Online aggression is becoming a serious problem among young people and adults who are avid of Internet and undertake risky behavior online that overexposing them to various attacks. Up to date, a lot of studies have been conducted to better evaluate and understand this social phenomenon. While most previous studies addressed the phenomenon among children and adolescents in Europe, America and Asia, our study has been undertaken among university students in Africa; those students intensively use Internet for both university work and entertainment such as sending and receiving text messages, publishing contents and other personal information on the web. Our study also put a specific focus on the students hyperconnectivity as the key driver of their involvement in cyberaggression context as suggested by many researchers. In the same vein, and contrary to many previous studies, we have built a clear conceptual framework which suggests that cyberaggression among students could be considered as an output of a combination of several and different factors including demographic factors, digital behavior risk, digital profile, parental control of students' internet use and psychological attributes that describe students personality. Finally, the study relies on an exploratory fieldwork; we developed an ad hoc questionnaire that we submitted to the participants who freely answered online without any human assistance, although they were not selected through a clear random sampling procedure.

First results do not necessarily confirm all the suggested hypotheses, while a few of them are consistent with the literature and our conceptual framework. Most of the previous researches emphasizes the gender differences, but we did not find any evidence of significant correlation between sex and online aggression. Hyperconnectivity that some researchers considered as key driver of online aggression find echo in our study. We found that students who undertake highly risky behavior online such as sharing the internet connection with others, publishing contents online or having dates with virtual friends, are among those who experience higher probability of being victim and/or perpetrator of online attacks. Parental control of students Internet use is also consistent with the literature, as students living within families where there is family communication around the internet have lower probability of being victim and/or author of online attacks; this finding echoes previous researches in the parental involvement in the cyberbullying reductions (Mendez-Baldwin et al., 2015). Psychological based attributes such as student self-esteem seem not to be significantly associated with the aggression in the Internet, although the clear theoretical mechanisms suggested in the literature and some empirical researches (Brewer and & Keslake, 2015; Palermiti et al., 2017). Beyond risk factors of cyberaggression, the results contribute to a better understanding of students use of Internet and social media sites. They also raise the need for deeper researches on cyberaggression among students in DRC to better understand mechanisms underlying the effects of different risk factors identified in this study; they have not been addressed here.

Although those contributions, some limitations should be considered. First, this study was not based on a random sample; this fact limits the inference of our results to all Congolese students. Second, methodology used to define victim and/or author of cyberaggression is prone to error of underestimation of cyberaggression authors/perpetrators; this should be improved in future research. Finally, the study was a cross-sectional design; such an approach makes it difficult to investigate causal relationships between risk factors and online aggression. Longitudinal approach would be more appropriate in this context, echoing what has been done by You and Ah Lim (2016) among Korean middle school students.

Based on our findings the following are some of the main recommendations that can help reduce the prevalence of online aggression among university students in the Republic Democratic of the Congo. First, parents should have open communications with their children about Internet usage and cyberaggression; they should also suggest them to limit time presence on Internet and social media networks. Second, Universities and colleges should regularly organize meetings, forums and conferences around the dangers of Internet and set up concrete measures to mitigate risk of online aggressions between students. Finally, professors and lecturers should have time in their teaching programs to talk students about cyberaggression.

REFERENCES

Alvarez-Garcia, D., Nûnez Perez, J.C., Gonzalez A.D. et Pérez, R.C. (2015). "Risks factors associated with cybervictimization in Adolescence", *International Journal of Clinical and Health Psychology*, Vol.15, pp 226-235.

Alvarez-Garcia, D., Barreiro-Collazo, A. et Nûnez Pérez, J.C. (2017). "Cyberaggression among Adolescents: Prevalence and gender Differences", Comunicar n.50, Vol. XXV, 9 pages.

Antipina, S. Bahkvalova, E. & Miklyaeva, A. (2020). "Psychological Determinants of cyber-aggression in institutionalized adolescents", https://ceur-ws.org/vol-2813/rpaper31.pdf

Athanasiades, C., Costanza A., Kamariotis T., Kostouli, M. & Psalti A. (2016). "The "net" of the Internet: Risk factors for cyberbullying among secondary-school students in Greece". *European Journal on Criminal Policy Research*, *22*(2), 301-317. https://doi.org/10.1007/s10610-016-9303-4

Baldwin, M., Cirillo, K., Matthew, M., Argento, V. (2015). "An examination of cyber Bullying and social media use in Teens: Prevalence, Attitudes and Behavior", *Journal of Bullying and Social Aggression*, Vol 1 Number 1.

Blaya, C. (2015). «Etude du lien entre cyberviolence et climat scolaire : enquête auprès des collégiens d'Ile de France», *Les Dossiers des Sciences de l'Education*, 33, pp. 69-90.

Blaya, C. (2011). « Cyberviolence et cyberharcèlement : approches sociologies », *La Nouvelle Revue de l'adaptation et de la scolarisation*, 2011/1, No53, pp. 47-65.

Bauman, S. (2009). "Cyberbullying in a rural intermediate school: an exploratory study". *Journal of Early Adolescence*, 30(6), pp. 803-833.

Baumon, S. & Baldasare, A. (2015). Cyberaggression among college students: Demographic Differences, Predictors, and the Role of the University. *Journal of college Students Development*, 56(4), pp. 317-330.

Beyazıt, U., im ek, & Bütün Ayhan, A. (2017). "An examination of the predictive factors of cyberbullying in Adolescents". *Social Behavior and Personality: An international Journal*, *45*, 1511-1522. doi: https://doi.org/10.2224/sbp.6267

Brewer, G. & Kerslake, J. (2015). "Cyberbullying, self-esteem, empathy and loneliness". *Computers in Human Behavior*, *48*, 255-260. https://doi.org/10.1016/j.chb.2015.01.073

Çakır, Ö., Gezgin D.M. & Ayas, T. (2016). "The Analysis of the relationship between being a cyberbully and cybervictim among adolescents in terms of different variables. *International Journal of Progressive Education, 12*(3), 134-154.

Chang, F., Chiu, C., Miao, N., Chen, P., Lee, C., Huang, T. & Pan, Y. (2015). Online gaming and risks predict cyberbullying perpetration and victimization in adolescents. *International Journal of Public Health*, *60*, 257-266. doi:10.1007/s00038-014-0643-x

Cho, S., Lee, H. Peguero, A.A. & Park, S. M. (2019). "Social-ecological correlates of cyberbullying victimization and perpetration among African American youth: negative binomial and zer-inflated negative binomial analyses". *Children and Youth Services Review*, 101 (March), 50-60. https://doi.org/10.1016/j.childyouth.2019.03.044

European Parliament (2016). *Cyber Bullying among Young People*, Study for the LIBE Committee, PE 571.67 (2016).

Even, C. (2019). *Cyberharcèlement chez les adolescents :impacts psychopathologiques, émotionnels et cognitifs : revue de littérature*, Thèse de Médecine, Faculté de Rouen. https://dumas.ccsd.cnrf.fr/dumas-02014346

Festl, R. & Quandt, T. (2016). "The role of online communication in long-term cyberbullying involvement among girls and boys", *Journal of Youth and Adolescence*, *45*, 1931–1945. doi: 10.1007/s10964-016-0552-9

Larrañaga, E., Yubero, S., Ovejero, A. & Navarro, R. (2016). "Loneliness, parent-child communication and cyberbullying victimization among Spanish youths". *Computers in Human Behavior, 65*, 1-8. https://doi.org/10.1016/j.chb.2016.08.015

Lee, C., & Shin, N. (2017). "Prevalence of cyberbullying and predictors of cyberbullying perpetration among Korean Adolescents». *Computers in Human Behavior*, *68*, 352-358. https://doi.org/10.1016/j.chb.2016.11.047

Les études du Centre Jean Gol (2017). *Le cyber – Harcèlement des enfants et des adolescents*, https://www.cjg.be.

Li, Q., Luo, Y., Hao, Z., Smith, B., Guo, Y. & Tyrone, C. (2020). "Risk factors of cyberbullying perpetration among school-aged children across 41 countries: a perspective of routine activity theory". *International Journal of Bullying Prevention*, https://doi.org/10.1007/542380-020-00071-6

Macilotti, G. (2019). « Violence et humiliation à l'ère numérique: une étude en milieu scolaire », *Médecine et Hygiène*, Vol. 43, pp. 299-328.

Malinoska-Cies'lik, M., Dzielska, A., Oblacin'ska, A. (2022). « Psychosocial Determinants of Adolescents 'cyberbullying involvement – The role of body satisfaction ». *International Journal of Environmental Research and Public Health*, 19(3), https://doi.org/10.3390/ijerph19031292

Mbaki Luzayisu, E. & Zamwangana Tungu, J.-P. (2023). *Cyberagression chez les étudiants en République démocratique du Congo (DRC). Essai de mesure et d'identification de facteurs de risque,* Survey Report, Kinshasa.

McCrae, R. R., & Costa, P. T. (1987). "Validation of the five-factor model of personality across instruments and Observers". *Journal of Personality and Social Psychology, 52*(1), 81–90. https://doi.org/10.1037/0022-3514.52.1.81

Mendez-Baldwin, M., Cirillo, K., Ferrigno, M. & Argento, V. (2015). "An examination of cyber-bullying and Social media use in Teens: prevalence, attitudes and behaviors". *Journal of Bullying & Social Aggression*, 1(1), 10 pages. http://sites.tamuc.edu/bullyingjournal/article/cyber-bullying-among-teens/

Menesini, E., Nocentini, A., Palladino, B.F., Frisen A., Berne, S., Ortega-Ruiz, R., Calmaestra, J., Scheithauer, H., Schutze-Krumbholz, A., Luik, P., Naruskov, K., Blaya, c., Berthaud, J. & Smith P.K. (2012). "Cyberbullying Definition Among Adolescents: A comparison Across six Europeans Countries", *CyberPsychology, Behavior and Social Network*, 15(9), https://doi:10.1089/cyber.2012.0040.

Merril, R. & Hanson, C. (2016). "Risk and protective factors associated with being bullied on school property compared with cyberbullied". *BMC Public Health*, *16*(145). doi: 10.1186/s12889-016-2833-3.

Mishina, F., Cook, C., et Gadalla, T., Daciuk, J. et Solomon, S. (2010). "Exploring factors associated with cyber bullying among middle and high school students", American *Journal of Orthopsychiatry*, 80(3), 362-374.

Palermiti, A., Servidio, R., Bartolo, G. & Costabile, A. (2017). "Cyberbullying and self-esteem: An Italian study". *Computers in Human Behavor*, *69*, 136-141. https://doi.org/10.1016/j.chb.2016.12.026

Park, S., Na, E.Y. & Kim, E.-M. (2014). "The relationship between online activities, netiquette and cyberbullying". *Children and Young Services Review*, 42, pp. 74-81. https://doi.org/10.1016/j.childyouth.2014.04.002

Petrosyan,A. (2022). *EU: internet penetration rate 2019-2022, by country*. https://www.statista.com/statistics/1246141/eu-internet-penetration-rate.

Peker, A. (2015). "Analyzing the risk factors predicting the cyberbullying status of secondary school students". *Education and Science*, *40*(181), 57-75. doi: 10.15390/EB.2015.4412

Rosenberg, M. (1965), *Society and the Adolescent Self-Image*, Volume 1979, Princeton, Princeton University Press. Runions, K. (2018). « Disentangling Functions of Online Aggression: The Cyber – Aggression Typology Questionnaire (CATQ) », *Agressive Behavior*, June 2016, online documentation.

Sasson, H. & Mesh, G. (2017). "The role of parental mediation and peer norms on the likelihood of cyberbullying". *The Journal of Genetic Psychology, 178*(1), 15-27. doi:10.1080/00221325.2016.1195330

Stahel L. and Weingartner, S. (2019). « Online aggression from a sociological perspective: An integrative view on determinants and possible countermeasures» . *Proceedings of the Third Workshop of Abusive Language Online*, Florence, Italy, Association for Computational Linguistics, pp. 181-187.

Tisseron, S., (2003), « Le désir d'extimité mis à nu », *Le Divan familial* 2003/2 (No11), pp. 53-62. UNICEF. (2019). *UNICEF poll: more than a third of young people in 30 countries report being a victim of online bullying*. U-Report highlights prevalence of cyberbullying and its impact on young people. from https://www.unicef.org/press-releases/unicef-poll-more-third-young-people-30-countries-report-being-victim-online-bullying.

UNICEF (online document). *Cyberbullying: what is it and how to stop it.* https://www.unicef.org/end-violence/how-to-stop- cyberbullying#2.

Vale, A. et al. (2018). « Cyber aggression in adolescence and internet parenting styles: a study with victims, perpetrators and victims perpetrators », *Children and Youth Services Review*, Vol. 93, pp. 88-99.

Wright, M. (2017). "Parental mediation, cyberbullying, and cybertrolling: The role of gender". *Computers in Human Behavior*, 71, 189-195. https://doi.org/10.1016/j.chb.2017.01.059

You, S. & Ah Lim. S. (2016). "Longitudinal predictors of cyberbullying perpetration: Evidence from Korean middle school students". *Personality and Individual Differences, 89*, 172-176. https://doi.org/10.1016/j.paid.2015.10.019

Yusuf, S., Al-Majdhoub, F.M., Mubin, N.N., Chaniago, R. H., Khan, F.R. (2021), « Cyberaggression - Victimization among Malaysians Youth », *Asian Journal of University Education*, online documentation.

Zang, Y., Chen, C., Zhaojun Teng, Z. et Gou, C. (2021), "Parenting Style and Cyber-Aggression in China Youth: The Role of Moral Disengagement and Moral Identity", *Front. Psychol*. Vol. 12:621878.

PROMISES OF DIGITAL IMAGINARIES AND RE-THINKING IDEALS OF DIGITAL YOUTH

Sari Tuuva-Hongisto and Kristiina Korjonen-Kuusipuro
Juvenia – Youth Research and Development Centre, South-Eastern Finland University of Applied Sciences (XAMK)
Patteristonkatu 3 D, 50100 Mikkeli, Finland

ABSTRACT

In this article we dismantle the digital imperative of youth and examine the practices and ideals of digital everyday life of Finnish youth, born in 2005-2006 in Finland. Our paper focuses on the different digital engagements in young peoples' everyday lives and what these engagements tell us about the ideals of digital citizenship and agency of young people. We ask, how these ideals shape the understandings of the use? To achieve these objectives, we draw on both qualitative and quantitative data in analyses of their interpretations of the use of digital technologies and their meanings. Our research reveals that it is important to elaborate how digitalisation is accessed, understood, and used, how it is constructed, and how people make sense of it.

KEYWORDS

Youth, Digital Ideals, Technological Imperatives, Finland

1. INTRODUCTION

Digital societies are developed around the ideology of technological imperatives (Talsi & Tuuva-Hongisto, 2009) and socio-technical imaginaries (Jasanoff & Kim, 2009). For example, in Finland, the national-level research of digitalization has followed these rationalities and leaned on a strong belief in them. Digital technologies have been kept apart from social structures and processes and positioned as a neutral good (Halford & Savage, 2010). These logics are visible and evident for example in an unquestioning push towards the digitalization of significant services, education, and political participation. Furthermore, digitalization has also been seen to provide equal opportunities for social and societal participation, especially among young people as "digital natives" and "cyber-children" (Holloway & Valentine, 2003; Prensky, 2001; Tapscott, 2008). Young people today are seen as a generation who find and adopt digital services first, and as such the youth-cultural meaning of digitalization is significant (Ito et al., 2010; Yalçın-İncik & İncik 2022).

As digitalisation strengthens, the use of various technologies and the related literacies are practically essential. Furthermore, variety of cultural and social structures, discourses and practices guide people to use these technologies. Such a necessity can be called a *technological imperative*, by which we mean the social and cultural pressure to exploit key technologies that have become a significant part of society and its ideology. (Talsi & Tuuva-Hongisto 2009.)

One of the key arguments of digitalization in various political debates is attached to the promise of equality of technology (Jasanoff & Kim 2015; boyd 2014). However, in critical reviews, digitalization is one more potential twist in the intersectional skein of a wide range of inequalities (e.g. Helsper 2021). The hope is that by strengthening everyone's digital participation, we will reach a situation where the promises of digitalisation can also be fulfilled and at the same time combat the potential threat of inequality.

The goal of this article is to dismantle the digital imperative of youth. To do this, we examine the practices and ideals of digital everyday life of Finnish youth, born in 2005-2006 in Finland. Our paper focuses on the different digital engagements in young peoples' everyday lives and what these engagements tell us about the ideals of digital citizenship and agency of young people. We ask, how these ideals shape the understandings of the use? To achieve these objectives, we draw on both qualitative and quantitative data in analyses of their interpretations of the use of digital technologies and their meanings. Youths' lives today are imbued with various kinds of uncertainties. However, everyday experiences and their experiential knowledge can be seen as a resource for coping with these uncertainties (Baillergeau & Duyvendak 2016).

2. THE PROMISES OF DIGITAL NATIVES

The discourse of the "digital generation" is precisely an attempt to construct the object of which it purports to speak. It represents not a description of what children or young people are, but a set of imperatives about what they should be or what they need to become. To some extent, it does describe a minority of young people who are actively using different technologies for social, educational and creative purposes, yet it seems very likely that most of these are already privileged in other areas of their lives and whose use of technology is supported by their access to other forms of social and cultural capital. (Buckingham 2008)

''Digital citizenship'' is the ability to participate in society online. The Internet has the potential to benefit society as a whole and facilitate the membership and participation of individuals within society. (Mossberg et al. 2007.) The networked, participatory potential of this new technology has been touted as creating new possibilities both for civic learning and for civic action. It provides a form of "networked citizenship" that is more inclusive and more participatory than the passive, dutiful citizenship of the past. Such arguments are frequently applied to the so-called digital generation of young people, who are apparently developing new forms of global political consciousness and activity as a result of their use of new media (Banaji & Buckingham 2013; Tapscott 2008). Far from dumbing down and disengaging young people, the new media are viewed as politically and personally empowering: they enable young people to become the agents or authors of civic action rather than merely the objects of adult interventions. (Banaji & Buckingham 2013.)

At the same time, if terms such as "digital youth" are taken to an extreme and analyzed in homogenous terms, people may assume that all kids of certain age are equally knowledgeable about and are willing to use all types of digital tools. Yet, studies show that young people engage with technologies in very different ways, ranging from the social to the intensely geeked-out (Ito et al. 2010). In addition, not all young people have the same access to technologized media. Scholars have observed "participation gaps" (Jenkins et al. 2006), which often exhibit familiar socio-cultural inequities based on sex, gender, ethnicity, and class. (Lange 2014.)

Despite the fore mentioned tendencies to universalize and idealize participation, there is increasing acknowledgement that youth's digital participation is both enabled and constrained by their particular social positions (boyd 2014). There is a broad "access rainbow" that patterns how young people engage with digital culture (Literat et al. 2018). It is partially reflective of their physical access to technological resources but also of the ways in which cultural capital and expectations influence forms of participation.

3. METHODS AND RESEARCH MATERIAL

The analysis is based on data collected in two phases in 2021 and 2022 from different parts of Finland with a target group of young people who are in the last grade of Finnish primary school, that is, about 15 years old. The ninth graders were chosen as a target group, because they are at the start of their transition years for their life-courses concerning education and growing up, approaching emerging adulthood in economically and socio-culturally different environments. We interviewed a total of 28 young people asking them, for instance, what meanings they give to digitalization in their daily lives and what role different digital devices and environments play in it.

A total of 418 young people during 2021 and 2022 quantitative data contextualizes and backgrounds the analysis we present in the article by rendering and structuring the research field and the "digital youth" under review in general. The questions asked in the online survey concerned the use and experiences of young people's various digital devices and environments. The survey also included some background information questions on, among other things, the respondent's gender, hobbies, family composition and livelihood, the level of education of parents, their place of residence, the use of public transport and the distance of various services essential to young people from the young person's home.

The qualitative data consists of 28 thematic individual or pair interviews and two focus group interviews with 6 participants. We have also conducted online observations and written fieldwork diaries of how these young people use digital media in their everyday lives. Our data collection was organized simultaneously to the emergence of the COVID-19 pandemic, and as face-to-face meetings with youth were cancelled due to various restrictions, we utilised digital ethnography and our interviews and observations were conducted online (about digital ethnography, see Korjonen-Kuusipuro, Tuuva-Hongisto & Berg 2022). The qualitative data includes the following themes and contents: young people's digital agencies, digital everyday practices, and meanings of places and contexts both online and offline (and the entanglements of these two).

Our analysis of qualitative data follows reflective thematic content analysis and has been carried out in a dialogic relationship with our research-based understanding of the effects of digitalisation on young people's lives (Braun & Clarke 2022). In our ethnographic study, the analysis phase of the data was entwined with the data collection. At the beginning of the classification of the data, two researchers encoded the transcribed interview data in a data-driven manner in the NVivo programme. We then reflected on the coding together and deepened our analysis by theming the material based on broader common themes and discussed the classification done with the entire research group. The names used in interview citations are pseudonyms to protect the privacy of the interviewees.

The analysis we present in this article is part of the research project Capturing Digital Social Inequality – Young Digi-Native's Asymmetrical Agencies Within Socio-Technical Imperatives and Imaginaries (DEQUAL, 2020–2024), funded by the Academy of Finland. The project is based on the idea of how physical places and environments in which different material structures open up to young people and where socio-cultural and youth cultural opportunities are different and unequal. The analysis of the ideals of digital youth is conducted through analytical frame of digital participation (cf. Literat 2018) and the dimensions of digital engagement: agency, practices, intensities and contexts (see Tuuva-Hongisto et al. 2022). In this article we focus on the restrictions of the use of digitalization that our interlocutors constantly produced in the interviews. They framed their use through the ideas of controlling the screen time, and picturing an ideal type of using digital devices and applications.

4. TECHNO-SOCIAL HIERARCHIES AND SHAME

Young people have a wide range of digital devices and skills at their disposal, and this is also true for the young people who participated in our research: for instance, almost all (96 %) of the youth who responded to the online questionnaire have smartphone at their disposal. The next most common digital devices used by young people are, according to the questionnaire data, a traditional personal computer (66 %) and a game console (47 %). In the questionnaire, we asked youth to take a stand on allegations concerning, for example, the respondent's digital skills. Most young people (90 %) somewhat or completely agreed with the statement 'I have sufficiently good skills to use different digital devices and services'. Only a few respondents (4 %) somewhat or completely disagreed with the statement. (Haverinen et al. 2023)

Based on our quantitative data, it generally appears that young people in general have access to a relatively comprehensive and diverse range of devices and applications that enable digital everyday life and 'hanging out', as well as other communication with friends. The most important online environments according to our online questionnaire were Snapchat (86,6%), TikTok (79,4%), Instagram (79,1%) and YouTube (75,4%), WhatsApp (73%). These results of the questionnaire communicate not only relatively diverse digital competences, skills and habits, but also the wider importance of digital communication and togetherness in the lives of young people. (See also e.g. Ito et al., 2009; Livingstone, 2010.)

Although the digital uses and skills were very mundane, the engagement to online participation varied. The individual differences in the digital participation, in its agencies and diverse roles, practices (playing, hanging, chatting, following), intensities with diversified engagement, screen times, control and online and offline contexts were huge. Despite these differences, most of our participants had the idea of using too much digital devices or spending too much time in different digital environments, even becoming addicted somehow. They were also envious to those who seemed more controlled and disciplined than them:

> *Sometimes during the weekend if you take the phone and the just stay on the couch and just watch YouTube videos, you get the feeling that you haven't done or reached anything. On the other hand, if you get out in the morning or go training or do something else useful, you realise that you are more energetic, that you have done something. The feeling that you have done something. (Juhani)*
>
> *Yes, I kind of envy people who can be apart from that phone for a really long time. But I could probably too, but I just have to make that change (laughs).* (Helmi)

Well, I use my phone quite a lot. I don't want to, but you do become addicted and it's not easy to get rid of. (Sofia)

Our research revealed that young people did not necessarily feel joy in their everyday use of digital devices, nor did they mention skills learned online. On the contrary, they often seemed to be ashamed and feel guilty for using too many digital devices and digital platforms. A feeling of shame is one of the primary negative affects, and in our analysis, we interpreted that shame was strongly connected to the social ideal of digital youth. Shame is intense and painful, and when we feel shame, we think we have done something that we feel is bad, and that something needs to be hidden. According to our interpretation, young people felt shame in situations where young people felt they had failed to live up to the social ideal. This way shame was strongly connected to the moral developments and reproduction of social norms. (See also Ahmed 2004.)

Hence, our research data do not contain only youth' s self-blaming but also efforts to locate oneself on the morally proper side of the line. In the interviews, young people's control over their own screen time and pursuit to keep it within certain limits of undefined normalcy became often justified by comparing how much time other young people or people 'in general' spend on different digital devices (cf. Salasuo, 2021). In this, the data contains many references to 'someone else': someone who uses a smartphone or plays with a computer more than themselves. In this shift, the young participants were also able to position themselves within suitable limits. In their reflections of time use, young people monitor and regulate not only their own but also their peers' use of digital media. (Haverinen et al. 2023.)

It is interesting how young people repeatedly told us that they need and want to do something "useful". Hanging out online did not seem to be useful, but when the interviewer asked what this useful might be, our participants often could not tell what they meant, or they mentioned for example that cleaning your room or wardrobe would be a good option as well as going out. This was encapsulated by one participant who said: "I have a feeling that I use a phone or computer all the time, and my life is being wasted", while another other argued how addictive digital devices were: "They are very addictive and take time from other important stuff".

And there's going to be shouting, like 'you're on that phone again, all this time, you're always on the phone, that's all you're ever going to do. (Emilia)

What is notable is the way our qualitative research material emerges the experience of overuse of digital communication devices described by youth themselves. However, they seem to be unable to define the limits of proper use more specifically – or what would be normal use, if such even exists: *'Well, I can't, actually, say what would be a proper time limit [in being on the phone]. But... I feel that I might use the phone a bit too much'* (Julia). As a result, the overriding feeling these young people have is that no matter what their actual screen time is – even if it was very minimal – they always spend a little too much time online, especially with their smartphones. Within this pattern we can recognize excuses for breaking something that is defined and recognized as morally wanted and demanding. In this, the discursive attitude climate obtrudes into the accounts around digitalization of the young ones' time use – showing that they are aware of the moral concern around their digital life contents. The blames are toned with expectations of more conventional everyday activities and with references where 'being on the phone' is seen as an irritative practice of doing nothing. (Haverinen et al. 2023; about doing nothing see Ehn & Löfgren 2010)

The ninth grades had a very clear idea of what kind of use of digital devices and applications were appropriate. They constantly referred to the time used on smart phones, they controlled, or wanted to control the used time with digital devices, especially on the phone. Especially those who had very limited screen time, had also very goal-oriented hobbies and their leisure time consisted of workout and their own special interests. They used digital media for school tasks, to read news and to follow current affairs. They said they did not post on social media, they were strict with their privacy; they did not have online friends and they were not actively communicating on any other channels. They used Youtube only to follow their own special tubers (usually connected to their hobbies) and they were following these mainly for learning purposes. They were heading to elite high schools, and they lived in a big city. This seems to illustrate a kind of ideal of today's digital youth: scarce, disciplined, controlled.

Matias is an example of what we call an "ideal digital youth", a disciplined, goal oriented and a media literate person. For him digital technologies are mostly for useful and educational purposes, he also follows news and current affairs, and his screen time is very limited.

And then at school. No. Not that much at the network. Of course, messaging and for this, but [...] Not really that much. I am one of the less using. I'm not that active there. The whole screen time is about an hour and a half. A while ago I kept a diary an my weekly training activities: amounts and such. In addition to soccer training, I spent about 30 hours a week in body maintenance and other sports. (Matias)

Matias wants to focus on school, and he is heading to an elite high school after the ninth grade. His hobbies are very goal-oriented: he plays football, and his plans include studying medicine or economics – or being a professional football player. His digital agency is controlled, passive and conscious, mostly for educational and targeted purposes. His digital practices are connected for studying, following news and current affairs and his hobbies: Matias uses social media platforms in order to look videos connected to football or other activities he wants to learn more about. The intensity of his use is very limited: in the interview he stresses that he doesn't have time to be on the phone or hang out online. He estimates his screen time is around one hour and half. The contexts of his digital participation are mostly Youtube and Instagram. Matias lives in the growing city centre, where he thinks he has all the possible opportunities for his future. The ideal of digital youth seems to be in motion: it is no longer connected to the potential of digitalization, its uses and skills. Thus, the discourse has shifted towards overuse, addictive and problematic use of digital devices, especially smart phones. The ninth graders commented on controlling their use; the idea of the freedom of mobile devices places new burdens and responsibilities on people as well as greater emotional stress. (Livingstone 2010.) We can also see how the "ethos of excellence" increasingly guides the content and practices of many children's and young people's leisure activities (Berg & Salasuo 2017).

The ethos of excellence is connected to the promise of a good life model that hard work, diligence and doing your best will be rewarded, at least at some point. Lauren Berlant (2011), a scholar of culture and politics, described this phenomenon by using the concept of cruel optimism. Cruel optimism requires citizens who focus on middle-class well-life goals that are not in the least possible or realistic for them.

5. CONCLUSIONS: SHIFTING IDEALS OF DIGITAL YOUTH

Today, self-control has become a sign of a good citizen; we must wage an endless battle to achieve the seemingly natural ideal norms of the elite and the stars. A failed and bad person is presented as not being able to take care of himself and fulfill these expectations of an ideal life dictated by no-one but shared by everyone. It is interesting that in a society that emphasizes freedom, we live amid so many silent but compelling norms and rules. (Jakonen 2020, 233-4.)

In our study we interpreted shame over the everyday use of digital devices, especially smartphones. When looking deeper into these tensions of controlling the use, we were able to interpret that shame and quilt young people felt for using too much digital devices and digital platforms were very visible. Even though, these feelings were "produced" in the context of the interviews, it was still surprising for us researchers how much shame was connected to the digital everyday.

The feelings of shame and quilt also echoes moral panic (Cohen 1972; Buckingham, 2008), where young people's ways to use digital devices are considered as a threat for a disciplined society. Furthermore, there seems to be evidence of demonizing talk around digitalisation: if you are not aware, controlled and disciplined, you will most probably be addicted, spend your time idly and learn nothing. We are not denying that disadvantages such as addictive behavior exist, but our data illustrates how young people seem to think that online activities are a waste of time. However, these ideas and feelings were "produced" in the context of the interviews, and those who participated in our study may have talked to the researchers in a way they think adults want to hear about their digital everyday life. We also noted how shame was not connected to a lack of skills or trendy devices, but clearly to a lack of control over their usage of digital devices and services.

Idealistic statements about digital culture stood in stark contrast to the ways that young people's digital lives are being pathologized in mainstream media, where parents are being taught to fear what young people might encounter if they spent too much time online. Medical authorities were urging parents to move the computer out of the teen's bedrooms and into publicly visible spaces to see if greater parental oversight might diminish their likelihood for mischief, cyberbullying, illegal downloads, violent gaming, or porn consumption. At the time, it was as if there were only two ways to perceive the Web— as a space of personal freedom and enlightenment or as a space of darkness and risk. (James 2014.)

These debates typically attribute an extraordinary power to technology and account for its role in highly deterministic terms: technology is seen to produce social change, irrespective of how and by whom it is used. Young people as emblems of the future are predictably invoked on both sides of this debate: they are at the same time "digital natives," whose facility with technology is creating new forms of social and cultural participation (Prensky 2001), and the "dumbest generation," stupefied and terminally distracted by the flickering screen (Bauerlein 2009). Such arguments tend merely to replay the binary logic that has historically characterized responses to all new technologies: either technology will liberate us, or it will enslave us; either it will expand our potential, or it will reduce us; either it will revitalize our social and cultural life, or it will take us all to hell. (Banaji & Buckingham 2013.)

Danah boyd (2014, 15.) states that a great deal of the fear and anxiety that surrounds young people's use of digital media stems from understanding or dashed hopes. Often, what emerges out of people's confusion takes the form of utopian and dystopian rhetoric. Sometimes misunderstandings result in moral panic. In other cases, such as the dystopian notion that teens are addicted to social media or the utopian idea that technology will solve inequality, the focus on technology simply obscures other dynamics at play. Both extremes depend on a form of technological determinism. Utopian and dystopian views assume that technologies possess intrinsic powers that affect all people in all situations in the same way.

Our research reveals that it is important to elaborate how digitalisation is accessed, understood and used, and how it is constructed. Furthermore, the dynamics of everyday lives – how people make sense of it and navigate its socio-technical contours are important (Webster et al., 2020). Feelings, emotions and affects can be analysed as hotspots showing us socially and culturally meaningful issues that need our attention (MacLure 2013; Ahmed 2004). Methodologically, understanding the everyday experiences and reading emotions and affects from research material is demanding. Therefore, diverse ways of exploring affect all need to be discussed further, particularly with reference to digitalisation, shame, and youth cultures. We hope the avenues opened in this paper will lead to additional research on the topic.

ACKNOWLEDGEMENTS

We want to thank all young people who participated in our research. DEQUAL project is funded by the Academy of Finland (330574).

REFERENCES

Ahmed, S., 2004. *The Cultural Politics of Emotion*. Edinburgh University Press.

Baillergeau, E. and Duyvendak, J.W., 2016. Experiential knowledge as a resource for coping with uncertainty: evidence and examples from the Netherlands. *Health, Risk & Society,* Vol. 18, No. 7–8, pp. 407–426. https://doi.org/10.1080/13698575.2016.1269878

Banaji, S. and Buckingham, D., 2013. *The Civic Web: Young People, the Internet, and Civic Participation*. MIT Press, Cambridge (MA) and London.

Bauerlein, T., 2009. *The Dumbest Generation.* New York: Penguin.

Berg, P. and Salasuo, M., 2017. Liikkuva luokka Liikunnan harrastaminen kunnon kansalaisuutena. *Yhteiskuntapolitiikka* 82 (2017) 3, 251-261.

Berlant, L., 2011. *Cruel optimism.* Duke University Press, Durham (NC).

Boyd, danah (2014) *It's complicated. The social lives of networked teens*. Yale University Press, New Haven & London.

Buckingham, D., 2008. Introducing Identity. In Buckingham, D. (Ed.) *Youth, Identity, and Digital Media*. MIT Press, Cambridge (MA) and London, 1–24. doi:10.1162/dmal.9780262524834.00

Braun, V., & Clarke, V., 2022. *Thematic Analysis. Practical guide.* Sage Publications.

Cohen, S., 1972. *Folk devils and moral panics: The creation of the mods and rockers*. MacGibbon & Kee, London.

Ehn, B. and Löfgren, O., 2010. *The secret world of doing nothing*. University of California Press Berkeley, Los Angeles and London.

Halford, S. & Savage, M., 2010. Reconceptualizing Digital Social Inequality. *Information, Communication & Society, Vol.* 13, No. 7, pp. 937–955.

Haverinen, V-S. et al. Forthcoming. Fear of missing out real life: from parental regulation to moral hierarchies of youth's screen time. Manuscript under review.

Helsper, E. J., 2021. *The Digital Disconnect. The social causes and consequences of digital inequalities*. Sage, London & Thousand Oaks (CA).

Holloway, S.L. and Valentine, G. (2003). Cyberkids. Children in the Information Age. Routledge/Falmer, London & New York.

Ito, M. et al., 2010. *Hanging out, Messing around and Geeking out: Kids Living and Learning with New Media.* MIT Press, Cambridge (MA) and London

Jakonen, M., 2020. *Konflikti. Talous ja politiikka*. Into, Helsinki.

James, C., 2014. *Disconnected: Youth, New Media, and the Ethics Gap.* MIT Press, Cambridge (MA) and London.

Jasanoff, S. and Kim, S-H. 2009. Containing the Atom: Sociotechnical Imaginaries and Nuclear Power in the United States and South Korea. *Minerva*, Vol. 47, No. 2, pp. 119–146. https://doi.org/10.1007/s11024-009-9124-4

Jasanoff, S. and Kim, S-H. 2015. Dreamscapes of Modernity. Sociotechnical Imaginaries and the Fabrication of Power. Chicago & London: Chicago University Press.

Jenkins, H. et al., 2006. *Confronting the Challenges of Participatory Culture: Media Education for the*
21st Century. The MacArthur Foundation, Chicago.

Korjonen-Kuusipuro, K., Tuuva-Hongisto, S. and Berg, P., 2022. Reflections on Digital Ethnography and Digital Realms of Young People. In Jantunen, J.H. et al. (Eds.) Diversity of Methods and Materials in Digital Human Sciences. Proceedings of the Digital Research Data and Human Sciences DRDHum Conference 2022, December 1-3, Jyväskylä, Finland, 83-94.

Lange, P. G., 2014. *Kids on YouTube: Technical Identities and Digital Literacies*. Taylor & Francis Group.

Literat, I. et al., 2018. Analyzing youth digital participation: aims, actors, contexts and intensities. *Information Society*, Vol. 34, No. 4. pp. 261-273.

Livingstone, S., 2010. Digital learning and participation among youth: critical reflections on future research priorities. *International journal of learning and media*, Vol. 2, No. 2-3. pp. 1-13.

MacLure, M., 2013. Researching without representation: language and materiality in post-qualitative methodology. *International Journal of Qualitative Studies in Education*, Vol. 26, No. 2, 658–667. https://doi.org/10.1080/09518398.2013.788755

Mossberger, K., et al. 2007. *Digital Citizenship: The Internet, Society, and Participation.* MIT Press, Cambridge (MA) and London.

Prensky, M., 2001. Digital Natives, Digital Immigrants. *On the Horizon,* Vol. 9 No. 5. pp. 1-6.

Salasuo, M., 2021. (ed.). *Harrastamisen äärellä. Lasten ja nuorten vapaa-aikatutkimus 2020.* Nuorisotutkimusseura & Nuorisotutkimusverkosto, Helsinki.

Talsi, N. and Tuuva-Hongisto, S., 2009. "Ei vietetty sinä jouluna tekniikan riemujuhlaa" Teknologinen imperatiivi teknologiaelämäkerroissa. *Kulttuurintutkimus*, Vol. 26, pp. 71–82. http://urn.fi/URN:NBN:fi:ELE-1551839

Tuuva-Hongisto, S., Kuusipuro-Korjonen, K. and Berg, P., 2022. *Tracing the digital landscapes of young people.* Presentation at NYRIS Conference 2022, Oslo.

Tapscott, D., 2008. *Grown up Digital! How the Net Generation is Changing your World.* The McGraw-Hill Companies, New York.

Webster, A., Svalastog, A. and Allgaier, J., 2020. Mapping digital landscapes. *Information, Communication & Society*, Vol. 23, No. 8, pp. 1100-1105. https://doi.org/ 10.1080/1369118X.2020.1784507

Yalçın-İncik, E., & İncik, T. 2022. Generation Z students' views on technology in education: What they want what they get. Malaysian Online Journal of Educational Technology, 10(2), 109-124. http://dx.doi.org/10.52380/mojet.2022.10.2.275

PUBLIC ATTITUDE TOWARD PARTICIPATION IN DISCUSSIONS ON ARTIFICIAL INTELLIGENCE IN JAPAN

Yukiko Maruyama
Tokai University
Hiratsuka Kanagawa, Japan

ABSTRACT

The increasing incorporation of artificial intelligence (AI) in our daily lives through various AI-enabled products and services has considerably changed the way our society functions. Therefore, public participation in discussions on AI development, use, and regulation, especially with AI becoming increasingly accessible to the public, is critical. However, few studies have focused on public attitudes toward participation in such discussions. This study focuses on understanding public attitudes toward participation in discussions and exploring factors influencing their intention. The results revealed that respondents tend to support public participation but are less willing to participate themselves. This trend was particularly evident among people 60 years and older.

KEYWORDS

Artificial Intelligence, Public Attitude, Public Awareness, Discussions on Artificial Intelligence, Online Survey

1. INTRODUCTION

Artificial intelligence (AI), through AI-enabled products and services, has increasing applications in the daily lives of people around the world, and has impacted the way our society functions. As mentioned in the OECD report, in addition to AI "contributes to better lives and helps people make better predictions and more informed decisions," "is also fueling anxieties and ethical concerns." (OECD 2019) Concerns remain regarding the development and societal implementation of AI and regulation as well as ethical issues. Studies have focused on public concerns and opinions because these play critical roles in discussions. Such studies have investigated the level of awareness and understanding of AI and perceptions toward AI. They also reveal expectations and trust in governmental corporations, multi-stakeholder institutions for developing and managing AI, and their opinions on the development of AI and the preferred direction of future development.

Cave et al. (2019) conducted a national survey of the UK population on the awareness and perceptions of AI and revealed that respondents had a negative perception of AI and felt that they had no control over its development. They stated that negotiating the deployment of AI requires overcoming these anxieties. Selwyn and Cordoba (2021) surveyed the Australian public regarding their level of awareness of AI development and the extent of change in AI support with additional exposure to information. The results revealed that the majority of respondents considered themselves to have limited knowledge of and familiarity with AI and were open to changing their opinions based on additional knowledge. This result suggests that public education on AI is critical. Kelly et al. (2021) surveyed 10,005 respondents from eight countries and six continents on their public perception of AI. They investigated sentiments toward and knowledge of AI, thoughts on its future and impact on society, and opinions on its responsible development and use. Regarding sentiments toward AI, differences were observed between countries. Although respondents perceived AI to be transformative and their expectations of AI-related change were not always positive in all countries, clear differences were observed between countries in terms of future effects. These studies suggest that negative public attitude and insufficient knowledge of AI had an indirect negative effect on AI development and regulation. This phenomenon suggests the necessity for public education on AI. Here, the public is not considered an active participant in the discussion, but as influencing the debate from the outside.

Zhang and Dafoe (2020) investigated Americans' perceptions of 13 types of AI governance and their trust in various actors to develop and manage the development and use of AI. The survey revealed that the respondents considered all AI governance challenges to have high issue importance and did not necessarily trust actors. Zhang and Dafoe mentioned, "As in these other policy domains, we expect the public to become more influential over time in impacting AI policy", and although the public is not a policy maker, they are expected to play a critical role in shaping policies. They considered the public to directly influence discussions.

Ema et al. (2016) stated that AI requires responsible research and innovation (RRI: Owen et al. 2012) with the collaboration of various stakeholders, including the public, from early stages. Therefore, they conducted a survey on the attitudes toward AI with various stakeholders, including the public, and aimed to suggest perspectives on future collaborations among stakeholders. Thus, public involvement is critical in discussions on AI development, use, and regulation, especially with AI becoming increasingly accessible to the public. However, few studies have focused on public attitudes toward participation in such discussions.

This study focused on understanding public attitudes toward participation in discussions and exploring factors influencing intention. The results can provide considerable insight into promoting collaboration among stakeholders in discussions.

2. METHOD

2.1 Sample

An online survey was conducted in December 2021. The respondents were members of an online market research panel with approximately 13 million members. A request to participate in the survey was sent to members of the online market research company until sufficient responses were received, following which people with similar answers to all questions were eliminated. A total of 1976 valid responses were collected. Based on responses to two questions regarding awareness of AI, we narrowed down the target population to those who had heard of AI but had never been involved in it in their daily lives or work. A total of 1334 responses were analyzed. Demographic data are presented in Tables 1 and 2.

Table 1. Sex of participants

Sex	Frequency
Male	634
Female	700

Table 2. Age of participants

Age	Frequency
20–29	210
30–39	245
40–49	264
50–59	289
60+	326

2.2 Questionnaire Design

The questionnaire consisted of sets of a series of multiple-choice and open questions. This study addressed multiple-choice questions related to the public awareness of AI and attitudes toward public participation in discussions on how to use AI in society. Regarding awareness, respondents were asked: "Have you ever heard about artificial intelligence (AI)?" and "Are you involved with artificial intelligence (AI) in daily life or your work?" These two questions were used to determine the target respondents. Furthermore, questions were posed regarding the sources of information on AI and self-assessment of AI-related knowledge. Regarding attitudes toward participation, respondents were asked to what extent they agreed with the following statements: "Public should be involved in determining how to use artificial intelligence (AI) in society" and "I wish to be involved in determining how artificial intelligence (AI) will be used in society."

To investigate factors influencing attitudes toward participation, six questions (see Table 3) were designed based on Omae et al. (2020). They evaluated AI education practices for promoting AI-related career development for elementary school students. They defined AI-related career development as "To acquire a desire to contribute toward society through AI." They assumed that educational practices do not promote career development directly but are mediated by improved motivation. Therefore, as stated in Eccles' expectancy-value theory (EVT; Eccles and Wigfield 2002), they adopted tree factors, namely intrinsic value, attainment/utility value, and expectancies for success as motivational factors influencing AI-related career development. EVT has been extensively used to examine motivational and achievement-related behaviors of elementary school students as well as to investigate parents' motivation related to involvement in their child's education. Omae et al. mentioned that career development is not equivalent to achievement motivation, but they adopted these three factors because another study (Omae 2016) revealed the usefulness of EVT in explaining career choices. Their results confirmed that these three factors positively influence AI-related career development.

Although attitudes toward participation in the discussions in this study are not equivalent to achievement motivation, they are similar to contributions to society, as described by Omae et al.'s definition of career development. Therefore, these three factors were included in this study.

Table 3. Questions pertaining to factors associated with attitude toward participation

Intrinsic value
I am interested in artificial intelligence.
I like artificial intelligence.
Attainment/utility value
I think artificial intelligence can be useful to society.
I think artificial intelligence is important to my life.
Expectancies for success
I am confident that I can influence the future development of artificial intelligence.
I am confident that I can learn about artificial intelligence.

3. RESULTS AND DISCUSSION

3.1 Awareness of AI

Among respondents, 84.8% respondents answered that they had heard of AI. By contrast, other surveys of the UK and Australian public report stated that 85% and 89.4% of respondents, respectively, claimed that they had heard about AI. In terms of awareness, this result was not different from that in other countries. Table 4 details the percentage of responses to questions and statements regarding respondents' involvement. The responses were multiple-choice. Of these responses, 1334 respondents who answered that they had never been involved with AI in their daily lives or at work were included in the analysis.

Table 4. Responses to questions and statements pertaining to "involvement in AI"

Involvement	percentage
I am/was involved in the development or research of AI technology	2.7
I am/was involved in the development and planning of AI-based products and services	3.4
I use/have used artificial intelligence in my work or activities	4.3
I use/have used artificial intelligence in my daily life	9.5
I have learned about AI	6.0
Other	2.0
I have never been involved with AI	80.0

Regarding self-assessment of AI-related knowledge, 26.8% of the Australian public investigated by Kelly et al. described themselves as knowing "a lot" or "a fair bit,". In contrast, the percentage of public respondents in Japan who marked themselves as "very knowledgeable," "knowledgeable," or "rather knowledgeable" was only 2.6%, a significantly low value.

The responses to questions about the sources of information on AI are presented in Table 5. Many respondents received information from TV/Internet news, TV shows, and newspapers. Few respondents answered questions related to SNS, friends, family, or work. According to Kelly et al. (2021), the top five were social media (45%), TV reports and commentaries (42%), movies or TV shows (40%), magazines or online articles (32%), and family and friends (31%). Compared with their results, the Japanese public does not seem to talk about AI on a personal level.

Table 5. Responses to questions pertaining to sources of information regarding AI

TV news	76%	Movie	15%	Others	7%
Internet news	38%	SNS	12%	Advertisement	4%
TV shows (except News)	32%	Friends and Family	11%	At work	4%
News Paper	29%	Books	11%		

3.2 Attitude Toward Public Participation in Discussion

3.2.1 Support and Intention

Figures 1 and 2 detail the responses to questions about the support and intention for public participation. More than half of respondents (52.8%) answered "strongly agree" or "agree" with "Public should participate in determining how to use AI in society." "Strongly agree" or "agree" was more prevalent in the age range of 60 years old and older (62.3%) and 20–29 years old (56.5%). By contrast, response was relatively low among 30–39-year-old participants (42.4%).

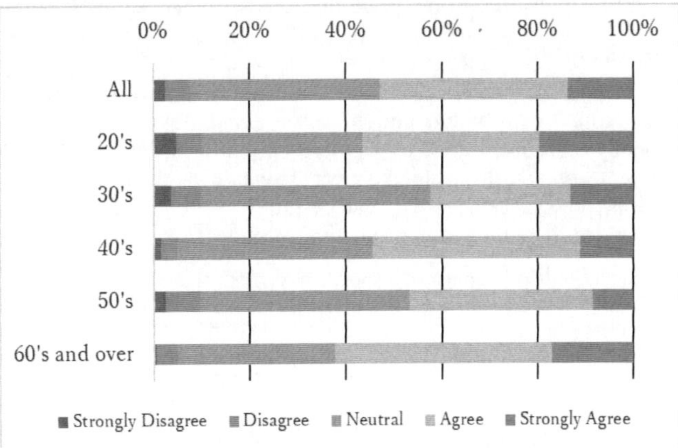

Figure 1. Responses to questions about support

Less than a quarter of respondents answered "strongly agree" or "agree" with "I would like to be involved in determining how AI will be used in society." Regarding support for public participation, positive responses were high among 20- to 29-year-old people (23%) and low among 60-year-old and older people (20.6%). This result indicates that respondents tend to support public participation but are less willing to participate themselves.

To examine the influence of age and sex on support and intention, the Mann–Whitney test and Kruskal–Wallis tests were conducted. The results indicated a statistically significant difference between age groups only for support ($p < 0.001$). Furthermore, multiple comparisons confirmed that statistically significant differences were observed between people in the age ranges 20–29 and 30–39, 30–39 and 60+, and 50–59 and 60 to 60+. This result suggests that the tendency to support public participation but to be less willing to participate by themselves is particularly strong among people aged 60 years and older. Although they revealed high levels of support for public participation, their keenness to participate was low.

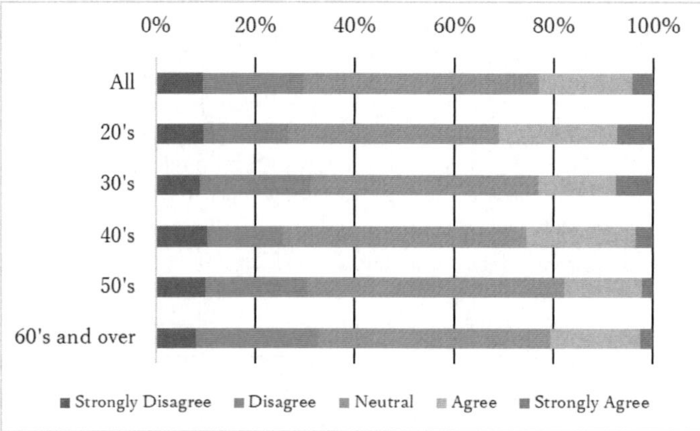

Figure 2. Responses to questions pertaining to intention

3.2.2 Factors Influencing Support and Intention

To investigate the differences in the three factors according to the extent of support and intention, respondents were categorized into three groups according to their answers to each question. The respondents who answered "strongly disagree," or "disagree" were grouped into "Negative," those who answered "neutral" were grouped into "Neutral," and those who answered "strongly agree" and "agree" were grouped into "Positive" respectively. Regarding items related three factors, respondents were asked to choose from five options, namely "strongly disagree," "disagree," neutral," "agree," and "strongly agree. "Each response was converted into a score on a Likert-type scale ranging from one to five. The average of the two items categorized for each factor was calculated as the score for each respondent. Tables 6 and 7 detail the number of respondents and average scores in each group.

The results of the Kruskal–Wallis test indicated statistically significant differences in the scores of the three factors between groups by opinion and intention. With regard to opinion, multiple comparisons confirmed that statistically significant differences were observed in the score of intrinsic value between all groups, in the score of attainment/utility value between "positive" and "negative," and "neutral", and in score of expectation for success between "negative" and "neutral," and "positive" respectively. In all cases, the scores of three factors tended to be high in the "positive" group.

Table 6. Support: Number of respondents and average of scores in each group

	Number of respondents	Average of scores		
		Intrinsic value	attainment/utility value	Expectancies for success
Negative	102	2.4	2.8	2.1
Neutral	525	2.8	3.2	2.6
Positive	702	3.2	3.6	2.7
p-value		< 0.001	< 0.001	< 0.001

Table 7. Intention: Number of respondents and average of scores in each group

	Number of respondents	Average of scores		
		Intrinsic value	attainment/utility value	Expectancies for success
Negative	394	2.5	3.1	1.9
Neutral	630	3.0	3.3	2.7
Positive	306	3.6	3.9	3.2
p-value		< 0.001	< 0.001	< 0.001

The results of multiple comparisons confirmed that statistically significant differences were observed in the scores of intrinsic value and expectation for success between all groups, in the score of attainment/utility value between "positive" and "negative," and "neutral". Similar to the responses for questions pertaining to support, in all cases, the scores of three factors tended to be high in the "positive" group. This result suggests that these three factors may be related to public participation support and intention.

The Kruskal–Wallis test was conducted to investigate the differences in the scores of the three factors between age groups. The results indicated statistically significant differences in intrinsic and attainment/utility values between the age groups. With respect to the intrinsic value, multiple comparisons confirmed statistically significant differences between the ages of 60 years old and older and 30–39, 40, and 50–59 years old respectively. In all cases, the scores of 60 years old and older was higher than those of other groups. Regarding the attainment/utility value, a statistically significant difference was observed between 50–59 years old and people aged 60 years and older. Individuals aged 60 years and older may find AI useful and important because of their interest in it.

Table 8. Average of scores in each group by age

	Average of scores		
Age	Intrinsic value	attainment/utility value	Expectancies for success
20–29	3.0	3.4	2.7
30–39	3.0	3.3	2.6
40–49	3.0	3.4	2.6
50–59	2.9	3.3	2.6
60-	3.2	3.5	2.5
p-value	0.001	0.007	0.300

4. CONCLUSIONS

An online survey was conducted in Japan to understand the public awareness of AI and attitudes toward public participation in discussions.

In terms of awareness, the result was not different from other countries. However, compared with the results of other studies targeting the public in other countries, knowledge of AI is the least, and AI does not seem to be a topic of personal conversation among Japanese respondents.

Regarding attitudes toward public participation in discussions, this result suggested that respondents tend to support public participation, but were less willing to participate. This trend was particularly evident among people 60 years and older. This result suggested that the three factors, namely intrinsic value, attainment/utility value, and expectancies for success, might be related to the support and intention of public participation. Individuals aged 60 years and older may find AI useful and important because of their interest in it.

While the results related the awareness of AI were compared the with the results of studies conducted in other countries, the results related the attitudes toward public participation in discussion were not compared. In the future, it is hoped that similar surveys will be conducted in other countries in the future.

REFERENCES

Cave, K. et al., 2019. Scary Robots: Examining Public Responses to AI. *Proceedings of the 2019 AAAI/ACM Conference on AI, Ethics, and Society*. Honolulu, USA, pp 331–337.

Eccles, J. S. and Wigfield, A., 2002. Motivational Beliefs, Values, and Goals. *Annual Review of Psychology*. Vol. 53, No. 1, pp 109–132.

Ema, A. et al., 2016. Future Relations between Humans and Artificial Intelligence: A Stakeholder Opinion Survey in Japan. *IEEE Technology and Society Magazine*. Vol. 35, No. 4, pp. 68–75.

Kelly, P. G., et al., 2021. Exciting, Useful, Worrying, Futuristic: Public Perception of Artificial Intelligence in 8 Countries. *Proceedings of the 2021 AAAI/ACM Conference on AI, Ethics, and Society*. Virtual Event, USA, pp. 627–637.

OECD., 2019. *Artificial Intelligence in Society*. https://doi.org/https://doi.org/10.1787/eedfee77-en

Omae, Y., 2016. Establishment of a Support Environment for Human Resource Development in Science and Technology Based on Machine Learning Analysis (translated by author). Nagaoka University of Technology, Doctoral dissertation. (in Japanese)

Omae, Y., et al., 2020. Artificial Intelligence Education in an Elementary School and its Evaluation by Career Development, Motivation and Rubrics. *Japan Journal Educational Technology*. Vol. 44, No. 2, pp. 213–223. (in Japanese)

Owen, R. et al., 2012. Responsible Research and Innovation: From Science in Society to Science for Society, with Society. *Science and Public Policy*. Vol. 39, Vo. 6, pp. 751–760.

Selwyn, N. and Gallo Cordoba, B., 2022. Australian Public Understandings of Artificial Intelligence. *AI & Society*. Vol. 37, pp. 1645–1662. https://doi.org/10.1007/s00146-021-01268-z

Zhang, B. and Dafoe, A., 2020. U.S. Public Opinion on the Governance of Artificial Intelligence. *Proceedings of the AAAI/ACM Conference on AI, Ethics, and Society*. New York, USA, pp. 187–193. https://doi.org/10.1145/3375627.3375827

SEARCHING FOR FALLS TRAJECTORIES OF THE ELDERLY

Ndiaye Amadou Maguette Djiogomaye[1,5], Michel Harel[2], Joseph Ngatchou-Wandji[3], Justine Trimouillas[5], Nassima Toumi[5], Caroline Gayot[5], Sophie Boyer[5,6] and Achille Edem Tchalla[1,4,5,6]

[1]*Laboratoire Vie-Santé, UR 24 134, Faculté de Médecine, 87025 Limoges, France*
[2]*INSPÉ de Limoges, Université de Limoges, 87036 Limoges Cedex, France; UMR5219 UPS, Institut de Mathématiques de Toulouse, 31062 Toulouse Cedex 9, France*
[3]*Sorbonne Paris Cité, Institut Elie Cartan de Lorraine, 54506 Vandoeuvre-les-Nancy cedex, France*
[4]*Service de Médecine Gériatrique, CHU de Limoges Dupuytren, Limoges, France*
[5]*Pôle Hospitalo-Universitaire de Gérontologie clinique, Service de Médecine Gériatrique, Unité de Prévention de Suivi et d'Analyse du Vieillissement, CHU de Limoges Dupuytren, Limoges, France*
[6]*Chaire d'excellence chez Fondation Partenariale de l'Université de Limoges I Institut Omega Health, Laboratoire VieSanté, UR 24134 Vieillissement Fragilité Prévention e-Santé, France*

ABSTRACT

The fall of the elderly is of great interest in geriatric. It can be linked to other diseases that make the elderly person frailer. Using Group Based Trajectory Modeling (GBTM) and K-means for longitudinal data (KML), we were able to find the trajectories that characterize our study population. Our participants were initially divided in two, randomly. We explored the trajectories that characterize the control group and the intervention group. We found five with the KML and four with the GBTM in the total study population and in each group. One of the groups is the intervention group and the other is the control group. The trajectories obtained in the intervention group testify to the impact of home-based interventions. The effectiveness of post-fall prevention is observed in both groups. However, the size of the sample and the limited number of falls recorded are a hindrance to the generalization of the results of the study.

KEYWORDS:

Geriatrics, Fall, Elderly, Mathematical Modeling, Longitudinal Data

1. INTRODUCTION

Just as children and adolescents have pediatrics as a medical specialty, the elderly have geriatrics. Geriatrics has become an increasingly popular field and has been a specialty in France since 2004 (Sénat, 2023). Regarding the study of aging, gerontology is the term used, which draws upon several fields including medicine, sociology, psychology, anthropology, economics, and many others (Bimou, 2019). This makes gerontology a rich field for collecting data that can help better understand patients and improve the quality of life for older adults. The objective is not to prevent aging, but to ensure that it goes well, which is referred to as "aging well" (US Preventive Services Task Force, 2018).

According to the World Health Organization (WHO), in 2019 out of the world's population of 7.7 billion people, more than 1 billion (13.2%) were 60 years old or older, compared to 382 million in 1980. This number is expected to increase to 2.1 billion by 2050 (WHO, 2023).

The narrowing of the birth rate in France and life expectancy, which is currently 85 years for women and 79 years for men as of 2021, has led to a considerable increase in the number of people over 65 years old (Insee, 2023). According to the National Institute of Statistics and Economic Studies (Insee), from 2016 to 2021, the number of inhabitants aged 65 or over increased by 1,379,699 in metropolitan France. In the group aged 75 or over, this number increased by 296,933; this strong increase will continue until 2035. This number will increase by 10 million by 2060, and one third of the population will be over 60 years old (Insee, 2023).

Studies aimed at finding ways to delay and improve aging are underway, with the goal of identifying ways to slow down the often irreversible consequences of aging (Park, 2018). Data collected from a Comprehensive Geriatric Assessment (CGA) on a cohort of people aged 75 and over are longitudinal and will help identify fall trajectories using statistical methods that have already produced promising results in this field. This statistical analysis will provide the physician with clues to improve their care (Tchalla et al., 2014).

2. GENERALITIES, MATERIALS, AND TOOLS

A fall is a complex event. To fully understand it, a comprehensive assessment of the older adult is necessary. The more information we have about them, the better it is. In this study, the protocol used allowed to collect as much information as possible from the participants.

2.1 Fall

A fall occurs when a person involuntarily ends up on the ground or at a lower level than their starting position (Moylan & Binder, 2007). Events caused by loss of consciousness, a violent push, stroke, or epileptic seizure are excluded. Falls can lead to disability, fractures, and even psychological trauma, making fall prevention a public health priority (SPF, 2023). The objective is to combat the rapid deterioration of the health of the population, especially the most affected: older people. To do this, it is essential to identify early warning signs of a first fall (Park, 2018). A first fall can lead to a reduction in physical activity, which is beneficial for maintaining the health of older adults, due to the fear of falling again (Mignardot et al., 2014). A fall may not be serious and may be repeated in some people, and this is referred to as "repeated falls" in geriatrics if a person falls at least twice in 12 months (WHO, 2023).

2.2 Comprehensive Geriatric Assessment (CGA)

The Comprehensive Geriatric Assessment (CGA) is a multidimensional and standardized approach that improves practices in the care of older people through a health assessment (Bimou, 2019).

2.3 Study Framework

Over a period of two years, the Unité de Prévention, de Suivi et d'Analyse du Vieillissement (UPSAV) collected data on 440 participants through the clinical research protocol GEROPASS during home visits conducted by a geriatrician and a nurse. The data from the Comprehensive Geriatric Assessment (CGA) were collected at the beginning of the study, which is the initial visit or V0 (it allows to take a snapshot of the participant's health at entry and to construct his/her Personalized Care Plan (PPS)), during the two visits that followed V1 (at 6 months) and V3 (at 12 months) but also at the end of 24 months, which is the visit V3 (it allowed to take a snapshot of the participant's health upon exiting the study).

Throughout the study, the intervention group participants had the four planned visits during the two years of the study. However, those in the control group only had the V0 and V3 visits (Figure 1). They didn't have the Personalized Care Plan (PPS).

The participants are men and women aged 75 and over.

To be included, participants must:

- provide written informed consent from the participant or their legal representative,
- not be participating in a clinical trial that modifies the patient's management,
- not have progressive pathologies that could affect short-term prognosis,
- not be in a Long-Term Care Unit (USLD) or in a Nursing Home for Dependent Elderly People (EHPAD),
- have a Mini Mental State Examination (MMSE) score of 10 or higher,
- be covered by social security at 100%.

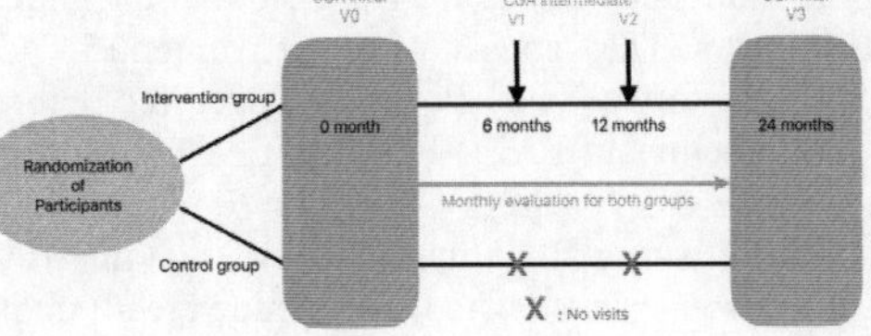

Figure 1. Study design

2.4 Database Formatting

The data is collected through home visits conducted by the Aging Prevention, Monitoring and Analysis Unit (UPSAV) and telephone calls made by the Clinical Research and Innovation Unit (URCI). In order to have a database with the necessary information, we joined the call database with the V0 visit database using Excel. This resulted in a database of 167 individuals, of which 79 are controls and 88 are interventions.

2.5 Descriptive Analysis

Table 1 shows the distribution of the 167 participants and their characteristics according to group. They have an average age of 83 years and are predominantly female (118 women and 49 men).

Table 1. Descriptive analysis of the study population by group at V0

		Over all	Both study groups		
			Intervention group,	Control group,	
Variables	N	N=167[1]	N= 88[1]	N=79[1]	p_{value}[2]
AGE	167	83.0 (80.0, 88.0)	83.0 (80.0, 89.0)	83.0 (80.0, 87.5)	0.98
Sex	167				0.34
Female		118 (71%)	65 (74%)	53 (67%)	
FALL HISTORY	167				**0.046**
No		42 (25%)	18 (20%)	24 (30%)	
Yes		107 (64%)	56 (64%)	51 (65%)	
Not specified		18 (11%)	14 (16%)	4 (5%)	
BMI	167				**0.022**
$\geq 21 kg/m^2$		147 (88%)	72 (82%)	75 (95%)	
$<21\ kg/m^2$		10 (6.0%)	9 (10%)	1 (1.3%)	
$<18\ kg/m^2$		10 (6.0%)	7 (8.0%)	3 (3.8%)	
OUTDOOR WALK	167				**<0.001**
No		10 (6.0%)	10 (11%)	0 (0%)	
Yes		71 (43%)	71 (81%)	0 (0%)	
Not specified		86 (51%)	7 (8.0%)	79 (100%)	
DAILY MEDICATION ≥ 4	167				0.35
Yes		124 (74%)	68 (77%)	56 (71%)	

1 Median (Interquartile Range) or Proportion (%)
2 Wilcoxon rank sum test; Pearson's Chi-squared test; Fisher's exact test

We see this average age in both study arms. We can note that the control group participants are younger than those in the intervention group. They consist of fewer women (53 versus 65) and fewer people with a history of falls (51 versus 56). A history of falls is a fall that occurred before the start of the study.

Study participants have a predominant FALL HISTORY. They represent more than 60% in each study group.

Most participants (147 or 88%) have a Body mass index (BMI) that is ≥ 21 kg/m², with 75 in the control group and 72 in the intervention group. Those who walk outside are all in the intervention group.

Those who take daily medication ≥ 4 are 124 (74% of participants), of whom 68 are in the intervention group and 56 are in the control group.

If we look at the variables with significant p_{values} (p_{value}< 0.05) among those presented in the table, we will retain: HISTORY OF FALLS (p_{value}= 0.046), BMI (p_{value} = 0.022), and OUTDOOR WALKING (p_{value} <0.001). The distribution of the proportions of these different variables between the two study arms is significant.

3. ANALYSIS OF TRAJECTORIES

In order to identify the fall trajectories of participants, we will use methods that allow us to identify individuals in a study population who have the closest profiles. This is referred to as clustering. For longitudinal data, SAS offers a mixed-parametric model developed by Nagin and his colleagues (Nagin, 2014) : Group Based Trajectory Modeling (GBTM). We also have a classification method for longitudinal data in R: K-means, developed by Genolini and colleagues, commonly referred to as KmL (Genolini & Falissard, 2011). These different models offer a range of choices for selecting one model over another based on validation scales, which solves the problem of the number of trajectories that should be retained.

3.1 Group Based Trajectory Modeling (GBTM)

Depending on the nature of our data, GBTM uses an appropriate approach (Nagin, 2014). A Poisson distribution is used if the response variable is a frequency, a censored normal model is used if it is continuous, and a logit model is used if it is dichotomous dependent. GBTM is implemented in SAS software and can be experimented with using the Traj procedure. In our case, we are studying the number of falls every 6 months, which is a quantitative variable, so we will use the censored normal model or CNORM in SAS software (Bimou, 2019).

3.1.1 Results

The fall event was collected in a binary manner. Let x be a participant in the study and $C_i(x)$ be a function that indicates the fall event in month i with $i \in [\![1,24]\!]$ for participant x. It is defined by:

$$C_i(x) = \begin{cases} 1\ if\ x\ had\ a\ fall \\ 0\ otherwise \end{cases} \tag{1}$$

This function allows us to construct our fall variable. However, using this variable did not allow us to find interesting clusters to analyze. This led us to explore other possibilities, including:

- replacing our categorical variable (which indicated whether the participant fell or not) with a quantitative variable that gives us the number of falls per month,
- using this quantitative variable to define a function that aggregates the number of falls observed every six months.

Let $D_j(x)$ be the function defined by:

$$D_j(x) = \sum_i C_i(x). \tag{2}$$

The function $D_j(x)$ lists the number of falls of participant x in month j with $j \in [\![1,24]\!]$.

Let $f(x)$ be a function defined by:

$$f(x) = \sum_{j=1}^{6} D_j(x). \tag{3}$$

The function $f(x)$ defined from equation (2), allows us to list the falls of participant x every six months.

The variable **FALL HISTORY** allowed us to have a fifth abscissa, giving us a snapshot of the patient before the start of the study.

We used the censored normal model (CNORM) since our variables are quantitative. We can already note that the algorithm represents the average falls over the study periods, which is not the objective of the study. We aim to identify clusters among our participants who will be grouped based on the number of falls.

The obtained results contain information that seems relevant. We chose the number of trajectories based on the distribution of participants in these trajectories (We set a threshold of 5% in each trajectory).

The obtained fall trajectories considering the entire study population allowed us to distinguish four fall trajectories.

Below are the trajectories characterizing the entire study population (Figure 2):

- Trajectory 1 in red consists of those who have never fallen
- Trajectory 2 in green consists of those with a fall history and who did not fall on average during the two-year study period.
- Trajectory 3 in purple consists of those with a fall history and who fell on average 1.2 times at six months of the study.
- Trajectory 4 in black consists of those with a fall history and who fell on average throughout the study period.

Throughout the study, we will keep the variable **number of falls** defined from the quantitative variable **number of falls per month**. Hence, for all the analysis methods we will use, we will work with the variable **number of falls every 6 months**.

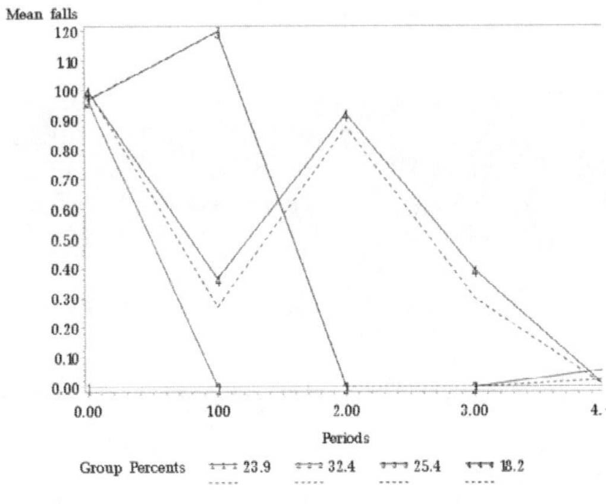

Figure 2. Fall trajectories with GBTM in the total study population.

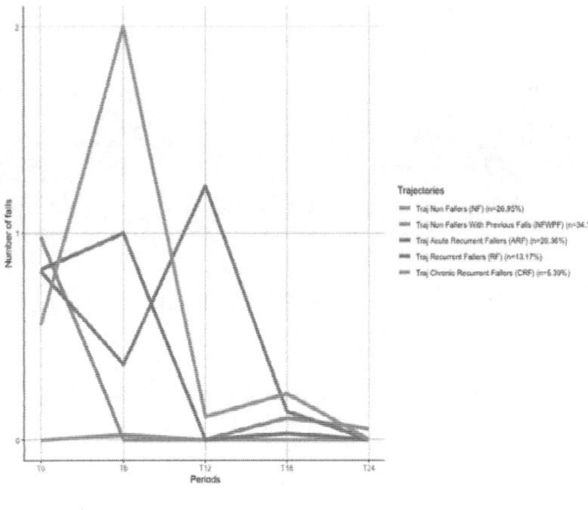

Figure 3. Fall trajectories with K-means in the total study population.

3.2 K-Means Classification (KML) for Longitudinal Data

The KML algorithm is a generalization of K-means clustering on longitudinal data (Genolini & Falissard, 2011). K-means applied to a data set partitions the individuals into subgroups that become increasingly homogeneous by maximizing inter-cluster distances and minimizing intra-cluster distances. It is recognized as an alternative to mixture models. KML performs this classification on data measured over a period, rather than at a single point in time. Implemented in R, KML can use either Euclidean or Manhattan distance to calculate intra-cluster and inter-cluster distances. The choice of model is made using the relevant inertia criterion: the Calinski-Harabasz criterion (Caliński & Harabasz, 1974).

3.2.1 Results

The choice of the five trajectories is made using the Calinski-Harabasz criterion.

In Figure 3, we have the trajectories of the total population falls in the study. We obtained five trajectories that are significant with a minimum threshold of 5% (each trajectory contains at least 5%) of the study population.

Unlike the CNORM of Group-Based Trajectory Modeling (GBTM), the K-means classification represents the number of falls over the study period.

Regarding the interpretation of the trajectories in Figure 3, we will start with the most stable:

- The trajectory of Non-Fallers (NF): it consists of participants who did not fall at the beginning (no history of falls) or throughout the study.
- The trajectory of Non-Fallers With Previous Falls (NFWPF): it consists of participants who fell before the study's beginning and did not fall during the study. Within the first six months of the study, it merges with the NF trajectory.
- The trajectory of Acute Recurrent Fallers (ARF): these are participants who have a history of falls and fell at least once in the first six months of the study. After one year, they no longer fall.
- The trajectory of Recurrent Fallers (RF): It consists of those who have a history of falls and will fall several times throughout the study.
- The trajectory of Chronic Recurrent Fallers (CRF): it consists of the largest fallers in the study. Each participant in this trajectory has a fall history, which did not prevent them from falling up to twice in the first six months. The falls decreased significantly throughout the study, although this trajectory never reaches the NF trajectory.

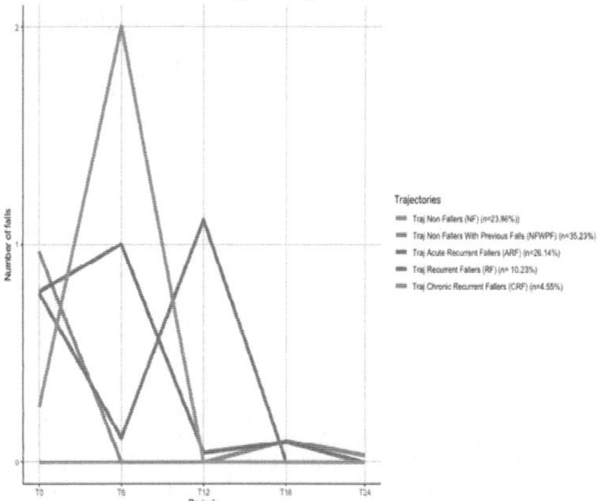

Figure 4. Fall trajectories with K-means in the Intervention Group

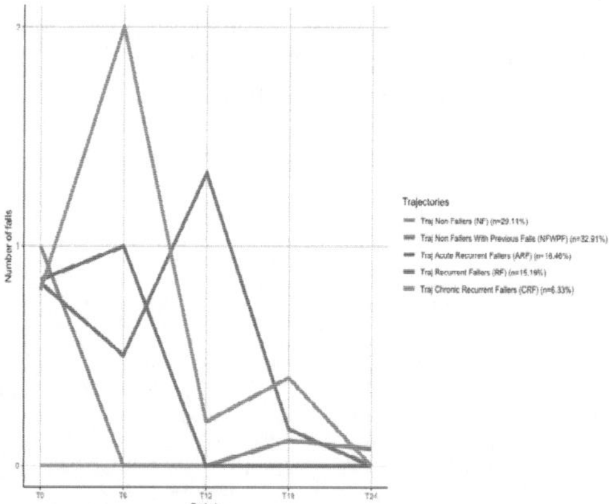

Figure 5. Fall trajectories with K-means in the Control Group

Observing the trajectories of the control and intervention groups allows us to see similar patterns in both Figure 4 and Figure 5 for all trajectories. The trajectories of NF and NFWPF are the same in both figures. If we take a closer look at the NFWPF trajectory in the control group, it starts from 1 at T0, unlike the intervention group where it starts from a lower value. Note that even though the algorithm places the number of falls on the ordinate, the center of a cluster (trajectories are constructed from clusters) is calculated by the average of the points around it. This justifies the fact that the trajectories do not necessarily pass through coordinates whose abscissas and ordinates are integers.

From the first six months, the NFWPF trajectory joins the trajectory of NF in both groups. In the intervention group, they coincide until the end of the study. From the twelfth month of the study, the NFWPF trajectory is distinguished from NF in the control group (Figure 5).

The ARF trajectory has the same pattern in both Figure 4 and Figure 5 even though its value at T0 in Figure 5 is closer to 1. The suspected difference in the first three most stable trajectories becomes apparent with the RF and CRF trajectories. At T6, the CRF trajectories reach a maximum of 2 falls. Meanwhile, the RF trajectories do not reach the same minimum value, indicating more falls in the control arm in the first 12 months.

This trend is confirmed in the second half of the study. The CRF trajectories (at the 12th month) and RF trajectories (at the 18th month) in the intervention group join the trajectory of NF. In the control group, these two trajectories do not stabilize. Here, the stability of trajectories is defined based on the trajectory of NF, which is the most stable (recall that the participants who make it up have never fallen)

4. DISCUSSION

The objective of the work assigned to us was to identify differences that could exist between two randomly defined groups. Describing the study population led us to scrutinize the characteristics that would appear. Using data collected at two different times, UPSAV provided data collected at home, and URCI provided data obtained through telephone calls. We were able to create a database that allowed us to describe the study population and find the trajectories of falls. In order to make the interpretation more fluid, we attempted to give names that summarized the trajectories.

By using the history of falls and transforming the categorical variable "fall or not" into a quantitative variable that records falls every six months for each participant, we found five trajectories of falls using KML. The similarity between the trajectories obtained for the total study population and those obtained for each arm required a comparison.

Until the sixth month of the study, the trajectories of the two groups had slight differences. From the twelfth month in the intervention group, the trajectories of the fallers (ARF, RF and the CRF) converged with those of the non-chronic and non-severe fallers (NF). In the control group, these fallers diverged from the NF. The fact that all trajectories tended towards zero at the end was due to the fact that there were few calls after the last six months.

Since participants in the intervention group had all protocol visits, we can conclude that UPSAV's intervention had an impact on reducing falls in the intervention group from the twelfth month of the study. All trajectories began to stabilize from the twelfth month of the study, except for the trajectory of RF, which started at the eighteenth month.

The effect of UPSAV's intervention was delayed in the trajectory of RF compared to the others. The decrease in falls in the control group is probably due to prevention by the treating physician after a fall. For those in the intervention group, whether the participant falls or not, prevention is provided, which explains the trajectories obtained. From two falls in a year, the participant begins to weaken. In order to provide good prevention, identifying risk factors becomes a priority.

In conclusion we have demonstrated the importance of closely monitoring older adults to prevent the first fall as well as repeat falls for those who have already fallen. This study will allow us to identify risk factors for falls in subgroups (within each trajectory of the study) of the total study population. The predictive factors for falls identified in each trajectory will enable specialists to better understand the profile of fallers and thus provide more targeted prevention strategies for patients. What are the predictive factors in each of our trajectories? Will these results be more meaningful with more data? The limitations of the study include the sample size and the recruitment of participants only in France. Telephone calls also posed challenges in data collection.

REFERENCES

Bimou, C. (2019). Analyse de trajectoires, perte d'autonomie et facteurs prédictifs : Modélisation de trajectoires [These de doctorat, Limoges]. https://www.theses.fr/2019LIMO0030

Caliński, T., & Harabasz, J. (1974). A dendrite method for cluster analysis. Communications in Statistics, 3(1), 1-27. https://doi.org/10.1080/03610927408827101

Genolini, C., & Falissard, B. (2011). KmL: A package to cluster longitudinal data. Computer Methods and Programs in Biomedicine, 104(3), e112-121. https://doi.org/10.1016/j.cmpb.2011.05.008

Insee. (2023). Espérance de vie à divers âges | Insee. https://www.insee.fr/fr/statistiques/2416631#graphique-figure1

Mignardot, J.-B., Deschamps, T., Barrey, E., Auvinet, B., Berrut, G., Cornu, C., Constans, T., & De, L. D. (2014). Gait disturbances as specific predictive markers of the first fall onset in elderly people : A two-year prospective observational study. Frontiers in Aging Neuroscience, 6, np. https://doi.org/10.3389/fnagi.2014.00022

Moylan, K. C., & Binder, E. F. (2007). Falls in older adults: Risk assessment, management, and prevention. The American Journal of Medicine, 120(6), 493.e1-6. https://doi.org/10.1016/j.amjmed.2006.07.022

Nagin, D. S. (2014). Group-Based Trajectory Modeling: An Overview. Annals of Nutrition and Metabolism, 65(2-3), 205-210. https://doi.org/10.1159/000360229

Park, S.-H. (2018). Tools for assessing fall risk in the elderly: A systematic review and meta-analysis. Aging Clinical and Experimental Research, 30(1), 1-16. https://doi.org/10.1007/s40520-017-0749-0

Sénat. (2023). Reconnaissance de la gériatrie comme une spécialité médicale—Sénat. https://www.senat.fr/questions/base/2003/qSEQ031009555.html

SPF. (2023). La surveillance épidémiologique des chutes chez les personnes âgées. Numéro thématique. Vieillissement et fragilité : Approches de santé publique. https://www.santepubliquefrance.fr/maladies-et-traumatismes/traumatismes/chute/la-surveillance-epidemiologique-des-chutes-chez-les-personnes-agees.-numero-thematique.-vieillissement-et-fragilite-approches-de-sante-publique

Tchalla, A. E., Dufour, A. B., Travison, T. G., Habtemariam, D., Iloputaife, I., Manor, B., & Lipsitz, L. A. (2014). Patterns, Predictors, and Outcomes of Falls Trajectories in Older Adults: The MOBILIZE Boston Study with 5 Years of Follow-Up. PLOS ONE, 9(9), e106363. https://doi.org/10.1371/journal.pone.0106363

US Preventive Services Task Force. (2018). Interventions to Prevent Falls in Community-Dwelling Older Adults: US Preventive Services Task Force Recommendation Statement. JAMA, 319(16), 1696-1704. https://doi.org/10.1001/jama.2018.3097

WHO. (2023). Chutes. https://www.who.int/fr/news-room/fact-sheets/detail/falls

DIGITAL CHALLENGES OF WORK MODEL

Katarina Šiber Makar[1], Anita Belak Bradač[2] and Ivana Ogrizek Biškupić[3]
[1]IN2 Company & University College ALGEBRA
[2]MBA student at University College ALGEBRA
[3]University College ALGEBRA
Gradišćanska ulica 24, 10000 Zagreb, Croatia

ABSTRACT

Nowadays we are facing global transformation driven by digital technology. A paradigm shift has been fostered by so many disruptions in almost all occupations and professions. Many companies that were present in the market for decades have digitalized parts of their business model, introduced new digital businesses, services or products and digitalized their processes, wherever applicable. Technology pace is increasing and is fundamentally changing the nature of job tasks, resulting in many new job roles that were created in the last decade, while others have become obsolete and fully automatized by robots, artificial intelligence, machine learning, chat bots and other digital technologies and solutions. Digital transformation has greatly affected all participants of a value change: service and product providers, their customers, and their employees. In this paper authors analyse challenges and benefits of remote/hybrid work for companies, employers, and employees, present the results of the conducted research on the employees' sample N:102 from one medium-sized company in the Republic of Croatia.

KEYWORDS

Work Model (Onsite, Hybrid, Online), Employees, Work Regulations, Workplace Disruption

1. INTRODUCTION

Business Digital transformation has created several significant changes in the way companies operate and in the way they provide their services and products. Technology has become an integral component of workspaces and is no longer viewed as an external support system (Venkatachalam, Parvathy et al., 2023). A greenfield digital companies have emerged that use fully digital processes and operate with their customers through digital channels only, often raving remote work model as only working option. Setting up digital company from scratch has proven to be much easier task than to transform existing business into digital, particularly when company has business processes operated with white collars employees. Many companies that have been present in the market for decades have digitalized parts of their business model, introduced new digital businesses, services or products and digitalized their processes, wherever applicable. Not only that products and services become globally available but the workers too. More and more companies have people employed across the globe and operate as one team regardless of their geographical dispersion. These companies have started applying remote work model as pioneers, but many more joined the same path because of Covid-19 pandemic (Ancillo, Antonio de Lucas et al., 2023) that significantly speeded up both digitalization and adoption of remote and hybrid work models. Technology pace is increasing and is fundamentally changing the nature of job tasks, resulting in many new job roles that were created in the last decade, while others have become obsolete and fully automatized by robots, artificial intelligence, machine learning, chat bots and other digital technologies and solutions. By 2025. machines and humans will be equal in terms of how tasks will be divided (Zahidi, S., 2020). Digital transformation is helping companies not only in customer interactions and in fulfilling emerging customers' expectations, many providing personalized customer experience and personalized services and products, but also in internal operations and process digitalization resulting in higher profits vs their competitors. Digital transformation has greatly affected all participants of a value change: service and product providers, their customers, and their employees.

2. WORKPLACE CHALLENGES

Some form of remote work model was here for more than 50 years, slowly developing as new technology were entering the workplace. First adopter of remote work were technology companies. CISCO in 1993 had one of the first systematic remote work programmes where employees could work either fully remotely or were offered flexible working hours and communicating via broadband technology. 10 years later, after a decade of working under remote and hybrid model, CISCO reported an increase in employee productivity and savings in 2003 of 195 mill $ (Neeley, Tsedal, 2021). Sun Microsystems, in the late 1990s, enabled remote work for 35% of their employees as part of their global expansion. Within a decade, just on reducing 15% of their real estate holdings, they reported savings of 500 mill $ (Neeley, Tsedal, 2021). Back in 1991. Peter F. Drucker declared that in the future "office work, rather than office workers, will do the traveling" (Drucker, Peter F. 2020). Since 2017 remote working started to be a common working option possibility, required by employees. Since the Covid-19 pandemic outbreak, telework (fully remote or hybrid) became the new normal, where almost 50% of all the employees all over the globe (during some point) worked remotely and continues to do so. "Flexible work is here to stay" according to findings done by Microsoft in 2021. Work Trend Index study of more than 30.000 people from 31 countries and trillions of Microsoft 365 and LinkedIn analyses of work signals (Microsoft, 2021). In 2015 19% of EU employees worked from home remotely or in a hybrid model using collaboration technologies, where Scandinavian countries were leading with 30%-38% working in remote or hybrid model (Samek Ladovici, M. European Parliament, 2021). According to a study by the International Labour Office from Geneva from 2019, 40% of US workers work remotely occasionally (Lister, Kate et al., 2020).

2.1 Challenges and Benefits for the Companies

Hybrid and remote work do not come without challenges. Organizations' dynamic capabilities play a major role in enabling this transformation of the workplace (Chatterjeea, Sheshadri et al. 2023). Management typically questions if employees will be fully engaged without continuous monitoring, while there is a challenge of work-life balance, sense of loneliness and isolation on employees' side, as well as lack of socialization with their colleagues. Chatterjeea, Sheshadri et al. 2023 developed a research model tested on a sample of 335 respondents from different types of organizations on the digital transformation. The study finds that the dynamic capabilities of organizations have a significant and positive influence on the digital transformation of the workplace, which in turn improves the employees' work-life balance resulting better employee performance and superior organization performance. The study also highlights the significant role that digital leadership plays in the digital transformation of the workplace. According to 2020 BCG research, success of hybrid/remote work depends on several factors like transparent remote-work vision, selecting right operating model, enhancing trust-based management, establishing the technological foundation and solutions (Ferreira., José et al., 2020). Leadership proved to have a key role in motivation and communication among employees. Additionally, without a company culture built on trust and clear goal settings, it is hardly possible to successfully introduce hybrid and remote working models. According to BCG survey from 2020, 24% of managers believe that it's harder to control and drive productivity while working remotely vs in-office work, 39% of managers believe it is harder to ensure team engagement and connection, and 30% of managers believe in less innovation while working remotely. Global Workplace Analytics survey (Lister, K. et al., 2019) points out that 35% of managers believe that main struggle of remote work is employee engagement, 34% believes that the main struggle is a lower performance of teamwork and cohesion among team members, followed by lower creativity and innovation and lower availability of manager's control over employees work performance. 2021 research from Gartner shows that employees were very engaged in remote/hybrid model even when their work had not been under constant company control and monitoring. According to research data of 2.865 participants, 70% of leaders confirmed that remote work performance was the same or better as on-site (in office) work (Kamouri, Anita et al., 2020). McKinsey Global Survey on digital transformation from 2018 lays down five characteristics of successful digital transformation as prerequisite for successful remote / hybrid work: leadership by having the right, digital-savvy leader in place, capability building for the workforce of the future (engaging integrator and technology-innovation manager who would support employees to bridge the gap between traditional and digital parts of business - they will translate and integrate new digital methods and processes into the existing way of working), empowering employees to work in new ways, continuous upgrading of tools used for collaboration and process digitalization, frequent communication via traditional and digital ways (McKinsey&Company, 2018). In a new remote and hybrid work model, relationships between manager and employees have become more important than ever, while through manager employee validate employer as well (Kropp, B. et al., 2022). Important criteria for successful remote work and productive teams are (Neeley, Tsedal, 2021): delivering results, individual growth facilitation, building team cohesion.

2.2 Employees Challenges and Benefits

Main benefits for employers/companies are (Florentine, Sharon, 2020): employee engagement and productivity increase, retention, morale, expanding talent pool, cost reduction. BCG survey from 2020 shows that companies have the following benefits from hybrid/remote work model: cost saving, talent attractiveness, productive workforce. Applying hybrid/remote working model allows companies to employ experts from any geography, allowing them to choose the best talents, but also serves as a way to expand their business in new geographies. What comes across all research and surveys investigated as common benefits of hybrid/remote working for employees and employers are following:

- work/life balance for employees,
- more personal and family time for employees,
- cost reduction for both employees and employers,
- increased flexibility and autonomy for employees,
- development of digital competences for employees,
- higher performance in well managed hybrid/remote work for both employees and employers,
- improved work satisfaction,
- expanding talent pool for employers.

BCG study from 2020 shows that two major employees benefit in hybrid/remote work are work-life balance (mainly related to less commute time which allows extra time to be spent for their personal needs and family) and cost reduction of living costs (mainly due to lower cost of commute, cost of meals, etc.). OWL Labs study from 2021 shows that more than 80% of respondents considers that main benefits of remote working are more time with their family 86%, that they are happier 84%, better able to handle work-life balance 83% and that hybrid work is better for their mental health 82% (OWL Labs, 2021). Main positive effects of remote/hybrid work according to European parliament study (Samek Ladovici, M. et.al., 2021) are: better balance of home and work life, more time to spend with a family 79%, more time for personal activities 66% and activities in local community 47%. Buffer – State of Remote report from 2022 shows that main disadvantages of working remotely are following: overwork and no clear boundaries between work and home time, too much of virtual communication and collaboration, social isolation, lack of support, inadequate equipment, career development challenges, inequalities between off site and on-site workers, difficulty focusing especially in fully remote work model, staying motivated especially in fully remote work model, feeling lonely and appropriate communication with managers or team members. The findings of the resarch made by (Van Zoonen, Ward et al. 2023) show that the negative relationship between isolation and trust is primarily established by reduced organizational communication quality and interpersonal information sharing quality compared to the frequency of communication. What comes across all research and surveys investigated as common challenges of hybrid/remote working for employees and employers are following:

- sense of loneliness and isolation on employees' side, as well as lack of socialization with their colleagues and losing team spirit,
- no clear boundaries between work and home time,
- too much of virtual communication,
- difficulty focusing on fully remote work model,
- staying motivated especially in fully remote work model,
- new colleagues onboarding,
- appropriate communication with managers and team members.

2.3 Management and Leadership Context of Challenges

In the "Neuroscience of Trust", published in Harvard Business Review 2017, after long years of research, university professor Paul Zak wrote: "*Compared with people at low-trust companies, people at high-trust companies report: 74% less stress, 106% more energy at work, 50% higher productivity, 13% fewer sick days, 76% more engagement, 29% more satisfaction with their lives, 40% less burnout.*" (Harvard Business Organisation, 2022). The best leaders will create a culture of flexibility and wellbeing of their employees. Culture is the result of how a company works and operates (Capgemini Research Institute and Capgemini Invent, 2021). Culture is composed of the collective experience of leadership and employees, what they believe in and what they value (Capgemini Research Institute and Capgemini Invent, 2021). Capgemini Research Institute defines digital culture based on seven key attributes: agility and flexibility, collaboration, customer centricity, data-driven decision making, digital-first mindset, innovation, and open culture. Top leaders play vital role in adopting culture playing a role-modelling desired mindset and behaviour. High trust is possible to achieve only if it's mutual, two-way street (Harvard Business Organisation, 2022). A leader's job is to give freedom to their employees, but also clarity what they need to do, including why and how their contribution matter supported by information they need to accomplish the task (Harvard Business Organisation, 2022). Leaders must give the ownership and credit to employees for their results, however, be there as a support if help is needed. True leadership is about building satisfied employee with high productivity for growing company results, and it's on leader to nudge employee to have reasonable balance between work and life. Company culture growing on trust is mandatory for successful hybrid/remote work. And when things are changing rapidly, there is no time to lead under known chain command (Meyer, P. P., 2020). Exactly how to approach in remote and hybrid work should functionate. Under big trust between e-leaders and employees, every employee should be a small individual entrepreneur of his jobs-to-be-done, with end-to-end mindset. E-leader should encourage such mindset and connect individuals into a team, under clarified decision rights.

For productive remote and hybrid team, there are tactics for successful leading remote workers (Neeley, Tsedal, 2021):

a) minimize location differences - leaders should not make differences in leading people who are working on a different location, no matter if locations are geographically different, or on site or off site working,
b) strengths, not status emphasize - leaders should emphasize individual strengths, not working status for remote workers, key for leaders is to promote group-level identity and stress out that each individual represents one part of a team (chain link),
c) create structure – predictability is important for remote workers, leaders should give clear direction to each team member with direct communication and clear job description and responsibilities,
d) give feedback – in order to motivate team members, leaders must give appropriate and constructive feedback as an individual support for career development no matter if working in office or remotely,
e) promote engagement, but not avoid conflicts – bonding of remote and hybrid workers through informal parts of the meetings in order to encourage engagement and fruitful discussion.

Taking all of the challenges employers and employees are facing in full remote work, hybrid work model seems to be more widely adopted by majority of employers, where applicable (Ancillo, Antonio de Lucas et al., 2023). It is considered that the optimum balance could be around two to three days of remote work a week (Samek Ladovici, M. et.al., 2021) giving possibility to the worker to organize work to maximize benefits of both options (in office and home office). Many companies are adopting much more aggressive hybrid models where majority of work is conducted remotely, while occasional physical in-office presence is used with intention to keep close relationships between co-workers and to nurture employees' sense of loyalty and belonging .

Hybrid/remote work model requires defining of measurable targets (Key Performance Indicators - KPIs) for achieving its purpose which will provide clarity about the goals and boost engagement of employees to achieve targets. The Performance Management must be adapted to a new way of working through (Ferreira., José et al., 2020):

a) Trust-based management – as a first enabler for successful hybrid/remote work, means shifting towards measuring workers' output (e.g., achievement of settled KPIs and quality and timeliness of the work). It's important that teams and leaders, no matter working remotely or onsite, share information, and focus on successful results delivery. They should behave in the best interest of the team, make themselves available for each other, ensure that everyone feel included, share difficult jobs, and show cultural sensitivity.
b) KPIs and OKRs (Objectives and Key Results) – companies should use measurable targets, which should be equally accessible and fair for both remote and onsite employees, by providing tools and information that enable reaching targets, and provide reporting for regular tracking of performance. OKRs are used for goals' setting and KPIs are used for tracking goal performance (Asana, 2022).
c) Coaching and feedback – important in driving employee's individual development and managing employee expectation, which could be very challenging in remote work model.

New flexible working models require new practices and routines which could improve efficiency in daily tasks. One of the proposed work practices is agile practice. Survey showed that companies which implemented at least two agile practices of named below, reported 40% increase in productivity (Ferreira., José et al., 2020). Agile practices proposed for hybrid/remote teams are (Ferreira., José et al., 2020): daily "stand up" sessions – to review backlog and align on process and priorities, structure work of the day, and discuss potential roadblocks, asynchronous communication – by using tools like Teams, Slack to ensure information flow, weekly team meetings – to review each week deliverables and to define upcoming priorities, demo days – sharing of results and collecting feedback from stakeholders.

3. CONDUCTED RESEARCH

3.1 Methodology

METRO AG (umbrella brand name) last year turnover was 25,6 billion € with strong investments worldwide of 627 million €. METRO C&C Croatia d.o.o. is a subsidiary of METRO AG, in Croatia operated since 2001 with headquarters located in Zagreb, and currently 172 employees, 10 wholesale stores in major cities in Croatia, strong Food Service Distribution with 24% share, and offering digital services through DISH platform to HoReCa customers, and is a strong and reliable B2B partner to HoReCa (hotels, restaurants, cafes, and catering), Trader (traditional trade shops, kiosks, bakeries) and SCO customers (Small companies, offices, and services). METRO C&C Croatia case study was conducted with the primary method of data collection research, which was through a survey questionnaire sent to HQ (people who work at the headquarters of an organization) aemployees (with a response rate of 88%), individual one-on-one interviews, including Board members, Senior managers, Middle managers and other "white collars" employees, as well as internal data received by authorized HR employees. One-on-one interviews were conducted with Board members, Senior managers, Middle managers, and white collars workers in HQ within different departments for fair coverage of the whole company. A Google Form questionnaire was conducted to examine wider HQ employees' responses of their expectations and preferences of work in the company located in Croatia (only in the office, only remote or hybrid), status of workplace disruption in the company, awareness of employees' expectations toward their workplace and the way they respond to it. METRO C&C Croatia belongs to retail/wholesale industry. This industry doesn't have the highest possibility for hybrid/remote work, as the nature of this sector is that there will always be a need to have employees in stores and warehouses (Retail Times, 2022). With new technologies emerging, rising trend of online shopping, staffing shortages etc. executives must consider the given situation and consider alternative ways of working and providing service.

COVID-19, like in many other companies worldwide, was the trigger to workplace disruption in METRO C&C Croatia. Before the pandemic, METRO C&C Croatia operated on-site work model. With the purpose to safeguard employee's health being, both in HQ and stores and warehouses, and to secure main business purpose to serve customers, METRO C&C Croatia leadership and crisis committee, following corporate guidance, decided to change the work model for HQ employees. At the beginning of the pandemic, it applied remote work model to maximum extent, with few exceptions with which administrative roles were able to operate due to legislation rules (e.g., reception was kept for scanning documents like invoices, contracts, etc.). Once the situation with the pandemic calmed down with vaccinations, the hybrid work model with limited number of people per room was chosen. Returning to the office full time was on the table, however, the survey among employees has shown great satisfaction with flexible working possibility and higher productivity of employees, confirmed by stores and warehouse employees who were continuously on-site working. At the beginning of this research, 3 hypotheses have been set:

H1 – Hybrid/remote work is currently the preferred work model for white collars employees. (confirmed)

Primary research method results, based on METRO C&C Croatia case study show that 92,5% of HQ employees preferred way of work is hybrid/remote, store and warehouse employees work onsite full time, as their job is not able to be done remotely according to interview statements of the Board members and decision of the Board that only HQ employees can work hybrid/remote work model. Hypothesis 1 is confirmed. Hybrid/remote work is preferred way of work for white collars.

H2 - Companies are adopting and applying hybrid/remote way of working.

Primary research method results, based on METRO C&C Croatia case study, show that 92,5% of HQ employees works hybrid/remotely, out of which 79,2% works 2 days per week remotely (home office), and 15,1% works 1 day per week remotely (home office). Hypothesis 2 is confirmed. Companies are adopting and applying hybrid/remote way of working.

H3 - Companies are coming up with new work process and tools to support the hybrid/remote work model.

Primary research method results, based on METRO C&C Croatia case study show that 98% of HQ respondents to the survey stated they are very satisfied with software's available for hybrid/remote work, 100% of HQ respondents to survey stated that technology they have for their hybrid/remote work like laptops, mobile phones, internet connections… is suitable for hybrid/remote work. Hypothesis 3 is confirmed. Companies are coming up with new processes and tools to support hybrid/remote work.

3.2 Results and Discussion

METRO C&C Croatia conducted survey among HQ employees (N=172), a few months after the pandemic started and when employees were home-office working for majority of the time, with the response rate of 59% (N=102) which has been the typical response rate for other METRO C&C Croatia surveys too. Results of conducted survey shown that 95% of respondents declared they are satisfied with new work model (54% "completely satisfied" and 41% "partially satisfied"), 75% of employees declared their level of productivity is "completely satisfied", 40% declared "completely satisfied" with their work-life balance, and 37% of employees declared their work-life balance in home-office working model as "partially satisfied" as shown in Figure 1 below. It's clear from the survey results how the biggest employees struggle in remote work is work-life balance.

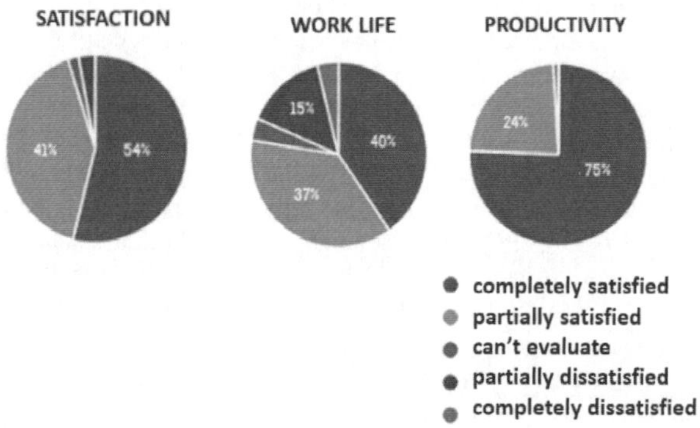

Figure 1. METRO C&C CROATIA survey results on employee productivity, satisfaction and work life balance

Survey has shown very positive results regarding communication with line manager and quality of feedback received from line manager, with 85% employees "completely satisfied" with communication with line manager and 82% of employees declared as "completely satisfied" with the feedback from line manager, in Figure 2.

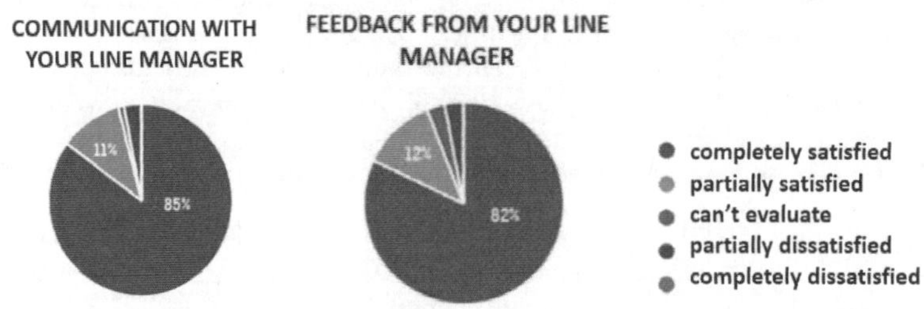

Figure 2. METRO C&C CROATIA survey results on communication and feedback from line manager

METRO C&C CROATIA leadership decided to keep the hybrid model as a new working model in the HQ, with fixed 3 days in the office (Monday-Wednesday) and 2 days home office (Thursday-Friday), and to keep stores and warehouse on-site working model. As research among Metro C&C Croatia HQ employees shown, 92,5% of employees prefer to work in a hybrid/remote work model which means they see many benefits working at least a few days at their home office. Main benefits, according to questionnaire among Metro C&C Croatia HQ employees are:

- Less commute time and more time for work and private needs (40% declared this benefit as the top 1 benefit).
- Financially better position (21% declared this priority as the top 2 benefit).
- Flexibility to choose location where to work (21% declared this benefit as top 2 benefit).
- More time for family (19% declared this benefit as top 2 benefit).
- Better focus on work (21% declared this benefit as top 3 benefit).

Additionally, 60,4% (32 employees) declared they are more productive at home-office, for 50% of those employees it is important and very important that the company co-finance home utility costs, 50% of them said they would probably or certainly leave the company if they did not offer hybrid work model, and >95% of them are age 41-56. One-on-one interviews with Board Members/Senior Managers and Middle Managers and other white collars employees working in HQ gave us an even deeper understanding about the benefits that hybrid model is offering, both for company and for employees. More than 90% of all interviewed express that hybrid work model is much better work option than fully remote, due to importance of onsite socialization, face-to-face meetings and discussions which are not able to be done productively via online meetings. Top benefits, according to interviews are less commute time and more time for work and private needs, higher productivity (in hybrid work model), higher satisfaction of employees and higher engagement (in hybrid work model), and higher availability of co-workers regardless of their distant location. Detailed overview of benefits collected during one-on-one interviews is shown in the Table 1 below.

Table 1. METRO C&C CROATIA: Benefits of working in a hybrid work model

	BENEFITS	
One-on-one interviews	**For company**	**For employees**
Board members/ Senior managers	– higher employees' satisfaction – higher productivity in >90% of employees – potential utility cost reduction (if some period completely closed HQ offices) – lower administrative and equipment costs (less printing paper and cartridges, less fixed printers - centralized printers installed vs printer in each room) – less postponing of the meetings (employees have chance to work from home when not feeling well, but able to work) – better competitiveness on the workforce market	– flexibility in 2 days of home office possibility – more time for their families – more time for themselves – less commute time – lower commute cost – healthier living – possibilities – less stress – higher productivity
Middle managers/ other white collars	– higher productivity – better focus on work – more time available for colleagues and managers – higher satisfaction of employees – happier and more engaged employees	– no commute time and cost – more time for work and private matters – higher productivity – less stress – better concentration due to less distraction during home office work (depends on situation at home) – better focus on work – fewer sick leaves – more satisfied and happier employees

Trust between Board Members, Senior Managers, Middle Managers, and other employees in Metro C&C Croatia HQ seems to be on a very high level, based on surveys and interviews conducted.

Based on survey results, Metro C&C Croatia has built a company culture on trust above authority, as prerequisite for hybrid/remote work. Employees were asked to grade their trust between managers and employees from no trust (1) to very high trust (5). 94% (50 employees) stated that there is a very high and high level of trust between managers and employees. The questionnaire survey among METRO C&C Croatia HQ employees has shown that the top 3 struggles are:

1. Overtime work (no boundaries between work and private time when home-office work) (26% said this is top 1 struggle). However, the employer is not able to apply time registration in home office and longer working time might occur also due to longer usage of private breaks or different distribution of working time.
2. Top 2 struggle (for 25%) was "in-office meetings (face-to-face) on home-office days (Thursday-Friday)".
3. Technical equipment not adequate or often technical problems was top 3 struggle (for 23% of employees).

Among additional struggles, employees stated (21% of them) that they feel inequality between in-office workers (the one that comes to the office in a day when home office is allowed) and the ones that stays at home office 2 days a week, also some of them (18%) struggle within line manager communication, 15% employees feel loneliness and some of them (8%) struggle with teamwork.

Table 2. METRO C&C CROATIA: struggles of working in a hybrid work model

STRUGGLES		
One-on-one interviews	**For company**	**For employees**
Board members/ Senior managers	– leading and maintaining of the meetings in hybrid mode (part of the team face-to-face and part of the team online) could be quite challenging and not so efficient – focus of the team is lower via Microsoft Teams meeting vs live meetings	– too many Microsoft Teams meetings without breaks – no clear boundaries between work and life lives during remote work – missing socialization and potential drop of motivation during remote work (feeling isolated) – lower concentration (depends on home situation) – self-neglecting (always in pyjama…)
Middle managers/ other white collars	– potentially exhausted employees if meetings are not properly managed – leaving the company if company doesn't offer flexible work model – harder onboarding if not offer flexible work	– too many Teams meetings without breaks – work-life balance disturbed (no clear boundaries between work and life in remote work) – meeting organized onsite during home-office days – meetings organized or mails sent by managers after regular work time – loneliness if fully remote work, in hybrid model not felt loneliness due to onsite socialization – in some cases, communication with managers could be improved (regular weekly meetings needed)

Questionnaire survey among METRO C&C Croatia HQ employees shows that communication and collaboration tool is very well accepted and helping them in daily collaboration with their line manager, their direct teams, and cross functional teams. According to one-on-one interviews, and internal data shared, many processes were changed across company and departments. Some of the examples are as follows:

- "Home office working rules and regulations" have been issued.
- Before "must printing files" were changed with e-documentation version (e.g., discount confirmation…) except for the Contracts and invoices that, according to Croatian Law, still must be signed and archived physically.
- Usage of OneDrive for sharing documentation, and real-time work on the same document within different departments has been introduced.
- MPOWER – software for digital vacation approvals, performance reviews, goal settings and development planning has been implemented.
- TalentLyft – talent search software for new employee selection process has been implemented.

- Tips and Tricks for on-line learning from home office prepared by METRO C&C HR team has been introduced.
- Wellbeing of hybrid/remote recommendations have been introduced.
- Mental health online booklet and body exercise guide for home office workers and their families has been issued by METRO C&C Croatia HR department.

Primary research method results, based on METRO C&C Croatia case study, are following: 92,5% of HQ employees preferred way of work is hybrid/remote, store and warehouse employees work onsite full time, as their job is not able to be done remotely according to interview statements of the Board members and decision of the Board that only HQ employees can work hybrid/remote work model. Primary research method results, based on METRO C&C Croatia case study, are following:

- 98% of HQ respondents to survey stated they are very satisfied with software's available for hybrid/remote work.
- 100% of HQ respondents to survey stated that technology they have for their hybrid/remote work like laptops, mobile phones, internet connections… is suitable for hybrid/remote work.
- METRO C&C Croatia is using following tools: Microsoft Office 365 platform with all available application, especially Microsoft Teams which is in everyday use, but also, Planner for project management, Viva Engage for international engagement, and many custom-built apps like Poll, Deskbooker B9, METRO.digital Chatbot, MAG HR Chatbot, I love learning, METRO Express…
- METRO C&C Croatia implemented many new processes for hybrid/remote work like "Home office working rules and regulations", e-documentation, OneDrive for sharing documentation and real-time work on the same documents/projects, new process for digital vacation approvals, performance reviews, goal settings and development planning via MPOWER, new process for talent search via TalentLyft software.

4. CONCLUSION

Research results have been analysed according to the sample of one middle-sized company in Croatia and probably have some limitation for wider conclusion but can serve as a solid basis for some other research. Nevertheless, the results clearly demonstrate that observed company has come up with new processes and tools to support hybrid/remote work. Hybrid work model, which is a blend of onsite and home-office work, and which this company is applying, seems to bring the best out of both models. Onsite work offer socialization, face-to-face meetings, direct communication of managers with direct reports for performance reviews and individual development, while home-office work offers focus on work and higher productivity, reduction in commute time and cost, better lifestyle, and happier employees. Hybrid work model became the mainstream in the observed company. Case study has shown that company can be transformed from fully onsite to hybrid work model organization, digitized certain processes to enable hybrid/remote working, invested into technology and digital tools, built company culture based on trust, settled clear OKRs and KPI's including feedback policy with achieving the best results ever in the current year. This research shows that the company has successfully adopted hybrid/remote work model, for both employees and company benefit from it. Observed example could also be applicable to other SMEs and that is something left for the further research.

REFERENCES

Ancillo, Antonio de Lucas; Gavrila Gavrila, Sorin; Nunez, Maria Teresa del Val, 2023. Workplace change within the COVID-19 context: The new (next) normal. *Technological Forecasting & Social Change* 194 (2023) 122673, 16 p. Available at https://doi.org/10.1016/j.techfore.2023.122673

Asana, 2022. *OKR vs. KPI: Which goal-setting framework is better?*: https://asana.com/resources/okr-vs-kpi

Capgemini Research Institute and Capgemini Invent, 2021. *Digital Mastery. Capgemini Research Institute and Capgemini Invent*. https://www.capgemini.com/wp-content/uploads/2021/02/Digital-Mastery-Report-1-4-1.pdf

Chatterjeea, Sheshadri; Chaudhurib, Ranjan; Vrontisc, Demetris; Giovando, Guido 2023. Digital workplace and organization performance: Moderating role of digital leadership capability. *Journal of Innovation & Knowledge* Volume 8, Issue 1, January–March 2023, 100334, Available at https://doi.org/10.1016/j.jik.2023.100334

Drucker, Peter F. 2020. *Practical leadership* (ISBN: 978-1-63369-931-1 ed.). (e. 978-1-63369-932-8, Ed.) Boston, Massachusetts: Harvard Business Review Press

Ferreira, José; Claver, Pablo; Pereira, Pedro and Thomaz, Sebastião 2020. *Remote working and the platform of the future.* The path to remote-working maturity: https://www.bcg.com/publications/2020/the-path-to-remote-working-maturity

Florentine, Sharon, 2020. *The keys to succesful remote work strategy.* https://www.cio.com/article/230829/the-keys-to-a-successful-remote-work-strategy.html: https://www.cio.com/article/230829/the-keys-to-a-successful-remote-work-strategy.html

Harvard Business Organisation, 2022: *Good Leadership: It all starts with trust*: https://www.harvardbusiness.org/good-leadership-it-all-starts-with-trust/

Kamouri, Anita; Lister, Kate, 2020. *Global Work-From-Home Experience Survey.* Globalworkplaceanalytics.com: https://globalworkplaceanalytics.com/global-work-from-home-experience survey#:~:text=About%20the%20Global%20Work-from-Home%20Experience%20Survey%20Nearly%203%2C000,the%20largest%20global%20post-COVID%20employee%20survey%20to%20date

Kropp, B., & McRae, E. R, 2022. *11 Trends that will shape work in 2022 and beyond.* Harvard Business Review: https://hbr.org/2022/01/11-trends-that-will-shape-work-in-2022-and-beyond

Lister, Kate; Kamouri, Dr. Anita, 2020. *Global work-from-home Experience Survey.* Global Workplace Analytics: https://globalworkplaceanalytics.com/global-work-from-home-experience-survey

Lister, K., & Harnish, T., 2019. *Telework and its effects in the United States.* In J. C. Messenger, Telework in the 21st Century (p. 128). Northampton, Massachusetts; Geneva: Edward Elgar Publishing, Inc., 2019; International Labor Office. doi:10.4337/9781789903751

McKinsey & Company, 2018. *Unlocking success in digital transformation.* McKinsey&Company. https://www.mckinsey.com/business-functions/people-and-organizational-performance/our-insights/unlocking-success-in-digital-transformations

Meyer, P. P., 2020. *Leading through VUCA* (Volatility, Uncertainity, Complexity, Ambiguity). https://medium.com/@pamelameyerphd/leading-through-vuca-volatility-uncertainty-complexity-and-ambiguity-769088de8815

Neeley, Tsedal, 2021. *Remote work revolution: Succeeding from anywhere.* New York: Harper Collins Publishers

OWL Labs, 2021. *State of remote work*,: https://owllabs.com/state-of-remote-work/2021/

Samek Ladovici, M. European Parliament, 2021. *The impact of teleworking and digital work on workers and society* (S. a. Policy Department for Economic, Ed.): doi:10.2861/691994 https://www.europarl.europa.eu/RegData/etudes/STUD/2021/662904/IPOL_STU(2021)662904_EN.pdf

Van Zoonen, Ward; Sivunen, Anu E.; Blomqvist, Kirsimarja, 2023. Out of sight – Out of trust? An analysis of the mediating role of communication frequency and quality in the relationship between workplace isolation and trust. *European Management Journal,* pp. 12, Available at https://doi.org/10.1016/j.emj.2023.04.006

Venkatachalam, Parvathy; Mishra, Rajhans, 2023. Future of workplace design from a socio-technical perspective. *IIMB Management Review*, Vol. 35, Issue 1, March 2023, p. 1-16, Available at https://doi.org/10.1016/j.iimb.2022.12.006

Zahidi, S., 2020. World Economic Forum. Retrieved from How COVID-19 can change work for the better - *the World Economic Forum's Saadia Zahidi on the Jobs Reset Summit*, https://www.weforum.org/agenda/2020/10/jobs-reset-saadia-zahidi/

DEVELOPMENT OF AN INTEROPERABILITY PLATFORM FOR INFORMATION SYSTEMS IN THE EHEALTH DOMAIN, FOR THE PORTUGUESE HEALTH SYSTEM

José D. Pereira, Miguel A. Brito and Ricardo J. Machado
Centro Algoritmi, University of Minho, Campus de Azurém 4800-058 Guimarães, Portugal
CCG/ZGDV Institute Campus de Azurém, University of Minho, 4800-058 Guimarães, Portugal

ABSTRACT

EHealth has emerged as a new field at the intersection of medical informatics, public health and business, referring to healthcare services and information provided or enhanced through information technologies. Today, it is very focused on improving the interoperability between Electronic Health Records (EHRs), Electronic Medical Records (EMRs), and between them and the applications installed on patients' devices (smartphones and IoT), with the development of new platforms. Developments of this type imply the use of standards such as Health Level 7 Fast Healthcare Interoperability Resource (HL7 FHIR), openEHR, and Clinical Data Interchange Standards Consortium (CDISC), which guarantee that health data transmission policies are respected.

The work presented in this article is developed within the framework of the Health for Portugal agenda (HfPT), of the "Agendas Mobilizadoras para a Inovação Empresarial" promoted by the "Plano de Recuperação e Resiliência portugês" (PRR), in collaboration with the Centro de Computação Gráfica (CCG/ZGDV), which is the organization responsible for development of this platform. This article is based on the initial phase of the development of a platform that allows the integration of different information systems, to solve the lack of interoperability of the EHRs of health stakeholders in Portugal and also to reduce the fragmentation of the HIS of the Portuguese SNS.

KEYWORDS

Ehealth, Interoperability, Architecture, Health Information Systems

1. INTRODUCTION

Today, eHealth is very focused on the interaction between patients and healthcare professionals, through improvements in EHRs and the development of new platforms that allow sharing and receiving health data directly from applications on patients' devices. To ensure interoperability, respecting privacy policies, and health data integrity, the development of this type of platform implies the use of standards such as HL7 FHIR, openEHR, and CDISC.

Within the scope of the HfPT agenda, promoted by the PRR, this work will be based on the development of a platform that allows the integration of different information systems, which guarantees compliance with standards, norms, and benchmarks for health data. In Portugal, there are several public and private health institutions that have their own information system operating in isolation, so that when a patient is transferred between hospitals, he needs to carry documentation with his clinical data in paper format and the medical examinations performed at the previous health institution are repeated at the hospital where the patient arrives.

Another situation that occurs, due to the lack of interoperability, is that the health information system (HIS) of the Portuguese SNS is also not capable of efficiently and effectively integrating the clinical information that is produced in private clinics and some public hospitals, which obliges the users' family doctors to manually update this system whenever there is new data. This is a problem that makes health services inefficient, increases the probability of medical errors, implies unnecessary expenses with repeating exams and forces family doctors to spend time on these tasks.

In this article the beginning of the development of a solution for this problem is documented. A literature review is carried out on interoperability platforms in the eHealth domain, which encompasses infrastructure, technologies and the application of standards related to health data transmission policies. It will serve as support for decisions on the architecture of the common platform, which allows interoperability between public and private hospitals, health centers, clinics, medical examination centers, the patient himself, the SNS, and in the future, other countries of the European community.

The work is based on a search carried out in the Scopus database, with the string "KEY (interoperability AND information AND systems AND eHealth)" and filtering of documents of the type "article" and "conference paper". With the aim of understanding the main theme of the project in a broader way and perceiving architectural, technological and regulatory trends, in order to make informed design decisions.

2. E-HEALTH

The designation eHealth emerged in 2001 by Eysenbach, designating it as an emerging field at the intersection of medical informatics, public health and business, referring to health services and information provided or enhanced through the Internet and related technologies (Eysenbach, 2001). It includes medical informatics, telemedicine, health telematics and health ICTs. It encompasses the interaction between patients and healthcare professionals and is seen as the most important revolution in healthcare since the advent of modern medicine or hygiene (Dinevski et al., 2010). The World Health Organization also defines eHealth as the safe and cost-effective use of information and communication technologies in support of health and health-related areas.

In today's society and also in eHealth, multichannel platforms, where patients are at the center, are gaining ground, but in this domain it is still systems such as EMR and EHR that play a central role. In Portugal, there is a scenario of lack of interoperability in the eHealth domain, related to the fact that the various health institutions have EHRs and EMRs that operate completely independently of each other, causing difficulties for patients, professionals, related industries and also to the main entity responsible for health in the country, the SNS.

3. E-HEALTH INTEROPERABILITY

Heterogeneous enterprise applications, at either the business or the manufacturing level, either within a single enterprise or among networked enterprises, need to share information and cooperate in order to optimize their performance. This information may be stored, processed, and communicated in different ways by different applications. The problem of managing heterogeneous information coming from different systems is referred to as the interoperability problem (Jaskó et al., 2020).

3.1 Developments on E-Health Interoperability Platforms

The Enterprise Electronic Cloud-Based Health Record System (E2CHRS) project for recording, retrieving, archiving, and updating medical records was presented by (Adebayo et al., 2014). It uses a unified cloud database for all hospitals, middleware that provides a common platform for all EHR systems in remote hospitals, an authentication server manages access to the system, and an e-web portal serves as the front end and connects the app to the cloud. This type of approach also facilitates the interoperability of HER systems with the pharmacy's computer systems and even with the applications of the users' electronic devices.
The availability of such a platform would later allow the development or integration of a solution such as the one proposed by (Pereira et al., 2018), which consists of a citizen-centered Electronic Prescription system. Where the patient enjoys a simplified and comfortable electronic prescription system, which will allow him to buy the medicines through an application, later the medicines will be delivered at home. As it is integrated with the RES, the physician simply writes down the prescription, normally, and the patient accesses it through this multichannel application.

In their article, (Lemus-Zúñiga et al., 2022) proposed a health monitoring system based on the integration of rapid prototyping hardware and interoperable software to build a system capable of transmitting biomedical data to health professionals. The system allows IoT devices and other systems to be interoperable, using standards for data exchange such as FHIR and the universAAL (universal Ambient Assisted Living) platform. They presented a new layer based on rapid prototyping sensors so as not to limit data acquisition with specific equipment suppliers. In this way, adopters or developers of this type of system can build the system to coexist with existing systems/platforms/services, as long as interoperability is guaranteed with the use of FHIR standards and the universAAL platform.

Several authors have already described attempts to incorporate blockchain technology and public cloud in this type of platforms, also (Carter1 et al., 2019) proposed a health data sharing platform that includes security through multilayer cryptography, data storage through of Amazon Web Services and data transfer using the Ethereum blockchain. For these authors, the combination of these technologies is promising, but they still do not consider it capable of solving the immediate problems in the eHealth area, as it will take some time to adapt these technologies to the standards and government policies required in the eHealth domain, such as the HITECH law applied in the United States.

Argentina's National Digital Health Strategy 2018-2024 aims to reduce quality gaps in healthcare through the implementation of interoperable information systems. In 2019, they launched the National Digital Health Network, using a service bus technology infrastructure to connect different systems across the country, which will enable new eHealth services and integrate existing ones. In an article by (Rizzato Lede et al., 2020), the work developed by the Ministry of Modernization of Argentina, together with the Ministry of Health, was presented, which consists of the Citizen Health Digital Portal (Mi Argentina / Mi Salud) for personal access of Argentine users. This allows them to access your information and change your consent to exchange health information for the National Digital Health Network. It has tools such as the Electronic Vaccination Card and intend to include the patient's international summary, laboratory results, and electronic prescriptions.

The work (Weber-Jahnke & Obry, 2012) presents a mechanism for peer-to-peer interoperable systems that restricts access to confidential medical data based on defined consent policies, allowing policies to be overridden when necessary. The project includes monitoring, audits, and is open source, designated CDAShip and available on SourceForge.

An interoperability solution for sharing data between heterogeneous data sources was presented by (Masud et al., 2012). Includes a metadata management framework for multimedia medical content, eg X-rays or ECG. The framework identifies resources, generates and represents metadata, and produces identifiers for multimedia content.

Also to achieve interoperability in the medical field, (Mazouz et al., 2017) proposed a system to integrate heterogeneous HIS using schema matching technique. In order to automatically find semantic correspondence between XML schemas to allow their integration into information systems. To achieve this, they used Metathesaurus in the Unified Medical Language System (UMLS).

A semi-distributed system architecture approach was proposed by (Fragidis et al., 2016) for an integrated national electronic health record system, incorporating the advantages of centralized architecture and distributed architecture. The architecture provides robust interoperability without healthcare providers changing their on-premises EHR systems. It is a pragmatic approach that takes into account the characteristics of the Greek national health system, together with the data communication network infrastructure of the national public administration, to achieve EHR integration with low implementation costs.

In Brazil, an architecture called SMART was developed to integrate the different telehealth platforms developed independently by the 9 Brazilian telehealth centers. The architecture aims to standardize information so that the Ministry of Health can monitor and evaluate the results of Telehealth actions (de Paiva et al., 2018). It has four components: a web tool for data manipulation, a web service to receive production data from the center, a component to convert the received data into decision support data, and a component that collects data from external sources to compose the data warehouse.

In an article published by (Id & Kim, 2018), an eHealth system with the Semantic Sensor Network (SSN) was presented, which is composed of an eHealth device, client, and server, which aims to solve the interoperability problem of the device using IETF YANG to model eHealth data semantics to represent sensor data. The developed ontology builds metamodels of eHealth sensors, for data that support different formats. The solution generates the broadcast message based on the semantic model, through a message generator. It includes YANG-to-JSON translator and eHealth client to present the meaningful SSN information to the user. The proposed approach assists in the self-configuration of eHealth sensors and in querying the sensor network, supporting semantic interoperability for the eHealth system.

In order to meet the urgent need for interoperability in EHR to improve healthcare delivery, the study by (Adel et al., 2017) recommends fuzzy ontology as an intelligent information system solution. These authors recommend the fuzzy ontology, involving current technology, such as a fuzzy database or Fuzzy Markup Language (FML).

Schiza et al. provided an article that articulates a framework for deriving a national health system based on interoperable EHRs. This eHealth ecosystem has been proposed for Cyprus and its most important components are the Central Citizen Data Warehouse, which uniquely links each citizen to the Central eGovernment Portal, the country's cloud-integrated EHR system is regulated by NeHA, ensuring interoperability, confidentiality and security. According to (Schiza et al., 2019), adopting EU principles following the relevant guidelines for a fully interoperable healthcare system is not simple. At social, political, and scientific levels, incorporating and developing modern eHealth IT technologies requires a major effort from all stakeholders.

The Evgeniy et. al., presented an article whose main objective is to investigate the most used definitions of clinical data structures and propose a model to adapt them to CEN 13606. More specifically, to develop a common methodology to transform existing information into XML documents, exchanged in the domain of health, using the CEN 13606 Archetype model. This research effort, combining medical and computer science professionals, resulted in a UML model. To develop the model, clinical documents generated during the provision of general medicine services were used. First, the global elements of the XML Schema definitions document (XSD) are converted into Java classes, and then these classes are used to extend the CEN 13606 reference model, preserving the original structure of the main classes, placing their relationships in the international standard. According to these authors, the model can support the construction of a prototype of an information system that allows semantic interoperability between clients in terms of RESTful web services provided by a NoSQL database (Evgeniy et al., 2020).

In any case, to achieve interoperable eHealth systems, platforms or applications that integrate EHRs, EMRs and even patient or wearable applications, interoperability standards are needed that meet the data protection standards and policies required by healthcare institutions and governments.

4. INTEROPERABILITY HEALTH STANDARDS

Reviewing the literature, some standards specially developed to be applied in HIS architectures and development are found.

With the aim of providing a comprehensive framework and standards for the exchange, integration, sharing, and retrieval of electronic health information Health Level Seven International (HL7) was born. It has members in over 50 countries, including Portugal, which could be relevant for this project. Developed the FHIR standard, intended to facilitate the exchange of health information between health professionals, patients, caregivers, payers, researchers, and anyone else involved in the health ecosystem (HL7 International). In the same vein, openEHR also publishes technical standards, including interoperability, for EHR platforms. It includes architectural concepts such as patient-centric shared lifetime health record, future-proofing data, and supporting clinical processes (openEHR). Another organization that develops data exchange standards is the CDISC, its standards facilitate the sharing of structured data between different information systems. Its standards transform incompatible formats, inconsistent methodologies, and diverse perspectives into a framework for generating clinical research data, making them accessible (CDISC).

4.1 Use of Standards in E-Health Interoperability Platforms

The intention to develop modern eHealth platforms, combined with the availability of standards and technologies that allow it, gave rise to the evolution of various solutions capable of meeting the needs of patients, professionals, health institutions, and governments. In this context, (Malaquias & Filho, 2021), developed a middleware based on HL7 FHIR. Also (Calamai & Giarré, 2012) proposed a unified structure for the interoperability of patient summary (SP), electronic prescription (with the involvement of the HL7 Italia Group) and the Chronic Care Model (CCM), based on the HL7 Version 3 Clinical Document Architecture Release 2.

The authors (Nayak et al., 2019) proposed a wearable medical device called Electronic Bracelet (E-BRACE) that allows secure temporary access to a patient's EHR in an emergency, which uses the FHIR protocol to make a call-to-service request and obtain the patient's health record.

In 2019 came GIRLS, a gateway to EHR interoperability that uses both the openEHR and FHIR standards. These authors started from a system that uses the openEHR standard to another that uses the FHIR standard, proposing an interoperability layered architecture, capable of exchanging data between different standards. It includes an API that connects different systems through endpoints, an intermediary server that maps data from different standards to the FHIR standard and manages microservices for data access and exchange, but the semantic interoperability between HIS using openEHR and FHIR is still a challenge. The openEHR standard is more complete than FHIR resources and can fully represent medical knowledge. However, the FHIR is much less disruptive for existing vendors due to its adherence to RESTful design principles (Gomes et al., 2019).

A new standards-based IT system has been proposed to help seniors navigate a safe path with public transport in a smart city for their routine medical checkups during epidemic outbreaks like COVID-19. The system supports evidence-based decision making for the elderly, integrating health and public transport data (provided by transport systems and using APIs such as Google Maps or Open Street Map), resorting to medical interoperability standards (HL7 FHIR standards can be used in other domains, for example smart city) and compliance tests were performed applying the HL7 FHIR 3.0.1 specification through the official HL7 FHIR validator (Urbauer & Forjan, 2020).

Using experiments, (Krastev et al., 2022) concludes that representing EHR using ISO/EN 13606 archetypes in conjunction with openEHR makes it possible to achieve a higher level of interoperability between HIS than with HL7 standards. According to these authors, using the dual information model of ISO/EN 13606 and openEHR, the new open platform architectures demonstrate cost efficiency in management and high quality of health services.

To achieve semantic interoperability in the eHealth domain, a terminology server provides information systems with the concepts needed to encode information correctly. The Semantic Competence Center (SCC) of ELGA GmbH has proposed an FHIR server, for the Swedish healthcare system, which will work together with a Git-based terminology server ("TerminoloGit"), the HL7 FHIR IG editor and some tools that were developed internally (Kleinoscheg et al., 2022).

The national eHealth systems implemented in Serbia as "MojDoktor" (MyDoctor) and in Macedonia as "MojTermin" (MyAppointment) are based on the same integrated health information platform, but a change in architecture towards a more formal approach is required to ensure their interoperability. To achieve this goal, the study by (Atanasovski et al., 2018) defined the Platform Independent Model (PIM), to represent the relationships and execution semantics, which allowed the development of a model to formalize data transformation to improve the system interoperability capable of transforming system data into HL7 messages.

4.2 E-Health Interoperability in The European Union

The EU Digital Agenda emphasizes the individual patient's right to access their own medical data in a secure manner, supporting patient empowerment and greater involvement in their own care. eHealth has experienced dynamic development across the European Union in recent years and enjoys the support of the European Commission, which seeks to achieve interoperability of national health systems to facilitate free movement, hampered by observed differences between member states on legal regulations, cultural approaches and technological solutions. It is necessary to reconcile differences in data protection policies between member states so that interoperability between European national eHealth systems is achieved (Kautsch et al., 2016).

In this sense, the European Commission funded the epSOS project and the KONFIDO project, to solve these problems of cross-border exchange of health data (Staffa et al., 2018). This new platform that we are going to develop, for the Portuguese health services, has to be developed taking these developments into account, so that it is fully interoperable with the other existing eHealth platforms in the other European member states.

Again, with the aim of making cross-border eHealth services available, the European Commission promoted the eHealth Digital Services Infrastructure (eHDSI), also known as "MyHealth @ EU". In this context, a selection and compilation of data elements for cross-border care was carried out to serve the eHealth Network Guidelines on Patient Summary as a format for health data exchange in Europe.

To be part of this network, countries must adopt international standards to transmit the content of the Patient Summary in a structured and codified way, understandable by the Health Professional at national and cross-border level. The CEF eHDSI Interoperability Infrastructure has currently adopted HL7 CDA Level 3 and Level 1 as the standard for document exchange. However, Member States are free to nationally adopt the same HL7 standard or another standard, e.g. HL7 FHIR (European Commission, 2022).

5. CONCLUSIONS

The results of the research carried out consist of a set of developments that had the main objective of achieving interoperability in the eHealth domain. It was found that there is a solution for recording, retrieving, archiving and updating medical records, which uses a unified cloud database for all hospitals, with middleware that provides a common platform for all EHR, an authentication server, and a web portal. A service bus technology infrastructure was developed to connect different systems across an entire country. There is a solution for sharing data between heterogeneous sources, which includes a metadata management framework for multimedia medical content. Another system integrates heterogeneous HIS using the schema matching technique to automatically find semantic matches between XML schemas.

The use of different architectural approaches was verified, such as a semi-distributed system architecture, to achieve EHR integration with low implementation costs, another architecture to integrate the different independently developed telehealth platforms, which aims to standardize information for of monitoring and evaluation, and the use of fuzzy ontology as an intelligent health information system solution.

Proposals were presented for a citizen-centered Electronic Prescription system and also a health monitoring system based on the integration of rapid prototyping hardware and interoperable software to transmit biomedical data, using standards such as FHIR and the universAAL platform.

Authors claim that the combination of technologies such as blockchain and public cloud, in this type of platforms, is promising, but it is not yet adequate to the standards and government policies required in the eHealth domain.

About eHealth developments, taking into account the references and standards for the exchange of health data, a middleware based on HL7 FHIR was developed, a unified structure was proposed for the interoperability of the patient summary, electronic prescription, and the model of chronic care, based on HL7 Version 3. A wearable medical device called the Electronic Bracelet (E-BRACE) has been proposed that allows safe temporary access to the EHR in an emergency, which uses FHIR. GIRLS is a gateway to EHR interoperability that uses the openEHR and FHIR standards capable of exchanging data between the different standards, includes an API that connects systems through endpoints, an intermediary server that maps data from different standards to the FHIR standard and manages microservices for data access and exchange. They developed a standards-based system to navigate a safe path on public transport in a smart city, using APIs such as Google Maps or Open Street Map, using HL7 FHIR. A model was developed to formalize data transformation to improve system interoperability capable of transforming system data into HL7 messages

Some authors claimed that the openEHR standard is more complete than the FHIR resources, one of these authors claims that the openEHR is able to fully represent medical knowledge, but the FHIR standard is much less harmful for the existing providers due to its adherence to the principles of RESTful design, another author claims that representing EHR using ISO/EN 13606 archetypes in conjunction with openEHR makes it possible to achieve a higher level of interoperability between HIS than with HL7 standards.

Another important observation is that to achieve semantic interoperability in the eHealth domain, a terminology server provides information systems with the necessary concepts to encode information correctly. It was also observed that architectures with a more formal approach are better at ensuring interoperability.

Therefore, to solve the problem that originated this work, the best solution will involve developing a structure to achieve a national health system for Portugal, based on interoperable HER system, with a central data warehouse, centered on the citizen, so that health data can be received, and accessed based on the identification of the citizen. But APIs must also be developed and made available that allow the future development of applications that citizens install on their connected personal devices, so that they can enjoy and produce their own health data, also APIs that allow access to data for statistical purposes and for feed prediction systems.

The architecture and development of this solution should take place in accordance with the unanimously accepted norms and standards that are as closely aligned as possible with the intended platform, so that the future work to be carried out next, in this project, is to make a detailed analysis and comparison between the existing standards, in order to make that decision.

Within the scope of the HfPT agenda, of the PRR, the preliminary phase of the development of an architecture of a platform that allows the integration of different information systems, to solve the lack of interoperability of the EHRs of the interested parties in health in Portugal and also to reduce the fragmentation of the HIS of the Portuguese SNS. The literature review has already been completed, which allowed observing trends and advances in the eHealth domain, analyzing and discussing the results of this work, which allowed reaching some conclusions and determining the path to follow for future work. The future work to be carried out next is to make a detailed analysis and comparison between existing standards, such as HL7, FHIR, openEHR or CDISC, in order to select the most suitable standards for this platform.

ACKNOWLEDGEMENT

This work has been supported by FCT – Fundação para a Ciência e Tecnologia within the R&D Units Project Scope: UIDB/00319/2020.

REFERENCES

About CDISC | CDISC. (2023). Retrieved February 11, 2023, from https://www.cdisc.org/about

About Health Level Seven International | HL7 International. (2023). Retrieved February 11, 2023, from http://www.hl7.org/about/index.cfm?ref=nav

About openEHR. (2023). Retrieved February 11, 2023, from https://www.openehr.org/about_us

Adebayo, A.-A., Ikuomola, A., Oluwakemi Abayomi-Alli, O., Abayomi-Alli, A. A., Ikuomola, A. J., Robert, I. S., & Abayomi-Alli, O. O. (2014). An Enterprise Cloud-Based Electronic Health Records System Learning with Technology using Facebook View project An Enterprise Cloud-Based Electronic Health Records System. *Journal of Computer Science and Information Technology*, *2*(2), 21–36. https://www.researchgate.net/publication/276027227

Adel, E., El-Sappagh, S., Barakat, S., & Elmogy, M. (2017). *International Journal of Computers and Applications Distributed electronic health record based on semantic interoperability using fuzzy ontology: a survey Distributed electronic health record based on semantic interoperability using fuzzy ontology: a survey*. https://doi.org/10.1080/1206212X.2017.1418237

Atanasovski, B., Bogdanovic, M., Velinov, G., Stoimenov, L., Dimovski, A. S., Koteska, B., Jankovic, D., Skrceska, I., Kon-Popovska, M., & Jakimovski, B. (2018). *Enterprise Information Systems On defining a model driven architecture for an enterprise e-health system on defining a model driven architecture for an enterprise e-health system*. https://doi.org/10.1080/17517575.2018.1521996

Carter1, G., Shahriar2, H., & Sneha1, S. (2019). Blockchain-Based Interoperable Electronic Health Record Sharing Framework; Blockchain-Based Interoperable Electronic Health Record Sharing Framework. *2019 IEEE 43rd Annual Computer Software and Applications Conference (COMPSAC)*, *2*. https://doi.org/10.1109/COMPSAC.2019.10248

Dinevski, D., Poli, A., Krajnc, I., Šušteršič, O., & Arh, T. (2010). E-health integration and interoperability based on open-source information technology. *Wiener Klinische Wochenschrift*, *122*(SUPPL. 2), 3–10. https://doi.org/10.1007/s00508-010-1354-9

European Commission. (2022). *Electronic cross-border health services*. European Commission. https://health.ec.europa.eu/ehealth-digital-health-and-care/electronic-cross-border-health-services_en

Evgeniy, K., Dimitar, T., Kalinka, K., Lyubomir, K., Petko, K., Simeon, A., & Nonka, M. (2020). *Standards based adaptation of clinical documents for interoperability of e-health services*. CEUR Workshop Proceedings 2656. https://resolver.ebscohost.com/openurl?custid=s1166128&groupid=main&profile=ftf&authtype=ip,shib&sid=Elsevier:Scopus&genre=proceeding&issn=16130073&ISBN=&volume=2656&issue=&date=2020&spage=14&pages=14-29&title=CEUR+Workshop+Proceedings&atitle=Standards+based+adaptation+of+clinical+documents+for+interoperability+of+e-health+services&aulast=Krastev&id=DOI:

Eysenbach, G. (2001). What is e-health? *Res*, *3*(2), 20. https://doi.org/10.2196/jmir.3.2.e20

Fragidis, L. L., Chatzoglou, P. D., & Aggelidis, V. P. (2016). Integrated Nationwide Electronic Health Records system: Semi-distributed architecture approach. *Technology and Health Care: Official Journal of the European Society for Engineering and Medicine*, *24*(6), 827–842. https://doi.org/10.3233/THC-161231

Gomes, F., Freitas, R., Ribeiro, M., Moura, C., Andrade, O., & Oliveira, M. (2019). GIRLS, a Gateway for Interoperability of electronic health Record in Low-cost System: ∗ Interoperability between FHIR and OpenEHR Standards. *2019 IEEE International Conference on E-Health Networking, Application and Services, HealthCom 2019*. https://doi.org/10.1109/HealthCom46333.2019.9009602

Id, W. J., & Kim, D. H. (2018). *Design and Implementation of e-Health System Based on Semantic Sensor Network Using IETF YANG*. https://doi.org/10.3390/s18020629

Jaskó, S., Skrop, A., Holczinger, T., Chován, T., & Abonyi, J. (2020). Development of manufacturing execution systems in accordance with Industry 4.0 requirements: A review of standard- and ontology-based methodologies and tools. *Computers in Industry*, *123*, 103300. https://doi.org/10.1016/J.COMPIND.2020.103300

Kautsch, M., Lichoń, M., & Matuszak, N. (2016). *Setting the scene for the future: implications of key legal regulations for the development of e-health interoperability in the EU*. https://doi.org/10.1002/hpm.2384

Kleinoscheg, G., Tanjga, N., Svec, N., Rainer-Sablatnig, S., & Sabutsch, S. (2022). TerminoloGit - An Open Source Terminology Server for Large Scale eHealth Environments. *Studies in Health Technology and Informatics*, *293*, 79–84. https://doi.org/10.3233/SHTI220351

Krastev, E., Abanos, S., & Tcharaktchiev, D. (2022). Health Data Exchange Based on Archetypes of Clinical Concepts. *CEUR Workshop Proceedings*, *3191*, 98–112.

Lemus-Zúñiga, L. G., Félix, J. M., Fides-Valero, A., Benlloch-Dualde, J. V., & Martinez-Millana, A. (2022). A Proof-of-Concept IoT System for Remote Healthcare Based on Interoperability Standards. *Sensors*, *22*(4). https://doi.org/10.3390/s22041646

Malaquias, R. S., & Filho, I. M. B. (2021). Middleware for Healthcare Systems: A Systematic Mapping. *Lecture Notes in Computer Science (Including Subseries Lecture Notes in Artificial Intelligence and Lecture Notes in Bioinformatics)*, *12957 LNCS*, 394–409. https://doi.org/10.1007/978-3-030-87013-3_30

Masud, M., Hossain, M. S., & Alamri, A. (2012). Data interoperability and multimedia content management in e-health systems. *IEEE Transactions on Information Technology in Biomedicine*, *16*(6), 1015–1023. https://doi.org/10.1109/TITB.2012.2202244

Mazouz, S., Malki, O. M. C., & Elloub, A. (2017). Towards a system for integrating heterogeneous health records. *ACM International Conference Proceeding Series*, 6. https://doi.org/10.1145/3175628.3175657

Nayak, S., Hossain, M. A., Mirza, F., Naeem, M. A., & Jamil, N. (2019). E-BRACE: A Secure Electronic Health Record Access Method in Medical Emergency. *Communications in Computer and Information Science*, *932*, 16–27. https://doi.org/10.1007/978-981-13-6052-7_2/COVER

Pereira, J., Beir, M., Teixeira, J., & Machado, R. J. (2018). *Patient-centric e-Prescription Services - An Integrated System Architecture Proposal; Patient-centric e-Prescription Services - An Integrated System Architecture Proposal*.

Rizzato Lede, D. A., Pedernera, F. A., López, E., Speranza, C. D., Maid, J. J., Gassino, F., Argibay, O., Castiglione, R., Pujadas, G., Gonzalez Alvaredo, E., Álvarez, R., Markiewicz, M. B., Alegre, F., Ferrareis, L., Rolandi, F., Ayala, F., Abadie, D. A., Cejas, C. A., López Osornio, A., & Rubinstein, A. (2020). Mi Argentina/Mi Salud: The Argentinian Citizen digital health portal. *Studies in Health Technology and Informatics*, *270*, 1011–1015. https://doi.org/10.3233/SHTI200314

Schiza, E. C., Kyprianou, T. C., Petkov, N., Schizas, C. N., Schiza, E. C., Petkov, N., & Kyprianou, T. C. (2019). Proposal for an eHealth Based Ecosystem Serving National Healthcare. *IEEE JOURNAL OF BIOMEDICAL AND HEALTH INFORMATICS*, *23*(3). https://doi.org/10.1109/JBHI.2018.2834230

Staffa, M., Coppolino, L., Sgaglione, L., Gelenbe, E., Komnios, I., Grivas, E., Stan, O., & Castaldo, L. (2018). KONFIDO: An OpenNCP-based secure ehealth data exchange system. *Communications in Computer and Information Science*, *821*, 11–27. https://doi.org/10.1007/978-3-319-95189-8_2/FIGURES/4

Urbauer, P., & Forjan, M. (2020). Integration of Health and Public Transport Data to Enable Decision Support for Seniors to Reduce Risk of Infection with Communicable Diseases. *ACM International Conference Proceeding Series*, 168–173. https://doi.org/10.1145/3439231.3439236

Weber-Jahnke, J. H., & Obry, C. (2012). Protecting privacy during peer-to-peer exchange of medical documents. *Information Systems Frontiers*, *14*(1), 87–104. https://doi.org/10.1007/S10796-011-9304-2

INNOVATIVE NUTRITIONAL DIGITAL BIOMARKERS BASED ON ACTIMETRIC BEHAVIOR MONITORING

Abderrahim Derouiche[1], Ghazi Bouaziz[1], Damien Brulin[2], Eric Campo[2] and Antoine Piau[3]
[1]LAAS-CNRS, University of Toulouse
CNRS, UPS, Toulouse, France
[2]LAAS-CNRS, University of Toulouse
CNRS, UT2J, Toulouse, France
[3]Center for Epidemiology and Population Health Research
CHU Toulouse, Toulouse, France

ABSTRACT

Older people represent a large part of the population. Focusing on this population in order to protect them from the health risks they may face during their lifetime, especially undernutrition, is a very important issue. Indeed, undernutrition is a disease state caused by an imbalance in nutritional intake in relation to the body's energy expenditure. With technological advances, biomarker measurement has become more efficient and specific, allowing us to better diagnose undernutrition in the elderly. This article presents a large study on healthy aging. Patients use a mobile application called ICOPE to screen for undernutrition and obtain biological data coupled with digital biomarkers obtained from a system for measuring lifestyle behavioral parameters. Actimetric measurements could be considered as a 'partial' parameter to be integrated with others, closer to physiological conditions.

KEYWORDS

Undernutrition, Biological Biomarkers, Digital Biomarkers, Elderly, Life Habits Monitoring, Health Data

1. INTRODUCTION

Worldwide, people aged 65 and over living alone represent a significant proportion of the population in the US (13%) and France (19.6%) (Piau, 2019). In our study, we are interested in these people living alone at home.Undernutrition is a pathological condition resulting from insufficient nutritional intake in relation to the body's energy expenditure. It is a major public health issue in the management of older people with chronic pathologies (up to 50% of elderly people) (Volkert, 2019). It is associated with increased morbidity and mortality, risk of loss of autonomy, falls, infections and emergency hospitalizations (Norman, 2021). In France, 400 000 elderly people at home suffer from undernutrition, so it is important to better understand the behaviors related to undernutrition. The use of digital biomarkers for the early detection of functional or cognitive decline is receiving increasing attention (Piau, 2020). For example, sensor-derived mobility indicators have been studied to predict an upcoming fall in older people (Piau, 2015). Currently, the assessment of an individual's eating behavior relies mainly on self-report measures and clinical tests. Although this methodology is necessary, it can be improved as it is limited by recall bias and by brief and mostly episodic assessments. Continuous real-time monitoring by a low-cost ambient sensors network can provide a more sensitive, objective and environmentally valid method to:

-Detect sub-clinical alteration in nutrition during chronic disease, prior to the onset of symptoms (allowing for targeted, personalized and more effective medicine (rather than one-size-fits-all medicine);

-Make a less biased assessment of treatment efficacy and side effects in a clinical trial after treatment modification;

-Remotely monitor the dynamics of recovery after hospitalization or surgery.

2. RELATED WORKS

Previous studies have concretely linked the physiological and the biological state of an individual, which is already an indication for diagnosing undernutrition.

Busnel et al. (Busnel, 2018) propose to estimate the diagnostic accuracy of the nutritional status score, body mass index and weight loss documented from the Resident Assessment Instrument – Home Care (RAI-HC). A sample of 267 home care recipients aged 65 years and older was assessed using the RAI-HC and the Mini Nutritional assessment short form (MNA-SF). The result reveals that the diagnostic accuracy of the RAI-HC indicators was not sufficient for optimal screening for undernutrition in elderly home care recipients. Additional assessment with the MNA-SF is recommended to optimize early detection of individuals at risk for undernutrition.

Birkan et al. (Birkan, 2018) propose to investigate the reliability and validity of the Turkish version of the Simplified Nutritional Appetite Questionnaire (SNAQ) in geriatric outpatients. The results indicate that in terms of validity of the SNAQ, Cohen's kappa analysis showed fair to moderate agreement between the SNAQ and the MNA ($\kappa=0.355$, $p<0.001$). Female gender, illiteracy, functional dependence in activities of daily living was significantly associated with poor appetite. The SNAQ score was weakly correlated with the MNA-SF and MNA-LF scores ($r=0.392$ and $r=0.380$, respectively, $p<0.0001$ for both).

Gambi et al. (Gambi, 2020) propose an application to monitor the food intake actions of individuals during a meal. Eating actions are estimated from the analysis of depth images. In order to optimize and to increase the performance of algorithms, the idea is to automate the process that allows us to identify the different actions of taking meals.

One of the main findings of previous studies is that weight loss is a fundamental defect identified in the elderly.Other studies also suggest that albumin levels in laboratory tests may be related to undernutrition. Finally, the identification of biomarkers (Panagoulias, 2021) is also proving to be tools to accurately prognosticate many diseases, classify them by type and stage, and personalize medical care.

3. PROPOSED SYSTEM

The proposed research is part of the regional Inspire program (Barreto, 2020). Inspire is a research platform dedicated to gerontological research on biological and healthy ageing to identify biomarkers, determine biological age and work on all marks of ageing. Inspire involves 1000 participants. The two main projects are:

3.1 ICOPE Mobile Application

As shown in Figure 1, Integrated Care for Older People (ICOPE) (Tavassoli, 2021) is a mobile application launched by the World Health Organization (WHO) in 2019 for the integrated care of older people. This app measures intrinsic capacity which is the combination of different abilities (physical, mental and psychological) and functional ability. There are 6 domains of intrinsic capacity that can be measured: cognition, mobility, nutrition, mood, vision and hearing. In order to maintain the health of older people and to better identify risks before the frailty stage is reached, the application is used as a monitoring tool. It is based on 5 steps: 1-Screening for decline in intrinsic capacity, 2-Specialized assessment, 3-Creating a personal care plan, 4-Regular monitoring and 5-Career and community integration. To monitor undernutrition, patients answered two questions about:

-Weight loss: have you unintentionally lost more than 3 kg over the last three months?

-Appetite loss: have you experienced a loss of appetite?

By answering these two questions, we can identify markers related to nutrition. We can also establish a link between nutrition and reduced mobility.

In order to better collect biological markers, we use an annual in-depth assessment and questionnaires.

Figure 1. ICOPE Mobile application

3.2 CART France

Collaborative Aging Research using Technology (CART) Sensor Platform is an initiative developed by Oregon Center for Aging & Technology (ORCATECH)(Thomas, 2021) to study various physical and physiological parameters. CART France is a living lab project based on one hundred connected homes. ORCATECH's sensor platform proposes new ways to assess people with cognitive impairment. By using a huge amount of information about changes in health status, activity and functioning in a real-world setting to improve the ability to assess and provide personalized care. Differences between the French CART project and the U.S. CART project are the number of sensors used in both projects and the fact that the French CART project is based on the nutritional status of individuals. As shown in Figure 2, the sensors used in CART France are physical and physiological sensors.

The project aims to combine the CART French data with clinical data to evaluate the relevance of numerical data to represent and predict an individual's health status.

The inclusion criteria for participants are as follows:

1- Over 75 years old

2- Living alone at home without assistance

3- Being robust

4- Having a computer and an internet terminal

To retrieve the physical parameters, we used passive infrared (PIR) motion and contact sensors installed at 3 different locations in the individual's home:

-the 1st PIR motion sensor is a wall sensor; it is used to detect the presence of the person in the room;

-the 2nd PIR motion sensor is a line sensor; 4 of them are placed in the hallway to calculate the walking speed when the person moves between two rooms. This parameter has already been studied to predict falls (Piau, 2020);

-the 3rd sensor is a contact sensor installed on the front door to detect the entry and exit of the person from the house.

All these function sensors are wireless sensors that use ZigBee radio communication. In order to retrieve physiological parameters, other types of sensors such as an impedance scale and a connected smart watch are added. Detailed descriptions of the hub computer and sensors have been published previously (Beattie, 2020) (Thomas, 2020). Figure 2 shows the different sensors used in the CART France project.

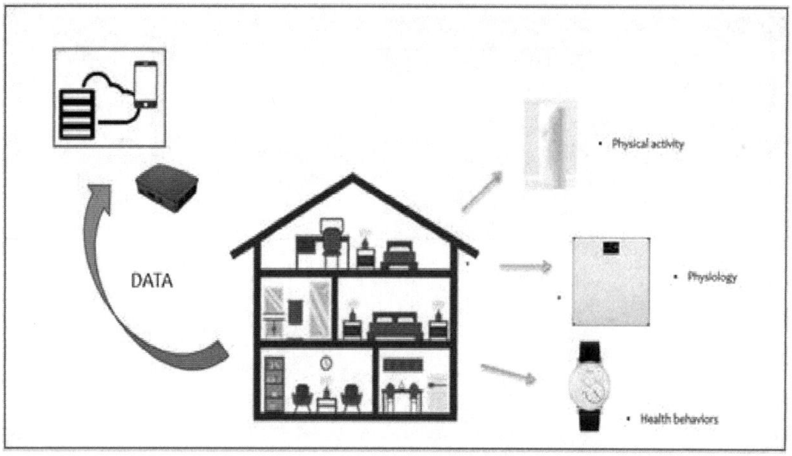

Figure 2. Overview of devices in the CART France project

3.3 System architecture

The French CART project was launched in July 2022. In this section we will present one of the first installations using only PIR sensors and a contact sensor to monitor physical activities. The architecture of the position of the sensors in the house can be visualized in Figure 3.

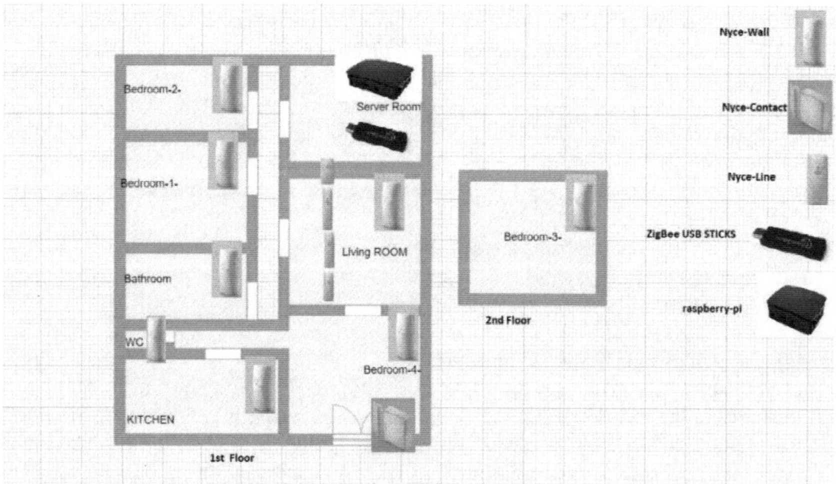

Figure 3. Architecture for the house

The exact number of sensors required, which is set at 13 sensors:
- 4 line sensors.
- 8 wall sensors.
- 1 contact sensor.

Figure 4 shows the location of the sensors in the house. Since the installation is done, we were able to see by using the Raspberry Pi (RPi) control panel that all the sensors are working correctly as shown in Figure 5. These sensors are connected to a hub via a personal area network (PAN).

The most important process is to map the location of the sensors in the house. This means that we will find the specific data for each room, as shown in Figure 6.

Figure 4. Example of sensors installation in the house

MAC	Inventory Name	Type	Last Checkin	Last Event	State	Active	Battery
...D6A232B		Contact Sensor	53m	7m	Closed		
...A993496		Contact Sensor	14m	4h 8m	Closed		
...0C204E8		Presence Sensor	58m	4h 51m			
...0C1D5EC		Presence Sensor	1h 3m	4m			
...0C19E9D		Presence Sensor	1h 3m	5m			

Figure 5. Example of sensors assigned in the console

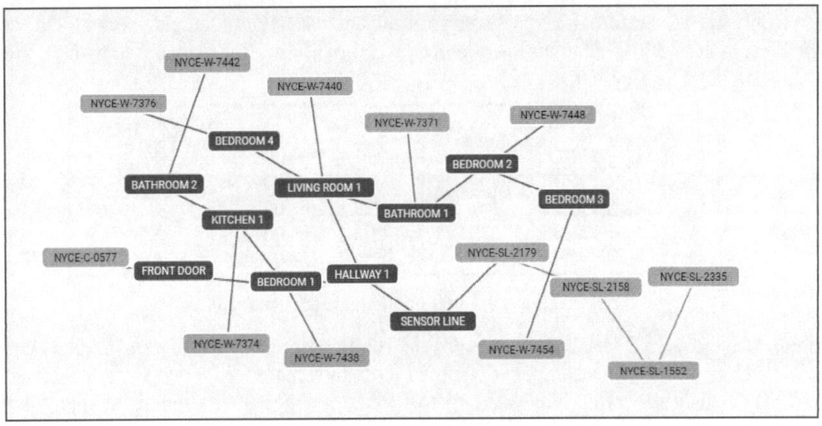

Figure 6. Location of sensors in the console

3.4 Data Collection

To date, we have collected 9 months of data. The objective of this data collection is to determine the life pattern corresponding to of this individual by correlating this data with the medical data from the weekly questionnaire and the ICOPE application. The collection and analysis of declarative data of people from the mobile application ICOPE are carried out in parallel to our study by the Hospital. They should make it possible to identify a change in the health state and appetite perception of the elderly.

The analysis described here is based on a naive approach that does not consider noise factors such as early triggering of sensors, transmission delay, or missing events, but it is sufficient to give a basic idea of their functionality. The data is stored in a database and includes all data set from all sensors. The resulting dataset was analyzed to find all walks in the walking line defined as a sequence of triggers of the 4 motion sensors (1,2,3,4 or 4,3,2,1) where the timestamps placed the sensors in a temporal order and where all sensors were triggered. Figure 7 shows the event logs in the database including homeid, stamp (date&time), areaid (room), itemid (sensors) and event (0: not in the room; 1: in the room).

homeid	study	stamp	itemid	areaid	event
2269	CART-FRANCI	17:08.1	27452	-1	0
2269	CART-FRANCI	17:12.4	27452	-1	1
2269	CART-FRANCI	17:15.9	27468	-1	1
2269	CART-FRANCI	17:21.1	23951	-1	1
2269	CART-FRANCI	17:21.1	23930	-1	1
2269	CART-FRANCI	21:05.4	23930	-1	0
2269	CART-FRANCI	21:05.6	23951	-1	0
2269	CART-FRANCI	21:16.9	27385	-1	0
2269	CART-FRANCI	35:13.8	27385	-1	0
2269	CART-FRANCI	35:50.1	27390	-1	0
2269	CART-FRANCI	36:44.9	24110	-1	0
2269	CART-FRANCI	37:13.8	27468	-1	0
2269	CART-FRANCI	37:41.2	27462	-1	0
2269	CART-FRANCI	38:05.0	27452	-1	0
2269	CART-FRANCI	38:35.8	27385	-1	1

Figure 7. Data Collection

If an areaid = -1, i.e. no sensors have been assigned to the room, as noted previously, to solve this problem, we need to build the sensors map (Figure 6). The first task is to preprocess the data by eliminating all insignificant areas.

	A	B	C
1	stamp	areaid	event
2	2022-07-27 10:34:03.357	29	1
3	2022-07-27 10:34:09.915	4	1
4	2022-07-27 10:34:14.157	29	0
5	2022-07-27 10:34:20.901	4	0
6	2022-07-27 10:35:56.845	4	1
7	2022-07-27 10:36:05.434	29	1

Figure 8. First Pre-Processing

Then, we only keep the information necessary for the analysis (stamp, areaid, events). On the other hand, in order to make correct and fast calculations, we need to convert time into Unix time, which allows us to filter events by area as shown in figure 9.

	A	B	C
1	stamp	areaid	event
2	1663658111	1	1
3	1663573521	1	1
4	1663573532	1	0
5	1663573544	1	1
6	1663073251	1	0

	A	B	C
24207	1659009059	4	0
24208	1659009054	4	1
24209	1659009045	4	0
24210	1659009040	4	1
24211	1659008207	4	0

Figure 9. Filtering of all events by areaid

Given the inter-individual variability in the physical activity intensity and duration, total energy expenditure can be estimated by multiplying resting energy expenditure by a factor reflecting the intensity of an individual's physical activity. This physical activity level factor (PAL) has been determined for many activities of daily living, including sedentary, occupational and sports activities (Westerterp, 2013).

$$PAL = \frac{TotalEnergyExpenditure}{BaselMetabolicRate}$$

The Basel Metabolism Rate (BMR) is obtained by the following basal metabolic rate formula:
-Men: BMR=88.362+ (13,397×weight in kg) + (4,799×height in cm) – (5,677×age in years)
-Women: BMR=447.593+ (9,247×weight in kg) + (3,098×height in cm) – (4,330×age in years)

Total energy expenditure=PAL×BMR

4. RESULTS AND DISCUSSION

Figure 10 shows an example of different events in each room of the house for 3 months of data. To better illustrate each of them, we represent each room by a histogram as shown in Figure 11. Each room contains events from which we can well distinguish the different activities of daily life, thus identifying a life pattern of that person. Finally, this will give us the ability to identify digital biomarkers.

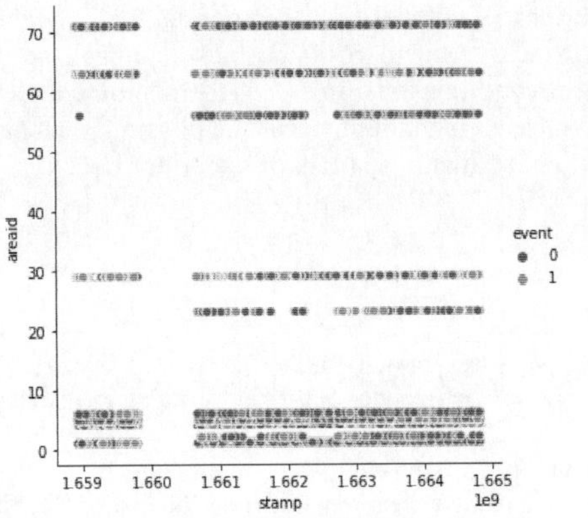

Figure 10. All the events that occurred in the house for 3 months

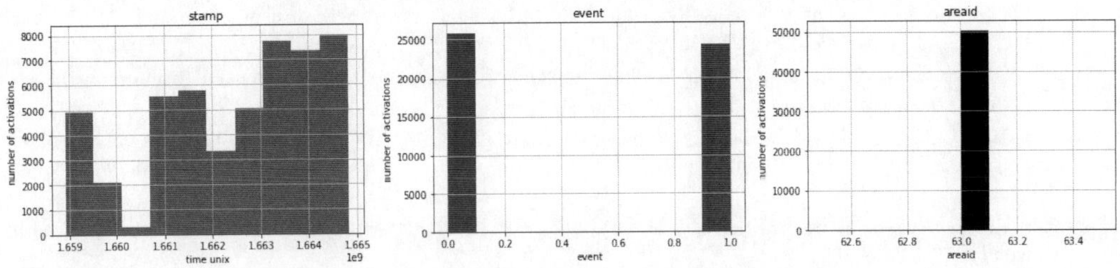

Figure 11. All events that occurred in area 63 for 3 months

We can also conclude that the use of PIR sensors not only detects the various activities of the person but it can also show the lifestyle which we can associate to other parameters to develop a very specific model for that person. It can also be used to calculate caloric expenditure. The data to calculate the BMR are for an individual with the following characteristics: age 77 years, sex: male, height 165 cm, weight 60 kg and work: writer.

$$BMR_{Men} = 88.362 + (13.397\times60) + (4.7988\times165) - (5{,}677 \times 77) = 1{,}246.855 \text{ cal}$$

Based on the lifestyle of the person in the house, we can classify this person in the PAL category between [1.4 and 1.69], which means he has a low-activity lifestyle.

$$Total\ Energy\ Expenditure = 1.4\times1\ 246.855\ cal = 1745.597 \text{ Cal.}$$

From the results of the weekly questionnaire, we can theoretically identify the lifestyle of the participants and validate this study by checking the activities records. The combination of the two studies (numerical and clinical) will explain and validate the results. However, the results of our analysis are approximately correct in the case where we measure the amount of daily energy spent in the home only. If we wish to calculate total daily energy expenditure, we must include other types of sensors, such as a Smartwatch, to monitor the total amount of activity inside and outside the home.

5. CONCLUSION

In this paper, we have presented a new system based on a regional research projects at the University Hospital Center in Toulouse-France. The CART initiative followed a principled design process to develop a platform capable of monitoring different fields of health and well-being. The data collected is typically used to explain, influence and/or predict health-related outcomes. In the case of CART France, the data collected must be used to identify nutrition-related behavior. To do this, we collect all the information related to the person's life habits such as the time, occurrences and duration of presence in each room, in our case the kitchen, as well as the use of household appliances such as the refrigerator. The use of digital biomarkers to detect subtle behavioral variations (undernutrition) prior to a clinical event is important to better understand individuals' adherence to recommendations. Actimetric measurements could be considered as a 'partial' parameter to be integrated with others, closer to physiological conditions. To this end, the combined use of clinical data (biannual nutrition assessment) and digital data (measurement of physical and physiological parameters, such as weight) will allow better diagnosis and prevention of situations at risk of undernutrition.

REFERENCES

Barreto De Souto, P., Guyonnet, S., Ader, I., Andrieu, S., Casteilla, L., Davezac, N., ... & Vellas, B. (2020). The Inspire Research Initiative: A Program for Geroscience and Healthy Aging Research going from Animal Models to Humans and the Healthcare System.

Beattie, Z., Miller, L. M., Almirola, C., Au-Yeung, W. T. M., Bernard, H., Cosgrove, K. E., ... & Kaye, J. (2020). The collaborative aging research using technology initiative: An open, sharable, technology-agnostic platform for the research community. *Digital Biomarkers*, *4*(1), 100-118.

Busnel, C., & Ludwig, C. (2018). Screening for undernutrition among older adults receiving home care: an estimation of the diagnostic accuracy of indicators available in the Resident Assessment Instrument-Home Care adapted for Switzerland. *Recherche en soins infirmiers*, *132*(1), 54-63.

Gambi, E., Ricciuti, M., & De Santis, A. (2020). Food intake actions detection: An improved algorithm toward real-time analysis. *Journal of Imaging*, *6*(3), 12.

İlhan, B., Bahat, G., Oren, M. M., Kilic, C., Durmazoglu, S., & Karan, M. A. (2018). Reliability and validity of Turkish version of the Simplified Nutritional Appetite Questionnaire (SNAQ). *The journal of nutrition, health & aging*, *22*, 1039-1044.

Norman, K., Haß, U., & Pirlich, M. (2021). Malnutrition in older adults—recent advances and remaining challenges. *Nutrients*, *13*(8), 2764.

Panagoulias, D. P., Sotiropoulos, D. N., & Tsihrintzis, G. A. (2021, November). Biomarker-based deep learning for personalized nutrition. In *2021 IEEE 33rd International Conference on Tools with Artificial Intelligence (ICTAI)* (pp. 306-313). IEEE.

Piau, A., Charlon, Y., Campo, E., Vellas, B., & Nourhashemi, F. (2015). A smart insole to promote healthy aging for frail elderly individuals: specifications, design, and preliminary results. *JMIR rehabilitation and assistive technologies*, *2*(1), e4084.

Piau, A., Mattek, N., Crissey, R., Beattie, Z., Dodge, H., & Kaye, J. (2020). When will my patient fall? Sensor-based in-home walking speed identifies future falls in older adults. *The Journals of Gerontology: Series A*, *75*(5), 968-973.

Piau, A., Wild, K., Mattek, N., & Kaye, J. (2019). Current state of digital biomarker technologies for real-life, home-based monitoring of cognitive function for mild cognitive impairment to mild Alzheimer disease and implications for clinical care: systematic review. *Journal of medical Internet research*, *21*(8), e12785.

Tavassoli, N., Piau, A., Berbon, C., De Kerimel, J., Lafont, C., De Souto Barreto, P., ... & Vellas, B. (2021). Framework Implementation of the INSPIRE ICOPE-CARE program in collaboration with the World Health Organization (WHO) in the Occitania region. *The Journal of frailty & aging*, *10*, 103-109.

Thomas, N. W. D., Beattie, Z., Marcoe, J., Wright, K., Sharma, N., Mattek, N., ... & Kaye, J. (2020). An ecologically valid, longitudinal, and unbiased assessment of treatment efficacy in Alzheimer disease (the EVALUATE-AD Trial): proof-of-concept study. *JMIR Research Protocols*, *9*(5), e17603.

Thomas, N. W., Beattie, Z., Riley, T., Hofer, S., & Kaye, J. (2021). Home-based assessment of cognition and health measures: The collaborative aging research using technology (CART) initiative and international collaborations. *IEEE Instrumentation & Measurement Magazine*, *24*(6), 68-78.

Volkert, D., Beck, A. M., Cederholm, T., Cereda, E., Cruz-Jentoft, A., Goisser, S., ... & Wirth, R. (2019). Management of malnutrition in older patients—current approaches, evidence and open questions. *Journal of clinical medicine*, *8*(7), 974.

Westerterp, K. R. (2013). Physical activity and physical activity induced energy expenditure in humans: measurement, determinants, and effects. *Frontiers in physiology*, *4*, 90.

PHYSICAL ACTIVITY PROMOTION AND DIGITAL INTERVENTIONS

Jennifer Santos[1] and Rita Oliveira[2]
[1]*Department of Communication and Art, University of Aveiro, 3810-193, Aveiro, Portugal*
Faculty of Arts and Humanities, University of Porto, 4150-564, Porto, Portugal
[2]*Digital Media and Interaction Research Centre (DigiMedia), Department of Communication and Art, University of Aveiro, 3810-193, Aveiro, Portugal*

ABSTRACT

There is countless evidence regarding the connection between levels of physical activity and obesity as an ongoing global pandemic. Likewise, as the body of evidence has grown, so has the need to take urgent action in finding potential solutions to increase physical activity levels. However, despite the clear benefits of physical activity and evidence associated to digital interventions' success, interventions approached so far have not been enough to pinpoint self-reported barriers to physical activity and tackle the issue successfully. This paper presents a literature review on the fundamental terminology, importance, and the current worldwide panorama of physical activity. In addition, reported barriers to physical activity and digital interventions to overcome barriers and promote physical activity are also reviewed.

KEYWORDS

Physical Activity, Health Communities, Online Communities, Digital Platforms

1. INTRODUCTION

According to Michie et al. (2017a), digital interventions are tools and programs based on digital technology that encourage or support behavioural change using smartphone apps, websites, social networks, text messages, or personal digital assistants. (Rose et al., 2018). These could be helpful in removing obstacles and promoting PA, particularly when taking into account the social support qualities and group-wide possibilities of digital interventions to promote PA (Manzoor et al., 2016; Zeng et al., 2018).

Although there is less proof of the effectiveness of solely digital interventions to promote a healthy diet, physical exercise, or weight control than there is for face-to-face interventions, evidence shows that proactive messaging (i.e., reminders to participate in behaviour change activities, feedback on progress, or advice) results in successful interventions (Rhodes et al., 2020). This suggests that interactivity may increase participation and increase effectiveness.

2. BRANCHES IN PHYSICAL ACTIVITY

The World Health Organization (WHO) describes PA as "any bodily movement produced by skeletal muscles that requires energy expenditure" (World Health Organization, 2020). Literature references different types of PA, such as aerobic and anaerobic PA, bone-strengthening activity, exercise, or fitness (Boratto et al., 2017; Guthold et al., 2018; Wu et al., 2015; Zeng et al., 2018). However, although these terms are frequently used interchangeably, there are individual differences among them.

Aerobic PA is movement in which the major muscles of the body move in a repetitive pattern over an extended length of time (World Health Organization, 2020). Aerobic exercise, often known as endurance exercise, promotes cardiorespiratory fitness. Walking, running, swimming, and biking are among examples of aerobic PA. Anaerobic PA, on the other hand, is referred to as short, intense bursts of movement, such as sprinting and weightlifting, where the demand for oxygen exceeds its supply (World Health Organization, 2020). Moreover, bone-strengthening activity is PA with the main goal of strengthening specific regions of the skeletal system, through force that impacts or exerts tension on the bones during exercises that encourage bone growth and vigour (World Health Organization, 2020). Activities that build bone strength include running, jumping rope, and lifting weights.

Exercise is a subcategory of PA where the goal is to maintain or enhance one or more aspects of physical fitness in an organized, timed, repeated, and intentional manner during leisure time. Whereas fitness encompasses physical and cardiorespiratory health and measures the body's capacity to operate effectively and efficiently in both work and leisure activities (World Health Organization, 2020).

Unlike work-domain PA, which occurs during paid or voluntary work, leisure-domain PA is conducted by a person at their convenience, as a non-necessary part of their daily life. Leisure-domain PA can include participation in sports, physical preparation for sports, and recreation activities like dancing, gardening, and strolling. The WHO (2020) also describes transport-domain PA, which includes walking, cycling, wheeling (the use of non-motorized wheels-equipped means of locomotion), and other similar activities performed with the intention of getting to and from places. The last type of PA is household-domain PA, which includes tasks carried out as part of daily domestic duties.

3. PHYSICAL ACTIVITY MATTERS

Despite evidence emphasising that PA is the cornerstone of a healthy lifestyle, global levels of PA have not been improving. Evidence has shown PA to be associated with lower rates of all-cause mortality obesity (Zhao et al., 2020), chronic diseases, such as cardiovascular diseases and hypertension (Carnethon, 2009; Hegde and Solomon, 2017; Lanier et al., 2016; O'Donovan et al., 2017), type-2 diabetes (Hamasaki, 2016), cancer (Brown, Justin C.; Kerri Winters-Stone; Augustine lee; Schmitz, 2014), mental health problems (Biddle, 2016, 1992), and better cognitive health and sleep (Swirski et al., 2019).

A physiological metric used to quantify the intensity of physical activities is the metabolic equivalent of task (MET), or simply, metabolic equivalent— which is the amount of energy a person expends while sitting still and at rest, expressed in METs (World Health Organization, 2020). The WHO (World Health Organization, 2020) uses METs to classify PA intensity into three main categories: vigorous, moderate, and light.

Vigorous-intensity PA refers to PA at 6.0 or greater METS and is usually rated a 7 or 8 on a scale of 0–10 relative to an individual's personal capacity. Moderate-intensity PA is defined as any exercise that is between three and six times as intense as rest. Moderate-intensity PA is typically rated a 5 or 6 on a scale of 0–10, depending on the person. Light PA is defined as consuming 1.5 to 3 METs, or less than three times the amount of energy used when that individual is at rest. This can involve easy movement, taking a shower, or other incidental activities that do not significantly raise heart rate or breathing rate.

However, some authors believe that time spent in sedentary behaviour is strongly and negatively associated with cardiometabolic health, cardiovascular disease, conditions such as type 2 diabetes, osteoporosis, and some cancers. This is why they make it a more significant indicator of poor health than moderate or vigorous PA (Gordon-Larsen et al., 2004; Owen et al., 2010; Thorp et al., 2011; World Health Organization, 2020).

The WHO Handbook for Guideline Development use the term "sedentary behaviour" to designate self-reported low movement sitting (during leisure, employment, and total time), television viewing, screen time, and low levels of movement as measured by equipment that measure movement or posture. Sedentary activities include those performed by those unable to stand—such as wheelchair users, most desk-based office work, operating a vehicle, and watching television.

Although sedentary behaviour is commonly used to refer to physical inactivity, it is not the same. Popularly, and from a one-dimensional standpoint, sedentary behaviour has been used to describe conducts "marked by much sitting" (Agnes, 2002). On a more complex perspective, the WHO Handbook for Guideline Development refers to physical inactivity as "an insufficient PA level to meet present PA recommendations" which refers to "any waking behaviour characterized by an energy expenditure of 1.5 METS or lower while sitting, reclining, or lying" (World Health Organization, 2020).

Some authors believe that although the suggested 1.5 MET threshold for sedentary behaviours seems reasonable, some sitting based activities may be classified as non-sedentary and on that account, further research is needed (Mansoubi et al., 2015). Moreover, the WHO reports that to help lessen the negative impact of excessive levels of sedentary behaviour on health, adults should strive to engage in more than the advised amounts of moderate-to-vigorous intensity PA (World Health Organization, 2020).

Obesity is a condition in which abnormal or excessive fat has accumulated to the extent that health may be impaired (World Health Organization, 2000). This condition is often related to sedentary behaviour because it is believed to be an important contributor to the growth in population average body weight (Aubert et al., 2021; Cabrera de León et al., 2007; Thorp et al., 2011; Vasques et al., 2012).

Besides being a chronic disease, obesity is an epidemic, frequently characterized as a potential health-impairing condition that comprises the accumulation of abnormal or excessive fat in the adipose tissue (World Health Organization, 2020). The World Health Organization (2000) recognizes that overweight and obesity are associated to some major health consequences, such as: NIDDM, CHD, hypertension, gallbladder disease, psychosocial problems, and certain types of cancer.

Over the past few decades, literature has defined the average PA recommendation time in a wide range of capacities. Cabrera de Léon et al. (2007) considered that an individual who engages in a PA daily average duration of 25-30 minutes can still be considered a sedentary person. Subsequently, the "Active Couch Potato" phenomenon, introduced a decade ago, explored the possibility of labelling individuals who engage in moderate-to-vigorous PA yet still have high sedentary behaviour time as sedentary individuals (Mansoubi et al., 2015; Owen et al., 2010).

The most recent PA and sedentary behaviour guidelines from the WHO report specific global PA recommendations complemented by reassuring evidence (World Health Organization, 2020). Children and adolescents (5-17 years old) should engage in at least an average of 60 minutes per day of aerobic PA of moderate to vigorous intensity over the course of the week. Of the recommended exercise period, at least three days a week should be dedicated to high-intensity aerobic exercise as well as activities that develop bone and muscle (Aubert et al., 2021).

Furthermore, for significant health advantages, adults (18-65 years old) should engage in at least 150–300 minutes of moderate–intensity aerobic activity, at least 75–150 minutes of vigorous–intensity aerobic activity, or an equivalent combination of moderate– and vigorous–intensity activity throughout the week. For additional health benefits, adults should also engage in muscle-strengthening exercises that use all the major muscle groups twice a week or more, at a moderate to high intensity (Watson et al., 2021; World Health Organization, 2020). As for older adults (65 years old and older), for significant health advantages, they should engage in at least 150–300 minutes of moderate–intensity aerobic PA, at least 75–150 minutes of vigorous–intensity aerobic physical exercise, or an equivalent combination of both during the week (World Health Organization, 2020).

4. OVERLOOKING PORTUGAL AND THE WORLD

The relationship strength between PA during the early stages of life and adulthood can predict PA's progression from youth to adulthood (Gordon-Larsen et al., 2004; Malina, 2001; Tamellin, 2003). Heredity is associated with PA, fitness, and health—the three main components of a relationship model designed by Bouchard, Shepard and Stephens (1994). Parent education, scholastic achievements, level of PA during childhood and adolescence are also among some of the factors that seem to be directly correlated with engaging in PA during adulthood (Tamellin, 2003).

Between adolescence and young adulthood there is a considerable decline in PA and stability of sedentary behaviour (Gordon-Larsen et al., 2004; Hayes et al., 2019). In view of that, the probability of becoming either overweight, or obese, increases with age. Accordingly, maintaining high levels of PA across the lifespan is an important challenge for public health promotion, especially since it is easier to maintain PA throughout adulthood with an active childhood than attempting to make major lifestyle changes in the future (Tamellin, 2003).

In longitudinal studies, significant rise in BMI, mean weight and body fat have been detected in higher education students. Pope, Hansen, and Harvey (2016) found that mean weight was higher at the end of their senior year than in the beginning of their freshman year, as students continuously gained weight over the course of the four years. In another four-year longitudinal study, Gropper et al. (2012) also noticed an increase in college students' weight, in overweight or obesity from 18% to 31%. This advocates that first-year bachelor's degree students are a population of concern for overweight and obesity, therefore a favourable research sample to examine throughout higher education experience. For this reason, Pope et al. (2016) believes that to combat the rising rates of overweight and obesity, college students should be a prevalent target-population.

Insufficient PA was prevalent globally among the 1.9 million people who self-reported their PA levels through the International Physical Activity Questionnaire in 2016, with a frequency of 27.5% compared to 28.5% in 2001 (Guthold et al., 2018). The burden of several chronic diseases and early mortality has been increased by 6% to 10% globally due to the high prevalence of physical inactivity (Ozemek et al., 2019). Additionally, the percentage of people that meet the WHO's Guidelines on PA recommendations varies across different age groups (Huber et al., 2020; Watson et al., 2021; Wilke et al., 2021). Correspondingly, approximately 81% of adolescents aged 11 to 17 years old and 28% of adults aged 18 years old and over are insufficiently physically active (Cardon and Salmon, 2020; Guthold et al., 2018).

In Portugal, over the last two decades (1995-1998 to 2003-2005), overweight and obesity prevalence has increased. Likewise, studies have found that more than half of the Portuguese adult population has unhealthy weight, with 54% of population from 18-64 years old being overweight, 14% obese and 30% of population from 10-18 years old overweight and 8% obese (Bordalo et al., 2015; Do Carmo et al., 2006; Ferreira, 2010).

In 2010, there was 30.4% overweight prevalence in adolescent Portuguese population (22.6% pre-obese and 7.8% obese) and, as projected, adolescents with a healthy weight had higher levels of PA (Ferreira, 2010). Also, in 2013, the central region of Portugal held the highest prevalence of overweight population (38.0%) (Rito and Graça, 2015). Likewise, Rito et al. (2015) found that the central region of Portugal presented the uppermost incidence of underweight (5.0%) and overweight (38.0%) individuals, as well as child obesity (17.8%).

The National Food, Nutrition and Physical Activity Survey (Lopes et al., 2017) reported that 43% of the Portuguese population over 14 years old are sedentary, considering that they do not meet any of the WHO Guidelines on PA. Moreover, only 36% of youth (15-21 years old), 27% of adults and 22% of the elderly (65-84 years old) are physically active, fulfilling the official Physical Activity Engagement for Health recommendations (Lopes et al., 2017).

Data associated with meeting the WHO Global PA recommendations during and after the COVID-19 pandemic (as of March 2020) must consider the resulting PA changes due to the government-imposed lockdowns, bans of social gatherings, and business closures. Recent literature has found that PA has significantly decreased since the COVID-19 pandemic, calling it a "pandemic within a pandemic" (Wilke et al., 2021).

According to the 2020 SummerStyles survey, 30.4% adults (18 years old or older) reported being less active, and 42.7% of adults reported no change in their amount of PA during the pandemic (Watson et al., 2021). Likewise, in Germany, 45% of young adults reported being less active during the pandemic, whereas 33% reported being more active (Huber et al., 2020).

In another study, compared to pre-restrictions self-reported PA, overall self-reported moderate-to-vigorous PA declined by 41% and vigorous PA declined by 42.2% (Wilke et al., 2021). Moreover, one study found that only one in five children met the WHO Global PA recommendations and that around 70% of participants used more screen time in weekdays than the advised 2 hours per day (Kovacs et al., 2021).

Consequently, the causes of variations in adults' levels of PA between studies, which are either higher or lower than recommended levels, are unknown. Knowing which subgroups were more active could assist and inform on the development of population-level tactics that may be applied to different groups or in non-pandemic situations (Hayes et al., 2019; Watson et al., 2021; Wilke et al., 2021). All things considered, these numbers suggest that more needs to be done, particularly in the younger groups, such as exploring new strategies and resources, to promote PA (Kovacs et al., 2021; Ozemek et al., 2019; World Health Organization, 2020).

5. REPORTED BARRIERS AND IMPEDIMENTS

Besides exploring the specificity of recommended intensity levels in PA, it is important to understand the underlying reasons for not meeting these criteria. Literature has identified a few multi-factorial categories for PA barriers: individual, psychological and environmental (Bantham et al., 2021; Daskapan et al., 2006).

In adolescence, individual and environmental barriers are among the most cited barriers (Bantham et al., 2021; Daskapan et al., 2006). Among them are factors such as lack of time due to busy lesson schedule or responsibilities related to the family and social setting, academic success priority over exercise, or lack of energy to do exercise. Whereas in adults, psychological barriers are, according to evidence, the biggest barriers that prevent adults from considering and engaging in PA (Bantham et al., 2021). All in all, although some authors believe that the importance of reported external barriers might outweigh that of perceived interior obstacles (Daskapan et al., 2006), recent literature reports otherwise (Bantham et al., 2021; Hamer et al., 2021; Toscos et al., 2010).

Illness or injury, and actual or anticipated body changes are among naturally occurring barriers reported by individuals considered under-represented by literature (Toscos et al., 2010). Overindulgence on PA has been associated with various illnesses and injuries, resulting in muscle soreness, difficulty getting back on track after illness (e.g., cold or flu), and inability to restart PA after a break (i.e., "falling off the wagon"). Actual or anticipated body changes, refers to complaints related to bulking up on the lower body or upper body, or losing weight in the "wrong" places due to PA.

Furthermore, several preliminary studies have shown that among middle-aged to older persons, greater levels of fear are linked to decreased activity levels (Hamer et al., 2021). This has resulted in the conceptualization of several fear types, including fears of experiencing pain, pain-related fears, fears of movement, fears of falling or injury, fears of cardiac events, fears of joint damage, fear of embarrassment, fears of enacted stigma, fear of harm, and fears of integrating community due to crime (Bantham et al., 2021; Daskapan et al., 2006; Hamer et al., 2021; Toscos et al., 2010).

Studies on fear among different age groups also vary. In other words, fears of injury, enacted stigma, and embarrassment have been observed across a range of age groups (Hamer et al., 2021). However, in younger age groups, the fears of pain, movement, and falling have rarely been reported.

Moreover, environmental barriers include but are not limited to having accessibility to safe spaces, healthcare inclusion and health conditions, financial constraints, social and cultural norms, family relationships (e.g., being married or having kids at home), work responsibilities, health and physical literacy levels, and/or access to information/educational resources. Access to safety-conscious places (i.e., recreational facilities, public parks) regard places areas that are risk-free, cool in the summer or warm in the winter, conveniently located near residences or places of employment, and inexpensive at no cost or cheap cost (Bantham et al., 2021).

Among the most frequent concerns are general safety, verbal abuse, gang-related activity, physical harm or assault, and gun violence (Andrade et al., 2018; Bantham et al., 2021; Toscos et al., 2010). Chronic health conditions have frequently been reported as challenges that prevent or make it difficult for individuals to be physically active, thus many individuals have identified that their current physical condition is a barrier for being unable to exercise (Hamer et al., 2021; Toscos et al., 2010).

The absence of informational and educational resources and lack of cultural considerations have also been linked to a higher risk of inactivity, according to observational research (Bantham et al., 2021; Sebastião et al., 2015). Some people believe they obtain enough PA from their daily activities because they are unaware of how much PA they should be performing to improve their health. This may be often brought on by an inadequate of knowledge availability, linguistic constraints, or trouble understanding offered information due to highly technical language and unfamiliarity with activities described in informational materials (Sebastião et al., 2015). Henceforth, if important cultural, social, political, historical, and economic elements within communities are not considered, the acceptability, viability, and sustainability of PA initiatives are jeopardized.

Without contemplating the role of cultural and societal value, the implementation of interventions may be inaccurate, as it risks being solely based on generalizations about a community or the researchers' own beliefs rather than objectively established, culture-centred hypotheses (Bantham et al., 2021; Sebastião et al., 2015). To overcome these barriers, literature suggests using methods that are linguistically and culturally suitable. Raising awareness is just one of them.

Others include creating PA communities, engaging in community-based participatory research, and forming interdisciplinary research teams (Bantham et al., 2021). In fact, addressing inequality in access to safe, inclusive, accessible, inexpensive, and suitable locations is a guiding principle of the WHO's Global Action Plan on PA for the 2018 to 2030. Therefore, multi-level intervention must be implemented to alleviate PA inequities in both general and marginalised population.

6. HOW DIGITAL CAN STAND IN A PHYSICAL DIMENSION

Digital interventions can be described as digital technology-based tools and initiatives that encourage or promote behavioural change (Michie et al., 2017b) using smartphone applications, websites, social network, text messaging, or personal digital assistants (Rose et al., 2018). Given their social support properties and communal potential to encourage PA, digital interventions may contribute positively to overcoming barriers and increasing PA (Manzoor et al., 2016; Zeng et al., 2018).

Although there is less proof of the effectiveness of solely digital interventions to promote a healthy diet, physical exercise, or weight control than there is for face-to-face interventions, evidence shows that proactive messaging (i.e., reminders to participate in behaviour change activities, feedback on progress, or advice) results in successful interventions (Rhodes et al., 2020). This suggests that interactivity may increase participation and increase effectiveness.

Evidence confirms that the internet may be a useful tool for extending the geographic scope of an initial, offline project and converting it into more pervasive public health practice (Mummery et al., 2006). Furthermore, literature suggests that using social networks rather than a PA monitoring system may encourage and support people in their regular PA (Elloumi et al., 2018; Resnick et al., 2010).

Since digital interventions' effectiveness seems to be dependent on socioeconomic status (Western et al., 2021), research recommends that the development of digital interventions considers literacy levels' inclusiveness (Rhodes et al., 2020). Moreover, majority of people have favourable opinions of most interactive and audio-visual elements' components and functionalities in digital interventions, independently of literacy levels (Rowsell et al., 2015).

Certain features, such as interactive individual challenges, are positively associated with increase of user engagement on websites (Davies et al., 2012). And, whereas interactive activity planners are perceived as difficult to complete and have biased responses, website features (i.e., interactive quiz, and tailored feedback), and content (i.e., images and audio-visual sequences) are perceived as pertinent, entertaining, motivating, and helpful (Rowsell et al., 2015).

However, notwithstanding the benefits in digital interventions, long-term engagement seems to be a frequently reported problematic. In the beginning, participants see a lot of potential for this type of online practice (García-Monge et al., 2019). Yet, subsequently, there is a decline in user engagement. Reported factors include issues such as lack of time, laziness, increased caution when uploading texts to an online forum, anxiety about being judged, and shyness around other more academically accomplished users who participate in discussion boards (Davies et al., 2012; García-Monge et al., 2019; Tague et al., 2014).

Essentially, using digital interventions could be a way to promote PA more dynamically and targeting a wider audience at reduced costs (Rhodes et al., 2020; Rose et al., 2018). This ought to be particularly applicable if initiatives prioritize adaptability in literacy levels and accessibility of resources (Bantham et al., 2021; Sebastião et al., 2015; Western et al., 2021). Nonetheless, despite the potential benefits of employing digital interventions to promote PA, several obstacles must be overcome (Malinen & Nurkka, 2015, Sanders et al., 2019). One of the most critical ones being: the long-term maintenance of user engagement (Tague et al., 2014, Toscos et al., 2010).

7. ONLINE COMMMUNITIES TO ENCOURAGE PHYSICAL ACTIVITY

It is undeniable that the public continues to struggle with satisfying WHO guidelines for healthy lifestyles in terms of PA (World Health Organization, 2020; Zhao et al., 2020). Similarly, we found in this study that OC has not been effectively employed in increasing PA. In reality, the current dynamics haven't been strong enough to continuously promote PA in a way that encourages long-term adoption of more desirable PA levels.

The combination of social relationships' pertinence and the undeniably significant component of efficient communication in DP in the modern world in online communities (Groenewegen et al., 2012; Resnick et al., 2010) and its significance within the context of science (Popp and Woratschek, 2016; Vandelanotte et al., 2019) provide a strong case for future PA interventions. This may present a chance for more effective PA promotion in online forums in an effort to lessen a currently recognized public health issue (Bordalo et al., 2015; Camões and Lopes, 2008; Do Carmo et al., 2008; Lopes et al., 2017).

Accordingly, it would be extremely helpful to undertake studies in online communities connected to PA digital interventions, to comprehend how they can serve as a tool to promote behaviours, and to discover characteristics that may affect long-term user participation in online communities. Particularly, discussions in online communities with the aim of promoting PA, to prevent lifestyles leading to in sedentarism and obesity (Gordon-Larsen et al., 2004; Zhao et al., 2020) and, subsequently, chronic diseases that significantly reduce quality of life and overall life expectancy (Biddle, 2016; World Health Organization, 2020).

ACKNOWLEDGEMENT

This work is financially supported by national funds through FCT – Foundation for Science and Technology, I.P., under the project UIDB/05460/2020. This work is supported by Fundação para a Ciência e Tecnologia (FCT) under the PhD grant SFRH/BD/144296/2019.

REFERENCES

Agnes, M. (Ed.), 2002. Webster's New World, 2nd ed. Wiley Publishing, Cleveland.

Andrade, E.L., Evans, W.D., Barrett, N., Edberg, M.C., Cleary, S.D., 2018. JMIR Public Health Surveill 4, e71.

Ardichvili, A., Page, V., Wentling, T., 2003. 7, 64–77.

Aubert, S., Brazo-Sayavera, J., González, S.A., Janssen, I., Manyanga, T., Oyeyemi, A.L., Picard, P., Sherar, L.B., Turner, E., Tremblay, M.S., 2021. Global prevalence of physical activity for children and adolescents; inconsistencies, research gaps, and recommendations: a narrative review. International Journal of Behavioral Nutrition and Physical Activity.

Bantham, A., Taverno Ross, S.E., Sebastião, E., Hall, G., 2021. Overcoming barriers to physical activity in underserved populations. Prog Cardiovasc Dis.

Biddle, S., 1992. Psychological Benefits of Exercise and Physical Activity.

Biddle, S., 2016. World Psychiatry 15, 175–176.

Boratto, L., Carta, S., Fenu, G., Manca, M., Mulas, F., Pilloni, P., 2017. Pervasive Mob Comput 36, 98–114.

Bordalo, A., de Carvalho, Á.A., Oliveira, A.L., Pedroso, A., Silva, A.J., Diniz, A., Al., E., 2015. Direção-Geral da Saúde 1–136.

Bouchard, C., Shephard, R.J., Stephens, T., 1994. Physical activity, fitness, and health: International proceedings and consensus statement., Physical activity, fitness, and health: International proceedings and consensus statement. Human Kinetics Publishers, Champaign, IL, England.

Brown, Justin C.; Kerri Winters-Stone; Augustine lee; Schmitz, K.H., 2014. Compr Physiol. 2, 2775–2809.

Cabrera de León, A., Rodríguez-Pérez, M.D.C., Rodríguez-Benjumeda, L.M., Anía-Lafuente, B., Brito-Díaz, B., Muros de Fuentes, M., Almeida-González, D., Batista-Medina, M., Aguirre-Jaime, A., 2007. Rev Esp Cardiol 60, 244–250.

Camões, M., Lopes, C., 2008. Revista Saúde Pública 42, 208–216.

Cardon, G., Salmon, J., 2020. J Sport Health Sci 9, 335–338.

Do Carmo, I., Dos Santos, O., Camolas, J., Vieira, J., Carreira, M., Medina, L., Reis, L., 2006. The International Association for the Study of Obesity 233–237.

Do Carmo, I., Dos Santos, O., Camolas, J., Vieira, J., Carreira, M., Medina, L., Reis, L., Myatt, J., Galvão-Teles, A., 2008. Obesity Reviews 9, 11–19.

Carnethon, M.R., 2009. Am J Lifestyle Med 3, 44S-49S.

Daskapan, A., Tuzun, E.H., Eker, L., 2006. J Sports Sci Med.

Davies, C., Corry, K., Van Itallie, A., Vandelanotte, C., Caperchione, C., Mummery, W.K., 2012. J Med Internet Res 14.

Elloumi, L., van Beijnum, B.J., Hermens, H., 2018. Health Technol (Berl) 8, 81–95.

Ferreira, J.S., 2010. Prevalência de obesidade infanto-juvenil: Associação com os hábitos alimentares, actividade física e comportamentos sedentários dos adolescentes escolarizados de Portugal Continental (Doctorate's). Universidade Nova de Lisboa.

García-Monge, A., González-Calvo, G., Bores-García, D., 2019. Open Learning 34, 257–272.

Gordon-Larsen, P., Nelson, M.C., Popkin, B.M., 2004. Am J Prev Med 27, 277–283.

Groenewegen, M., Stoyanov, D., Deichmann, D., Van Halteren, A., 2012. Lecture Notes in Computer Science (including subseries Lecture Notes in Artificial Intelligence and Lecture Notes in Bioinformatics) 7710, 96–109.

Gropper, S.S., Simmons, K.P., Connell, L.J., Ulrich, P. V., 2012. Applied Physiology, Nutrition, and Metabolism 37, 1118–1123.

Guthold, R., Stevens, G.A., Riley, L.M., Bull, F.C., 2018. Lancet Glob Health 6, e1077–e1086.

Hamasaki, H., 2016. World J Diabetes 7, 243.

Hamer, O., Larkin, D., Relph, N., Dey, P., 2021. Obesity Reviews 22.

Hayes, G., Dowd, K.P., MacDonncha, C., Donnelly, A.E., 2019. Tracking of Physical Activity and Sedentary Behavior From Adolescence to Young Adulthood: A Systematic Literature Review. Journal of Adolescent Health.

Hegde, S.M., Solomon, S.D., 2017. Physiol Behav 176, 139–148.

Huber, B.C., Steffen, J., Schlichtiger, J., Graupe, T., Deuster, E., Strouvelle, V.P., Fischer, M.R., Massberg, S., Brunner, S., 2020. Alteration of physical activity during COVID-19 pandemic lockdown in young adults. J Transl Med.

Kovacs, V.A., Starc, G., Brandes, M., Kaj, M., Blagus, R., Leskošek, B., Suesse, T., Dinya, E., Guinhouya, B.C., Zito, V., Rocha, P.M., Gonzalez, B.P., Kontsevaya, A., Brzezinski, M., Bidiugan, R., Kiraly, A., Csányi, T., Okely, A.D., 2021. Eur J Sport Sci.

Lanier, J.B., Bury, D.C., Richardson, S.W., 2016. Am Fam Physician 93.

Lopes, C., Torres, D., Oliveira, A., Severo, M., Alarcão, V., Guiomar, S., Mota, J., Teixeira, P., Rodrigues, S., Lobate, L., Magalhães, V., Correia, D., Pizarro, A., Marques, A., Vilela, S., Oliveira, L., Nicola, P., Soares, S., Ramos, E., 2017. Inquérito Alimentar Nacional e de Atividade Física.

Malina, R.M., 2001. J. Hum. Biol 13, 162–172.

Malinen, S., Nurkka, P., 2015. International Journal of Web Based Communities 11, 153–169.

Mansoubi, M., Pearson, N., Clemes, S.A., Biddle, S.J.H., Bodicoat, D.H., Tolfrey, K., Edwardson, C.L., Yates, T., 2015. BMC Public Health 15.

Manzoor, A., Mollee, J.S., Aráujo, E.F.M., Halteren, A.T. van, Klein, M.C.A., 2016. Online Sharing of Physical Activity: Does It Accelerate the Impact of a Health Promotion Program?, in: 2016 IEEE International Conferences on Big Data and Cloud Computing (BDCloud), Social Computing and Networking (SocialCom), Sustainable Computing and Communications (SustainCom) (BDCloud-SocialCom-SustainCom). pp. 201–208.

Michie, S., Yardley, L., West, R., Patrick, K., Greaves, F., 2017a. Developing and evaluating digital interventions to promote behavior change in health and health care: Recommendations resulting from an international workshop. J Med Internet Res.

Michie, S., Yardley, L., West, R., Patrick, K., Greaves, F., 2017b. Developing and evaluating digital interventions to promote behavior change in health and health care: Recommendations resulting from an international workshop. J Med Internet Res.

Mummery, W.K., Schofield, G., Hinchliffe, A., Joyner, K., Brown, W., 2006. J Sci Med Sport 9, 424–430.

O'Donovan, G., Stensel, D., Hamer, M., Stamatakis, E., 2017. Eur J Epidemiol 32, 559–566.

Owen, N., Healy, G.N., Matthews, C.E., Dunstan, D.W., 2010. Exerc Sport Sci Rev 38, 105–113.

Ozemek, C., Lavie, C.J., Rognmo, Ø., 2019. Global physical activity levels - Need for intervention. Prog Cardiovasc Dis.

Pope, L., Hansen, D., Harvey, J., 2016. J Nutr Educ Behav 49, 137-141.e1.

Popp, B., Woratschek, H., 2016. Sport Management Review 19, 183–197.

Resnick, P.J., Janney, A.W., Buis, L.R., Richardson, C.R., 2010. J Med Internet Res 12, e72p.1-e72p.11.

Rhodes, A., Smith, A.D., Chadwick, P., Croker, H., Llewellyn, C.H., 2020. Exclusively digital health interventions targeting diet, physical activity, and weight gain in pregnant women: Systematic review and meta-analysis. JMIR Mhealth Uhealth.

Rito, I., Graça, P., 2015. Surveillance Childhood Obesity Initiative - COSI:2013.

Rose, T., Barker, M., Jacob, C., Morrison, L., 2018. 61, 669–677.

Rowsell, A., Muller, I., Murray, E., Little, P., Byrne, C.D., Ganahl, K., Müller, G., Gibney, S., Lyles, C.R., Lucas, A., Nutbeam, D., Yardley, L., 2015. J Med Internet Res 17.

Sanders, I., Short, C.E., Bogomolova, S., Stanford, T., Plotnikoff, R., Vandelanotte, C., Olds, T., Edney, S., Ryan, J., Curtis, R.G., Maher, C., 2019. JMIR Mhealth Uhealth 7, e12484.

Sebastião, E., Chodzko-Zajko, W., Schwingel, A., 2015. BMC Public Health 15.

Swirski, M., Shaikh, N., Chinner, A., Gaaikema, E., Coulthard, E., 2019. bioRxiv.

Tague, R., Maeder, A.J., Vandelanotte, C., Kolt, G.S., Caperchione, C.M., Rosenkranz, R.R., Savage, T.N., Van Itallie, A., 2014. Stud Health Technol Inform 206, 84–92.

Tamellin, T., 2003. Physical Activity From Adolescence To Adulthood and Health- Related Fitness At Age 31, Public Health.

Thorp, A.A., Owen, N., Neuhaus, M., Dunstan, D.W., 2011. Sedentary behaviors and subsequent health outcomes in adults: A systematic review of longitudinal studies, 19962011. Am J Prev Med.

Toscos, T., Consolvo, S., McDonald, D.W., 2010. IHI'10 - Proceedings of the 1st ACM International Health Informatics Symposium 346–355.

Vandelanotte, C., Duncan, M.J., Kolt, G.S., Caperchione, C.M., Savage, T.N., Van Itallie, A., Oldmeadow, C., Alley, S. J., Tague, R., Maeder, A.J., Rosenkranz, R.R., Mummery, W.K., 2019. More real-world trials are needed to establish if web-based physical activity interventions are effective. Br J Sports Med.

Vasques, C., Mota, M., Correia, T., Lopes, V., 2012. Revista Portuguesa de Cardiologia 31, 783–788.

Watson, K.B., Whitfield, G.P., Huntzicker, G., Omura, J.D., Ussery, E., Chen, T.J., Fanfair, R.N., 2021. International Journal of Behavioral Nutrition and Physical Activity 18.

Western, M.J., Armstrong, M.E.G., Islam, I., Morgan, K., Jones, U.F., Kelson, M.J., 2021. The effectiveness of digital interventions for increasing physical activity in individuals of low socioeconomic status: a systematic review and meta-analysis. International Journal of Behavioral Nutrition and Physical Activity.

Wilke, J., Mohr, L., Tenforde, A.S., Edouard, P., Fossati, C., González-Gross, M., Ramírez, C.S., Laiño, F., Tan, B., Pillay, J.D., Pigozzi, F., Jimenez-Pavon, D., Novak, B., Jaunig, J., Zhang, M., van Poppel, M., Heidt, C., Willwacher, S., Yuki, G., Lieberman, D.E., Vogt, L., Verhagen, E., Hespanhol, L., Hollander, K., 2021. Int J Environ Res Public Health 18, 1–12.

World Health Organization, W., 2000. Obesity: Preventing and Managing the Global Epidemic.

World Health Organization, W., 2020. WHO Guidelines on physical activity and sedentary behaviour, World Health Organization.

Wu, Y., Kankanhalli, A., Huang, K.W., 2015. 2015 International Conference on Information Systems: Exploring the Information Frontier, ICIS 2015 1–12.

Zeng, L., Almquist, Z.W., Spiro, E.S., 2018. Lecture Notes in Computer Science (including subseries Lecture Notes in Artificial Intelligence and Lecture Notes in Bioinformatics) 10914 LNCS, 137–147.

Zhao, M., Veeranki, S.P., Magnussen, C.G., Xi, B., 2020. The BMJ 370.

DECISION-MAKING REGARDING LABOR AND BIRTH: AN ANALYSIS OF ONLINE SYSTEMS FOR EXPECTANT PARENTS

Carla V. Leite[1,2], Rita Oliveira[1], Catarina Lélis[1] and Ana Patrícia Oliveira[1]
[1]*DigiMedia, Department of Communication and Art, University of Aveiro*
Campus Universitário de Santiago, 3810-193 Aveiro, Portugal
[2]*Department of Nursing Science, Faculty of Medicine, University of Turku*
Medisiina B6, FI-20520 Turku, Finland

ABSTRACT

Decision making during the perinatal period is a complex, multidimensional and individual process. A Birth Plan, as a communication tool, can help promote perinatal literacy, to trigger discussions between expectant parents and healthcare professionals, and clarify decisions, needs, and expectations. To understand what topics are included in online decision-making systems, also referred as birth plan generators, and how complete the options provided are, this research was conducted. Using a web search procedure, four plans were selected, and their content analyzed. From the emerged data, 13 Categories and 59 Subcategories were clustered, and a score from 2 to 8 was assigned depending on completeness. The most common Subcategories found were *Water* (Type of birth) and (Freedom) *To move*, which scored the highest; while the least common were: *Updates, First Language, Religion, Privacy, Urinary catheterization, End of Placenta,* (Freedom) *To breath* and *To vocalize*, which scored the lowest possible. Several inferences were discussed regarding the subcategories results, and in conclusion, it is suggested that the clustered data be used as an information architecture for future work, to develop a better and more comprehensive decision support solution for labor and birth.

KEYWORDS

Decision-Making, Birth Plan, Perinatal, Labor, Content Analysis, Reproductive Health Literacy

1. INTRODUCTION

A Birth Plan (BP) is a self-care intervention for health, and a tool for comprehensive sexuality education (WHO, 2019), since the process of its creation is a meaningful communication trigger between expectant parents and health care professionals. It requires the discussion of options and clarification of perceptions, beliefs, fears, and motivations, leading to awareness, knowledge building, and ultimately, informed decision- making (Whitford et al., 2014). It is thought to have a positive impact on the expectant parents' satisfaction towards labor and birth, by encouraging their participation and supporting their decisions (Lothian, 2007). In addition, studies show that expectant parents who have created a BP tend to be better prepared, to feel being more in control, and to be more involved in the process, which helps reduce anxiety and stress during labor and birth (Hidalgo-Lopezosa et al., 2017).

Although several prestigious institutions around the world advocate for childbirth education and the creation of a BP (AAP and ACOG, 2012), there is a well-documented negative perception and controversy about its use among health care professionals (Afshar et al., 2017), given that the course of labor and birth is uncertain and can become complicated at any time, and unforeseen events can occur that can lead to a "breach" of the previously established birth plan (Hidalgo-Lopezosa et al., 2017).

Considering that not every detail during the perinatal period can be predicted, a BP is more accurately described as a birth preferences' document (Whitford et al., 2014); a practical guide, so that the midwifery team can respond as humanely and holistically as possible, following a person-centered approach (WHO, 2019; Suárez-Cortés et al., 2015). It could be said, that a generator of Birth Plans is a decision-making system that ideally includes multiple scenarios and all the issues regarding the entire process of birthing.

Birth Plans emerged in 1970 due to criticism of overmedicalized child birthing practices (Yam et al, 2007); in the 1980s, WHO classified them in the top category of recommended practices for safety reasons (WHO, 2006); however, no template has been made available worldwide, so some health systems provide the option at the country level to use a birth plan created from scratch, while others provide a local template (Suárez-Cortés et al, 2015).

This study is a follow-up of a previously published benchmarking analysis (Leite and Almeida, 2022), so this time a content analysis was conducted in order to understand the online decision support solutions related to labor and birth that were found earlier. The goal is to establish how complete the options offered are, to pave the way to the creation of a new complete BP generator, as it may affect how expectant parents exercise their fundamental right to know and to decision-make, their expectations, participation and autonomy, along with how midwifery care is provided (Inkeroinen et. al, 2022; WHO, 2019).

2. METHODS

A Content Analysis was conducted with the goal of identifying and acknowledge how the Internet is supporting the creation of birth plans, in order to answer the following question: *What topics are mentioned in the birth plans, and how complete are the options provided?*

Options related to labor and birth were analyzed; while options related to infants and puerperium were considered out of the scope of this work. The purpose of this analysis is to show the occurrence and completeness of key themes (Marshall and Rossman, 1989).

The solutions included in this study resulted from the procedure established on the previously published work (Leite and Almeida, 2022), regarding a benchmarking analysis of e-health services to support the perinatal period that made use a Web search procedure for the key words/phrases: "birth plan generator", "create birth plan" and "birth preferences", that better ranked and were considered as contributing to this study according to the following selection criteria: (a) available online; (b) pregnant women as the target audience; (c) labor and/or birth plan creation features; (d) in English. Redundancies were eliminated, so the similar ones were not included in this list. The selected plans were coded using the capital letters of the website or institution that made them available online: National Health Service from the United Kingdom (NU); The Bump (TB); She Knows (SK); Mama Natural (MN).

To examine the material, the raw data were organized and clustered in order to highlight their characteristics, as already acknowledged on table 1 from the previous work. The Categories and Subcategories that emerged from a single plan or from the combination of two or more plans, are mutually exclusive, homogeneous, relevant, productive, objective and reliable (Bardin, 2008), and are aforementioned on the figures 1-13 of the Results Section of this article.

The presence of each Subcategory was addressed, and a score was assigned according to its completeness, in comparison to each BP. Since a binary analysis was not possible, three different classifications were established: a) Complete (2 points); b) Incomplete (1 point); c) Severely incomplete or Not mentioned (0 points). The Subcategory was classified as Complete if the majority of options were covered by the respective BP; it was considered Incomplete if some of options were covered but not all; it was considered Severely incomplete or Not mentioned if the topic was either not mentioned or it was mentioned but not presented as an option and no decision was possible to state.

The scoring allowed an overview of each plan, and also an analysis of each Subcategory itself. The maximum possible score for each Subcategory was 8 points, meaning that all four plans contained Complete options, while the minimum score was 2 points if only one plan mentioned that Subcategory or if two plans contained complementary options.

3. RESULTS

The "Introduction" Category emerged, clustering four Subcategories (figure 1) that clarify: how to name the document; who the expectant parents and soon-to-be born child are; their intentions regarding the place of birth; the clinical history of the pregnant person.

The *Headline* is present in all BPs, and indicates how to address the document and may clarify or imply the expectations, intentions, and approach of the expectant parents. Only SK plan was considered Complete since it allows to choose one of the six options for titles rather than a rigid title; it also allows a brief description of the plan's purpose with a sample text that can be customized.

Identification refers to clarifying who is involved in the plan. In addition to the name of the pregnant person, the planned name for the baby can also be entered at SK, and the name will be then automatically used in the document when referring to the baby.

In the *Clinical history* sub-category, in addition to the Due/Induction date, TB and SK ask for other clinical details, namely: Group B streptococcus status, Rh compatibility and, gestational diabetes.

Location was highly explored by NU plan, since it lists possible places for the birth to occur, namely: home, midwifery station, hospital maternity ward, as well as the option of not yet being sure about the place.

Introduction	**NU**	**MN**	**TB**	**SK**	**Total**
Headline	1	1	1	2	**5**
Identification	1	0	1	2	**4**
Clinical History	1	0	2	2	**5**
Location	2	0	1	1	**4**

Figure 1. "Introduction" Category results, by Sub-categories and by Plan

The "Consent and Information" category emerged (figure 2), which clustered options on how expectant parents wanted to deal with information during labor and birth.

The *Update* Subcategory received the lowest possible score. TB offers the option to be kept informed at all times in the event of a C-section, or only upon request; while, SK offers the option to request information in advance related to any situation, but does not offer the option to avoid updates. Those two plans shared the minimum score since they offer complementary options.

Regarding *Informed of options* Subcategory, SK offers three options for the need of discussion and active participation of expectant parents on the decisions. TB only offers that option in case a C-section deems necessary, adding: "*I would like: A second opinion / To make sure all other options have been exhausted*".

The *Professional's decision* Subcategory refers to the possibility that expectant parents may decline to decide under certain circumstances and leave it to the professional to do it to the best of their knowledge. Although all four plans mentioned it, they were considered Incomplete, since they only offer options such as *"if the provider deems necessary"* in 1 to 4 cases each along the full plan.

Consent and Information	**NU**	**MN**	**TB**	**SK**	**Total**
Update	0	0	1	1	**2**
Informed of options	0	0	1	2	**3**
Professional's decision	1	1	1	1	**4**

Figure 2. "Consent and Information" Category results, by Sub-categories and by Plan

Factors that are not close related to the perinatal period may influence the labor and birth experience too, so "Particular Requirements" Category emerged from the analysis (figure 3).

NU was the only plan that considered language as a potential barrier to communication, and therefore it offers the option of asking for someone to interpret English into the person's first language. In terms of *Accessibility*, together the plans offer: sign language interpreter, impaired vision, hard hearing, and "other special needs".

The option for "certain religious customs to be observed" appears only on the NU plan, not specifying which ones can be offered, so it is left to the expectant parents to fill them in the text box.

Regarding *To Bring something from home* Subcategory, both NU and TB offer the possibility of bringing equipment from home; NU has a text box to specify what, while TB suggests four ideas of what to carry.

Particular requirements	**NU**	**MN**	**TB**	**SK**	**Total**
First language	2	0	0	0	**2**
Accessibility	1	0	1	1	**3**
Religion	2	0	0	0	**2**
To bring something	1	0	2	0	**3**
To record/photograph	0	0	2	2	**4**

Figure 3. "Particular Requirements" Category results, by Sub-categories and by Plan

The Category "Freedom" is related to various needs and behaviors during the perinatal period (figure 4). *To move* was a widely explored Subcategory, besides body movement, some plans also offered the opposite: laying down, or/and just standing up. Being able *To breathe* freely and without guidance was mentioned on two plans in a complementary way, whereas *To vocalize* was explored on only one as: "Please allow me to vocalize as desired during labor and birth without comment or criticism", making it one of the Subcategories with the lowest score. Two of the plans explore the options regarding *To push*: follow the instinct to push; no time restrictions; to have guidance/coaching; to have support or freedom while pushing.

Regarding *To change position* Subcategory, different approaches emerged from the analysis: six positions are suggested for labor, and twelve for birth on TB plan; an explanation that labor position may be dynamic process and seven birth positions are suggested on NU; and, SK offer the choice of freedom to change position during labor, and the position for birth.

To eat/drink options are offered on three of the plans; and *What to wear* their own clothes, instead of a hospital gown is an option explored in only two plans.

Freedom	**NU**	**MN**	**TB**	**SK**	**Total**
To move	2	2	2	2	**8**
To breathe	1	0	1	0	**2**
To vocalize	0	0	0	2	**2**
To push	0	0	2	2	**4**
To change position	2	0	2	2	**6**
What to wear	0	0	2	2	**4**
To eat and drink	0	2	2	2	**6**
Others	0	0	1	1	**2**

Figure 4. "Freedom" Category results, by Sub-categories and by Plan

"Ambience and Equipment" category is related to different options to set the perinatal milieu along with relaxation and birth equipment (figure 5).

Privacy is one of the Subcategories with lowest score, and the options offered are complementary between two plans, one allowing to request a special room for the perinatal period, and the other allowing to request the door to be closed and people to knock first.

Requests regarding *Silence or Music* playing, and the usage of a *Mirror* (Vaginal birth) or a *Low screen* (gentle c-section) were considered in some plans. Despite NU offering only one option regarding Ambience, it offers a great amount of options for Equipment (e.g., beanbag, mat, birthing pool, equipped room and brought from home). Also, TB adds other possibilities: foot pedals, squatting bar, birthing tub, chair and stool.

Ambience and Equipment	**NU**	**MN**	**TB**	**SK**	**Total**
Light	0	2	2	2	**6**
Mirror or Low screen	0	1	2	2	**5**
Silence or Music	0	0	2	2	**4**
Equipment	2	0	1	0	**3**
Privacy	1	0	0	1	**2**

Figure 5. "Ambience and Equipment" Category results, by Sub-categories and by Plan

Regarding "People" involved or present during labor and birth, four Subcategories emerged (figure 6), with TB plan standing out by covering almost completely all Subcategories.

Companion or Partner is mentioned in three of the four plans; however, it was considered incomplete when the option did not count with a textbox to name the person. *Support (doula/coach)* was fairly mentioned with only one plan leaving the option behind, while *Students/interns/in training* achieved a high score of 7.

TB asks for the Medical Doctor's name, not specifying whether this was intended for the physician following the pregnancy or the one planned to assist the birth to be named. On the other hand, SK considers the possibility of various health care professionals to be involved on the birthing process.

People	NU	MN	TB	SK	Total
Companion/Partner	1	0	2	1	**4**
Support (doula/coach)	2	0	2	2	**6**
Students/interns/In training	1	2	2	2	**7**
Health care professionals	0	0	1	2	**3**
Others (siblings, visitors, etc)	0	1	2	0	**3**

Figure 6. "People" Category results, by Sub-categories and by Plan

"Type of birth" Category clusters four sub-categories (figure 7), and TB plan achieved the highest score possible. Clearly, a *VBAC* and a *Water birth* are also Vaginal births, but the set of options describing each one of these routes, and the fact that they imply different direct and indirect choices, were reasons enough to be considered as distinct Subcategories.

Vaginal birth options are only fully explored in one of the plans, while the others fail to include both instrumental (forceps or vacuum extraction) and non-instrumental routes.

Water birth is related to the option of being on a tub, pool or shower during the second stage of birth - from the complete dilatation of cervix to total expulsion of the new-born (Hutchison et al, 2020), and this Subcategory is highlighted in all plans scoring the maximum.

Options regarding *C-section* (elective or emergency surgery performed to deliver the new-born), are mentioned in all plans, but fully offered in only half of them.

Type of birth	NU	MN	TB	SK	Total
Vaginal	1	1	2	1	**5**
Water	2	2	2	2	**8**
C-section	1	1	2	2	**6**
VBAC	0	2	2	0	**4**

Figure 7. "Type of birth" Category results, by Sub-categories and by Plan

"Early Interventions" category cluster seven sub-categories (figure 8) that emerged from options related to procedures related to admission and to the labor period.

The contrast between NU plan and TB and SK plans, is noticeable, with the latter two considering early interventions worth mentioning. They also offered conditional options based on clinical status (e.g., *Stripping membranes performed only if baby is in distress*). *Urinary catheterization* Subcategory scored the lowest possible, with only TB plan offering both its request and refusal (offering "*to use the bathroom as needed or desired*" as the other choice).

Early Interventions	NU	MN	TB	SK	Total
Trichotomy (shaving)	0	0	2	2	**4**
Enema (clyster)	0	0	2	2	**4**
Perfusion	0	2	2	2	**6**
Cervical exams	0	2	2	2	**6**
Rupture/Stripping Membranes	0	2	2	2	**6**
Urinary Catheterization	0	0	2	0	**2**
Alternative methods	0	0	2	2	**4**

Figure 8. "Early Interventions" Category results, by Sub-categories and by Plan

"Fetal Monitoring" Category clusters three Subcategories related to the different ways to perform it (figure 9). The most common option found was related to the usage of *Doppler* device, scoring 7 points.

Others sub-category is related to different ways to approach the question, namely the wished frequency, and which situations are deemed necessary to perform the monitoring.

Fetal monitoring	NU	MN	TB	SK	Total
Cardiotocography	1	0	2	0	**3**
Doppler	1	2	2	2	**7**
Others	1	0	2	1	**4**

Figure 9. "Fetal monitoring" Category results, by Sub-categories and by Plan

"Pain management" category is related to different means to deal with and to relief pain during labor and birth and was fairly mentioned in all plans (figure 10). Options regarding *Water* usage to manage pain during labor, through shower, pool or tub, was considered on this Subcategory, since they do not necessarily imply a *Water birth*. *Other* sub-category includes several alternative therapies.

Pain management	**NU**	**MN**	**TB**	**SK**	**Total**
Anesthesia	2	1	2	2	**7**
Analgesia	2	1	2	2	**7**
Water	2	2	2	1	**7**
Others	1	1	2	1	**5**

Figure 10. "Pain management" Category results, by Sub-categories and by Plan

"Medication" category includes options for requesting medication administrated for *Induction or augmentation* and *Other* purposes (figure 11).

The contrast between NU and the other plans is again striking, with the *Induction/Augmentation* Subcategory mentioned in detail in the TB and SK plans.

Medication	**NU**	**MN**	**TB**	**SK**	**Total**
Induction/Augmentation	0	1	2	2	**5**
Others	0	1	1	1	**3**

Figure 11. "Medication" Category results, by Sub-categories and by Plan

"Late and post interventions" category is related to procedures and interventions related to the second and third stage of birth - expulsion and immediate post-partum (figure 12).

Perineal relaxation through different techniques (*massage with oil* or *hot compresses*) is offered in both TB and SK, contrasting with NU and MN that don't mention any.

Options regarding the usage of *Forceps or Vacuum extraction* are displayed on MN and TB plans, while SK fail to mention it, and on NU they are not offered as Consent/Refuse clinical intervention, but rather the preference of who can stay in the room while the intervention is being carried out.

Although *Episiotomy*, the surgical incision through the perineum is referred in all the plans, only TB and SK plans offer opposite preferences and conditional decisions. If an *Episiotomy* is performed or if the woman suffers a second-degree tear, since damage of muscle layer is involved, *Perineal suture* is needed to repair it (Kettle et al., 2010). TB only offers a decision regarding the use of local anesthesia during repair, while SK plan also offers an option to avoid stitching in case the tear is not severe. Even though *Episiotomy* was rated 6, *Perineal suture* received only 3 points because it was not mentioned in two of the plans.

Late and post interventions	**NU**	**MN**	**TB**	**SK**	**Total**
Perineal relaxation	0	0	2	2	4
Forceps or Vacuum extraction	1	2	2	0	5
Episiotomy	1	1	2	2	6
Perineal suture	0	0	1	2	3

Figure 12. "Late and post interventions" Category results, by Sub-categories and by Plan

"Umbilical cord and Placenta" category clusters two Subcategories focused on umbilical cord: *Clamp/Cut* and *Blood/Steam cells*; and three on placenta: *Discharge procedure*, its *Observation* and its *End* (figure 13), since they are physically interconnected.

Regarding *Clamp/Cut* of the umbilical cord, SK plan stands out by offering options related to waiting time and to name the person to perform the cut. It also offers options regarding the collection for banking, storage or donation of *Blood/Stem cells*.

Despite *Discharge* options being somehow offered in all plans, they are in fact incomplete and complementary, each one being related to different topics: maneuvers, guidance or timing.

NU and MN fail to mention any option regarding the *Observation* of the placenta from expectant parents, while TB and SK take it in consideration. The options for the *End* of the placenta are only offered by MN plan, through the inclusion/exclusion of a *"Save placenta"* icon. This Subcategory scored the lowest possible.

Umbilical cord and placenta	NU	MN	TB	SK	Total
Clamp or Cut	0	1	1	2	4
Blood or Stem cells	0	1	1	2	4
Discharge	1	1	1	1	4
Observation	0	0	2	2	4
End	0	2	0	0	2

Figure 13. "Umbilical cord and placenta" Category results, by Sub-categories and by Plan

4. DISCUSSION & CONCLUSIONS

Overall, the most common Subcategories were *Water* (Type of birth) and (Freedom) *To move*, scoring the highest, while the least common were *Updates, First language, Religion, Privacy, Urinary catheterization, End of Placenta,* freedom *To breath, To vocalize, and Others*, scoring the minimum possible. Also, Subcategories from "Consent and Information" Category had remarkably low scores. These results may reflect the most and least common requests and expectations from pregnant people.

Despite the high importance of "Type of birth" for expectant parents according to several studies (Figueiredo et al., 2002; Zakerihamidi et al., 2015), *Vaginal* birth seems to be adopted as the first option in NU, SK and MN plans. This may be due to the fact that vaginal birth is considered the safer path for low-risk pregnancies (WHO, 2016). Nonetheless, the TB plan stands out since it begins with a section in which the pregnant person can indicate what Type of Birth they plan to have; later in the document, it is possible to add details in case unforeseen events lead to a breach of the initial plan (Hidalgo-Lopezosa et al., 2017). Offering alternatives, mentioning risks and benefits of different paths based on scientific evidence, and giving the floor for pregnant people to express their views and communicate their feelings to health experts is considered the best approach to support informed decisions and manage expectations (Stirling et al., 2017).

To breath and *To vocalize* and *Other* Subcategories from "Freedom" scored the lowest, while *To move* achieved the maximum. Indeed, solid studies recognize evidence that freedom of movements during labor and choice of birth position is anatomical and physiologic beneficial (WHO, 2016), as it is a chance to release the muscle tension that increases pain, it allows a sense of greater personal freedom, and can be a distraction from discomfort (Shilling, 2007). Notwithstanding, recent studies on mindfulness-based childbirth consider promising the usage of innovative strategies to address fear and acute pain through meditation techniques, including sitting and walking meditation, body scan, mindful movement, mindful eating, and pain coping strategies, such as: mindfulness of breath, partner touch, and sounding (i.e. using low/loud vocal tones during periods of intense physical sensation) (Duncan, 2017).

When NU requests about settings in the event of Early, Late or Post interventions happening, consent for the event to happen seems to be assumed, and as already established this goes against what is advocated in Beijing Declaration and Platform for Action (United Nations, 1995) and as human right standards: Right to information and Right to make informed-decisions as a patient (WHO, 2019). Moreover, offering the option of any topic to be discussed with the midwife, instead of requesting from the expectant parents their view point, consent, refusal or to ask for alternatives, does not really add up to a BP, since the communication between them and the health professionals is one of the main goals of a BP.

By mentioning several, and even opposing, possible positions to answer the same question, and the inclusion of decisions regarding interventions, it reflects the BP approach as following a patient-centered care, which is preferable than standard protocol (WHO, 2018; Rocha and Novaes, 2010).

Furthermore, not all Subcategories from "Late and Post interventions" category were widely offered by all the plans, with *Perineal suture* scoring only 3, even though a high number of people experiencing some degree of perineum damage after a vaginal birth, and the fact that the damage and following repair may result in perineal pain for two weeks after the birth, or even long-term pain and discomfort during sexual intercourse (Kettle et al., 2010). Moreover, the type of material (e.g.: synthetic absorbable stitches or catgut), could also be relevant options to include for discussion on a plan, since catgut might be refused based on religion or ethics.

It is known that delaying the clamp/cut has benefits for the new-born, however it affects drastically the chances of obtaining clinically useful content and volume for banking (Ciubotariu, 2018). Despite these two options being co-related, no information of this conflicting position is clarified in any of the plans.

In addition, placentas are considered medical waste by the majority of the hospitals worldwide, however a rising number of families are requesting to take the placenta home instead of being disposed, so they can do a ceremony, a special burial or to use it processed as a puerperal remedy (Hayes, 2019; Johnson et al, 2018), and this was only reflected on one of the plans. The same plan offers *Lotus birth* as an option, so the decisions regarding both *Umbilical cord* and *Placenta* should be interconnected, since it the baby remains attached from the baby's naval to the placenta, until the cord naturally detaches, without any cut (Gönenç et al, 2019).

By being a visual plan, MN adds another layer of information and might facilitate readability. However, several Subcategories were not offered, and no real explanation about the topics was provided, which are considered limitations. A BP that covers all relevant issues to be discussed in advance can be an empowering tool for expectant parents and facilitator tool for the birth itself, considering that it is critical to be in quiet surroundings for the pregnant person neocortex stop working, away from all stimulants, including communication, feeling observed, so their primitive brain structures can more easily release the necessary hormones for the labor and birthing process (Odent and Moraloğlu, 2015). To be used as a Consent and Information system by itself, it needs to be approached from both health professionals, pregnant person, and companion's perspective, considering different dimensions: by request, by need, or by frequency. As future work, a proposal for a Decision System related to labor and birth topics could be developed using these findings to build the information architecture, to define the approach and to acknowledge its completeness.

ACKNOWLEDGEMENT

This work is financially supported by national funds through FCT – Foundation for Science and Technology, I.P., under the project UIDB/05460/2020. CVL would like to acknowledge the fund support through the Doctoral Programme in Technology Enhanced Learning and Societal Challenges (PD/00173/2013) by FCT, through the research grants PD/BI/135241/2017, doctoral grant PD/BD/143077/2018 and exceptional grant COVID/BD/152551/2022.

REFERENCES

Afshar, Y., Mei, J. Y., Gregory, K.D., Kilpatrick, S.J. and Esakoff, T.F., 2017. Birth plans-Impact on mode of delivery, obstetrical interventions, and birth experience satisfaction: A prospective cohort study. *Birth*, 45(1), pp.43–49. doi: 10.1111/birt.12320.

American Academy of Pediatrics & The American College of Obstetricians and Gynecologists, 2012. *Guidelines for Perinatal Care* (7th ed.).

Bardin, L., 2008. *Análise de conteúdo (L'analyse de contenu)*. Lisbon, Portugal: Edições 70. Original document in French published by Presses Universitaires de France.

Ciubotariu, R., Scaradavou, A., Ciubotariu, I., Tarnawski, M., Lloyd, S., Albano, M., Dobrila, L., Rubinstein, P. and Grunebaum, A., 2018. Impact of delayed umbilical cord clamping on public cord blood donations: can we help future patients and benefit infant donors? *Transfusion*, 58(6), pp.1427–1433. doi: 10.1111/trf.14574.

Duncan, L.G., Cohn, M.A., Chao, M.T., Cook, J.G., Riccobono, J. and Bardacke, N., 2017. Benefits of preparing for childbirth with mindfulness training: a randomized controlled trial with active comparison. *BMC Pregnancy and Childbirth*, 17(1). doi: 10.1186/s12884-017-1319-3.

Figueiredo, B., Costa, R., and Pacheco, A., 2002. Experiência de parto: Alguns factores e consequências associadas. *Análise Psicológica*, 20(2), pp.203–217. doi: 10.14417/ap.306.

Gönenç, İ.M., Aker, M.N. and Ay, E., 2019. Qualitative Study on the Experience of Lotus Birth. *Journal of Obstetric, Gynecologic & Neonatal Nursing*, 48(6), pp.645–653. doi: 10.1016/j.jogn.2019.08.005.

Hayes, E. H., 2019. Placentophagy, Lotus Birth, and Other Placenta Practices. *The Journal of Perinatal & Neonatal Nursing*, 33(2), pp.99–102. doi: 10.1097/jpn.0000000000000402.

Hidalgo-Lopezosa, P., Hidalgo-Maestre, M. and Rodríguez-Borrego, M.A., 2017. Birth plan compliance and its relation to maternal and neonatal outcomes. *Revista Latino-Americana de Enfermagem*, 25. doi: 10.1590/1518-8345.2007.2953.

Inkeroinen, S., Virtanen, H., Stolt, M. and Leino-Kilpi, H., 2022. Patients' right to know: A scoping review. *Journal of Clinical Nursing*. pp.1-14. doi: 10.1111/jocn.16603.

Johnson, S. K., Pastuschek, J., Rödel, J., Markert, U. R., and Groten, T., 2018. Placenta - worth trying? human maternal placentophagy: Possible benefit and potential risks. *Geburtshilfe Frauenheilkd*, 78, pp. 846–852. doi: 10.1055/a-0674-6275.

Kettle, C., Dowswell, T., and Ismail, K. M. K., 2010. Absorbable suture materials for primary repair of episiotomy and second-degree tears. *Cochrane Database of Systematic Reviews*, (6), p.CD000006. doi: 10.1002/14651858.CD000006.pub2

Leite, C.V. and Almeida, A.M., 2022, e-Health Services to Support the Perinatal Decision-making Process: An Analysis of Digital Solutions to Create Birth Plans. *Proceedings of the 15th International Joint Conference on Biomedical Engineering Systems and Technologies (BIOSTEC 2022)* - Volume 5: HEALTHINF, pp. 405-412, doi: 10.5220/0010814600003123.

Lothian, J. A., 2007. Listening to Mothers II: Knowledge, Decision-Making, and Attendance at Childbirth Education Classes. *The Journal of perinatal education*, 16(4), pp. 62–7. doi: 10.1624/ 105812407X244723.

Marshall, C., and Rossman, G., 1989. *Designing qualitative research*. Newbury Park: SAGE Publications.

Odent, M. and Tekin Moraloğlu, O., 2015. Neocortical Inhibition During Labor and Stress Deprivation Associated with Cesarean Section Without Labor. *The Journal of Gynecology - Obstetrics and Neonatology*. 12(3), pp. 123-125.

Rocha, J. A., and Novaes, P. B., 2010. Uma reflexão após 23 anos das recomendações da organização mundial de saúde para o parto normal. *Femina*, 38(3), pp. 119-126.

Shilling, T., Romano, A. M., and DiFranco, J. T., 2007. Care practice 2: Freedom of movement throughout labor. *The Journal of Perinatal Education*, 16(3), pp. 21-24. doi: 10.1624/105812407X217101.

Stirling, D., Vanbesien, J., and McDougall, R., 2017. *Informed Decision-making for Labour and Birth*. Ontario Public Health Association.

Suárez-Cortés, M., Armero-Barranco, D., Canteras-Jordana, M., and Martínez-Roche, M. E., 2015. Use and influence of delivery and birth plans in the humanizing delivery process. *Revista Latino-Americana de Enfermagem*, 23(3), pp. 520–526. doi: 10.1590/0104-1169.0067.2583.

United Nations, 1995. *Beijing Declaration and Platform for Action: Beijing+5 Political Declaration and Outcome*. Reprint, New York: UN Women, 2014.

Whitford, H., Entwistle, V., van Teijlingen, E., Aitchison, P., Davidson, T., Humphrey, T., and Tucker, J., 2014. Use of a birth plan within woman-held maternity records: a qualitative study with women and staff in northeast scotland. *Birth* 41(3), pp. 283–289.

WHO, 2006. Birth and emergency preparedness in antenatal care. *Intergrated management of pregnancy and childbirth (IMPAC)*.

WHO, 2016. *WHO Recommendations on antenatal care for positive pregnancy experience*. Geneva, Switzerland: World Health Organization Press.

WHO, 2018. *WHO recommendations intrapartum care for a positive childbirth experience*. Geneva, Switzerland: World Health Organization Press.

WHO, 2019. *WHO consolidated guideline on self-care interventions for health and well-being: Sexual and Reproductive Health and Rights*. Geneva, Switzerland: World Health Organization.

Yam, E., Grossman, A., Goldman, L., & Garcia, S., 2007. Introducing birth plans in mexico: an exploratory study in a hospital serving low-income mexicans. *Birth* (34), pp. 42–48. doi: 10.1111/j.1523-536X.2006.00124.x.

Zakerihamidi, M., Latifnejad Roudsari, R., and Merghati Khoei, E., 2015. Vaginal Delivery vs. Cesarean Section: A Focused Ethnographic Study of Women's Perceptions in The North of Iran. *International journal of community-based nursing and midwifery*, 3(1), pp. 39–50.

EXPLAINABLE ARTIFICIAL INTELLIGENCE IN THE DIAGNOSIS OF CARDIOVASCULAR DISEASES IN SMALL SAMPLES

Natalia Gusarova, Ivan Tomilov, Danil Zmievskii, Vladimir Shilonosov, Tatiana Polevaya and Aleksandra Vatian
ITMO University,
Saint Petersburg, Russia

ABSTRACT

In this work several DNN-based solutions aimed at classifying different of CVDs in conditions of small training samples are proposed. Using GradCAM technology, the main areas of attention of DNNs in these conditions are indentified, and then experimentally being compared with the areas of attention of cardiologists when making a clinical decision. It is shown that even on small samples it is possible to achieve the efficiency of SOTA solutions obtained on large datasets, which should increase the confidence of medical staff, including less qualified ones, in DNN as a means of supporting rapid diagnosis of CVD in acute cases. In turn, combining the results of the physician and AI tools can improve the quality of diagnosis and, therefore, the patient's chances for a speedy and successful cure.

KEYWORDS

Cardiovascular Disease, Explainable Artificial Intelligence, XAI, Small Samples

1. INTRODUCTION

1.1 Nature of the Problem

Cardiovascular disease is the cause of death worldwide, and this trend is only increasing in the 21st century. Although AI-based decision support applications in such areas have large enough datasets for training, which are the basis for a quite high level of accuracy, however, practitioners are in no hurry to use such tools. The reason is largely due to the fact that machine learning methods based on deep neural networks (DNNs), which are the de facto standard for use in medical applications, are fundamentally "opaque". Without being able to follow the decision-making process performed by the DNN, the doctor, who is personally responsible for the patient, has no reason to fully trust its results and cannot fully take them into account when making his decisions.

The problem is even more exacerbated when it comes to life-threatening conditions and rapidly developing diseases such as myocardial infarction. In this case, on the one hand, the level of responsibility of the doctor for the clinical decision for a particular patient increases. On the other hand, the DNN should quickly retrain on a small set of data, which can be obtained in a short time of the development of the disease in this patient. Under these conditions, the performance accuracy of the DNN expectedly falls, which further reduces the doctor's confidence in its results. As a rule, in such cases, the results formed by DNN, which could act as a source of "second opinion" for the attending physician, are completely disavowed, which negatively affects the diagnosis and further course of the disease.

The way out of this collision may be to involve the concept of explainable artificial intelligence (XAI) [Vilone, Speith]. Namely, it is necessary to transfer the process of work of the DNN to the state of a "transparent box", that is, to enable the doctor to track and interpret all stages of this process. For example, when diagnosing a heart attack, a sign that is traditional for clinical practice is elevation of the ST segment of the ECG. If, when analyzing this ECG, the DNN pays the most attention to the ST segment, then for the doctor this will be evidence that the network is working correctly, and its results can be trusted.

1.2 Previous Works

The literature [Barz, Olson, Brigato] presents a number of ways to improve the efficiency of DNN in the context of small samples, including: transfer learning [Xu], data augmentation [Pan], GAN's [Li], Bayesian neural networks [Madan], etc. However, as a detailed analysis shows, to support the operational clinical decisions, which is typical for CVD, two approaches seem to be the most promising.

The first one is to reduce the dimension of the model. According to [Brigato], model complexity is a critical factor when only a few samples per class are available. Here, measures such as the transition to lower-dimensional DNNs while maintaining the basic architecture, dropout regularization, and others that help reduce the number of adjustable model parameters can be applied.

The second approach uses architectures specialized for learning from small samples, such as Siamese neural networks (SSN) [Bromley] and few-shot learning (FSL) [Santoro].The SNN inputs consist of paired signals (ECG in our case), with each signal passed through an identical deep-convolutional subnetwork. The SNN is typically trained using a contrastive loss function, thereby highlighting the difference in input signals. The FSL is based on a similarity function which is formed by pre-training the deep model on a large-scale labeled dataset. From a technological point of view, it is important that in order to represent the similarity function a SNN can be used. Then, using this similarity function, the FS model can determine the closest class for each newly seen item.

Both SSN and FSL, alone and in combinations, are widely used to detect differences in artifacts of various domains [Liu D., Wang, Figueroa-Mata], and in particular in medicine, including clinical language processing [Li (2023), Oniani], high-tech medical imaging [Li (2020), Deepak], sleep staging based on single-channel EEG [You], ECG time series classification [Gupta, Pałczyński].

The main processed artifact in the classification of CVD using DNN is the electrocardiogram (ECG) (Figure 1), which consists of a sequence of QRS complexes [Wagner, Liu F.F., Moody]. From the point of view of the formation of the DNN feature space, CVD can be divided according to the type of manifestation on the ECG into CVD associated with displacements of the main (R) wave, and CVD associated with local changes in other components of the QRS complex. The first type includes such CVDs as supraventricular premature beat, premature ventricular contraction, fusion of ventricular and normal beat, etc. The most important representative of CVD of the second group is myocardial infarction (MI), which can manifest itself on the ECG as an elevation (STE) and (or) depression (STD) of the ST segment. Obviously, the classification of CVD of the second type is a much more difficult task than the first, regardless of the nature and type of classifier used, which must be taken into account when comparing the effectiveness of classifiers.

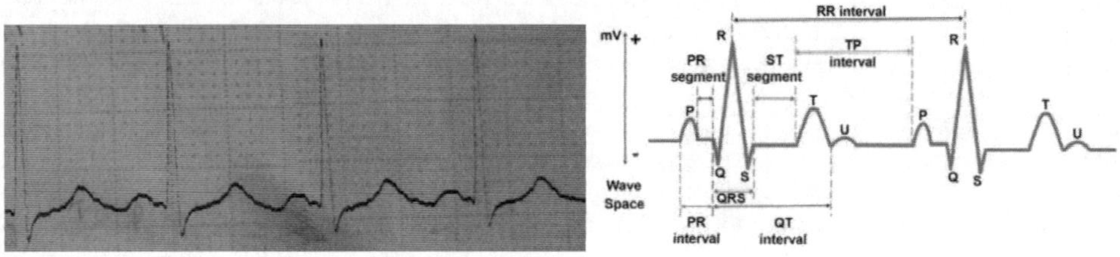

a b

Figure 1. Fragment of ECG (a) and its characteristic intervals (b)

An analysis of the literature allows us to select as SOTA-level for the classification of CVD by ECG the works [Gupta, Pałczyński, Jeong, Strodthoff], where the achieved level of accuracy was 57–92%, depending on the type of diseases classified (more detailed data are given in section 3.1). It should be emphasized that the same level is typical for cardiologists when making operational decisions on the ECG. According to the results of a meta-analysis [Cook], the pooled accuracy in ECG interpretation performed by medical personnel of various skill levels ranges from 56% for residents to 69% for practicing physicians and 75% for cardiologists.

In this regard, the explainability of AI-based models for the performed ECG diagnosis acquires a special role. The literature [Paralic] notes the lack of explainability of DNN-based methods in relation to the medical field and to ECG interpretation, in particular. For example, the review [Jeyakumar] considers XAI methods for

ECG data within the framework of general approaches to multiple types of data, with an emphasis on the explainability of the ECG as a whole. To do this, explanation-by-example methods are offered as the optimal way of XAI. As a means for generating such examples, it is proposed to use a pair of encoder-decoder [Attia]. A similar problem is solved in [Paralic] – to mark parts of ECG the most important for manual disease diagnosis, and perturbation-based XAI methods turned out to be optimal.

For visual explanations of DNN predictions of CVD different XAI methods are used. For example, [Anand] applied SHapley Additive exPlanations (SHAP) for ECG containing CVD of arrhythmia group. [Strodthoff] employed the Gradient×Input method. However, in ECG analysis performed by DNN, backpropagation methods of XAI dominate, among which GradCAM [Taniguchi, Raza, Ganeshkumar] ranks first.

A number of works have proposed methods for testing the effectiveness of the developed XAI tools. In [Paralic, Cabitza] for this purpose the questionnaire offered to doctors is used. [Strodthoff] highlights areas of the ECG curves that do or don't contribute to the correct assessment of myocardial infarction. [Siddiqui] proposes a framework for comparative evaluating of XAI methods accomplished using expertly annotated ECG signals. The technique uses two metrics - similarity and stability.

Let's summarize. Building AI-based CVD classification tools with small training samples is an open area of research. The main ways here are recognized as reducing the dimension of the model and using FSL-based architectures. The accuracy of existing systems varies greatly depending on the type of CVD and is comparable to that of practitioners. To increase the confidence of doctors in the results of AI systems, it is advisable to use XAI. However, in the subject area under consideration, XAI solves either the problem of improving the usability of ECG analysis in the manual diagnosis of CVD, or the problem of assessing the effectiveness of the AI system itself in diagnosing CVD. Evaluation of the possibilities of combining the results obtained by doctors and the AI system in the development of clinical decisions on CVD is not presented in the literature available to the authors.

1.3 Our Contribution

1. We have proposed DNN-based solutions to classify different CVDs having small training samples, based on two approaches - reducing the dimension of the model and SNN + FSL-complexing, which provide efficiency on small samples that is superior to SOTA-level.

2. Using GradCAM technology, we identified the main areas of attention of DNNs in these conditions and compared them with the areas of attention of cardiologists when making a clinical decision;

3. We have shown that interpretations of CVD by typical ECG performed by specialized DNN-based AI tools and medical personnel are adequate in terms of the accuracy achieved and comparable in terms of the distribution of areas of attention when making a decision.

2. METHOD AND MATERIALS

To develop a solution that uses the principle to reduce the dimension of the model, we used the ResNet-18 architecture [Jeong] as the initial schema. To adapt the network to work on small samples, the following measures were used: decrease in the number of network parameters due to the transition to a smaller number of layers (from 18 to 10); replacement of the activation function from ReLu to Gelu; combining regularizers L1 and L2; optional use of the dropout regularizer. To develop a few-shot solution, we used the architecture [Gupta] as the initial schema, with the following changes: after the last convolutional layer, a dropout layer was added for regularization in order to identify the most relevant features, and the vector proximity function was changed from L2-distance to cosine similarity as more robust to the direction of embeddings. Both solutions use the Grad-CAM [Selvaraju] technique, which is compatible with any DNN architecture, as a means of implementing attention. F1-measure and accuracy were used as performance evaluation metrics.

To verify the zones of attention allocated by DNN, they were compared with the zones of attention used in the diagnosis of relevant diseases in clinical practice. For this purpose, expert cardiologists (three specialists) were involved, who performed manual marking of their areas of attention on ECG records extracted from the dataset and visualized in the form equivalent to native paper ECG. For all types of classified CVD, zones of attention were manually allocated in the diagnosis of characteristic QRS complexes. The total number of labeled data for cardiovascular diseases was 70 cases. For a comparative assessment of the areas of attention allocated by medical experts and the DNN, the Jaccard measure was used:

$$K_J = \frac{S(A \cap B)}{S(A \cup B)}, \quad (1)$$

where the numerator is the measure of the intersection of the compared objects, and the denominator is the measure of their union, respectively. In the case of comparing the classification by cardiovascular diseases, the length of the corresponding zone along the time axis was used as such a measure.

To improve the comparability of the results, the training and testing of the developed solutions was carried out on the same datasets as the SOTA sources. Characteristics of datasets and specific application in each case are presented in Table 1.

Table 1. Datasets characteristics

Name, source	Characteristics	Where applied	Application specifics
CPSC2018 [Liu F.F.]	6877 records, 8 classes	[Jeong]	Classification into 5 classes out of 8
		Present work	Decimation by 4 times (up to 1741 records), 5 classes used
PTB-XL [Wagner]	21837 records, 5 classes, 71 types of heart diseases	[Pałczyński]	Classification into 5 and(or) 20 classes
		Present work	Classification into 2 classes
ECG5000 [Chen F.]	Length - 140, 5 classes	[Gupta]	for training, classification into 5 classes
MIT-BIH Arrhythmia Database [Moody]	109,000 heart-beats, 47 classes	[Gupta]	for testing, classification into 5 classes
PTB [Bousseljot}	549 records of 12-lead ECG, 9 classes	[Strodthoff]	Classification into 2 classes

3. RESULTS AND DISCUSSION

3.1 Results

Efficiency estimates of the developed solutions in comparison with SOTA models are presented in Table 2 for FCL-based schemes and in Table 3 for ResNet-based schemes. The following designations are used: N – the number of classes, K –- the number of labeled samples for each of the N classes, n/d – no data available; I-AVB – first-degree atrioventricular block, RBBB – right bundle branch block, PVC – premature ventricular contraction, STD – ST-segment depression, STE – ST-segment elevated, MI – myocardial infarction.

Table 2. Efficiency estimates of the developed solutions in comparison with SOTA models for FCL-based schemes

Source	Arcitecture	Dataset	CVD type	N	k	Accuracy	F1-score
[Gupta]	FSL +SSN	ECG5000	normal / arrytm	5	1	0.848	0.852
					2	0.905	0.907
					3	0.920	0.921
					4	0.920	0.922
					5	0.923	0.924
[Pałczyński]	FSL + SVM with linear kernel	PTB-XL	normal / not normal	2	5	0.912	0.903
			normal / arrytm / MI	5	5	0.788	0.717
			normal / arrytm / MI	20	5	0.653	0.320
Present work	FSL +SSN	PTB-XL	normal / MI	2	1	0.712	0.703
					2	0.691	0.705
					3	0.829	0.831
					4	0.832	0.834
					5	0.832	0.834

Table 3. Efficiency estimates of the developed solutions in comparison with SOTA models for ResNet-based schemes

Source	Arcitecture	Dataset	CVD type	N	Accuracy	F1-score
[Jeong]	ResNet-18	CPSC2018 (full dataset)	I-AVB	5	0.837	0.803
			RBBB		0.860	0.850
			PVC		0.613	0.636
			STD		0.822	0.760
			STE		0.571	0.525
[Strodthoff]	ResNet-based	PTB	normal / MI	2	n/d	0.932
Present work	ResNet-10	CPSC2018 (decimation by 4 times)	I-AVB / normal	4	0.853	0.834
			RBBB / normal		0.890	0.852
			PVC		0.702	0.647
		PTB-XL	normal / MI	2	0.715	0.683

Figure 2 shows an example of the implementation of XAI in the developed solutions (a solution based on the ResNet-10 architecture, the input is characteristic ECG leads, labeled by experts as acute myocardial infarction). The attention zones of the DNN are marked with vertical stripes, and the intensity of the color corresponds to the degree of attention of the neural network. A fragment of such an ECG, marked up by an expert, and the corresponding DNN attention zones are shown in Figure 3, a, b.

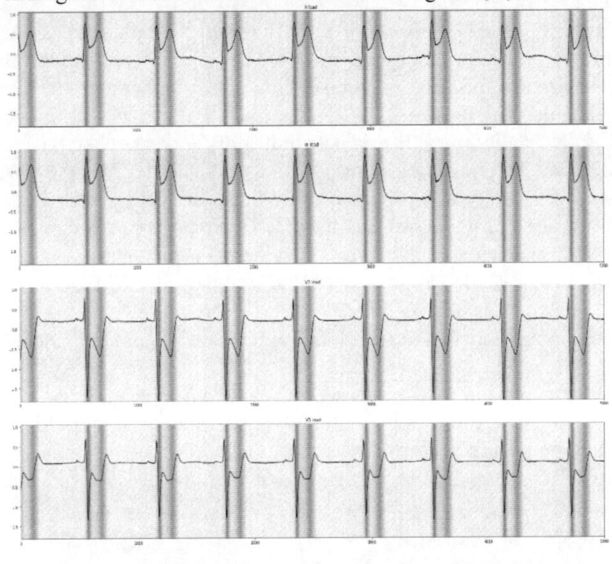

Figure 2. DNN attention zones in classification of ECG with MI

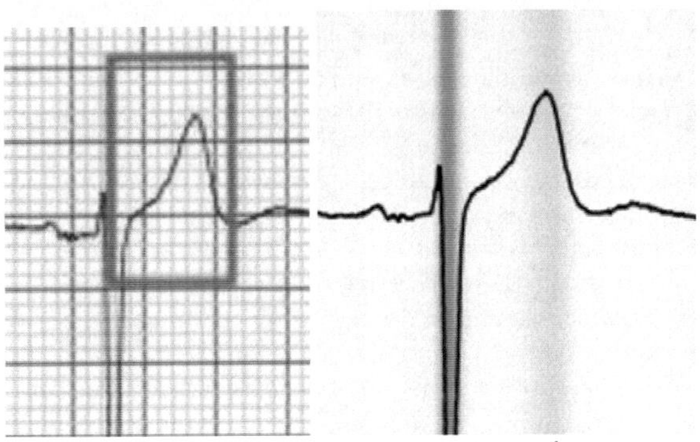

a b

Figure 3. Comparison of attention zones of an expert (a) and of DNN (b) in the classification of ECG with MI

The mean value and standard deviation of the Jaccard measure (1), averaged for both proposed solutions, were: mean = 0,461, std = 0.076 respectively.

3.2 Discussion

As can be seen from Figure 2, the attention of the neural network is stably localized on the characteristic elements of the ECG - this is the peak of the R wave and the ST segment. In both cases, the first "burst" of attention of the neural network falls on the peak of the R wave, which is the main indicator of heart rate, i.e. the nature of the work of the heart muscle as a whole. At the same time, the second "burst" of attention of the neural network is localized in that area of the ST segment, which determines a specific type of pathology. Thus, we can say that the distribution of attention zones of the neural network corresponds to the physiological mechanism of the manifestation of certain pathologies of the cardiovascular system on the ECG.

Small std value of the Jaccard measure with a sufficiently high mean value indicate that the tactics of expert doctors and the "tactics" of DNN work are stable and quite comparable. This conclusion is well confirmed by comparison of Figures 3 (a, b): DNN steadily localizes attention to two ECG zones corresponding to meaningful signs of the disease; the same zones are included in the area of attention of the doctor-expert; at the same time, the expert doctor, considering these zones as the main ones, also distributes his attention over the entire ST-segment of the ECG.

An analysis of Tables 2 and 3 shows that the effectiveness indicators of the developed solutions are generally consistent or even somewhat higher compared to SOTA level, even with a 4-fold reduction in the training set. We emphasize that this level is comparable or even higher than the level of medical personnel of various skill levels when deciphering the ECG.

At the same time, in all studies, the effectiveness indicators depend on the type of CVD, in particular, when classifying MI and its potential manifestations, such as STD and STE, they are usually less than when classifying diseases with a purely arrhythmic manifestation on the ECG. In this case, the stability of the coincidence of the "opinions" of the expert and the DNN is important, which can increase the confidence of the experts in the DNN data, even at relatively small values of the F-measure, and vice versa. For example, in [Strodthoff], with a high value of the F1-measure (Table 3), on 4 out of 12 ECG leads, network attention varies, and it identifies dominant signs against the presence of MI, and on 8 – for it. At the same time, in our work, at moderate values of the F1-measure (Tables 2, 3), this agreement is quite high (std = 0.076).

Finally, comparing the results of Figure 3 and Tables 2, 3 gives grounds to treat DNNs and expert doctors as equally powerful classifiers, but taking into account different aspects of the manifestation of CVD on the ECG. This, in turn, opens the way for formal ensembling of such decisions, which has proven to be effective in various decision-making problems [Ali].

4. CONCLUSION

We have proposed DNN-based solutions aimed at classifying different of CVDs in conditions of small training samples. The effectiveness of the developed solutions depends on the CVDs classified, but in all cases is at the level of the best DNN-based SOTA models, as well as with the best levels of healthcare experts..

Using GradCAM technology, we identified the main areas of attention of DNNs in these conditions, experimentally compared them with the areas of attention of cardiologists when making a clinical decision, and obtained a match at std = 0.076. Such a coincidence, even at relatively low levels of F-measures (F1=0.834) can serve as a basis for increasing physicians' confidence in using DNN as a support tool for making clinical decisions in CVD, and vice versa.

Thus, we have shown that even on small samples it is possible to achieve the efficiency of SOTA solutions obtained on large datasets, which should increase the confidence of medical staff, including less qualified ones, in DNN as a means of supporting rapid diagnosis of CVD in acute cases. In turn, combining the results of the physician and AI tools can improve the quality of diagnosis and, therefore, the patient's chances for a speedy and successful cure.

Our research also revealed a number of limitations of the considered approaches. As noted in Section 1.2, diagnosis of various diseases based on ECG can be associated with both relatively simple (albeit somewhat non-linear) signs characteristic of disorders such as arrhythmias, and rather subtle and complex signs, as in the case of myocardial infarction and its subtypes. . The doctor, analyzing the ECG, necessarily draws on a fairly extensive context, both in the form of previous experience and in the form of additional signs. At the same time, the neural network model, which is trained based on the detection of ECG similarity, relies only on the features it has identified. Due to the specifics of training based on minimizing the l2-distance between ECG embeddings, the complexity of the identified features and the resolution associated with the adjusted parameters may be insufficient. Partially, this problem is solved by adding mechanisms for evaluating the attention of the model, however, additional research is required on the influence of architecture, advanced regularization methods, the chosen method for generating embeddings and calculating the distance between them, as well as other parameters on the properties listed earlier.

ACKNOWLEDGEMENT

This work was supported by Russian Science Foundation, Grant № 23-11-00346.

REFERENCES

Ali S., Tirumala S.S., Sarrafzadeh A. (2015) Ensemble learning methods for decision making: Status and future prospects. *14th International Conference on Machine Learning and Cybernetics (ICMLC)*, July 2015. doi: 10.1109/ICMLC.2015.7340924.

Anand A., et al. (2022) Explainable AI decision model for ECG data of cardiac disorders. *Biomedical Signal Processing and Control*. V. 75, May 2022, 103584

Attia Z.I., Friedman P.A. (2023) Explainable AI for ECG-based prediction of cardiac resynchronization therapy outcomes: learning from machine learning? *European Heart Journal*, V. 44, Is. 8, 21 Feb. 2023, Pp. 693–695, https://doi.org/10.1093/eurheartj/ehac733

Barz B., Denzler J. (2020) Deep Learning on Small Datasets without Pre-Training using Cosine Loss. IEEE Winter Conference on Applications of Computer Vision (WACV) 2020. arXiv:1901.09054v2 [cs.LG] 11 Dec 2019

Bousseljot R, Kreiseler D, Schnabel, A. (1995) Nutzung der EKG-Signaldatenbank CARDIODAT der PTB über das Internet. *Biomedizinische Technik*, Band 40, Ergänzungsband 1 (1995) S 317

Brigato L., Iocchi L. (2021) A Close Look at Deep Learning with Small Data. 25th International Conference on Pattern Recognition (ICPR), 2021, pp. 2490-2497, DOI: 10.1109/ICPR48806.2021.9412492

Bromley J., et al. (1994). Signature verification using a "Siamese" time delay neural network. *Advances in Neural Information Processing Systems*. 6: 737–744.

Cabitza F., et al. Painting the black box white: experimental findings from applying XAI to an ECG reading setting. arXiv:2210.15236 [cs.AI]. https://doi.org/10.48550/arXiv.2210.15236

Chen Y., et al. (2015) The UCR time series classification archive, July 2015. www.cs.ucr.edu/~eamonn/time series data/.

Cook D.A., Oh S.Y., Pusic M.V. (2020). Accuracy of Physicians' Electrocardiogram Interpretations. *JAMA Intern Med.* 2020 Nov; 180(11). P. 1–11.

Deepak S., Ameer P.M. (2021). Brain tumour classification using siamese neural network and neighbourhood analysis in embedded feature space. *IMA*, V.31, Is.3, September 2021, Pp. 1655-1669. https://doi.org/10.1002/ima.22543

Figueroa-Mata G., Mata-Montero E. (2020) Using a Convolutional Siamese Network for Image-Based Plant Species Identification with Small Datasets. *Biomimetics (Basel)*. 2020 Mar; 5(1): 8. doi: 10.3390/biomimetics5010008

Ganeshkumar M., et al. (2021) Explainable Deep Learning-Based Approach for Multilabel Classification of Electrocardiogram. IEEE Transactions on Engineering Management, doi: 10.1109/TEM.2021.3104751.

Gupta P., Bhaskarpandit S., Gupta M. (2021) Similarity Learning based Few Shot Learning for ECG Time Series Classification. *2021 Digital Image Computing: Techniques and Applications (DICTA)*, Gold Coast, Australia, 2021, pp. 1-8, doi: 10.1109/DICTA52665.2021.9647357.

Jeong D.U., Lim K.M. (2021) Convolutional neural network for classification of eight types of arrhythmia using 2D time–frequency feature map from standard 12-lead electrocardiogram. *Sci Rep* 11, 20396. 2021. https://doi.org/10.1038/s41598-021-99975-6

Jeyakumar J.V., et al. How can i explain this to you? An empirical study of deep neural network explanation methods. *Adv. Neural Inf. Process. Syst*. 2020, 33, 4211–4222.

Li D.-C., et al. (2021) A Generative Adversarial Network Structure for Learning with Small Numerical Data Sets. *Appl. Sci.* 2021, 11(22), 10823; https://doi.org/10.3390/app112210823

Li M.D., et al. (2020) Siamese neural networks for continuous disease severity evaluation and change detection in medical imaging. *npj Digit. Med.* 3, 48 (2020). https://doi.org/10.1038/s41746-020-0255-1

Li, Q., He, S. (2023) Similarity matching of medical question based on Siamese network. *BMC Med Inform Decis Mak* 23, 55 (2023). https://doi.org/10.1186/s12911-023-02161-z

Liu D., et al. (2022) Design of Siamese Network for Underwater Target Recognition with Small Sample Size. *Appl. Sci.* 2022, 12(20), 10659; https://doi.org/10.3390/app122010659

Liu F.F., et al. (2018) The China physiological signal challenge 2018: automatic identifcation of the rhythm/morphology abnormalities in 12-lead ECGs. In Te 7th International Conference on Biomedical Engineering and Biotechnology (ICBEB 2018) http://2018.icbeb. org/Challenge.html (2018).

Madan A., Prasad R. (2021) B-Small: A Bayesian Neural Network Approach to Sparse Model-Agnostic Meta-Learning, *IEEE International Conference on Acoustics, Speech and Signal Processing (ICASSP)*, Toronto, ON, Canada, 2021, pp. 2730-2734, doi: 10.1109/ICASSP39728.2021.9414437.

Moody GB, Mark RG. (2001) The impact of the MIT-BIH Arrhythmia Database. IEEE Eng in Med and Biol 20(3):45-50 (May-June 2001). (PMID: 11446209)

Olson M., Wyner A.J., Berk R. (2018) Modern Neural Networks Generalize on Small Data Sets. *32nd Conference on Neural Information Processing Systems (NeurIPS 2018), Montréal, Canada*. 10 p.

Oniani D, et al. (2023) Few-Shot Learning for Clinical Natural Language Processing Using Siamese Neural Networks: Algorithm Development and Validation Study. *JMIR AI* 2023;2:e44293. doi: 10.2196/44293

Pałczyński K., et al. (2021) Study of the Few-Shot Learning for ECG Classification Based on the PTB-XL Dataset. *Sensors* 2022, 22(3), 904; https://doi.org/10.3390/s22030904

Pan Y., et al. Knowledge-based data augmentation of small samples for oil condition prediction. *Reliability Engineering & System Safety*, Volume 217, January 2022, 108114. https://doi.org/10.1016/j.ress.2021.108114

Raza A., et al. (2022) Designing ECG Monitoring Healthcare System with Federated Transfer Learning and Explainable AI. arXiv:2105.12497v2 [cs.LG] 10 Jan 2022

Santoro A., et al. (2016). One-shot Learning with Memory-Augmented Neural Networks. arXiv:1605.06065v1 [cs.LG] 19 May 2016

Selvaraju R.R., et al. (2016) Grad-CAM: Visual Explanations from Deep Networks via Gradient-based Localization. arXiv:1610.02391v4 [cs.CV].

Siddiqui K., et al. (2022) Trust Metrics for Medical Deep Learning Using Explainable-AI Ensemble for Time Series Classification. 2022 IEEE Canadian Conference on Electrical and Computer Engineering (CCECE), 18-20 September 2022. DOI: 10.1109/CCECE49351.2022.9918458

Speith T. (2022). A Review of Taxonomies of Explainable Artificial Intelligence (XAI) Methods. *2022 ACM Conference on Fairness, Accountability, and Transparency* (FAccT '22), June 21–24, 2022, Seoul, Republic of Korea. 12 p. https://doi.org/10.1145/3531146.3534639

Taniguchi H., et al. (2021). Explainable Artificial Intelligence Model for Diagnosis of Atrial Fibrillation Using Holter Electrocardiogram Waveforms. Int Heart J 2021;62(3):534-539. doi: 10.1536/ihj.21-094.

The top 10 causes of death (Dec. 2020). https://www.who.int/news-room/fact-sheets/detail/the-top-10-causes-of-death.

Vilone G., Longo L. (Oct. 2020) Explainable Artificial Intelligence: a Systematic Review. arXiv:2006.00093 [cs.AI]. https://doi.org/10.48550/arXiv.2006.00093

Wagner P., et al. (2020) PTB-XL, a large publicly available electrocardiography dataset (version 1.0.1). *Sci. Data* 2020, 7, 1–5.

Wang Z.-M., et al. (2021) A Few-Shot Learning-Based Siamese Capsule Network for Intrusion Detection with Imbalanced Training Data. *Comput Intell Neurosci*. 2021; 2021: 7126913.

Xu Y., et al.(2022) Iterative self-transfer learning: A general methodology for response time-history prediction based on small dataset, *Journal of Computational Design and Engineering*, Volume 9, Issue 5, October 2022, Pp. 2089–2102, https://doi.org/10.1093/jcde/qwac098

You Y., et al. (2023) A Siamese Network-Based Method for Improving the Performance of Sleep Staging with Single-Channel EEG. *Biomedicines* 2023, 11(2), 327; https://doi.org/10.3390/biomedicines11020327

DEVELOPMENT OF A MOBILE APP DIAGNOSTIC SYSTEM FOR TROPICAL FEBRILE DISEASES

Okure Obot[1], Daniel Asuquo[1], Kingsley Attai[2], Ekemini Johnson[2], Kiirya Arnold[3], Mercy Edoho[1], Moses Ekpenyong[1], Christy Akwaowo[4], Owoidoho Udoh[5], Victoria Usen[6] and Faith-Michael Uzoka[7]

[1]*University of Uyo, Uyo, Akwa Ibom State, Nigeria*
[2]*Ritman University, Ikot Ekpene, Akwa Ibom State, Nigeria*
[3]*Kyambogo University, Kampala, Uganda*
[4]*University of Uyo Teaching Hospital, Uyo, Akwa Ibom State, Nigeria*
[5]*Primary Health Centre, Ukana Ikot Ide, Essien Udim, Akwa Ibom State, Nigeria*
[6]*Primary Health Centre, Mbak Etoi, Uyo, Akwa Ibom State, Nigeria*
[7]*Mount Royal University, Calgary, Canada*

ABSTRACT

Febrile diseases present confusable symptoms that can be highly challenging to inexperienced physicians during diagnosis. This is in addition to the acute shortage of experienced medical experts in rural communities, especially in low-to-middle-income countries (LMICs). To address these challenges, there is a need for the development and adoption of mobile health (mHealth) application - a medical decision support system that is both effective and easy to use by inexperienced physicians and frontline health workers (FHWs) who conduct primary patient diagnosis in rural communities in a number of LMICs. An app developed for the diagnosis and treatment of febrile illnesses, if properly utilized by this category of health workers can serve as a DSS to save many lives lost due to inaccurate diagnosis and self-medication. With data gathered from 3253 patients in four states in Nigeria, several models were developed from the data leading to the use of one of the models (the analytic hierarchy process [AHP] for systems development). This paper presents the methodology of the developed system, based on the AHP model, which showed 90% diagnosis accuracy and 91% precision. A major component of our methodology was the use of elements of design thinking (involving physicians and FHWs) to ensure excellent user experience with functional app features and components usable across a range of devices and platforms.

KEYWORDS

Febrile Disease, Malaria, Multi-criteria Decision Analysis, Analytic Hierarchy Process, Medical Decision Support System

1. INTRODUCTION

Febrile diseases thrive well in the tropical and sub-tropical regions of the world because of high humidity, high temperature, and heavy rainfall, which create a favorable environment for their agents (Nunthavichitra et al, 2020). Poor vector control, poor sanitation, drug resistance, and self-diagnosis have complicated the situation, especially in low-to-middle-income countries (LMICs). These diseases present confusable and sometimes, overlapping symptoms that make it difficult for an inexperienced physician to perform accurate differential diagnosis (Attai et al, 2022). There is a paucity of experienced physicians in LMICs, which has resulted in the reliance on Frontline Health Workers (FHWs) to provide needed health services. According to WHO (2016), the overall global acute shortage of physicians has led to growing interest in the training and use of FHWs to render healthcare services. Most times, patients present multi-diseases and multi-symptom situations, thereby complicating the work of an FHW in the diagnosis process.

Multi-criteria decision analysis (MCDA) requires techniques that include a large number of criteria to guide rational decisions in order to minimize errors. Generally, the procedure presents complexity often due to conflicting criteria that raise the level of uncertainty in the final decision output (Asuquo and Onuodu, 2016). In health care, these procedures are even more complex involving numerous specialists with diverse subjective preferences/opinions, which might hinder the final decision in optimizing health care systems for diverse applications. Though the responsibility for the decision rests on a designated decision maker (DM), the final decision is often a product of an interaction between the DM's preferences and those of other stakeholders to meet economic, social, moral and budgetary expectations (Angelis, Tordrup and Kanavos, 2015). Consequently, this study recognizes the need for the computer analyst who provides methodological support for the decision process, the client who acts as an intermediary between the DM and the analyst, and the medical professionals that know the mechanisms of the behavior of the object of study. The specialists help in identifying the symptoms (criteria) related to each febrile disease (alternative) and provide pair-wise comparison ratings based on their experiential knowledge.

This paper aims at developing a decision support system (DSS) for the diagnosis of febrile diseases by FHWs. The Android app is composed of functional features and components usable across a range of devices and platforms with or without internet access. We focused on domain expertise and end-user involvement, following the principles of design thinking. The specific objectives of the study are to; a) obtain experiential knowledge of medical experts in the diagnosis of febrile diseases; b) apply agile software engineering methodology in the design of an app for differential diagnosis of febrile diseases by FHWs; c) develop and validate an AHP-based MCDA model for the differential diagnosis and d) develop functional graphic user interfaces in collaboration with the FHWs who are the eventual users of the app. The rest of the paper is organized as follows: Section 2 presents related literature while Section 3 describes the methodology of the study. In Section 4, the system implementation results are presented, while the conclusion and recommendations of the study are presented in Section 5.

2. RELATED WORKS

Adehor and Burrell (2008) developed an intelligent DSS for prompt diagnosis of malaria and typhoid fever. Their system diagnosed illness based on users' responses/answers to physical examinations for symptoms which were used to compute the degree of certainty of the disease. Sharma et al. (2013) designed a symptom-based DSS for malaria and dengue diagnosis using fuzzy logic. The system used symptoms of patients, to diagnose dengue fever and malaria. In Patra, Sahu, and Mandal (2010), a diagnostic expert system was developed for diagnosing diseases using data and knowledge from medical experts and secondary sources. Symptoms were categorized into groups and mapped against observed symptoms to indicate the presence of a disease. Similarly, Abiola et al. (2017) developed a rule-based Lassa fever diagnosis system (LFDS) using medical records. Patient symptoms were used to build rules and inputs for intelligent diagnosis. However, these studies did not support predictive learning, thereby limiting the efficacy of their models to predict confusable symptoms.

Oguntimilehin (2020) developed a mobile application for malaria diagnosis using 19 conditional symptoms, with instances of datasets corresponding to selected patients' medical records, using non-nested generalized exemplar (NNGE) was to label and classify the symptoms. Rules were then generated by the NNGE to construct the inference engine for a user-friendly mobile diagnosis application. The datasets in the study lacked verification of symptoms through confirmatory laboratory tests.

Adebayo et al. (2013) developed an expert system for diagnosing and treatment of malaria based on medical expert information collected through structured interviews, and an extensive literature review, the design adopted the waterfall software development method. Fatumo, Adetiba, and Onolapo (2013) proposed an expert System called XpertMalTyph for diagnosing the complications of malaria and typhoid using MySQL, Java, and Java Expert System Shell (JESS). The system was to enable patients with mild cases of malaria and typhoid complications to get treatment without visiting the hospital. However, these systems had no element of intelligent modeling/prediction. Maidabara et al. (2021) developed an expert system using secondary data, for diagnosing malaria and typhoid. They employed three ML algorithms including support vector machine (SVM), artificial neural network (ANN), and Naïve Bayes (NB) but SVM and NB gave the best classification accuracies. However, their work solely relied on patients' medical history and physical examination without confirmatory (laboratory) tests to verify physician suspicion or address confusable symptoms. Furthermore, the authors' datasets were binary transformations (0 and 1) and most unlikely to effectively model the degree or severity of the disease(s). Ozkan, Koklu, and Sert (2018) applied decision tree, SVM, random forest, and ANN in the diagnosis of urinary tract infection (UTI) using examination data and diagnosis results of 59 UTI patients and ANN gave the highest accuracy of 98.3%. Shabut et al. (2018) presented a mobile-enabled plasmonic ELISA-based tuberculosis (TB) detection scheme using smartphones with an ensemble classifier (random forest) and the results showed an accuracy of 98.4%.

From the foregoing literature, our research integrates collaborative crowdsourcing of input symptoms or data–a community-based approach that captures both patient and physician perspectives before a symptom is rendered, hence, eliminating diagnostic biases. Further experiential knowledge in the form of disease-specific thresholds defining the universe of discourse (UoD) or cutoffs to disease burden estimation is also captured, to address disease-specific ecosystems.

3. METHODOLOGY

The development of the application for diagnosing febrile diseases utilized an Agile software methodology that emphasized iterative user involvement. In the Agile model, the processes are interleaved with a significant level of iterative user involvement, based on the principles of empathy in design thinking. Empathy (in design thinking) helped us to: appreciate potential users' (FHWs) needs, gain deep insight into how the FHW interacts with the health system, the Standing Order, and patient in the diagnosis process, realize how patients' lives could be impacted through the diagnosis, and examine the motivations and mental processes employed by physicians and FHWs in the diagnosis process. The functional requirements of the system were iteratively specified at every incremental phase by the FHWs who are the end-users of the system.

3.1 Agile Software Development Approach

The Agile software methodology steps deployed in the app development of our diagnostic system are shown in Figure 1 and are described as follows.

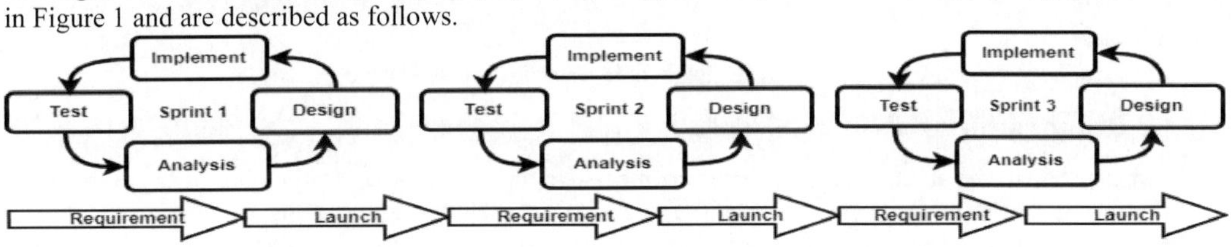

Figure 1. Iterative Process in Agile software development

3.1.1 Requirements Gathering

For a thorough understanding of the proposed system functionalities, relevant stakeholders and users were integrated into the iterative process. From the various meetings held during this stage, the functionalities were elicited from the stakeholders. Three major and overlapping sub-processes were employed for the elicitation, documentation, and understanding of the system functionalities by the stakeholders and users. During requirement elicitation, major and minor use cases were collected from stakeholders and users. Next, requirement documentation was carried out where the elicited requirements were organized and documented in appropriate software engineering format. The use case diagram in Figure 2 depicts the graphical interaction between the febrile disease diagnosing system and the Users, FHW, Facility Head, Record Officer, Patient, and Administrator. The use case model is segmented into Patient Assessment, Appointment and Follow-up, Report Module, Patient Clinical Information, and Personnel information. The four major stakeholders (FHW, Facility Head, Record Officer, and Patient) share various use cases in three of the four modules.

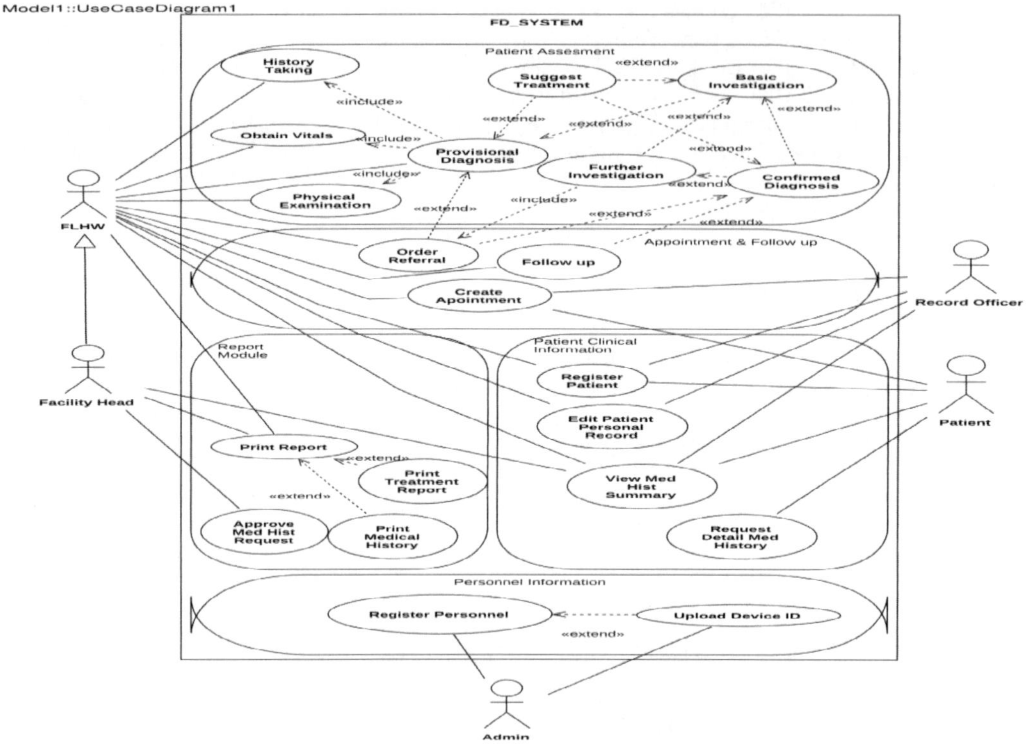

Figure 2. Use Case Diagram of the Febrile Disease Diagnostic System

3.1.2 Requirement Analysis and Validation

To ensure that all the stakeholders and team members share a common understanding of the software to be built, the gathered requirements were analyzed and validated through various convened meetings. The meetings involved additional elicitation and revision of the requirement documentation obtained by considering the strengths and weaknesses of each requirement. The deliverable from these meetings is a unanimous requirement for the proposed system that is documented graphically in a context diagram as shown in Figure 3. The context diagram models the boundary between the system and its environment showing entities that interact with it.

Furthermore, the elements of a Standing Order for the index diseases were integrated to develop the app. A Standing Order is a written protocol that guides the FHWs in the diagnosis and treatment of illnesses at the primary healthcare level. The procedure for using standing order for the treatment of a disease is as follows: Welcome the client and give him /her a seat; introduce yourself and reduce tension by encouraging the patient; collect history; carryout physical examination and record the needful; carryout further investigation to ascertain your findings, if necessary; Treat accordingly.

3.1.3 Design

The user interface, knowledge base, and diagnostic engine are the major aspects designed in this study, and consideration is given to the interactions of the software (comprising the data storage, access logic, application logic, and presentation logic) and the hardware (mobile device and the server). The architecture is presented in Figure 4. The architecture comprises, the medical experts, patients, user interface, knowledge base, a diagnostic system with its sub-components (AHP diagnostic engine, decision support filters), and diagnosis and treatment. The knowledge base contains extracted experiential knowledge from sixty-two (62) physicians in secondary and tertiary health facilities, which have expertise in the domain of diagnosis of febrile diseases including interrelated symptoms.

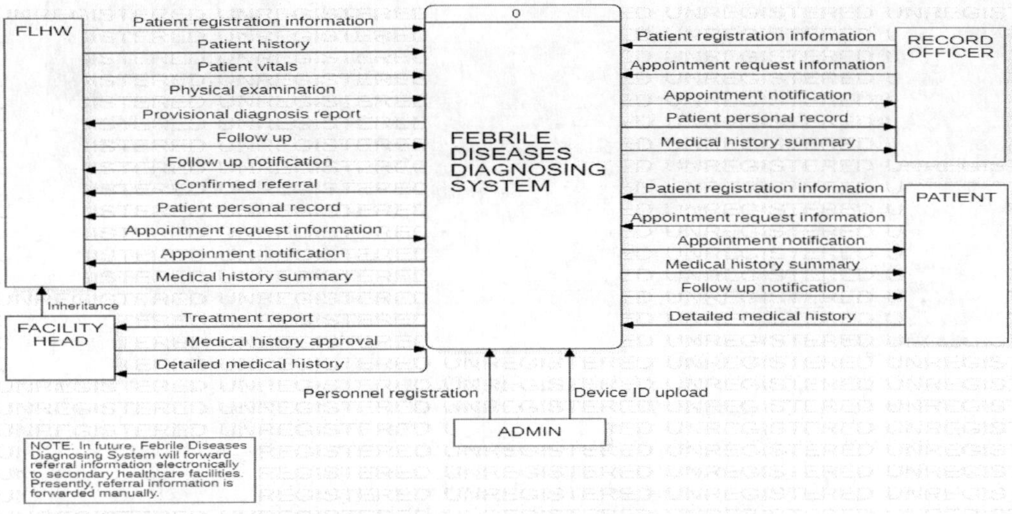

Figure 3. Context Diagram of the Febrile Disease Diagnostic System

It also contains sufficient patient examination and test results including risk factors that predispose a patient to these febrile diseases. This information helps in the formulation of a multi-disease multi-symptom model necessitating their representation in an AHP structure to form the AHP diagnostic engine. The contents of the knowledge base are sent to the diagnostic system, where the decision support filters utilize such information to simulate the behavior of an experienced physician using the AHP diagnostic engine to trigger appropriate individual febrile disease model(s) for further analysis. The AHP engine contains linear consensus models that specify which disease to be triggered by the decision support filters. The AHP diagnostic engines distinguish the symptoms of febrile illnesses and facilitate the identification of the suspected illness based on the patient's symptoms and risk factors to suggest the most suitable form of treatment. The FHW can also visualize the predicted diagnosis on the user interface for timely and accurate healthcare decision-making in terms of medical advice and treatment.

The systematic approach used by the decision support filter of the diagnostic system for malaria (MAL) is presented as follows. The system performs intelligent diagnosis based on the contribution of each symptom to a disease using physicians' consensus preferences and confidence levels. The criteria-by-criteria pair-wise comparison for each disease alternative was evaluated using a typical AHP comparison matrix, with each leading diagonal equal to 1. Later, the AHP method determines the relative weight of each criterion and aggregates the eigenvectors for each comparison matrix until the composite weight (final score) for each alternative is obtained. The values of the final weight vector indicate the relative importance of each alternative with respect to the goal of the decision problem. A DM then uses the eigenvectors to rank the alternatives for appropriate decision-making. The normalized vector matrix is obtained and the relative weight or weighted eigenvector is computed as the row average of the resulting normalized matrix. The consistency ratio (CR) of each performance matrix was estimated to ascertain the level of inconsistency due to the subjective judgment of the DM. If $CR \leq 0.1$, the level of inconsistency is acceptable else, the inconsistency is high and the DM is advised to revise the elements of the matrix to realize a more consistent matrix. The probability of occurrence of each febrile disease was computed from the aggregate diagnostic factor index (ADFI) based on the linguistic score of its symptoms: 1=absent; 2=very low; 3=low; 4=moderate; 5=high; 6=very-high and the weighted Eigenvector (priority score) generated from the AHP process. The ADFI for MAL is expressed in equation (1) while the probability of occurrence (P_i) is expressed in equation (2). Table 1 shows occurrence probability ranges generated from the computed ADFI (Uzoka et al., 2017).

ADFI_MAL=(CHLNRIG_Score*0.115693613)+(FTG_Score*0.108827326)+(FVR_Score*0.132092708)+(GENBDYPN_Score*0.114760331)+(HDACH_Score*0.120826663)+(JONPN_Score*0.107823379)+(BITAIM_Score*0.088720975)+(PRTN_Score * 0.086728546)+(HGDFVR_Score * 0.124526459) (1)

$$P_i = \frac{ADFI}{5} * 100 \quad (2)$$

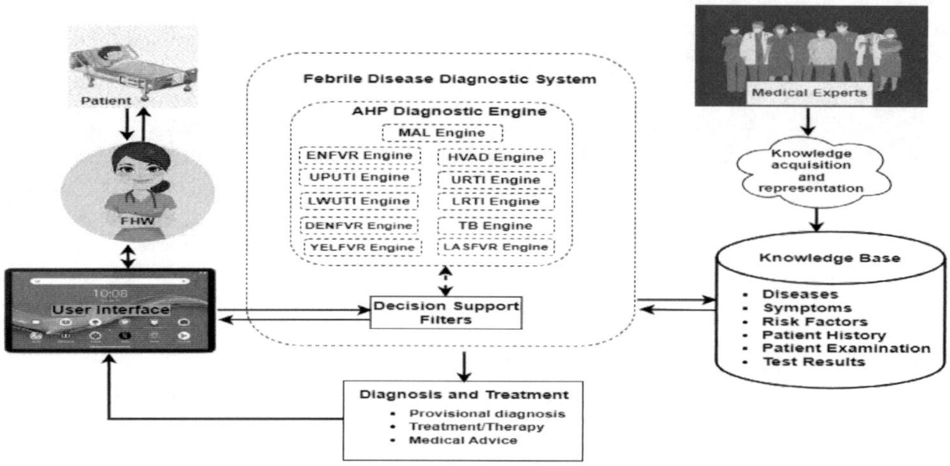

Figure 4. System Architecture

Table 1. Occurrence probability ranges

Uniform Rating	ADFI Range	Probability range (%)
1	0.000 - 1.000	0.01-20.00
2	1.001 – 2.000	20.01 – 40.00
3	2.001 – 3.000	40.01-60.00
4	3.001 – 4.000	60.01 – 80.00
5	4.001 – 5.000	80.01– 100.00

4. SYSTEM IMPLEMENTATION RESULTS

The software tool used for the front-end development was eXtensible Markup Language (XML), to build the mobile layouts – graphical user interface (GUI) while the back-end had SQLite for local database storage, Python and SQL for application programming interface (API) development. Others include; Android Studio Dolphin 2021 v3.3 as the main Integrated Development Environment (IDE) used for building, compiling, and debugging the system app using Java as the programming language. Android Software Development Kit (SDK) Version 28 for the binaries that are used to build, debug, run, and test the App. MPAndroidChart: v 3.0.3 is an Android library used to create graphical representations of computations and data. Git, a source code management tool used to track changes in the code and enable programmers to work together on this nonlinear development, and Adobe Illustrator, a vector-based graphics tool used to create icons and images that are placed in the app layouts to portray more information to users.

Figures 5-8 shows the various screen of the developed app. In the login screen, the FHW is required to input his or her worker identification (ID) and password to control access and prevent unauthorized users into the system. Output from a correct ID entry permits the FHW to view the medical ID screen in Figure 5, and either register a new patient, search for existing patients, or view registered patients in the system. Figure 6 shows the system dashboard which provides the working space for FHW to access different modules (vital signs, diagnosis, follow-ups, and appointments) of the system. Figure 7 shows how patient's perceived symptom is captured using sliders on a 5-point scale. A slider is a GUI control element that uses a lever moved horizontally to control a variable. It simply allows FHWs to select a range from a fixed set of options by simply sliding the bars corresponding to each sign and symptom. Our approach ensured that the visual design of this element does not hinder its usability. The aesthetic control also ensures that users can make selections correctly without having to struggle to hit a precise value. In addition, the slider labels are displayed below in order to remain visible while the user is selecting a value. These values are combined with the eigenvalues for each symptom in the AHP inference engine for evaluation of ADFI and the probability of each disease diagnosis. Figure 8 shows the diagnosis report depicting the initial diagnosis without risk factors and final diagnosis with considered risk factors for every confirmed febrile illness. Results with risk factors assessment indicate that the AHP model closely mimics domain experts' diagnosis accuracy performance.

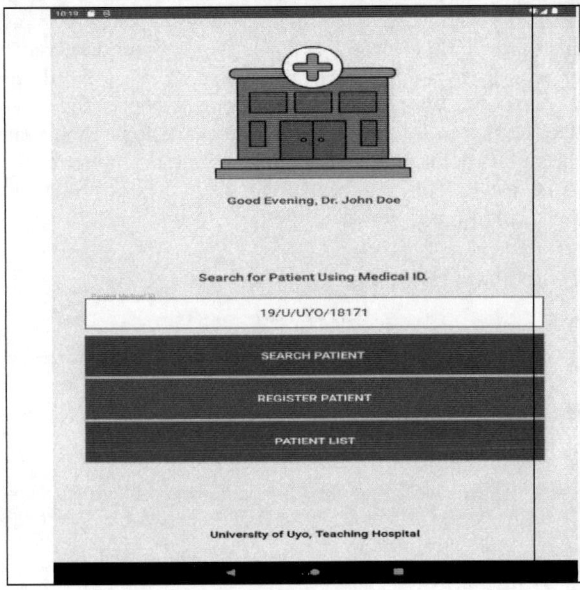

Figure 5. Medical Search ID

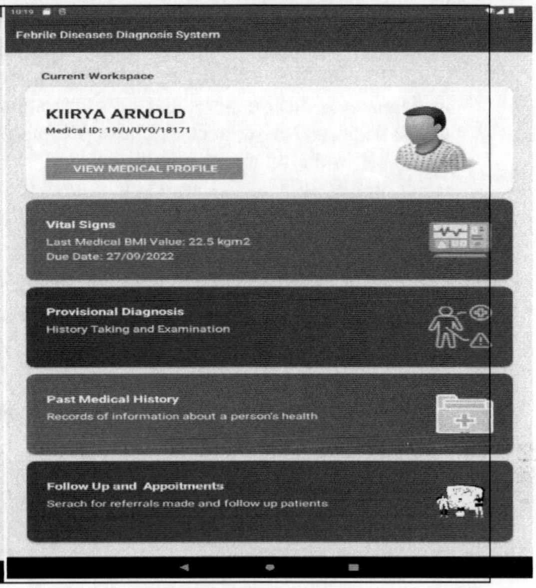

Figure 6. System Dashboard

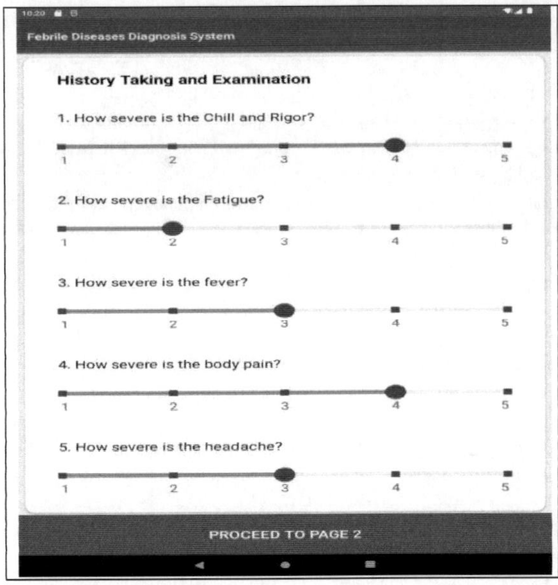

Figure 7. History Taking and Examination

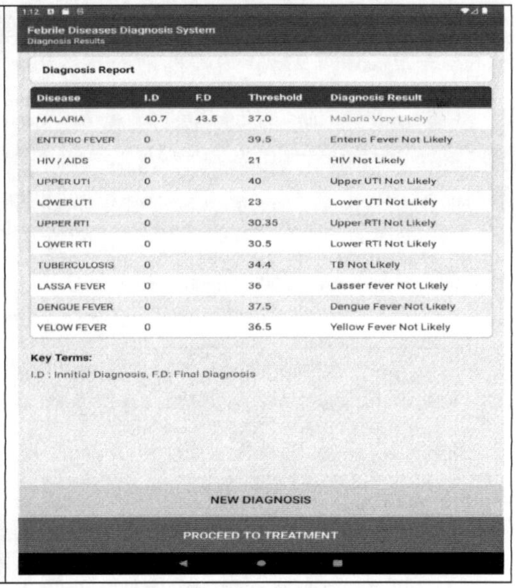

Figure 8. Diagnosis Report

5. CONCLUSION

Medical diagnosis is a complex task and the paucity of experienced physicians in LMICs where febrile diseases seem to be more prevalent presents a huge challenge. Therefore, an app to assist FHWs in diagnosing febrile diseases with overlapping symptoms was developed for this purpose. With a multi-disease multi-symptom analytical model for eleven (11) febrile diseases, AHP, an MCDA methodology was found the most suitable for this task and the results from the AHP-based medical DSS indicate a high accuracy value of 90% when compared to the physician's prediction performance. Thus, this decision tool can be effectively utilized for differential diagnosis of febrile diseases in rural settings and LMICs faced with inadequate medical experts and poor operating conditions. We are presently testing the app in four states where datasets were collected. The app is therefore limited to these four states; so, we are hoping to expand the scope in the future. We have noticed that the results of diagnosis are only accepted as an intermediate result until a rapid test is carried out to give the final results. This will facilitate FHWs in making prompt and accurate decisions regarding the diagnosis and treatment of febrile disease.

REFERENCES

Abiola, H.M., Alaba, A.A. and Luka Joy, D., 2017. Expert system for Lassa fever diagnosis using rule based approach. *Ann. Comput. Sci. Ser*, 15(2), pp.68-74.

Adebayo, A. O., Fatunke, M., Nwankwo, U., & Odiete, O. G. (2013). The design and creation of a malaria diagnosing expert system. School of Computing and Engineering Sciences, Babcock University, PMB, 21244. COMPUSOFT, An international journal of advanced computer technology, 2 (12)

Adehor, A.B. and Burrell, P.R., 2008, An intelligent decision support system for the prompt diagnosis of malaria and typhoid fever in the malaria belt of Africa. In IFIP International Conference on Artificial Intelligence in Theory and Practice (pp. 287-296). Springer, Boston, MA.

Angelis, A., Tordrup, D. and Kanavos, P., 2015. Socio-economic burden of rare diseases: a systematic review of cost of illness evidence. *Health Policy*, 119(7), pp.964-979.

Asuquo, D.E. and Onuodu, F.E., 2016. A fuzzy AHP model for selection of university academic staff. *International Journal of Computer Applications,* 141(1), pp.19-26.

Attai, K., Amannejad, Y., Vahdat Pour, M., Obot, O. and Uzoka, F.M., 2022. A Systematic Review of Applications of Machine Learning and Other Soft Computing Techniques for the Diagnosis of Tropical Diseases. *Tropical Medicine and Infectious Disease*, 7(12), p.398.

Fatumo, S.A., Adetiba, E. and Onolapo, J.O., 2013. Implementation of XpertMalTyph: an expert system for medical diagnosis of the complications of malaria and typhoid. *IOSR Journal of Computer Engineering*, 8(5), pp.s34-40.

Maidabara, A.H., Ahmadu, A.S., Malgwi, Y.M. and Ibrahim, D., 2021. Expert System For Diagnosis Of Malaria and Typhoid. *Computer Science & IT Research Journal*, 2(1), pp.1-15.

Nunthavichitra, S., Prapaso, S., Luvira, V., Muangnoicharoen, S., Leaungwutiwong, P. and Piyaphanee, W., 2020. Case report: COVID-19 presenting as acute undifferentiated febrile illness—a tropical world threat. *The American journal of tropical medicine and hygiene*, 103(1), p.83.

Oguntimilehin, A., 2020. A mobile malaria fever clinical diagnosis system based on non-nested generalized exemplar (NNGE). *International Journal of Emerging Trends in Engineering Research*, 8(2), pp.259-64.

Ozkan, I.A., Koklu, M. and Sert, I.U., 2018. Diagnosis of urinary tract infection based on artificial intelligence methods. *Computer methods and programs in biomedicine*, 166, pp.51-59.

Patra, P.S.K., Sahu, D.P. and Mandal, I., 2010. An expert system for diagnosis of human diseases. *International Journal of Computer Applications*, 1(13), pp.71-73.

Shabut, A.M., Tania, M.H., Lwin, K.T., Evans, B.A., Yusof, N.A., Abu-Hassan, K.J. and Hossain, M.A., 2018. An intelligent mobile-enabled expert system for tuberculosis disease diagnosis in real time. *Expert Systems with Applications*, 114, pp.65-77.

Sharma, P., Singh, D.B.V., Bandil, M.K. and Mishra, N., 2013. Decision support system for malaria and dengue disease diagnosis (DSSMD). *International Journal of Information and Computation Technology*, 3(7), pp.633-640.

Uzoka, F.M.E., Nwokoro, C., Debele, F., Akinnuwesi, B. and Olaniyan, M., 2017, December. AHP model for diagnosis of tropical confusable diseases. In *2017 International Conference on Computational Science and Computational Intelligence (CSCI)* (pp. 1758-1763). IEEE.

WHO, 2016. *Global Strategy for Human Resources for Health: Workforce 2030*. Accessed on October 15, 2018 from https://apps.who.int/iris/bitstream/handle/10665/250368/9789241511131-ng.pdf?sequence=1&ua=1 2.

EXAMINATION OF UNREMITTING KIDNEY ILLNESS BY UTILIZING MACHINE LEARNING CLASSIFIERS

Fareeha Sarwar[1, 2], Nuno Garrido[1], Pedro Sebastião[1] and Akmal Rehan[2]
[1]*Instituto de Telecomunicações (IT-IUL)*
Ed. Armadas, 1649-026, Lisbon, Portugal
[2]*Faculty of Science, Department of Computer Science*
University of Agriculture, Faisalabad, Pakistan

ABSTRACT

Chronic kidney disease is a rising health issue that affects millions of people worldwide. Early detection and characterization of this disease is essential for effective management and control. This disease is associated with several serious health risks, such as cardiovascular disease, increased risk of stroke, and end-stage renal disease, which can be effectively prevented by early detection and treatment. Medical scientists rely on machine learning algorithms to diagnose the disease accurately at its outset. Recently, adding value to healthcare is being accomplished through the integration of machine learning algorithms into mobile health solution. Considering this, this paper proposes a predictive model of three machine learning classifiers, including Support Vector Machine, Decision Tree, and Multilayer Perceptron for chronic kidney disease prediction. The performance of the model was assessed using confusion matrix and executed in popular machine learning software tools such as WEKA and Rapid Minor. The study found that support vector machine yielded the highest accuracy rate of 98% in predicting chronic kidney disease in WEKA among other standard classifiers by using 10-fold cross validation. In addition, the proposed prediction model has been compared with existing models in terms of accuracy, sensitivity, and specificity. The experimental results indicate that the proposed predictive model shows promising results. These findings could integrate with the development of mobile health solution and other innovative approaches to prevent and treat this debilitating condition.

KEYWORDS

Machine Learning Classifiers, Chronic Kidney Disease, WEKA, Rapid Minor, Mobile Health Solution

1. INTRODUCTION

Worldwide, non-communicable diseases (NCDs) have replaced communicable diseases as the leading cause of morbidity and premature death. In low- and middle-income countries, 80% of the burden occurs, and 25% occurs in people under 60. NCDs are expected to have a significant impact on the global economy: by 2015, just two diseases (cardiovascular disease and diabetes) will account for 5% of global GDP. Heart disease, stroke, and peripheral vascular disease account for approximately half of the economic burden, causing more deaths than HIV/AIDS, malaria, and tuberculosis combined. Among diabetes and cardiovascular disease (including hypertension), kidney disease is an essential determinant of poor health outcomes, and the World Health Organization recommends that national NCD programs focus on preventing kidney disease, particularly at the primary care level (Couser, Remuzzi, Mendis, & Tonelli, 2011). As chronic kidney disease (CKD) does not follow any age limit, it can appear at any age. Further, if one has already developed CKD, one is more likely to experience sudden deterioration of kidney function. Procrastination can cause severe kidney damage. This disease requires early detection in order to be treated successfully. However, CKD doesn't show symptoms in the initial stages, making it impossible to identify it without testing. A person with milder chronic kidney disease may do not show any disease-related symptoms, and that can make it hard to predict. The effected person will show symptoms that are very common, like nausea, high blood pressure and blood in your urine. At this stage if CKD is detected, it can be cured and monitored to prevent prolonged damage to the kidney. Urine and blood test can be conducted to detect it an early stage. When chronic kidney disease worsens or at a high stage, it starts showing symptoms like anemia, low immune response, blood pressure, nerve damage, and weak bones. But it is also observed that some people with chronic kidney disease do not show any symptoms at all. Glomerular filtration rate can be obtained from simple blood tests. The stages of kidney disease are shown in Table 1.

Machine learning algorithms are used widely for classification and prediction in the medical field. Because of the dynamic, progressive, and heterogeneous nature of chronic kidney disease, it has become necessary to predict the presence and progression of chronic kidney disease with good accuracy and precision. If the patient has chronic kidney disease, then the severity level is checked by measuring at which stage of chronic kidney disease the patient is. Chronic kidney disease stage measure is very crucial for diagnosis recommendation as treatment decisions are highly influenced by chronic kidney disease stages. Nowadays, electronically availability of medical data makes it easy for researchers to analyze, classify and predict with the help of machine learning algorithms and predictive models. Some mathematical computational models developed in recent years for accurate prediction and support decisions of healthcare providers (Itani, Rossignol, Lecron, & Fortemps, 2019).

Mobile system development in health care system for data gathering and monitoring purpose is an emerging technology that becomes essential especially for emergency situations such as spinal cord and acute trauma. To handle huge amount of data cloud storage-based solutions have been introduced in compliance with mobile application in healthcare. From past years most of the focus was on the development of mHealth Solution, that can manage patients remotely. Many studies focus on the mHealth Solution model, technology, data gathering techniques, and communication. This solution has evolved exponential over time with standard and regulation (Zhang, et al., 2019). The motive of this study is to improve machine learning model performance in predicting chronic kidney disease so that in future it can be integrated with mobile Health Solution.

Table 1. Stages of kidney disease

Stages	GFR Scale	Description
1st	>=90	Kidney function normally working, little bit problem in urinating
2nd	60-89	At this stage kidney function mildly disturbed
3rd	30-59	Moderate level of kidney function working
4th	15-29	At this stage kidney function severely reduced
5th	<15	Depict end stage of kidney functioning, patient will be on dialysis

2. REVIEW OF LITERATURE

Numerous researchers have worked in healthcare domain using Machine Learning and Data Mining classifiers. Various studies delineated that Machine Learning and Data Mining algorithms are potent for prediction and prognosis in medical field. In this research study, author achieved high classification accuracy in prediction using Machine Learning techniques on specified problem. In past few years mHealth System has been introduced that has ability to predict and monitor diseases via mobile applications.

Different Machine Learning classifiers have been applied in prediction of chronic kidney disease.(Jongbo, Adetunmbi, Ogunrinde, & Badeji-Ajisafe, 2020) employed Bagging and Random Subspace approaches on three data mining classifiers including k-Nearest Neighbor, Naïve Bayes, and Decision Tree for improving the performance of these classifiers. The data set used in this research was the data of 400 CKD patients collected from the UCI repository. After preprocessing, data was split into two parts, consisting of 70% data to train the model and the remaining 30% data to test the performance of the model. After that, researchers trained three classifiers (Decision Tree, KNN, and Naïve Bayes) build models and checked their performance. The predictions of the classifiers were gathered utilizing the gathering method (Bagging) with greater part casting a ballot rule and the results of this research explored that 97% accuracy of prediction can be attained by utilizing a random subspace ensemble on k-Nearest Neighbors classifier.

A study conducted by (Islam, Akter, Hossen, Keya, Tisha, & Hossain, 2020) on the risk factor prediction of chronic kidney disease using machine learning classifiers. They collected 1032 patient data through questionnaire and then formatted it into CSV files format. In the preprocessing stage they remove noisy data values and inputted missing values. They proposed model using 6 machine learning algorithms, namely Random Forest, Simple Linear Regression, Naïve Biased, Linear Regression, Simple Logistic Regression and

Decision Stump. The researchers train model with 63% data and test with 37% dataset. The results show that for this real time data set this model gave satisfactory accuracy in prediction of risk of chronic kidney disease. They also concluded that the fundamental driver of kidney ailment is hemoglobin factor.

Mobile Health solution is an emerging technology in terms of cost and time saving proposed mHealth solution architecture. (Qureshi, b, Jeon, & Piccialli, 2020) collected data through mobile application, clinics, and healthcare data sets. For handling bulk data cloud-based storage model was incorporated using two tier architecture. Permission framework was used to implement access control system. In the application end security and privacy was achieved through user authentication interface. A dataset of 400 patients containing 25 attributes was taken from the UCI repository. They proposed dynamic and predictive model using 4 machine learning classifiers named Support Vector Machine, Decision Tree, KNN Model, and Naïve Byes Model to predict cardiovascular disease prediction. For this model data was derived from cloud storage and several feature selections techniques were used to obtain dataset. Afterwards dataset was stratified into 75% of training and 25% of testing. This proposed predictive model provided good accuracy with compared to previous predictive models. This model has some limitation also. They suggested some further treatment decisions and early preventive measures about cardiovascular disease.(Estonilo & Festijo, 2022) also advanced mHealth Solution using deep learning classifiers to predict and monitor Diabetes Mellitus remotely. They also improved the quality of mHealth Solution architecture in terms of security and responsiveness. Complied MAG guidance to maintain standard of proposed architecture. A Google Form was used to collect data from November 2021 to December 2021. An online evaluation followed after the data collection. People with diabetes or without diabetes, familiar or not familiar with m-health applications, were the target participants. They named their model DMChecker which gave 93% accuracy in predicting Diabetes Mellitus.

Many research stated that data mining is useful for inconsequential clinical data for discovering hidden patterns from kidney disease patient's test report.(Brito, Esteves, Peixoto, Abelha, & Machado, 2022) worked specifically on serum creatinine value to identify those patients who were at the last stage of the kidney disease. The patients who undergo continuous ambulatory peritoneal dialysis (CAPD) treatment at the last stage of kidney disease were at the subject of this research. As CAPD is not a permanent heal from kidney disease, there is a need for continuous monitoring of these patients' blood tests. Through blood tests, a patient's evolution, therapeutic drug monitoring, and another prognosis can be observed. Creatinine level is used to determine the working of kidney function, through this value kidney disease stages can be determined.

3. MATERIALS AND METHODOLOGY

The main objective of this research is to improve machine learning classifiers accuracy in prediction chronic kidney disease in the proposed architecture of mHealth Solution. For this, dataset was preprocessed and analyzed for generating legitimate results. Proposed methodology specifies three classifiers used to predict chronic kidney disease presence with accuracy shown in Figure 1.

3.1 Data Set Description

Dataset plays a vital role during predictive research. Your results are totally dependent on it so that is really very important to collect accurate data. The data set used in this research was the data of 400 CKD patients collected from the UCI repository (Polat, Mehr, & Cetin1, 2017).The data set contains a total of 14 attributes having 6 nominal values and 8 numeric values. The data were classified into two classes: Yes (containing the data of patients suffering from Chronic Kidney Disease) and No (containing the data of healthy people).A description of each feature can be found in Table 2.

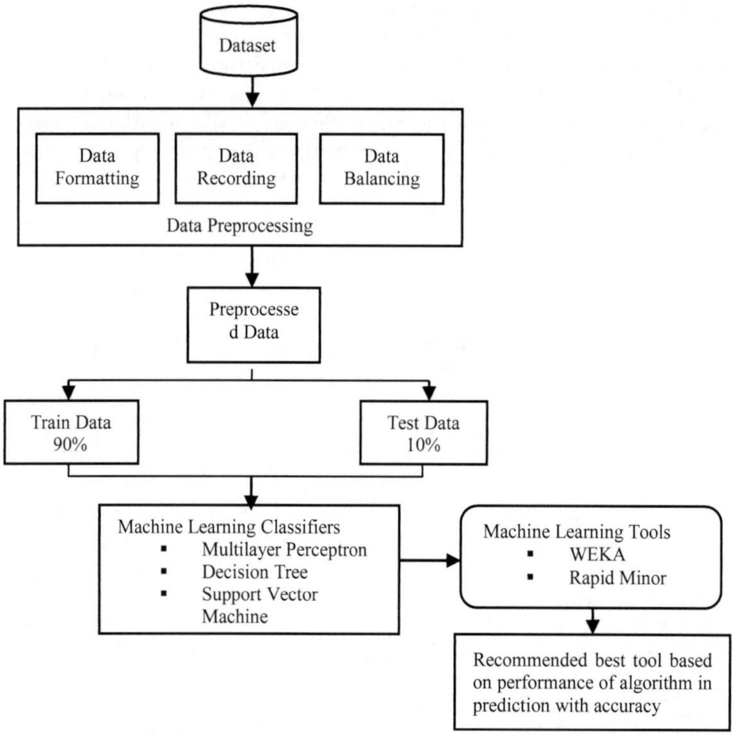

Figure 1. Workflow of proposed methodology

Table 2. Data set description

CKD dataset	Attributes meaning	Category	Scale	Missing values
bp	Blood pressure	Numerical	Mm/Hg	9
sg	Specific gravity	Numerical	1.005 to 1.025 (the higher value means higher risk)	11
al	Albumin	Nominal	0 to 5	47
su	Sugar level	Nominal	0 to 5	49
rbc	Red blood cells	Nominal	Abnormal or Normal	152
bu	Blood urea	Numerical	mgs/dl	19
sc	Serum creatinine	Numerical	mgs/dl	17
sod	Sodium	Numerical	mEq/L	89
pot	Potassium	Numerical	mEq/L	88
hemo	Hemoglobin	Numerical	gms	17
wc	White blood cell count	Numerical	cells/cumm	112
rc	Red blood cell count	Nominal	millions/cmm	131
htn	Hypertension	Nominal	yes, no	2
classification	Class	Nominal	CKD, not CKD	0

3.2 Data Preprocessing

There were outliers and noise in the dataset. Since certain measures may be lost when patients are being tested, the preprocessing stage incorporated the estimation of missing values and noise elimination, including outliers, normalization, and unbalanced data checking. Data was preprocessed in multiple steps. In first step rename the columns in broaden form. Then data types of all columns were corrected. At that moment ablution of categorical columns were performed and to manage the missing values in the data set median value method was used. The noisy values were soothed by using binning technique. The original dataset contained 25 attributes like age, blood pressure (BP), sugar, red blood cells, white blood cell, red blood cells, sodium, potassium, hemoglobin, etc. during the phase of preprocessing, we eliminated few attributes based on importance and change the type of few attributes from discrete values to nominal.

3.3 Mobile Health Solution Architecture

Due to increase support in chronic diseases prediction and monitoring Mobile Health solution has gained a significant attention. Refine diagnostic procedure in outlying and rural areas where medical facilities are not available or at a distant or medical care is not possible, a self-diagnostic application will be developed to predict chronic kidney disease and its severity level. The model will be connected with application using Flask API considered as end points between model and application. Micro services will be used to authenticate user, predict disease, and monitor patients on daily basis. Application will input data through sensors, user interface and clinical history to predict presence of CKD and monitor patient. For huge data storage, cost effective and stability, cloud-based storage solution will be employed. The data from the mobile application and previous history of patient will be stored in this storage model. To ensure privacy and confidentiality of patient data isolation strategies can be implemented. The wearable sensors will monitor the patient and transmit the data to mobile application automatically. Data will go under preprocessing stage where outliers, noisy data will be removed. Missing values will be inputted to enhance the quality of the model. By importing all the libraries feature selection method will be implemented. After preprocessing, three machine learning classifiers named Multilayer Perceptron, Decision Tree and Support Vector Machine will be applied to optimally reduced dataset. The accuracy of each classifier will be evaluated for the prediction of CKD. Based on this prediction, further recommendation, treatments, and continuous monitoring will be suggested through mobile application. With this mobile architecture solution, kidney disease can be detected at an early stage. Patients can be monitored according to their severity level. Figure 2 depicts the system architecture.

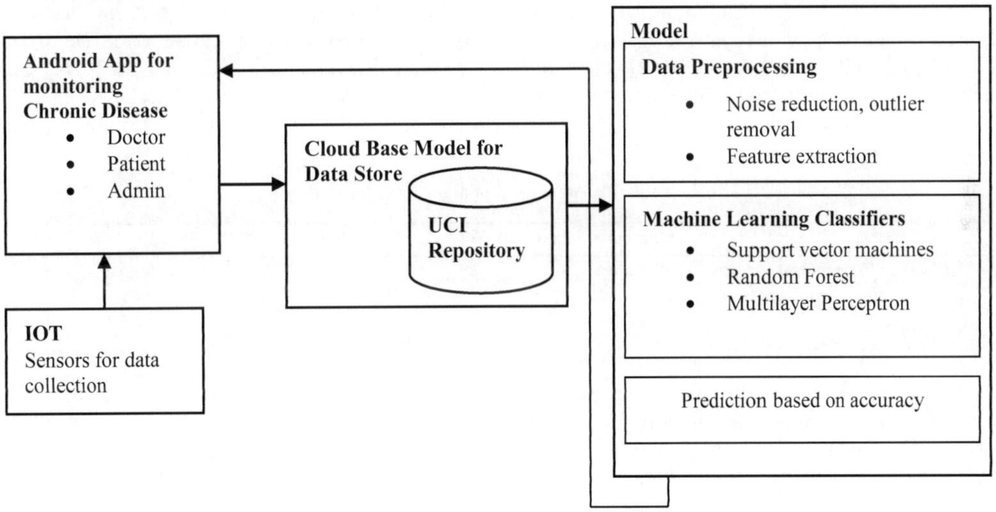

Figure 2. System architecture

3.4 Evaluating the Model

In this peculiar analysis, value of CKD marked as positive and value of not CKD marked as negative. With the help of this confusion matrix the performance of machine learning classifiers was assessed. With the help of this confusion matrix the specific results were displayed. From the CKD samples the true positive (TP) specifies that the disease was accurately identified whereas false negative (FN) specifies that the disease was incorrectly identified. For the non-CKD samples false positive (FP) indicates that the disease was wrongly diagnosed. While on the contrary, true Negative (TN) highlights the disease was accurately diagnosed. Model performance was measured using precision, recall, sensitivity, accuracy, F1 score, and specificity.

4. EXPERIMENTAL RESULTS

Using two machine-learning tools, three classifiers were applied to a dataset. In this research, two phases of machine learning will be employed. First is the training phase, where a test dataset is trained to fit the classifier parameters. Second is the Validation Phase, which estimates how well the model has been introduced. In order to train and test the data, we used 10 cross-validation methods. Machine learning algorithms Decision Tree, Support Vector Machine, and Multilayer Perceptron will be compared. Machine learning tools WEKA and Rapid Miner will be used to train and test the dataset. The comparative analysis will be performed among the selected classifiers based on accuracy and precision in the prediction of kidney disease (Rady & Anwar, 2019).

As a result of previous research, the Multilayer Perceptron provided an accuracy of 92.5% when detecting kidney disease. In order to train and test their model, researchers used a dataset containing 25 attributes. Based on multilayer perceptrons applied to a reduced dataset, this study achieved 96.5% accuracy on a 10-fold validation with an error rate of 3.5%. A total of 386 instances are correctly classified, while 14 instances are incorrectly classified. A 10-fold validation divides the dataset into ten parts, with one part being used to train the model and the other part being used to evaluate whether the model is correct (Vashisth, Dhall, & Saraswat, 2020). Previously, SVM was applied to a dataset with 25 attributes in order to detect kidney disease with an accuracy of 70%. In this study, Support Vector Machines were applied to a chronic kidney disease dataset with 14 attributes and gave 98% accuracy with a 2.5% error rate on a 20-fold validation. In a set of 400 instances, 391 are correctly classified while 9 are incorrectly classified. 20-fold validations allow the dataset to be divided into 20 parts, one of which is used to train the model and the other part is used to evaluate whether or not the model generates correct results (Vashisth, Dhall, & Saraswat, 2020). Previous research has found that decision trees are 95% accurate when there are 25 attributes in the dataset. On a dataset with 14 attributes, Decision Tree provided 97.75% accuracy with an error rate of 2.25% after 10-fold validation. In a 10-fold validation, 400 instances are divided into 10 parts and one part is used to train the model, and the other part is used to determine which model is reliable. 391 instances have been correctly classified while 9 have been incorrectly classified. Results can be found in Table 3 (Baidya, Umaima, & Islam, 2022).

Table 3. Comparison of algorithms in weka

Classifiers	Correctly Classified %	Precision	F measure	Error Rate
Multilayer Perceptron	96.5	0.967	0.977	3.5
Support Vector Machine	98	0.981	0.98	2
Decision Tree	97.75	0.978	0.978	2.25

As in Rapid Miner, three classifiers are applied to the dataset and the results are presented separately in Table 4. First, the results are presented one by one, and then they are presented collectively in a tabular format. The Decision Tree provided 94% accuracy when applied to the dataset. The accuracy of the Support Vector Machine on the chronic kidney disease dataset was 94.50% while the accuracy of Multilayer Perceptron on the chronic kidney disease dataset was 63.50%.

Table 4. Comparison of algorithms in rapid miner

Classifiers	Accuracy %
Decision Tree	94.17
Support Vector Machine	94.50
Multilayer Perceptron	63.75

Table 5 compares the performance of the same algorithm on two different machine learning tools. Three machine learning algorithms were applied to a kidney disease dataset, and the results indicate that Support Vector Machine is the most accurate in predicting chronic kidney disease, with 98% accuracy in WEKA. The multilayer perceptron did not perform well in Rapid Miner, but the accuracy in WEKA was satisfactory in predicting chronic kidney disease. Compared to previous research, this predictive model shows good accuracy with a reduced dimension dataset. There is evidence that most previous work shows good accuracy with a dataset with 25 attributes, but this accuracy can be achieved using a dimensionally reduced dataset with 14 features, which will result in a reduction of processing time and space.

Table 5. Comparison of algorithms in both tools

Tools	Algorithm	Accuracy %
WEKA	Decision Tree	97.75
	Multilayer Perceptron	96.50
	Support Vector Machine	98.00
Rapid Minor	Decision Tree	94.17
	Multilayer Perceptron	63.75
	Support Vector Machine	94.50

5. CONCLUSION

The advancement of technology makes healthcare services more advanced, and M-Health is also integrated with such technologies to monitor, process, and store patient data. Further, machine learning techniques are being employed to accurately predict disease analysis and classify disorders. In this study, three machine learning classifiers were used to predict chronic kidney disease, including support vector machine, decision tree, and multilayer perceptron. These classifiers were executed in machine learning software tools such as WEKA and Rapid Miner. The results of this study showed that support vector machine had the highest accuracy rate of 98% in predicting kidney disease in WEKA, while multilayer perceptron did not show good accuracy in Rapid Miner, but showed good accuracy in WEKA. It seems that the proposed machine learning model would be an effective tool for predicting kidney disease, based on an analysis of patient features from a dataset. Compared to existing state-of-the-art models, it outperformed them in accuracy, sensitivity, and specificity. In conclusion, the results of this study demonstrate the potential of machine learning for improving the accuracy of kidney disease prediction that can be integrated with mobile health solution architecture. Future research in this area could focus on the development of hybrid machine learning approaches for predicting chronic diseases, including kidney disease with mobile health solution integration. Additionally, dimension reduction techniques could be applied to minimize computational resources and time in the prediction of kidney disease and other chronic diseases.

ACKNOWLEDGEMENT

This work was supported by Instituto de Telecomunicações, Portugal.

REFERENCES

Baidya, D., Umaima, U., Islam, M. N., Shamrat, F. J. M., Pramanik, A., & Rahman, M. S. (2022, April). A deep prediction of chronic kidney disease by employing machine learning method. In *2022 6th International Conference on Trends in Electronics and Informatics (ICOEI)* (pp. 1305-1310). IEEE.

Brito, C., Esteves, M., Peixoto, H., Abelha, A. and Machado, J., 2022. A data mining approach to classify serum creatinine values in patients undergoing continuous ambulatory peritoneal dialysis. *Wireless Networks*, pp.1-9.

Couser, W. G., Remuzzi, G., Mendis, S., & Tonelli, M. (2011). The contribution of chronic kidney disease to the global burden of major noncommunicable diseases. *Kidney international*, *80*(12), 1258-1270.

Estonilo, C.G. and Festijo, E.D., 2022, January. Evaluation of the Deep Learning-Based M-Health Application Using Mobile App Development and Assessment Guide. In *2022 IEEE 12th Annual Computing and Communication Workshop and Conference (CCWC)* (pp. 0435-0440). IEEE.

Islam, M.A., Akter, S., Hossen, M.S., Keya, S.A., Tisha, S.A. and Hossain, S., 2020, December. Risk factor prediction of chronic kidney disease based on machine learning algorithms. In *2020 3rd international conference on intelligent sustainable systems (ICISS)* (pp. 952-957). IEEE.

Itani, S., Rossignol, M., Lecron, F. and Fortemps, P., 2019. Towards interpretable machine learning models for diagnosis aid: a case study on attention deficit/hyperactivity disorder. *PloS one*, *14*(4), p.e0215720.

Jongbo, O.A., Adetunmbi, A.O., Ogunrinde, R.B. and Badeji-Ajisafe, B., 2020. Development of an ensemble approach to chronic kidney disease diagnosis. *Scientific African*, *8*, p.e00456.

Kulkarni, E.G. and Kulkarni, R.B., 2016. Weka powerful tool in data mining. *International Journal of Computer Applications*, *975*, p.8887.

Pandit, A. K. (2020). Chronic Kidney Disease Prediction by Machine Learning.

Polat, H., Danaei Mehr, H. and Cetin, A., 2017. Diagnosis of chronic kidney disease based on support vector machine by feature selection methods. *Journal of medical systems*, *41*, pp.1-11.

Qureshi, K.N., Din, S., Jeon, G. and Piccialli, F., 2020. An accurate and dynamic predictive model for a smart M-Health system using machine learning. *Information Sciences*, *538*, pp.486-502.

Rady, E.H.A. and Anwar, A.S., 2019. Prediction of kidney disease stages using data mining algorithms. *Informatics in Medicine Unlocked*, *15*, p.100178.

Tahmasebian, S., Ghazisaeedi, M., Langarizadeh, M., Mokhtaran, M., Mahdavi-Mazdeh, M. and Javadian, P., 2017. Applying data mining techniques to determine important parameters in chronic kidney disease and the relations of these parameters to each other. *Journal of renal injury prevention*, *6*(2), p.83.

Vashisth, S., Dhall, I., & Saraswat, S. (2020, January). Chronic kidney disease (CKD) diagnosis using multi-layer perceptron classifier. In *2020 10th International Conference on Cloud Computing, Data Science & Engineering (Confluence)* (pp. 346-350). IEEE.

Webster, A.C., Nagler, E.V., Morton, R.L. and Masson, P., 2017. Chronic kidney disease. *The lancet*, *389*(10075), pp.1238-1252.

Zhang, Y., Liu, C., Luo, S., Xie, Y., Liu, F., Li, X. and Zhou, Z., 2019. Factors influencing patients' intentions to use diabetes management apps based on an extended unified theory of acceptance and use of technology model: web-based survey. *Journal of medical Internet research*, *21*(8), p.e15023.

CONNECTED PUERPERIUM IN PORTUGAL: A DIGITAL SYSTEM FOR AN INTEGRATIVE POSTPARTUM CARE

Carla V. Leite[1,2], Tala Al Shrbaji[3], Hannakaisa Niela-Vilen[2], Ana Patrícia Oliveira[1] and Rita Oliveira[1]

[1]*DigiMedia, Department of Communication and Art, University of Aveiro*
Campus Universitário de Santiago, 3810-193 Aveiro, Portugal
[2]*Department of Nursing Science, Faculty of Medicine, University of Turku*
Medisiina B6, FI-20520 Turku, Finland
[3]*School of Health Sciences, University of Aveiro*
Campus Universitário de Santiago, 3810-193 Aveiro, Portugal

ABSTRACT

The goal of this research was to conceptualise a system that answers to an emerging health care need: connecting women during puerperium with different users, namely: midwife, obstetrician, paediatrician, breastfeeding consultant, physical therapist, family nurse and doctor, psychologist, and informal caregiver. Those are the stakeholders, along with her family and the Portuguese Health System, since the implementation of this system could improve women's health and well-being, reduce costs with obstetric, psychiatric and general healthcare, and contribute to the positive after birth experience and breastfeeding rates, since: it enables the communication of women with different users of the system and between them; it allows them to upload up to date data/results/exams, accessible by different users who might need it for a customised and integrative diagnosis/treatment; and, it is accessible online, allowing women´s constant and physically distant follow-up. The use cases diagram and flows of the system, along with the interface design of a mobile app were created.

KEYWORDS

Postpartum, Puerperium, Digital, Mobile App, eHealth

1. INTRODUCTION

Pregnancy is considered a time of great transformation, but both physiological and psychological changes continue after birth in the postpartum period. A woman's body begins to recover from the pregnancy-related changes, while many health problems may also occur. Pain, abnormal discharge, urinary complications, or mood disorders, for example, are common symptoms for new mothers (Yonemoto et al., 2021). The duration of the postpartum period is not consensual, sometimes being defined as six weeks, and other times as the 12 weeks period after the birth, also referred to as the "fourth trimester" (Womersley et al., 2021; Verbiest et al., 2018). The early postpartum period appears to be the most stressful time concerning the transition to motherhood (Almalik, 2017; Tully et al., 2017). Yet, antenatal and intrapartum care are prioritised over the postpartum period (Womersley et al., 2021), with the latter being reported for a decade as the most neglected period for the provision of quality care, especially in resource-limited settings from low- and middle-income countries (WHO, 2014).

The care of a woman and her infant in the first weeks after birth is regulated differently among distinct countries. In general, the postpartum hospital stay is short and little information is provided regarding that period. The current situation in Portugal, to the best of our knowledge, is that no track of which information regarding postpartum is being delivered to soon to be parents; after birth, followed by discharge, there is no interaction or follow up from the public hospital nursing team. When no complications or risk situations are diagnosed, the first and single puerperium appointment with a Family Physician and Nurse Practitioner will happen between the 4th and 6th week after the birth (SNS, 2015). Moreover, there is no online system to facilitate the communication and interaction between new parents and public healthcare professionals regarding questions that might arise after birth (e.g., newborn care, breastfeeding, puerperium care, etc.).

Although Doula and Breastfeeding Consultant support might be necessary to mother-baby bond and to establish breastfeeding (Brooks et al., 2016), they are not regulated professions by the Portuguese Healthcare System, but there are established networks and online lists and databases of private providers, some with International Certification and some doulas specialised in the Postpartum period (e.g. for doulas: https://www.redeportuguesadedoulas.com; https://www.doulasdeportugal.com; https://doular.org; e.g. for Breastfeeding Consultants: https://www.apclc.pt; https://www.camsdeportugal.pt).

Currently, the generalised options to attempt to find answers through the public sector are: calling, sending an email or going in person to the health centre they belong to in order to schedule an appointment; going to the hospital or maternity units in person for an emergency consultation - which might not the best policy if it's not an emergency due to the risk of exposing the newborn and recovering mom, and to keep the focus and priority in the services; or, calling the 24/7 national call centre to talk to a nurse (DGS, 2023; ERS, 2021). Clearly, these services are conditioned by waiting times, service schedules and/or staff availability. Notwithstanding, there is a 24/7 video call and web chat service for the deaf population in order to be able to communicate with nurses or administrative staff (SNS 24, 2023).

Furthermore, there is no digital system assuring the connection between, and among, the tertiary sector (e.g. private medical physicians, physical therapists, doulas, nurses, breastfeed consultants, psychologists, etc.) and the patient information collected by the national healthcare system over the years. The new parents need to fill in/tell their clinical history in each private practice/office, and to carry their exams and analysis results, medicine prescriptions and guides of treatment.

Effective follow-up systems for postpartum women should be considered essential to be put in place, as many pregnancy-associated problems, such as gestational diabetes, anxiety, or depressive symptoms can continue to affect women's health in the long term (Womersley et al., 2021). Mobile health apps are suggested as a complementary support for parents in the postpartum period. However, the quality of commercially available apps varies and not all provide adequate health information, nor are customised (Tucker et al., 2021). Postpartum care should be based on a woman's individual needs and concerns. Proactive health services are needed after hospital discharge to ensure continuous support (Almalik, 2017). Tailored and timely information using different methods, such as handouts, texts, and discussions, would support mothers better than a single health care visit (Verbiest et al., 2018).

2. METHODS

The aim of this research was to conceptualise a system to comply with the *Rights to Information and to Healthcare Access* (WHO, 2019; United Nations, 1995) for the Portuguese residents, in order to facilitate the interaction of new parents with both physical and emotional care professionals. Consequently, the system can potentialise their self-care engagement during the postpartum period, in order to avoid commonly known physical and mental complications (Borghei et al., 2017).

The conceptualisation process counted with two phases: firstly, an exploratory phase which included an unstructured interview of two midwives, one breastfeeding consultant, two doulas, and three new mothers in Portugal, o understand how they are solving the underlying issue of lacking a digital platform focused on promoting the interaction between stakeholders during the puerperium period. Moreover, they also contributed to identify who could be the stakeholders. It was also asked about the data they would like to send/receive. With those inputs, a first draft of the use cases was created and analysed by an expert in healthcare systems. With the feedback obtained from the validation process, a final proposal of the system model was created using Use Cases, and they will be all presented far ahead.

Use cases (UC) are a structured representation of tangible interactions between users to achieve a goal (Douglass, 2016), inspired by real-life scenarios to which the digital world must provide an answer. The aim is to show the subsequent goals of the application and the possible roles of the user or other systems, making UC a highly relevant tool for user research and user experience design (Matz and Germanakos, 2016). This early step in the development process enables to involve human factors that will impact the requirements of the application (Liaghati et al., 2020). Use cases help answer fundamental questions: *What are the users of the system trying to do? What's the user experience?* (Rosenberg and Stephens, 2007), considering the research scope it can be translated as: *What does the digital system need to do for puerperium women to be able to reach and get support?*. A single use case may group similar or related uses in only one diagram, including activities, sequences, and system states (Delligatti, 2013). By representing the system in diagrams, it is possible to start with a simple, user-based view, and build that in order to create all the necessary diagrams as complex as needed (Liaghati et al., 2020). UCs are not expected to unambiguously define into detail how the application is intended to be developed, but rather provide a simple, high-level representation of what the application is designed to accomplish (Liaghati et al., 2020). Software can be used to be effective, accurate, readable, and understandable to all project stakeholders, sponsors, and end users (Matz and Germanakos, 2016), so the online collaborative software Lucid (https://lucid.app/) was used. Moreover, UCs facilitate the creation of interfaces that can lead later to a highly usable prototype (Matz and Germanakos, 2016), so the last step of the project was to design the interfaces based on the proposed UC diagrams, OmniGraffle software (https://www.omnigroup.com/omnigraffle) was used.

3. RESULTS

It became clear that the interviewed participants were finding ways through email, messages and phone calls to keep in contact individually, but also their personal social media (i.e. Facebook groups) and group messages (i.e. Messenger, WhatsApp) were used so the new mothers could support each other and be in touch with their peers - met during the birth and parenthood preparation courses, events or workshops targeting pregnant women. They also wished to involve their family and the other parent/partner, highlighting branches of the issue initially hypothesised.

In order to achieve the proposed goals, the primary and secondary stakeholders of the system were defined, to be able to find the system actors and their behaviours through UML Use-case diagrams (UC1-6). The first ones were identified as: a) Postpartum women (UC1); b) Postnatal Specialists (UC2), such as: nurse, midwife, physical therapist, physician, breastfeeding consultant, doula, psychologist, etc., mainly who focus rather on prevention (ex: risk posture and behaviours), promotion (ex: active lifestyle), diagnosis, treatment, recovery (ex: musculoskeletal, neurological, respiratory), mental, emotional and/or physical (ex: breastfeeding techniques) support. The Secondary as: c) the Portuguese Health System, since better outcomes on postpartum care could mean improvements on women's general health, therefore less costs involved for different sectors on the long term; d) Family members & Close Friends (UC3), including the other parent of the newborn; e) General population, since it could affect the public perception of afterbirth, and ultimately the family planning potentiating birth rates. Furthermore, Online services from the Portuguese Health System such as RSE (Electronic Health Records) and Patient's Portal were services considered relevant to be connected and used as data bases for the proposed system (UC4 and 5, respectively). RSE can only be accessed by Nurses and Physicians (and no other profession) who work on the Portuguese Health System from the Ministry of health (public sector). They are also the agents that can access and update the Patient's Portal clinical data, and each citizen can consult it, although no specialised area for pregnancy, birth or postpartum exists or is made available when the time comes. The sixth identified actor is the system administrator, who controls and manages whatever deemed necessary in the back office (UC6).

3.1 Postpartum Woman (UC1)

This actor of the system is the woman experiencing the puerperium phase (UC1 - Figure 1), and she will be able to view, edit and add data related to the user account, profile and status. It will start with the creation of the account, the customization of the profile, and it will be possible to update their emotional, physical and psychological changes. The actor will also be able to manage Postnatal specialists in order to connect, contact and request an appointment. It will also be possible to manage appointments, for example, reschedule or cancel them. In addition, she will receive notifications about messages received, upcoming appointments and new user connections. This user will also be able to manage their family and close friends in the system, by connecting accounts and deciding on their permissions, i.e. what data they can view, edit and add about her.

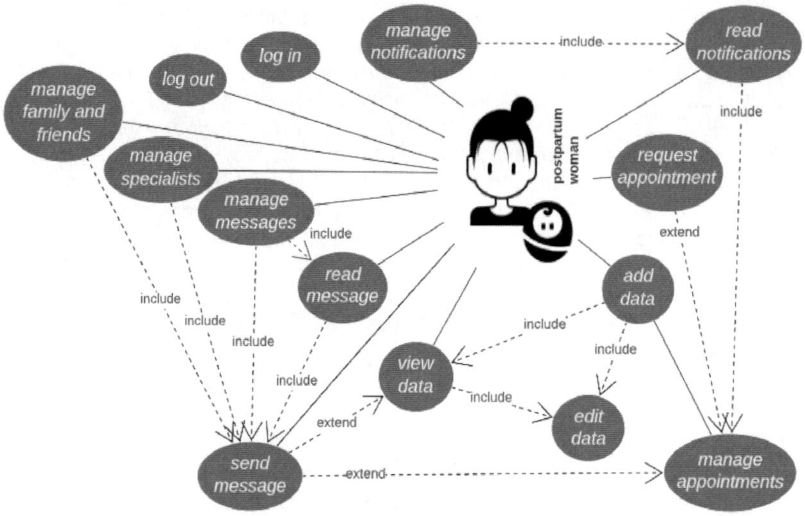

Figure 1. Use case of Woman during puerperium period (UC1)

3.2 Postnatal Specialist (UC2)

This actor is the Postnatal specialist (UC2 - Figure 2) focused on the care of women during the puerperium phase. The actor will create an account, view and edit their profile, which should include a description of its speciality, approach and care model they currently follow, experience, accomplishments and statistics (activity in the system, number of postpartum women under their care, their ratings, etc.). Although this role may be common to users from different academic and practical backgrounds, they are all professionals responsible for the emotional, physical, and psychological care of postpartum women. Since they are able to care for multiple postpartum women at the same time, they must have the ability to see them all, their statistics and status, and also select each one under their care to view, edit and add new data. This user will be able to exchange messages with other users, including postpartum women, along with their connected family and friends. It will be possible to schedule and manage their appointments, which will trigger notifications about the changes made in order to alert all the users involved.

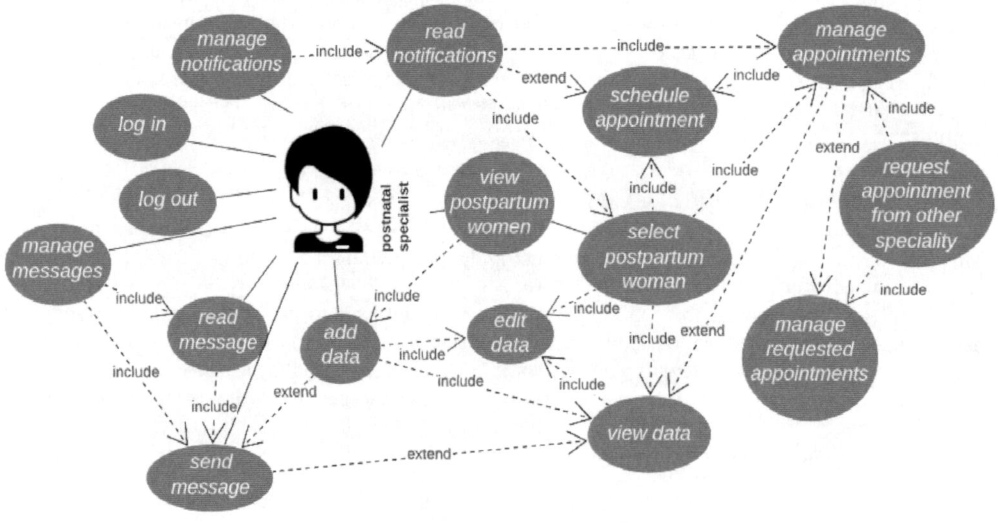

Figure 2. Use case of Postnatal specialist (UC2)

3.3 Family and Close Friends (UC3)

This actor represents a family member (i.e., the other parent, the partner, a parent, etc.) or a close friend of the postpartum woman (UC3 - Figure 3), so they can be included as her informal support network. They can check on her scheduled appointments that are shared with them, and report their availability to join her. They can also see who the postnatal specialists involved are, and can request appointments for themselves in case they need information to better support the puerperium woman (no clinical or private information should be shared, but rather general scientific knowledge, possible techniques, etc.). The actor will also be able to see the reported emotional, physical, and psychological changes that the puerperium woman has shared with them, and also report their support. They will be able to exchange messages with the users they are connected to, and receive notifications about changes or upcoming appointments.

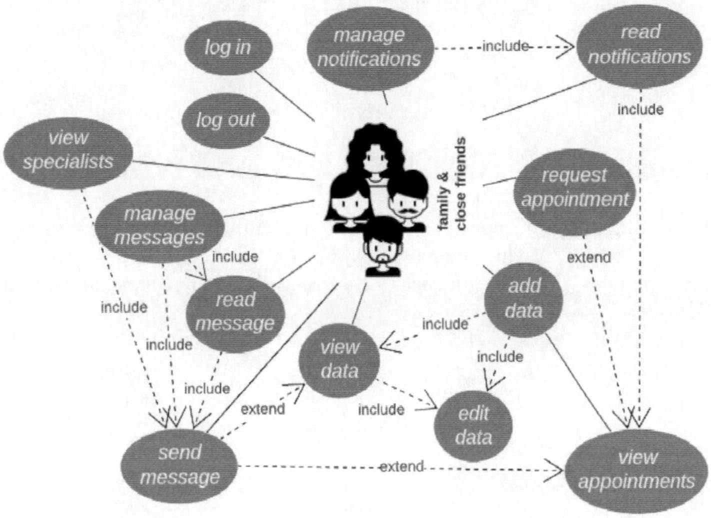

Figure 3. Use case of Family and Close Friends (UC3)

3.4 RSE (Electronic Health Records) (UC4)

This actor is a digital system used by the Portuguese Health System (UC4 - Figure 4), and its database gathers all the clinical information of each patient registered, including diagnosis, treatments, exams and analysis results. This information was created by or communicated to medical doctors and nurses from the public sector - currently the only ones with access to view, edit or add information to the system. Its approach is walled garden since patients or other professionals, including from private care, do not have access to view/edit/add data. However, this research takes into account the progress of adding Big Data to it, and strengthening the pipeline of digital healthcare services (Dash et al., 2019). With this visionary approach, this actor could feed the proposed application with valuable clinical information of each patient, and even relevant data from healthcare professionals, such as statistics.

Figure 4. Use case of Electronic Health Records (UC4)

3.5 Patient's Portal (UC5)

This actor is an online system currently being used by Patients of the Portuguese Healthcare service (UC5 - Figure 5) in order to check the data updated by medical doctors and nurses upon appointments on the public service (i.e. health centres and hospitals), prescriptions from medical doctors from the private sector and work entries from clinical analysis/exams laboratories. However, currently there is no specific section regarding pregnancy or postpartum on the patient's portal (Leite and Almeida, 2021).

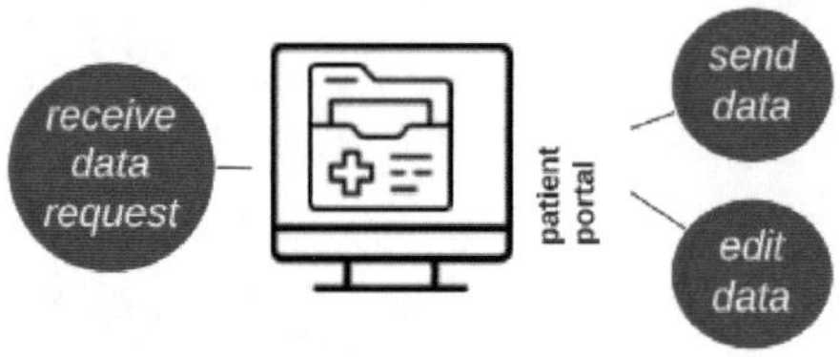

Figure 5. Use case of Patient's Portal (UC5)

3.6 System Administrator (UC6)

This actor (UC6 - Figure 6) is responsible for managing settings, messages, notifications, and also users when its intervention is needed. This actor is also who will be reached when there are any issues to solve within the system, and who will update the back office according to new developments and changes that may affect the system.

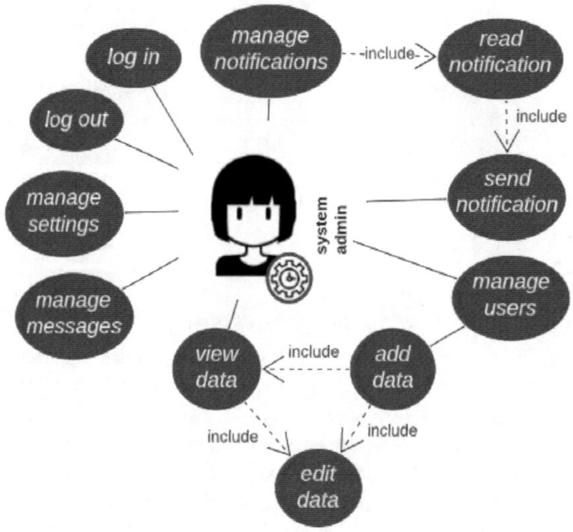

Figure 6. Use case of System Administrator (UC6)

3.7 Interfaces – Proof of concept

In order to illustrate how the use cases can be materialised visually to the users, interfaces of the mobile application were created, and took in consideration the predicted flux of events and usability principles, Figures 7-9 were included next.

Figure 7-9. Proof of concept of the action to connect to Postnatal Specialist as a Puerperium Woman user

4. DISCUSSION AND CONCLUSION

After extensive monitoring during pregnancy, a shift towards newborn care occurs after birth, and the women seem to face the postpartum period challenges without adequate support. They have reported learning needs regarding baby care, episiotomy care, and breastfeeding, and reported that maternal health services are not meeting them (Almalik, 2017), so they feel unprepared for postpartum (Martin et al., 2014). The single appointment with health care professionals is planned to happen only a few weeks after birth, and studies show the inadequacy of the postnatal check, and the dissatisfaction of women, expecting a more comprehensive approach as reproductive health patients (Womersley et al., 2021; Tully et al., 2017).

Future developments to this proposal could take into consideration that pregnant women and their caregivers' needs during the perinatal period were already reported as unmet regarding logistical, social, emotional, and informational (Robinson et al., 2018; Leite and Almeida, 2022), the proposed model could be modified to include both prenatal and postnatal periods - a possible contribution that was out of the scope of this project.

According to Ventola's (2014) framework, clinical practice can be transformed with the usage of technology that facilitates tasks such as information and time management, health record maintenance and access, communications and consulting, reference and information gathering, patient management and monitoring, clinical decision-making, and education and training. Mobile devices, such as smartphones and tablets, are highlighted since they are powerful tools from hardware to operating systems, and allow features regarding sound recording, voice and/to text, web search, location, and high-quality cameras. More recently, these increasingly sophisticated tools can also be part of the Internet of Things, connected to wearables, gadgets, ubiquitous devices, and smart home and interactive television sets - all sources of records and data that could be useful for health and wellbeing purposes (Grym et al., 2019).

However, to comply with clinical practice standards and to be integrated with health care services, the conceptualisation and validation processes need to be established, and measures to increase the quality and safety of the applications. The most common benefits of online health support to patients are convenience, reliability, health care availability, and cost savings (Almathami et al., 2020). Further, they can reduce travel time, waiting time in the clinics, and absences from work. It is also suggested that digital healthcare services may empower patients to participate more in their care (Ghimire et al., 2023).

Notwithstanding, healthcare providers may sometimes have ambivalent and sceptical attitudes toward digital health services, namely when including telemedicine approaches. Young healthcare professionals seem to recognise the potential of modern technology for care provision (Grassl et al., 2018; Niela-Vilen et al., 2020). Taking this into consideration, it became relevant to conceptualise a service not based on real-time interaction and for diagnosis purposes, but to enable the possibility of a timely interaction, that circumvents bureaucratic obstacles between stakeholders during the postpartum period. This way, postnatal specialists could be focusing on the quality of the encounter with new parents instead of tasks that technology can facilitate (Niela-Vilen et al., 2020).

By having a virtual space to meet puerperium women often, brief opportunistic advice to tackle issues like nutrition or tobacco consumption can also take place aiming behaviour changes. Visionary developments could make use of big data to detect behaviour patterns of puerperium women as users in order to provide them with timely and adequate suggestions aiming for support. The application can also be connected to artificial intelligence systems that are trained for postnatal topics and to provide predictive answers to their questions. Moreover, puerperium women seem to wish to share their postpartum experience (Verbiest et al., 2018), so a social approach inside the application itself or social media connections might need to be integrated. Clearly indicate advantages, limitations and possible applications.

ACKNOWLEDGEMENT

This work is financially supported by national funds through FCT – Foundation for Science and Technology, I.P., under the project UIDB/05460/2020. CVL would like to acknowledge the fund support through the Doctoral Programme in Technology Enhanced Learning and Societal Challenges (PD/00173/2013) by FCT, through the research grants PD/BI/135241/2017, doctoral grant PD/BD/143077/2018 and exceptional grant COVID/BD/152551/2022. CVL and HN-V would like to acknowledge the University of Turku for providing an interdisciplinary collaborative environment for the development of this work. CVL and TAS would like to acknowledge Nelson Rocha and Alexandra Queirós for their expert review of the proposal, under the scope of Health Information Systems Course from the School of Health Sciences, University of Aveiro.

REFERENCES

Almalik, M.M., 2017. Understanding maternal postpartum needs: A descriptive survey of current maternal health services. *Journal of Clinical Nursing*, 26(23-24), pp.4654–4663. doi: 10.1111/jocn.13812.

Almathami, H.K., Win, K.T. and Vlahu-Gjorgievska, E. (2020). Barriers and facilitators that influence telemedicine real-time online consultation at patients' home: A Systematic Literature Review. *Journal of Medical Internet Research*, 22(2). doi: 10.2196/16407.

Borghei, N. S., Taghipour, A., Latifnejad Roudsari, R., Jabbari Nooghabi, H., 2017. Investigating the Determinants of Maternal Empowerment During Pregnancy: A Strategy for Prenatal Healthcare Promotion, *Journal of Midwifery and Reproductive Health*, 5(3), pp. 988-997. doi: 10.22038/jmrh.2016.7980.

Brooks, E.C., Genna, C.W. and Mannel, R., 2016. The Lactation Consultant: Roles and Responsibilities, in Wambach, K. and Riordan, J. (eds.) *Breastfeeding and Human Lactation*, 5th ed. Burlington MA: Jones & Bartlett Learning, pp. 3-36.

Dash, S., Shakyawar, S.K., Sharma, M. and Kaushik, S., 2019. Big data in healthcare: management, analysis and future prospects. *Journal of Big Data*, 6(1), pp.1–25. doi: 10.1186/s40537-019-0217-0.

Delligatti, L., 2013. *SysML Distilled: A Brief Guide to the Systems Modeling Language*. 1st ed, Addison-Wesley.

DGS, 2023. *Norma DGS, Norma: 001/2023, Organização dos cuidados de saúde na preconceção, gravidez e puerpério*. Retrieved: https://normas.dgs.min-saude.pt/wp-content/uploads/2023/03/norma_001_2023_org_cuidados_preconcecao_gravidez_puerperio.pdf.

ERS, 2021. *Direitos e Deveres dos Utentes dos Serviços de Saúde*. Porto: Entidade Reguladora da Saúde. Retrieved: https://www.ers.pt/media/sfbd4x2h/publica%C3%A7%C3%A3o-ers_direitos-e-deveres.pdf

Ghimire, S., Martinez, S., Hartvigsen, G. and Gerdes, M., 2023. Virtual prenatal care: A systematic review of pregnant women's and healthcare professionals' experiences, needs, and preferences for quality care. *International Journal of Medical Informatics*, 170, p.104964. doi: 10.1016/j.ijmedinf.2022.104964.

Grassl, N., Nees, J., Schramm, K., Spratte, J., Sohn, C., Schott, T.C. and Schott, S., 2018. A Web-Based Survey Assessing the Attitudes of Health Care Professionals in Germany Toward the Use of Telemedicine in Pregnancy Monitoring: Cross-Sectional Study. *JMIR mHealth and uHealth*, 6(8), p.e10063. doi: 10.2196/10063.

Grym, K., Niela-Vilén, H., Ekholm, E., Hamari, L., Azimi, I., Rahmani, A., Liljeberg, P., Löyttyniemi, E. and Axelin, A., 2019. Feasibility of smart wristbands for continuous monitoring during pregnancy and one month after birth. *BMC Pregnancy and Childbirth*, 19(1). doi: 10.1186/s12884-019-2187-9.

Liaghati, C., Mazzuchi, T.A. and Sarkani, S., 2020. A method for the inclusion of human factors in system design via use case definition. 2(1-4), pp.45–56. doi: 10.1007/s42454-020-00011-1.

Leite, C. V. and Almeida A. M., 2021, The Current Status of Online Resources from the Portuguese Health System for Childbirth Education. *2021 16th Iberian Conference on Information Systems and Technologies (CISTI)*, Chaves, Portugal, 2021, pp. 1-4, doi: 10.23919/CISTI52073.2021.9476461.

Leite, C.V. and Almeida, A.M., 2022, e-Health Services to Support the Perinatal Decision-making Process: An Analysis of Digital Solutions to Create Birth Plans. *Proceedings of the 15th International Joint Conference on Biomedical Engineering Systems and Technologies (BIOSTEC 2022)* - Volume 5: HEALTHINF, pp. 405-412, doi: 10.5220/0010814600003123.

Martin, A., Horowitz, C., Balbierz, A. and Howell, E.A., 2014. Views of Women and Clinicians on Postpartum Preparation and Recovery. *Maternal and Child Health Journal*, 18(3), pp.707–713. doi: 10.1007/s10995-013-1297-7.

Matz, A. and Germanakos, P., 2016. Increasing the Quality of Use Case Definition Through a Design Thinking Collaborative Method and an Alternative Hybrid Documentation Style, in Zaphiris, P., Ioannou, A. (eds) *Learning and Collaboration Technologies*, Lecture Notes in Computer Science vol 9753, Switzerland: Springer, Cham, pp. 48-59, doi: 10.1007/978-3-319-39483-1_5.

Niela-Vilen, H., Rahmani, A., Liljeberg, P. and Axelin, A., 2020. Being 'A Google Mom' or securely monitored at home: Perceptions of remote monitoring in maternity care. *Journal of Advanced Nursing*, 76(1), pp. 243–252. doi: 10.1111/jan.14223.

Robinson, J.R., Anders, S.H., Novak, L.L., Simpson, C.L., Holroyd, L.E., Bennett, K.A. and Jackson, G.P., 2018. Consumer health-related needs of pregnant women and their caregivers. *JAMIA Open*, 1(1), pp. 57–66. doi: 10.1093/jamiaopen/ooy018.

Rosenberg, D. and Stephens, M., 2007. Use Case Modeling, in: Use Case Driven Object Modeling with UML. Apress, pp 49-82, doi: 10.1007/978-1-4302-0369-8_3.

SNS, 2015. *Programa Nacional para a Vigilância da Gravidez de Baixo Risco*. Lisboa: Direção-Geral da Saúde, ISBN: 978-972-675-233-2. Retrieved from https://www.dgs.pt/em-destaque/programa-nacional-para-a-vigilancia-da-gravidez-de-baixo-risco-pdfl1.aspx.

SNS 24, 2023. *Contacto acessível para cidadão surdo*. Available at: https://www.sns24.gov.pt/contacto-acessivel-cidadao-surdo/ [Accessed 28 May 2023].

Tucker, L., Villagomez, A.C. and Krishnamurti, T., 2021. Comprehensively addressing postpartum maternal health: a content and image review of commercially available mobile health apps. *BMC Pregnancy and Childbirth*, 21(1), 311. doi: 10.1186/s12884-021-03785-7.

Tully, K.P., Stuebe, A.M. and Verbiest, S.B., 2017. The fourth trimester: a critical transition period with unmet maternal health needs. *American Journal of Obstetrics and Gynecology*, 217(1), pp.37–41. doi: 10.1016/j.ajog.2017.03.032.

United Nations, 1995. *Beijing Declaration and Platform for Action: Beijing+5 Political Declaration and Outcome*. Reprint, New York: UN Women, 2014.

Ventola C. L., 2014. Mobile devices and apps for health care professionals: uses and benefits. *P & T: a peer-reviewed journal for formulary management*, 39 (5), pp. 356–364.

Verbiest, S., Tully, K., Simpson, M. and Stuebe, A., 2018. Elevating mothers' voices: recommendations for improved patient-centered postpartum. *Journal of Behavioral Medicine*, 41(5), pp.577–590. doi: 10.1007/s10865-018-9961-4.

WHO, 2019. *WHO consolidated guideline on self-care interventions for health and well-being: Sexual and Reproductive Health and Rights*. Geneva, Switzerland: World Health Organization. Retrieved from: https://apps.who.int/iris/bitstream/handle/10665/325480/9789241550550-eng.pdf

WHO, 2014. *WHO recommendations on postnatal care of the mother and newborn*. Geneva, Switzerland: World Health Organization. Retrieved from: https://apps.who.int/iris/bitstream/handle/10665/97603/9789241506649_eng.pdf

Womersley, K., Ripullone, K. and Hirst, J.E., 2021. Tackling inequality in maternal health: Beyond the postpartum. *Future Healthcare Journal*, 8(1), pp.31–35. doi: 10.7861/fhj.2020-0275.

Yonemoto, N., Nagai, S. and Mori, R., 2021. Schedules for home visits in the early postpartum period. *Cochrane Database of Systematic Reviews*, 2021(7). doi: 10.1002/14651858.cd009326.pub4.

TOWARDS A DATA-DRIVEN CITY: AN ANALYSIS OF CITIES' DATA STRATEGIES

Yusuf Bozkurt[1,2], Simon Flaisch[2], Dieter Hertweck[2] and Alexander Rossmann[2]
[1]*School of Computing, Engineering & Physical Sciences, University of the West of Scotland Paisley, PA1 2BE, Scotland, United Kingdom*
[2]*Department of Computer Science, Reutlingen University 72762 Reutlingen, Germany*

ABSTRACT

Smart cities are considered data factories that generate an enormous amount of data from various sources. In fact data is the backbone of any smart services. Therefore, the strategic beneficial handling of this digital capital is crucial for cities. Some smart city pioneers have already written down their approach to data in the form of data strategies, but what should a city's data strategy include, and how can the goals and measures defined in the strategies be operationalized? This paper addresses these questions by looking closely at the data strategies of cities in Germany and the top three countries in the EU Digital Economy and Society Index. The in-depth analysis of 8 city data strategies has yielded 11 dimensions that cities should consider in their data strategy. These are (1) relevance of data, (2) principles, (3) methods, (4) data sharing, (5) technology, (6) data culture, (7) data ethics, (8) organizational structure, (9) data security and privacy, (10) collaborations, (11) data literacy. In addition, data governance is a concept to put these 11 strategic dimensions into practice through standardization measures, training programs, and defining roles and responsibilities by developing a data catalog.

KEYWORDS

Smart Cities, Data Governance, Urban Data Strategy, Urban Data Governance

1. INTRODUCTION

Digital transformation is not feasible without data. Therefore, proper governance of data assets is essential for any organization to create value (Anthony Jnr, 2021; Dremel et al., 2017; Vial, 2019). Nowadays, there is no denying the importance of data, with the EU considering it a vital source of "growth, competitiveness, innovation, job creation, and societal progress" (European Commission, 2022b), benefiting citizens and businesses in many ways through data-driven applications. Today, cities are considered data factories, representing an enormous, rapidly growing, but also heterogeneous data landscape (Barns, 2018; Moustaka et al., 2018). A great deal of research addresses smart services used in different city dimensions, as shown in a literature review by (Bozkurt et al., 2020). Data is the backbone of many services, but what is the right way to deal with it? Cities address this question by publishing data strategies. In this work, we aim to analyze the openly published data strategies of EU cities in the top three countries of the EU Digital Economy and Society Index (DESI) (European Commission, 2022a) (Finland, Denmark, and the Netherlands) and Germany to get an integrated overview of the dimensions of urban data strategies, providing a holistic view of cities' data domains. The results aim to contribute to the body of knowledge of smart cities by providing an overview of the strategic dimensions of data to help practitioners approach the topic of data strategy and provide guidance on the areas in which they should examine their data. In doing so, we explored the following two research questions (RQ) in this paper:

RQ 1: What are the dimensions of cities' data strategies?

RQ 2: How can data governance in cities contribute to achieving the goals of data strategies?

RQ 1 is answered using content analysis of data strategies. The results provide relevant dimensions for a data-driven city. These dimensions allow data governance to be further defined and aligned with the data strategy. RQ 2 views the whole from the perspective of data governance, as data governance can be a promising concept for developing a data-driven smart city.

An analysis of previous scientific publications on urban data strategy yielded no results. We searched the IEEE Xplore, Web of Science, Emerald, and ACM scientific databases for peer-reviewed articles with the search terms "data strategy" AND ("smart city" OR "city"). In total, we found 139 hits, but after analyzing the titles and abstracts, none of the articles met the criteria of an analysis of city data strategies or guidance for developing city data strategies.

The rest of the paper is organized as follows. The next section explains the background concepts of smart cities, data governance, and data strategy. Section 3 describes the research approach to illustrate the data collection and analysis. Section 4 presents the results of the data strategy analysis by describing the dimensions of data strategies. Next, section 5 discusses how data governance contributes to achieving urban data strategies' goals, followed by the implications and limitations of this work. The final section provides a conclusion.

2. BACKGROUND

Increasing urbanization poses new challenges for cities. The UN estimates that ¾ of the world's population will live in cities (United Nations, 2018). To meet the challenges of urbanization, e.g., public transport, waste management, and urban utilities, the concept of smart cities is of great relevance (Ahvenniemi et al., 2017; Jin et al., 2014; Zanella et al., 2014). Smart city research has yielded many definitions, yet providing a unified definition is difficult as perspectives range from information and communication technology (ICT) focus to socioeconomic focus (Albino et al., 2015; Manville et al., 2014). Although there are many definitions, a common denominator seems to be the use of ICT for the operation of smart cities, as shown in several papers (Bozkurt et al., 2020; Caragliu et al., 2011; Lombardi et al., 2012; Washburn & Sinduh, 2010). Nevertheless, just using ICT is not enough; instead, the focus must be on how ICT uses data to drive growth and guide urban development processes (Caragliu et al., 2011). Thus, the success of smart cities is highly dependent on how sound data can be used to decide and shape urban development and growth (Albino et al., 2015). That leads to the need for appropriate handling of data.

Data governance is one concept that addresses the data asset in a way consistent with an organization's overall strategy by defining principles, standards, responsibilities, and processes for the entire data lifecycle. However, it should not be confused with the technical management of data. Instead, it is the decision-making in all data-related issues, which is then implemented in data management (Mahanti, 2021b, 2021a; Seiner, 2014; Weber & Klingenberg, 2021). The literature on data governance in the urban context is sparse (Bozkurt et al., 2022; Kvalvik et al., 2022), although there is already a lot of practical and scholarly work on data governance in the enterprise context (Khatri & Brown, 2010; Mahanti, 2021b, 2021a; Otto, 2011; Seiner, 2014; Weber et al., 2009; Weber & Klingenberg, 2021). In the smart city context, however, only two literature reviews on urban data governance analyze only a handful of relevant literature (Bozkurt et al., 2022; Kvalvik et al., 2022). There is also little work with primary data, e.g., Paskaleva et al. (2017) developed a framework for data governance in smart city initiatives based on interviews with Manchester, Eindhoven, and Stavanger, finding that stakeholder collaboration is a key aspect in urban data governance. In addition, data governance in the city can define the direction for the city's generation, collection, and utilization of data consistently aligned with the city strategy and across all stakeholders (Paskaleva et al., 2017).

Maximizing the business value of data requires a cohesive, enterprise-wide data strategy that addresses how data is collected, stored, managed, and used (Adelman et al., 2005). Data is a cross-cutting issue, especially in smart cities, as many stakeholders and actors are involved in the data ecosystem. Thus, a city must know its data infrastructure and act accordingly to set a strategic direction. The lack of a data strategy can lead to reactive rather than proactive initiatives or to a poorly planned and designed system, which later leads to interoperability issues or vendor lock-ins, and to repeatedly solving the same organizational and technical problems. Therefore, the need for an overarching data strategy that involves all sectors is a collaborative effort to ensure that goals and technology capabilities are aligned to reap the benefits of data (Fleckenstein & Fellows, 2018).

3. RESEARCH APPROACH

In data collection, we followed the basic outline of the gray literature analysis of Garousi et al. (2017), as cities themselves publish data strategies and do not fall under the category of peer-reviewed scientific literature.

For the data analysis, used an inductive coding approach (Chandra & Shang, 2019; Corbin & Strauss, 1990; Gioia et al., 2013) and captured concepts, ideas, and action items that emerged in individual articles through full-text reading.

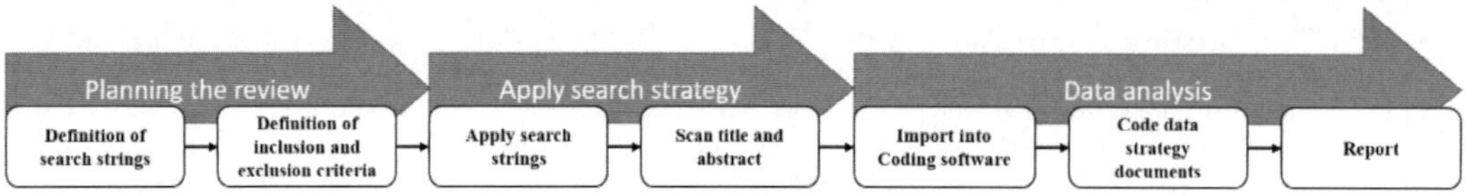

Figure 1. Research Process

Figure 1 shows the research process. The first phase is planning the review. That includes the step of defining the search words. We defined our search words in a brainstorming session within the author team. First, we created a list of terms that could be relevant to our search. Then, we iteratively adjusted these search terms to obtain relevant search results and cover our research questions. As mentioned in the introduction, our search was limited to the top three countries according to DESI (Finland, Denmark, and the Netherlands) and Germany. In the end, the following search strings emerged:

- *<Denmark/Finland/Germany/The Netherlands>:* ***"data strategy*** *<country>* ***city"*** *OR* ***"data governance*** *<country>* ***city"*** *OR* ***"urban data governance*** *<country>****"*** *OR* ***"smart city data strategy*** *<country>****"***

The next step in the planning phase of the review is the definition of inclusion and exclusion criteria. We applied the inclusion and exclusion criteria to distinguish the relevant search results from the irrelevant ones, as shown in Table 1.

Table 1. Inclusion and exclusion criteria

Inclusion	The article contains a data strategy of a city that belongs to Germany or to the top three countries, according to DESI.
	The article is published in English or German.
	The article was published by or on behalf of the cities.
	The URL is working and freely available.
Exclusion	The source is non-text based.
	The article contains the duplicated content of a previously examined article.

After the planning phase of the review, applying the search strategy was the next phase of the research process. We applied the search terms in the Google search engine during the period March-April 2023. To conduct a proper analysis, we included the yielded results relevant to our research questions in an Excel file according to our inclusion and exclusion criteria. In this way, we identified a total of 15 relevant articles. We then reviewed them for duplicates and analyzed the articles in detail for their contribution to our research questions. Thus, at the end of the study selection process, eight relevant articles (Table 2) remained for in-depth analysis, which were imported into MAXQDA qualitative data analysis and mixed methods software (MAXQDA, 2022).

Table 2. Identified cities

Country	City	Strategy document
Germany	Cologne	(Blauhut & Stadt Köln, 2021)
Germany	Soest	(Stadt Soest, 2021)
Germany	Monchengladbach	(Tillmanns et al., 2022)
Germany	Berlin	(Land Berlin, 2018)
Germany	Hamburg	(Stadt Hamburg, 2020)
Finland	Helsinki	(The city of Helsinki, 2023)
The Netherlands	Amsterdam	(The city of Amsterdam, 2021)
The Netherlands	The Hague	(The city of The Hague, 2020)

Data analysis was performed using MAXQDA software, in which each document was coded, and the codes created were discussed within the team. In this way, data analysis iteratively produced a series of codes that were eventually combined into broader categories of similar themes in a workshop with the author team, resulting in the 11 dimensions presented in section 4.

4. RESULTS

The following section presents the analysis results of the data strategies, which are synthesized into 11 dimensions of urban data strategies.

(1) Relevance of data:
The relevance of data finds an introductory place in the data strategies. A common understanding is established, and the goals, motivation, and vision are described. That includes communicating the goals of optimizing urban services using data, creating new innovative solutions, making urban planning data-driven, and being an attractive location for businesses. The relevance of data protection, data sovereignty, data sharing, and data quality are addressed in that context (Blauhut & Stadt Köln, 2021; Land Berlin, 2018; Stadt Hamburg, 2020; Stadt Soest, 2021; The city of Amsterdam, 2021; The city of Helsinki, 2023; The city of The Hague, 2020; Tillmanns et al., 2022).

(2) Principles:
The Principles address the fundamental aspects of handling data. (1) Citywide standards for high data quality along the entire data lifecycle must be established with a focus on harmonizing specific data models and visible interface alignment and guidelines. Master data management, including reference data, must be revised to develop administrative and citywide standards for elementary attributes of data sets (Blauhut & Stadt Köln, 2021; The city of Helsinki, 2023; Tillmanns et al., 2022).

Next, (2) Data quality must be ensured over the long term, including syntactic and semantic data quality, and redundancies and non-meaningful data must be minimized. KPIs are a tool to measure data quality based on criteria such as timeliness, control intensity, etc. (Stadt Hamburg, 2020; Stadt Soest, 2021; The city of Helsinki, 2023; The city of The Hague, 2020; Tillmanns et al., 2022).

Ensuring (3) sovereignty is another principle: The data about the city must be available to the city and remain so. Citizens and visitors have a right to know how their data is being used. They must always have easy access to their data and be able to use the opportunities offered by laws and regulations. Control over data must be well-organized and transparent. In doing so, partners working with the city must be transparent about using technology and data to solve urban challenges. Dependence on specific technologies and data must be reduced, and the public sector must be able to decide on access rights to the holistic urban data space (Blauhut & Stadt Köln, 2021; The city of Helsinki, 2023; The city of The Hague, 2020).

Finally, the 4) data in and about the city must be classified. The goal of data classification is to identify data sets characterized by standard rules and working methods in order to finally define data usage guidelines adapted to the data classes, improve the usability of the data and design data architectures tailored to the application (The city of Helsinki, 2023).

(3) Methods:
The methods address approaches and tools in handling data. Thus, a roadmap defines the next steps for realizing the strategic goals and provides a transparent common understanding. This ensures that the strategy is put into action and avoids unclarity due to the overwhelming volume of goals (Stadt Soest, 2021; The city of Helsinki, 2023). The goals are to be addressed step by step in an agile working method with all stakeholders in the city. Data governance is seen as a dynamic and continuously evolving framework for implementing the data strategy. This approach improves operational procedures, and the data strategy adapts and scales (Stadt Hamburg, 2020; Tillmanns et al., 2022). Furthermore, success must be measurable by defining KPIs and regular benchmarking (The city of Helsinki, 2023).

A data catalog is a central tool in the data strategy, which contains the data with descriptive metadata and evolves constantly. The goal is to create a citywide overview of the city's data assets, what quality parameters it has, and who the data owners are. Cities can improve data quality, clean up redundant data sets, know about the current data sets in various departments, and provide all interested stakeholders with an overview of the data that is relevant to them and its sources using the data catalog (Stadt Soest, 2021; Tillmanns et al., 2022).

(4) Data exchange:
The data-exchange aspect is divided into open data and non-open data. Cities want to make more data available to the public, e.g., by making frequently requested data available online by default. Open data is the basis for

information and participation and for civil society to develop its own applications. In addition, open data lays the foundation for transparency and trust in city government. In addition to raw, machine-readable data, services based on open data must also be made openly accessible, i.e., applications created with public funds must also be accessible to the public. Cities have many creative businesses and residents that could use open urban data to develop practical data-based applications for the urban ecosystem. This is how to sustain the vitality of the urban ecosystem (Blauhut & Stadt Köln, 2021; Land Berlin, 2018; Stadt Soest, 2021; The city of Amsterdam, 2021; The city of Helsinki, 2023; Tillmanns et al., 2022).

Non-open data, on the other hand, refers to data that other units can use within the city administration, following legal considerations. By combining different data sources, new insights are possible, and at the same time, the effort for further data collection can be reduced and redundancies minimized. Data should not be considered in their departmental silos but rather across departmental boundaries. The classification of data sets as open data or internal use only, as well as access control, must be decided through data governance processes (Stadt Hamburg, 2020; Stadt Soest, 2021; The city of Helsinki, 2023; Tillmanns et al., 2022).

(5) Technology:

The technological dimension is primarily about modernizing the data infrastructure regarding homogeneity, security, reliability, and integrability. A city's data infrastructure must be designed so that there is no lock-in to individual vendor solutions by relying on open source. In the long term, scalability must be ensured to deal with ever more and larger volumes of data. In addition, modern analytics environments and tools are needed to process data faster and more securely (Blauhut & Stadt Köln, 2021; The city of Amsterdam, 2021; Tillmanns et al., 2022).

In terms of analytics environments and data-driven decision-making, machine learning methods are applied to develop forecasting models and urban planning simulations visualized in user-centric dashboards (The city of Helsinki, 2023). Generally, an urban data platform is a key concept in urban data infrastructure. It is considered the backbone of the urban data infrastructure and facilitates data-driven decision-making through an integrated view and standardized interfaces. As a system of systems, an urban data platform connects various data sources through appropriate interfaces and enables regulated access to each data set. Based on this infrastructure, for example, city subsidiaries and the administration can share data, and an open data portal can be connected to the urban data platform to publish specific data sets (Stadt Hamburg, 2020; Stadt Soest, 2021; Tillmanns et al., 2022).

(6) Data culture:

Data is not merely about processes or technology but also people and culture. This is reflected in cities' data strategies. The focus is on communicating the benefits of data use and sharing in the course of data-driven applications and decisions to citizens, employees, and all other stakeholders in the city. It is vital to demonstrate real-world example projects to make them tangible for everyone, as data projects are abstract for many people, therefore it is often not directly apparent why one should engage with data. Communication of the benefits of data must be incorporated into city activities (The city of The Hague, 2020), so prototypes and showcases are essential means of communication (Tillmanns et al., 2022).

Further, thinking in domain silos must be broken down. The reluctance to share one's data with others must be overcome, and the understanding must increase that data and information often improve when shared with others. Combining different data sources creates new insights and reduces the need to recreate data. Clarifying the benefits of data at all levels of the hierarchy fosters motivation and a shared sense of long-term direction for data-driven work (The city of Amsterdam, 2021; Tillmanns et al., 2022).

(7) Data ethics:

Dealing ethically with data is equivalent to being ethical in the physical world, so cities want to ensure peace, justice, and security in the digital world (The city of The Hague, 2020). Cities must respect societal values when implementing data-driven activities, as not everything legal is legitimate (Tillmanns et al., 2022). The intention is to prevent unintended negative consequences of data use (Stadt Soest, 2021). Ethics committees ensure the ethical handling of data by reviewing the entire data lifecycle from openness, transparency, and fairness aspects (Blauhut & Stadt Köln, 2021; Stadt Hamburg, 2020; Stadt Soest, 2021; The city of Amsterdam, 2021; The city of Helsinki, 2023; The city of The Hague, 2020; Tillmanns et al., 2022). In this way, cities aim to ensure the traceability of results with high transparency and away from data-based applications being a black box (Tillmanns et al., 2022). The results and workflows must be transparently traceable and reproducible for third parties. This will minimize potential disadvantages or discrimination against groups or individuals (Stadt Soest, 2021).

The data ethics dimension is mainly guided by the FAIR principles, which consist of the principles of (1) findability, meaning that users can easily find the data they need. The (2) accessibility, by being retrievable by their identifier through a standardized communication protocol. The (3) interoperability, when the data use commonly accepted identifiers, data models, and ontologies and finally the (4) reusable use of the data, in which the attributes are precisely described, use licenses are clearly defined, and the source is given (The city of Helsinki, 2023).

(8) Organization structure:
Dealing with data requires structural and organizational changes. Data governance and management teams within existing structures are required, and responsibilities and roles must be defined (Blauhut & Stadt Köln, 2021; Stadt Hamburg, 2020; Stadt Soest, 2021; The city of Helsinki, 2023; The city of The Hague, 2020; Tillmanns et al., 2022). For example, roles such as data owner and data manager within departments must be defined, allowing for decentralized implementation, but data matters must be governed centrally (Stadt Hamburg, 2020). For this reason, a steering group made up of cross-departmental representatives from each department derives specific measures from the data strategy centrally (Soest, 2021). Thus, data governance defines how strategic use can be ensured regarding organization and processes (Tillmanns et al., 2022).

Data governance controls data management, which executes data processes operationally. Data governance forms the basis for coordinated data management through rules, organization, and processes on a functional and technical level (Stadt Soest, 2021; The city of Helsinki, 2023; Tillmanns et al., 2022). In contrast, the city's data management primarily ensures that the data processed in the city is available and provided in the desired form through predefined rules and quality controls (Stadt Soest, 2021; The city of Amsterdam, 2021; The city of Helsinki, 2023; Tillmanns et al., 2022).

(9) Data security and privacy:
Data security and privacy are basic requirements for gaining citizens' trust. Cities must ensure that data collected as private is also processed as confidential (Blauhut & Stadt Köln, 2021; Stadt Hamburg, 2020; Stadt Soest, 2021; The city of Amsterdam, 2021; The city of The Hague, 2020). The principle of "privacy by design" ensures that all hardware and software used in collecting and processing personal data is designed to keep the data secure during its entire lifecycle (Stadt Soest, 2021). However, data security and data protection should not be seen as an obstacle to innovation but rather as a driver of it. There is a high degree of uncertainty about who may use which data and how, and what formal requirements must be met. This uncertainty leads to data being shared and used at a lower rate than it could be (Tillmanns et al., 2022).

Furthermore, not all data need the same level of protection and must be treated differently. Data protection classes can help define a coherent set of rules that determine which data needs which level of security, considering the criteria of confidentiality, integrity, and availability (Tillmanns et al., 2022).

(10) Cooperations:
A stable and sustainable data ecosystem relies on coopeations with all internal and external stakeholders (The city of The Hague, 2020). Stakeholders must be identified through iterative dialog processes, and the city's data ecosystem must be strengthened through collaborative processes, ultimately leading to the basis for better services (Blauhut & Köln, 2021). This is not only to create a good foundation for successful project implementation but also to ensure a high level of acceptance and participation of all interested parties so that projects can be deeply embedded in the city's overall development (Blauhut & Stadt Köln, 2021).

Inter-municipal cooperations are a great way to benefit from the experiences of others, as cities often face the same challenges, saving time and costs. Moreover, data is not bound by city boundaries, so the urban data ecosystem must also provide the ability to implement data-driven solutions across cities. In the end, collaboration with neighboring cities helps tackle complex problems that cannot be solved alone due to limited resources (Stadt Soest, 2021; The city of Amsterdam, 2021; Tillmanns et al., 2022).

(11) Data literacy:
Building data literacy is critical for cities as data becomes prevalent in traditional areas that were not previously data-driven. Building data literacy aims to ensure that the knowledge and skills needed to gain insights from data and use data consciously and responsibly are in place (The City of The Hague, 2020; Tillmanns et al.,

2022). Data literacy applies to all disciplines but also aims to provide citizens with a basic understanding of using technology to make better decisions and know the risks and consequences of sharing their data (The city of The Hague, 2020).

Cities must invest in staff training and empowerment, both in individual training and in the development of a learning organization (Tillmanns et al., 2022). Training programs must be established divided into two categories. The (1) basic training program for all employees builds a common understanding and knowledge about the use of data. Only on the basis of this common ground will it be possible to develop the city government into a data-driven organization (Stadt Soest, 2021; Tillmanns et al., 2022). In addition to the basic training program, employees require specific data skills, depending on their position and task. Employees responsible for data governance and management processes must complete training in the specific areas of responsibility by (2) professional advanced training programs. Investing in their employees will increase the cities' attractiveness as an employer in the long term (Stadt Soest, 2021; The city of Helsinki, 2023; Tillmanns et al., 2022).

5. DISCUSSION

Urban data strategies divide into the 11 dimensions detailed previously. The relevance of data is at the forefront of data strategies. It establishes a shared commitment to data utilization, lists possible scenarios for data utilization, and identifies how the city, citizens, businesses, and other stakeholders can benefit from data. Data utilization must be deeply rooted in the culture and mindset to be successful in a sustainable manner. All stakeholders must be made aware of the benefits that proper use of data brings to the public good. The data strategy can only be accepted if all have a common understanding. To this end, it is necessary to communicate fundamental knowledge and present comprehensible application scenarios from everyday practice demonstrating the benefits. This also includes a transparent portrayal and educational campaign concerning data security and protection, as a large amount of sensitive data is in municipal hands – incorrect handling of data damages the trust of stakeholders. The city needs to ensure that data is handled properly throughout its entire lifecycle but also that the right measures are taken for the further innovative development of data ecosystems, e.g., through open data, collaboration with municipal stakeholders, and the development of innovative technologies, including data portals and digital twins. Therefore, data governance is the executive hand of the data strategy, assuming a data governance organizational unit exists in the city (Stadt Hamburg, 2020; Stadt Soest, 2021; The city of Amsterdam, 2021; Tillmanns et al., 2022).

Bozkurt et al. (2022) define urban data governance as a unit that deals with all data-related issues from a holistic, overarching perspective to pursue sustainable urban development by governing data assets in the interest of citizens and promoting business and services. The focus is on sustainable urban development, which is consistent with the overall strategy of cities. The city's data governance unit can operationalize the dimensions listed in the data strategy in collaboration with relevant stakeholders (internal and external) through defined actions. For example, the data governance unit can define city-wide standards and data quality principles according to the city's needs to ensure the highest level of interoperability and integrability. In addition, given its data access scope (Bozkurt et al., 2022), the data governance unit aims to establish policies and mechanisms for data exchange in an internal sense, external sense, and open data. To this end, it plays a crucial role in data classification and the creation of a citywide data catalog to promote the sharing of data of different uses and, through the data catalog, provide a standardized view of the city's data resources, promoting not only technological and semantic compatibility, but also data ethics issues such as trust, ownership, and transparency. This ultimately benefits the implementation of data platforms and other digital services. Bozkurt et al. (2022) eight dimensions of urban data governance have many connecting elements and parallels to the, in this paper, identified dimensions of data strategy. Thus, data governance drives the city's data strategy and hinders the strategy from becoming a dusty document on the shelf. Data governance is a concept that needs to be practiced continuously by all stakeholders and requires a cultural change and transformation process. This can be achieved by regularly educating the staff and building data literacy in the long term.

With this work, we intend to provide an overview of how urban data strategies are constructed (RQ1) and how they can be operationalized through data governance (RQ2). Data has immense importance for the development of cities, be it for data-driven urban planning or sensor-based services in the city. This paper aims to contribute to this critical smart city area, which has received little attention from the academic community concerning data strategies and data governance in smart cities. The identified 11 dimensions of data strategies show that data must be strategically considered from a multifaceted perspective. This ranges from technology and data protection to cultural factors in dealing with data. Data governance as a tool for implementing data strategies can help cities determine gaps to ensure their data strategy is not just a strategic paper dusting away in a drawer. In this way, practitioners can use the results of this work as initial guidance on how to develop the data strategy and understand its implementation. In addition to this practical contribution, there is also a scientific contribution in further exploring the body of knowledge about smart cities from the data perspective and integrating data governance further into the smart city discussion. Scholars can use the results of this work, for example, to develop a maturity model of the data aspects of smart cities and transfer concrete data governance measures to cities. Still, we do not consider the identified 11 dimensions of data strategy for smart cities as final and take it as a starting point for further development in further research, such as broadening the sources by expanding the search criteria. In addition, it is quite possible that the data strategies are not publicly available and may be requested directly from cities; ideally, this can be expanded through primary data collection in the form of interviews with cities. Finally, smart city strategies, which are more accessible openly than data strategies, can be further analyzed to identify other strategic directions related to data, as the data aspect may also be present in some smart city strategies.

6. CONCLUSION

This paper has examined 8 data strategies of cities in Germany, Finland, Denmark, and the Netherlands to contribute to the knowledge of smart cities by providing an overview of the strategic dimensions of urban data and clarifying how data governance in cities can contribute to data strategy. The analysis revealed the data strategy dimensions of (1) relevance of data, (2) principles, (3) methods, (4) data exchange, (5) technology, (6) data culture, (7) data ethics, (8) organizational structure, (9) data security and privacy, (10) collaboration, (11) data literacy. In addition, it shows that data governance is a concept to put these 11 strategic dimensions into practice through standardization efforts, training programs, defining roles and responsibilities, and developing a data catalog. Data governance drives the city's data strategy and prevents the strategy from becoming a dusty document on the shelf.

Ultimately, the findings help practitioners approach the topic of data strategy and provide guidance on areas where they should examine their data. Further, the scientific community is expanding knowledge about urban data and the strategic dimensions of cities seeking to be data-driven and providing a basis for developing further work in areas such as the maturity model for data-driven cities and data governance concepts for smart cities.

REFERENCES

Adelman, Sid., Abai, Majid., & Moss, L. Terpeluk. (2005). *Data strategy*. Addison-Wesley.

Ahvenniemi, H., Huovila, A., Pinto-Seppä, I., & Airaksinen, M. (2017). What are the differences between sustainable and smart cities? *Cities*, *60*, 234–245.

Albino, V., Berardi, U., & Dangelico, R. M. (2015). Smart Cities: Definitions, Dimensions, Performance, and Initiatives. *Journal of Urban Technology, 22*(1), 3–21.

The city of Amsterdam. (2021). *City of Amsterdam Data Strategy*. https://assets.amsterdam.nl/publish/pages/1017819/data_strategy_city_of_amsterdam_2021-2022.pdf

Anthony Jnr, B. (2021). Managing digital transformation of smart cities through enterprise architecture–a review and research agenda. *Enterprise Information Systems*, *15*(3), 299–331.

Barns, S. (2018). Smart cities and urban data platforms: Designing interfaces for smart governance. *City, Culture and Society*, *12*, 5–12.

Land Berlin. (2018). *Gründbuch für die Digitalisierungsstrategie des Landes Berlin*.

Blauhut, D., & Stadt Köln. (2021). *Daten "machen" Strategie*. https://buergerinfo.stadt-koeln.de/getfile.asp?id=821964&type=do

Bozkurt, Y., Braun, R., Rossmann, A., & Hertweck, D. (2020, July 21). The Advent of Smart Cities: Status Quo and Future Research Directions. *6th IADIS International Conference on Connected Smart Cities (CSC2020)*.

Bozkurt, Y., Rossmann, A., & Pervez, Z. (2022). A Literature Review of Data Governance and Its Applicability to Smart Cities. *Proceedings of the 55th Hawaii International Conference on System Sciences 2022*, 2680–2689.

Caragliu, A., del Bo, C., & Nijkamp, P. (2011). Smart Cities in Europe. *Journal of Urban Technology*, *18*(2), 65–82.

Chandra, Y., & Shang, L. (2019). Inductive Coding. In *Qualitative Research Using R: A Systematic Approach* (pp. 91–106). Springer Nature Singapore.

Corbin, J., & Strauss, A. (1990). Grounded Theory Research: Procedures, Canons, and Evaluative Criteria. In *Qualitative Sociology* (Vol. 13).

Dremel, C., Herterich, M., Wulf, J., Waizmann, J.-C., & Brenner, W. (2017). How AUDI AG Established Big Data Analytics in Its Digital Transformation. *MIS Quarterly Executive*, *16*(2).

European Commission. (2022a). *Digital Economy and Society Index (DESI) 2022*. https://digital-strategy.ec.europa.eu/en/policies/desi

European Commission. (2022b). *A European Strategy for data*. https://digital-strategy.ec.europa.eu/en/policies/strategy-data

Fleckenstein, M., & Fellows, L. (2018). Modern Data Strategy. In *Modern Data Strategy*. Springer International Publishing.

Garousi, V., Felderer, M., & Mäntylä, M. V. (2017). Guidelines for including grey literature and conducting multivocal literature reviews in software engineering. *Information and Software Technology*, *106*, 101–121.

Gioia, D. A., Corley, K. G., & Hamilton, A. L. (2013). Seeking Qualitative Rigor in Inductive Research. *Organizational Research Methods*, *16*(1), 15–31.

The city of Helsinki. (2023). *The City of Helsinki Data Strategy - Digitaalinen Helsinki*. https://digi.hel.fi/english/helsinki-city-data-strategy/

Jin, J., Gubbi, J., Marusic, S., & Palaniswami, M. (2014). An information framework for creating a smart city through internet of things. *IEEE Internet of Things Journal*, *1*(2), 112–121.

Khatri, V., & Brown, C. V. (2010). Designing data governance. *Communications of the ACM*, *53*(1), 148–152.

Kvalvik, P., Sánchez-Gordón, M., & Colomo-Palacios, R. (2022). Beyond technology in smart cities: a multivocal literature review on data governance. *Aslib Journal of Information Management*.

Lombardi, P., Giordano, S., Farouh, H., & Yousef, W. (2012). Modelling the smart city performance. *Innovation: The European Journal of Social Science Research*, *25*(2), 137–149.

Mahanti, R. (2021a). Data Governance and Compliance. In *Data Governance and Compliance*. Springer Singapore.

Mahanti, R. (2021b). *Data Governance and Data Management Contextualizing Data Governance Drivers, Technologies, and Tools Data Governance*. Springer Nature.

Manville, C., Cochrane, M., Cave, J., Liebe, A., & Massink, R. (2014). *Mapping Smart Cities in the EU*. https://www.europarl.europa.eu/RegData/etudes/etudes/join/2014/507480/IPOL-ITRE_ET(2014)507480_EN.pdf

MAXQDA. (2022). *MAXQDA | All-In-One Tool for Qualitative Data Analysis & Mixed Methods - MAXQDA* (2022.4). https://www.maxqda.com/

Moustaka, V., Vakali, A., & Anthopoulos, L. G. (2018). A Systematic Review for Smart City Data Analytics. *ACM Computing Surveys (CSUR)*, *51*(5).

Otto, B. (2011). Data governance. *Business and Information Systems Engineering*, *3*(4), 241–244.

Paskaleva, K., Evans, J., Martin, C., Linjordet, T., Yang, D., & Karvonen, A. (2017). Data governance in the sustainable smart city. *Informatics*, *4*(4).

Seiner, R. (2014). *NON-INVASIVE DATA GOVERNANCE: The Path of Least Resistance and Greatest Success* (1st ed.). Technics Publications.

Stadt Soest. (2021). *DATENSTRATEGIE SOEST*. https://digital-soest.de/images/AIDW/Datenstrategie_Final_2021-05-14.pdf

Stadt Hamburg. (2020). *DIGITALSTRATEGIE FÜR HAMBURG*. https://static.hamburg.de/fhh/epaper/digitalstrategie/#0

The city of The Hague. (2020). *Towards a data-driven and responsible city - Data strategy 2020-2022*.

Tillmanns, K., Platzer, M., Lenzen, I., & Mönchengladbach. (2022). *Leitlinien Datensouveränität*. https://www.moenchengladbach.de/fileadmin/moendico/Leitlinien_Datensouver%C3%A4nit%C3%A4t_MG.pdf

United Nations, D. of E. and S. A. (2018). *World Urbanization Prospects: The 2018 Revision*. https://population.un.org/wup/Download/Files/WUP2018-F21-Proportion_Urban_Annual.xls

Vial, G. (2019). Understanding digital transformation: A review and a research agenda. In *Journal of Strategic Information Systems,* 28(2), pp. 118–144. Elsevier B.V.

Washburn, D., & Sinduh, U. (2010). Helping CIOs Understand "Smart City" Initiatives. *FORRESTER*.

Weber, K., & Klingenberg, C. (2021). *Data Governance - Der Leitfaden für die Praxis*. Hanser.

Weber, K., Otto, B., & Österle, H. (2009). One Size Does Not Fit All---A Contingency Approach to Data Governance. *Journal of Data and Information Quality*, *1*(1).

Zanella, A., Bui, N., Castellani, A., Vangelista, L., & Zorzi, M. (2014). Internet of things for smart cities. *IEEE Internet of Things Journal*, *1*(1), 22–32.

SMART CITY CONSTRUCTION AT ITS BEST: THE URBAN CASE STUDY OF SMART TUNNEL LIVELIHOOD SYSTEM

Zeyi He[1, 2]* and Haidong Wu[1]
[1]*Changsha Urban Development Group Co.LtD, China*
[2]*Zhongnan University of Economics and Law, Changsha, China*
**Dr.*

ABSTRACT

This paper explores the development of Smart City Construction that prioritizes the needs of citizens. We highlight the Smart Tunnel Livelihood System as a case study, which provides an innovative and practical solution to improve safety, traffic efficiency, and emergency response in urban tunnels. In today's cities, smart tunnels should offer a revolutionary way to manage emergency vehicle traffic. Our system consists of three crucial components: traffic flow monitoring, closed-loop data sharing applications, and real-time traffic monitoring and guidance. By integrating these three components, we can address the challenges faced by smart tunnels, enhance their safety, and improve traffic efficiency in urban areas. Through our multi-layered approach to data management and real-time traffic monitoring, we demonstrate how to design a solution that meets the demands of citizens in the modern Smart City Construction.

KEYWORDS

Smart Tunnel, IoT system, Smart City, Real-Time Control, Tunnel Livelihood System

1. INTRODUCTION

The rise of smart cities has been driven by technological advancements and the urgent need to address environmental and urbanization issues. Many cities worldwide are adopting smart city technologies to improve the quality of life for citizens, enhance urban infrastructure, and promote sustainable development, such as IoT technology is used for supply chain management (Nagarajan et al., 2022), and blockchain technology is used in intelligent transportation system (Ullah et al., 2023). Moreover, smart city design involves much more than the merely physical structures but includes the entire framework as well.

Yet, the rapid development of smart cities has brought to light concerns about the potential negative impact on vulnerable groups. Therefore, it is essential to ensure that smart city technologies cater to the needs of all citizens, including those with urgent medical requirements (Kamtam et al., 2022). Issues concerning emergency medical services' efficiency and quality have become increasingly pressing, given the life-threatening consequences of delayed response times. An efficient system for the implementation of smart city technologies is vital to achieving the desired results in the "last-mile" delivery of services.

In this paper, we contend that smart tunnel systems have immense potential in mitigating the urgent challenge of emergency medical transportation in developing smart cities. To begin, we introduce the fundamental components of smart tunnel systems, as well as the challenges posed by emergency medical transportation to the transportation system. Subsequently, we present an assessment of the hurdles associated with implementing smart tunnel systems in urban areas that are primed for smart city infrastructure. Then, we propose solutions for the control and integration of data systems in smart tunnels. Lastly, we highlight the implications and significance of this experimental approach. Overall, this paper showcases that intelligent tunnel systems possess the capability to address the exigencies of emergency medical transportation and represents a crucial step towards smart city development.

1.1 Development of Smart Cities

Smart cities are a cutting-edge urban development paradigm, driven by the urgent need to overcome diverse challenges stemming from the rapid pace of urbanization (Mohanty and Kumar, 2021). They involve the integration of various urban systems, such as energy, transportation, and government services, as well as the implementation of innovative technological solutions that enable comprehensive and effective planning, management, and service delivery. Smart city concepts are founded on more sustainable, inclusive, and resilient approaches to urban living, nurturing a symbiotic relationship between the physical and digital environments. The ultimate goal of smart cities is to enhance people's quality of life by enhancing their living conditions in both physical and digital spaces. By forging a closer connection between these two dimensions, smart cities strive to deliver a more comfortable, convenient, and sustainable living environment that fulfills the needs and expectations of present-day urban communities.

A multitude of smart cities have already appeared around the world, including Singapore, Barcelona, and London, among others (Joss et al., 2019). The development of smart cities is a continually evolving process. Academics have discussed extensively the possible benefits and challenges related to smart city technology. Some scholars assert that the use of smart city technology can improve urban living standards through effective and thorough planning and management. For example, Deakin and Al Waer (2011) affirm that smart cities can promote social and economic development through the application of ICT solutions. On the other hand, other researchers have highlighted potential issues related to smart city technology, such as data privacy, security, and accessibility. Zanella et al. (2014) warn about the requirement for ethical policies and regulations to prevent data breaches and cyberattacks in the development of smart cities. Furthermore, scholars have emphasized the importance of engaging stakeholders and fostering public-private partnerships to guarantee equitable distribution and participation in the development of smart cities (Nam and Pardo, 2011).

However, the fast-paced development of smart cities has also raised concerns about potential negatives impacts on vulnerable groups, particularly those in urgent need of medical attention. While smart city technologies can enhance the efficiency and quality of emergency medical services, it is crucial to ensure that they cater to the needs of all citizens, including those who may not have access to the latest technologies or who may face other healthcare-access barriers.

1.2 Focus on Urgent Needs

In order to ensure the safety and well-being of vulnerable groups, such as those in need of emergency medical attention, prioritizing their needs should be a vital consideration for smart city technologies. For instance, it is essential to ensure that emergency medical services are accessible to all citizens, irrespective of their socioeconomic status or location. The fast and efficient transportation of patients to medical facilities is of utmost importance, especially in time-sensitive cases, such as heart attacks, strokes, and trauma injuries. Therefore, smart tunnels play a crucial role in addressing emergency medical traffic challenges in urban traffic systems, which often face heavy traffic loads, frequent accidents, and extensive operational and maintenance restrictions. By linking various hospitals and emergency departments, smart tunnels offer a dedicated and streamlined transportation route for emergency medical services to reduce response time and ultimately improve patient outcomes.

1.3 IoT Technology

The Internet of Things (IoT) is an innovative technology that comprises of tangible devices connected to the internet via sensors and protocols. It allows for the exchange and communication of information among these connected devices. The IoT technology enables functions such as intelligent recognition, positioning, tracking, supervision, and control, making it a valuable asset for a variety of fields, including transportation, medical care, and energy management (Li et al., 2014) [9]. By utilizing IoT sensors, it is possible to obtain real-time monitoring of environmental parameters and facility status information, which can be processed, analyzed, and controlled to detect anomalies and enable the automatic activation of emergency procedures. The integration of IoT technology in transportation infrastructure marks a significant step towards smarter transportation systems, which can enhance urban mobility and improve the quality of life of city-dwellers.

Furthermore, by monitoring traffic patterns, reducing congestion, and minimizing energy consumption, IoT-powered smart tunnels could promote the creation of greener transportation networks, thereby improving environmental sustainability.

1.4 Smart Tunnel Systems

Smart tunnel infrastructure is a novel approach to managing tunnel operations and enhancing transportation efficiency. By leveraging advanced sensing and control technologies such as IoT, smart tunnels enable a wide range of applications such as tunnel accident detection systems (Kim et al., 2019), tunnel crack detection systems (Zhong et al., 2019), tunnel brightness detection systems (Li et al., 2019), as well as traffic monitoring, management, and ventilation control. Additionally, smart tunnels represent a groundbreaking solution to the challenge of emergency vehicle traffic in cities. With the aid of sophisticated sensors and control systems, smart tunnels can detect approaching emergency vehicles and adjust traffic flow automatically, allowing seamless passage without interruption. This approach can substantially diminish emergency response times and enhance patient medical outcomes. Smart tunnel systems offer a promising solution to the problems of emergency vehicle traffic in urban areas and provide an opportunity to optimize the efficiency of emergency medical services.

2. PROBLEMS AND CHALLENGES

This article aims to identify an appropriate case study for the deployment of a smart tunnel system, which meets certain criteria such as addressing specific societal needs, having adequate space for installation across two lanes, and having essential infrastructure and information technology to enable the creation of a complete IoT system. The integration of a smart tunnel system in a practical context will offer insight into the challenges and opportunities that accompany such infrastructural changes. The chosen case study must have urgent requirements for enhancing emergency response and traffic management, as well as fundamental infrastructure and information technology crucial for the implementation of essential IoT systems. One instance could be a bustling urban highway plagued by frequent traffic congestion and high reliance on emergency medical services. Nevertheless, executing this implementation process demands substantial investment in both financial and logistical resources.

2.1 Selected Case

The case study of Changsha City, China satisfies the three criteria mentioned in this article for implementing the smart tunnel system. Firstly, with a significant proportion of the population over the age of 65, accounting for over 15% of the permanent residents, Changsha City requires advanced healthcare infrastructure to address medical emergencies. The smart tunnel system can assist with this by improving traffic flow and enabling quicker emergency medical response times.

Secondly, Changsha City's location, with one river and two banks, provides ample space for the installation of the smart tunnel system. The Xiangya Road Cross River Tunnel, connecting multiple medical institutions, Xiangya Hospital and Third Xiangya Hospital, two Chinese famous top hospitals, is currently under construction, while nearby healthcare facilities like Hunan Maternal and Child Health Hospital and Hunan Cancer Hospital further emphasize the need for efficient emergency response. China Ambulatory Surgery Alliance announced that in 2021, with an annual total of over 3.473 million outpatient and emergency visits and 102,000 surgeries in Xiangya Hospital, respectively, Changsha City has a large number of patients requiring medical attention. Geographic map of Xiangya Road Cross River Tunnel as shown in Figure 1.

Lastly, the selected case study should have basic infrastructure and information technology in place, enabling the implementation of necessary IoT systems. The Xiangya Road Cross River Tunnel is planned to be the main road of the city, which is a two-way six-lane road. The length of the North line tunnel is 3320m, the length of the south line tunnel is 3648.24m, and the design speed is 50km/h for the main line tunnel and 35km/h for the on-ramp tunnel. The entrance and exit are equipped with video analysis cameras for vehicle detection. With such condition, IoT technology can create green channels for medical institutions to transport critically ill patients effectively, although ensuring safety, efficiency and normal operation under traffic loads, traffic accidents, and safety accidents is a major challenge.

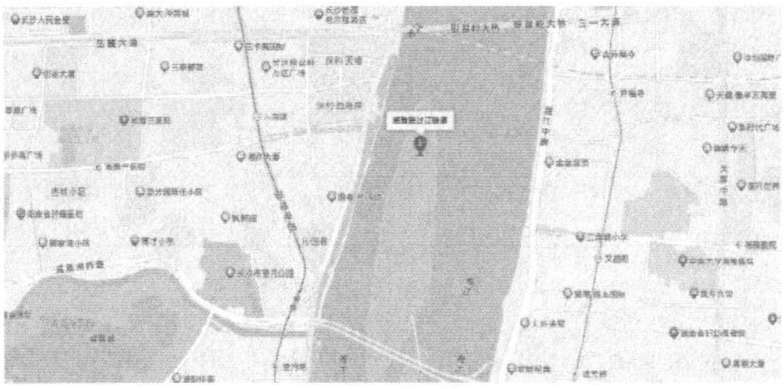

Figure 1. Geographic map of Xiangya Road Cross River Tunnel

Thus, the case of Xiangya Road Cross River Tunnel provides a compelling example for implementing a smart tunnel system that can cater to specific societal needs. This innovative solution not only makes efficient use of available installation space but also enables the seamless transportation and treatment of a large number of patients in need of medical attention, as outlined in this article.

2.2 Challenges and Difficulties

There are several challenges that need to be overcome in order to implement the livelihood system into the smart tunnel. These include managing traffic in congested urban tunnels; integrating data and communication systems among hospitals, emergency departments and the tunnel management system; and ensuring ambulance priority management with the real-time emergency vehicle navigation.

First, managing traffic flow in urban tunnels requires innovative solutions that can seamlessly integrate with existing transportation systems while prioritizing emergency vehicle traffic. Urban tunnels are notoriously congested, so implementing comprehensive monitoring and control systems for smart tunnels is a complex task. The management and monitoring systems of a tunnel are typically part of different communication protocols and control systems, adding an extra layer of complexity. Therefore, designing a new smart tunnel system requires careful consideration to avoid disrupting the intricate network of existing systems and protocols, as even the slightest misstep could potentially disrupt an already complex system.

Integrating different data systems and platforms used by hospitals, emergency departments, and the tunnel management system presents another significant challenge. The healthcare industry has struggled with fragmented technology adoption due to the varying electronic health record (EHR) systems, communication protocols, and data formats used by different hospitals and emergency departments. Additionally, the data collection and analysis technologies and standards used by the tunnel management system may differ, adding another layer of complexity. Therefore, integrating different data sources while ensuring data integrity and security is crucial.

Lastly, real-time ambulance priority management presents a challenge. Traditional traffic control systems may not be adequate to ensure ambulances are given priority in congested tunnels. A smart city tunnel system must be equipped with real-time sensors and algorithms to detect ambulance presence and adjust traffic signals accordingly. Further, the system should provide real-time feedback to ambulance drivers about the estimated time of arrival to the destination hospital.

3. SOLUTION

To address the challenges facing smart tunnels in cities, enhance their safety, and improve traffic efficiency, this paper proposes a comprehensive set of solutions for a smart tunnel livelihood system.

The proposed program addresses the following three issues:

- Creating a complete smart tunnel livelihood system to monitor and control traffic flow;
- Opening the closed loop of multi-point data sharing and analysis applications;
- Enabling real-time traffic monitoring and guidance scenarios.

Intelligently managing and controlling urban tunnels presents significant challenges, particularly when it comes to achieving interoperability between multiple devices, sharing data, coordinating management, enabling collaboration, and facilitating rapid responses across different IoT platforms, public service entities (including medical institutions), and government control centers situated in multiple locations. To tackle these challenges, this top-level design approach employs three design dimensions: IoT cluster management, integrated data scheduling, and scenario-based command and dispatch systems.

The overall design aims to integrate new technologies with existing resources to enhance system efficiency. The system architecture, as illustrated in Figure 2, provides a comprehensive solution:

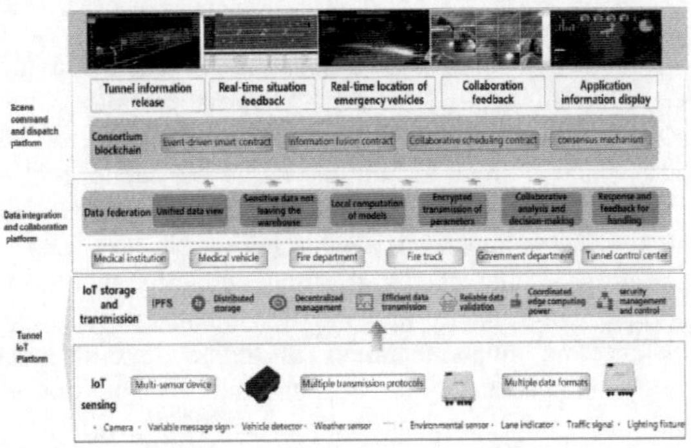

Figure 2. Overall Architecture of Smart Tunnel Livelihood System

3.1 Constructing a Tunnel Edge IoT Platform

Through analysis and evaluation, it is found that the current tunnel monitoring and communication systems consist of five systems, namely, the central computer system, fire alarm system (FAS), traffic monitoring system, environmental equipment monitoring system (BAS), and communication system, as well as CCTV and video detection systems. Each of the systems operates independently, and Programmable logic controller (PLC) is adopted as a mature communication solution. These five systems respectively meet the requirements for basic information collection and management in the tunnel, which constitute a classic scheme for tunnel monitoring and communication and ensure the completeness of electromechanical functions, reliability of technology, and safety of tunnel management, as shown in Figure 3(a).

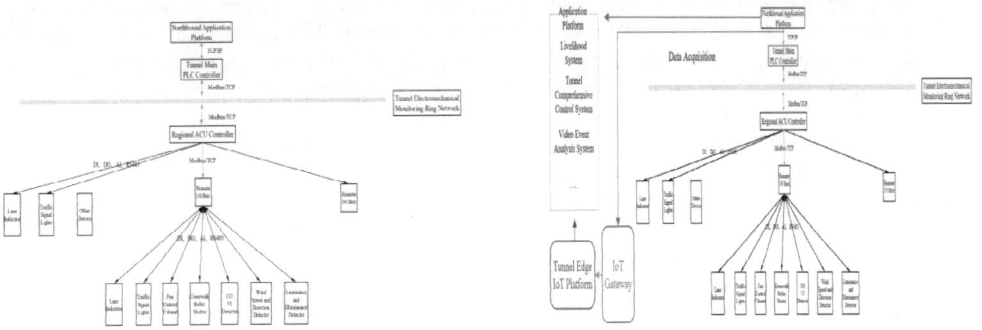

(a) Current Tunnel Information and Control System (b) Tunnel Edge IoT Platform

Figure 3. IoT Technology Application in Tunnel Information and Control System

However, the entire system also faces several shortcomings: The traditional electromechanical system suffers from control separation and data isolation issues, with inadequate sharing and utilization of data and resources throughout the tunnel; the limited front-end sensing capabilities and absence of real-time sensing, automatic early warning, and rapid disposal are cause for concern; the PLC architecture's centralized control mode frequently experiences multiple faults and cannot provide equipment status point sensing, resulting in high troubleshooting costs; Finally, the overall architecture lacks the logic of intelligent IoT, necessitating the application of AI technology to attain visualized, measurable, controllable, and intelligent management.

To overcome these limitations, we propose establishing a tunnel edge IoT platform that facilitates integrated data collection, analysis, and sharing, sensor deployment for comprehensive event sensing, and AI-based smart management. Specifically, an IoT gateway will be installed on the main PLC controller in the tunnel, enabling data collection across different systems and devices. To unify device identifiers, a standardized tunnel state model will be established, addressing the problem of data and information isolation. The tunnel edge IoT platform will be built on the tunnel side, with application cases and models tailored to specific scenarios as depicted in Figure 3(b).

The proposed approach aims to enable a distributed control mode that enhances the integrated access and sharing of data resources, laying a conceptual foundation for advanced tunnel behaviours analysis and prediction. The solution will also deploy advanced sensors capable of comprehensive event sensing, facilitating automatic early warning and rapid disposal. The PLC architecture of the current system belong to a centralized control mode, which is prone to multiple faults and cannot realize point sensing of equipment status, resulting in high cost of troubleshooting; in contrast, this architecture allows for a more precise point sensing of equipment status, reducing the cost of troubleshooting.

By incorporating the logic of intelligent IoT, the tunnel edge IoT platform will utilize AI technology to enable visualized, measurable, controllable, and intelligent management, and will play a crucial role in supporting the development and implementation of tunnel intelligentization. The platform boasts compatibility with multiple communication protocols and data compression technologies to maximize data exchange efficiency, and thus reduce operational costs. With increasingly complex tunnel data and equipment, the platform's analytical capabilities will prove essential in identifying trends, anomalies, and potential hazards in tunnel operations.

Overall, the tunnel edge IoT platform represents a promising approach for addressing the shortcomings of the current tunnel monitoring and communication systems, allowing for enhanced and cost-effective tunnel management and maintenance. The platform's advanced sensing equipment, extensive data processing capabilities, and smart management logic will facilitate more proactive and data-driven decision-making and improve tunnel safety.

3.2 Data Integration Scheme Design

To enable efficient and reliable data sharing across multiple points, including healthcare institutions, government departments, tunnel control centers, and medical vehicles, a data integration platform based on data federation has been established. The platform facilitates data analysis, sharing, and application while ensuring that sensitive data remains local. A unified data view is provided for logical data integration to avoid security risks from data leaving the local environment. The integration platform also establishes different endpoint nodes for local computing, enabling local model training, deployment, computation, and secure transmission of encrypted parameters between different points.

Moreover, the integration platform provides coordinated control to implement related tasks in various scenarios, synchronous model computation, on-site response, and feedback coordination, all of which support the development of advanced applications. Serving as the core of the civil system's data operation, the data integration and collaboration platform is the foundation for secure and reliable data application among multiple endpoint nodes.

The above introduces the overall architecture and main functions of each part of the system. In this section, we mainly introduce how to break through the data loop of perception, decision-making, and control, and achieve efficient and reliable data sharing applications. The overall architecture of system data integration is shown in Figure 4. In the integration scheme illustrated in Figure 4, multiple endpoints are involved, including tunnel IoT devices, tunnel on-site management control centers, medical rescue institutions, and government management departments. Given the different data storage and processing capabilities among these endpoints, the heterogeneous and non-uniform data situation must be addressed in the system design, with a focus on resolving the storage, analysis, and transmission of massive IoT data. Furthermore, the security, timeliness, and reliability of data sharing among multiple endpoints must be ensured, along with the realization of a closed-loop business for data perception analysis, decision-making, and coordinated control among multiple endpoints.

In the case of this tunnel, the data required by the civil system comes from various data silos in the field of social governance. The internal data and system management of each party are complex, and significant variances exist in the types, attributes, and structures of the data. This makes data sharing and interaction nearly impossible in such a multi-party autonomous data environment. This article proposes a solution for data federation to facilitate data integration and cooperate learning. Various departments and institutions, as well as IoT devices, can use data federation to achieve layered integration, as shown in Figure 5. The data on the IoT side, the people's livelihood side, and the government side usually come from various departments and institutions and on-site sensing equipment. The internal data and system management of each party are more complicated, and the types, formats, and scales of data between each other are different. In order to solve problems such as difficult data sharing, high data privacy requirements, and slow response of data analysis applications, the data of various departments, institutions, and IoT devices can be layered and integrated in the form of data federation, as shown in Figure 5:

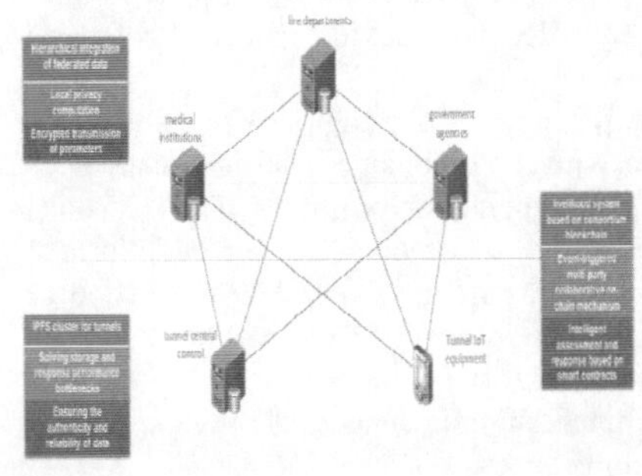

Figure 4. Overall Architecture of Data Integration

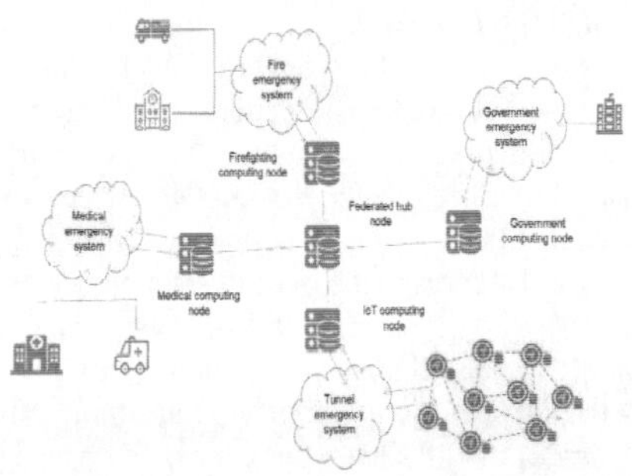

Figure 5. Data Federation Layered Integration Architecture

Data federation refers to a scheme for data sharing among multiple entities while protecting data privacy and security. It is a method for integrating multiple sources of big data in a decentralized data governance environment. To achieve effective data querying and analysis, various methods for data integration and query analysis are used to integrate distributed data sets located in different places into a unified data view. In data federation, each entity's data sets remain isolated and not visible to others. For example, patient data held by hospitals should not be visible to tunnel management, and tunnel traffic data should not be visible to hospitals. Therefore, this project proposes using data federation to solve the problem of data availability and invisibility. The goal of data federation is to improve data availability while protecting data privacy and confidentiality. It allows different entities to share data without centralizing it in one place. Data federation is often closely integrated with federated learning technology and uses secure multi-party computation, encryption technology, data sharing protocols, etc. to ensure data privacy and security.

The data federation in this project belongs to vertical federation of relational data. The essence of vertical federation is combining different feature data, which is suitable for scenarios with overlapping users but fewer overlapping features. For example, through ambulance information passing through the tunnel, they are all ambulance information and have a unique license plate identifier (same sample). However, hospitals have information about the vehicle and patient conditions, while tunnel management has information about vehicle entry and exit, and emergency medical services have information about vehicle and driver information (different features). Therefore, the three data entities have the same user information, but the intersection of their user features is relatively small. Vertical federation learning is to aggregate these different features in an encrypted state to enhance the model's federated learning capabilities.

The methods used to achieve data federation are not discussed in this article, and there will be a dedicated article after the project is completed to elaborate on it. This article only discusses how federated computing contributes to the function of the smart tunnel system. Taking an ambulance passing through the tunnel as an example, data federation can connect hospital data, local medical emergency system data, and tunnel management system data through certain methods to obtain data query results, obtain emergency vehicle information, and determine the level of medical assistance. It opens up a green channel for data including emergency assistance, vehicle location, preliminary diagnosis (judgment), and traffic congestion feedback, triggering data collection and push tasks in real-time. The real-time situation data is pushed to locally deployed computing nodes, and the tunnel civil scenario vehicle navigation model is launched.

Between local computing nodes of medical, firefighting institutions, government departments, tunnels, and other different types of entities, the triggering of analysis tasks and the transmission of model parameters are coordinated through the federation hub node. The federation hub node itself does not perform model calculations but interacts with smart contracts on the alliance chain and various aggregation nodes to trigger local model calculations, encrypt parameter transmission between entities, and synchronize result data. This helps the alliance chain achieve privacy computation and encrypted communication.

3.3 Implementation of Real-Time Intelligent Vehicle Navigation

Through the IoT platform at the tunnel end and the two-phase work of federated data layered integration, the perception function and data integration of the livelihood system have been initially achieved. In this section, we mainly discuss how to realize the ambulance perception, decision-making, and control parts of the livelihood system through the control method.

The livelihood system comprises three main functions: perception, decision-making, and control. The perception function serves to connect the livelihood system with the ambulance and information exchange systems in the tunnel, enabling it to receive emergency passage demands promptly. The decision-making function involves the system analyzing and judging the federated data module's received information and activating the response plan to open up a green channel if conditions are met. The control function generates broadcasts, large-screen information, and advance warning alerts for emergency avoidance scenarios and opens up green channels for ambulances, providing necessary navigation.

In this part, for the control phase of opening up green channels for ambulances and navigation, two traffic organization paths have been designed based on specific circumstances.

The first is the Green channel traffic organization (mainline entry): Ambulances enter the tunnel through the mainline, and social vehicles are prohibited from using the lanes in the mainline (control strategy: the variable message sign at the entrance of the tunnel shows "No Entry"). Vehicles entering the tunnel from the ramp are prohibited from changing lanes (control strategy: the variable message sign on the ramp shows "Stay in Lane and Prohibited from Changing Lanes"), while supporting broadcasts inform and CCTV systems monitor the status of green channel vehicles. This lane setting will reducethe mainline capacity to some extent, but the ramp is hardly affected, showed as the Figure 6(a).

The second is Green channel traffic organization (entry from the ramp): Ambulances enter the tunnel from the ramp, and social vehicles are prohibited from using the lanes on the mainline (control strategy: "No Entry" is displayed on the variable message sign at the entrance of the tunnel). Vehicles entering the tunnel from the ramp are prohibited from changing lanes (control strategy: "Stay in Lane and Prohibited from Changing Lanes" is shown on the variable message sign on the ramp), while corresponding broadcasts are made for information dissemination, and the CCTV system is employed to monitor the status of green channel vehicles. This lane setting may decrease the capacity of the ramp to some extent while having hardly any impact on the mainline, showed as the Figure 6(b).

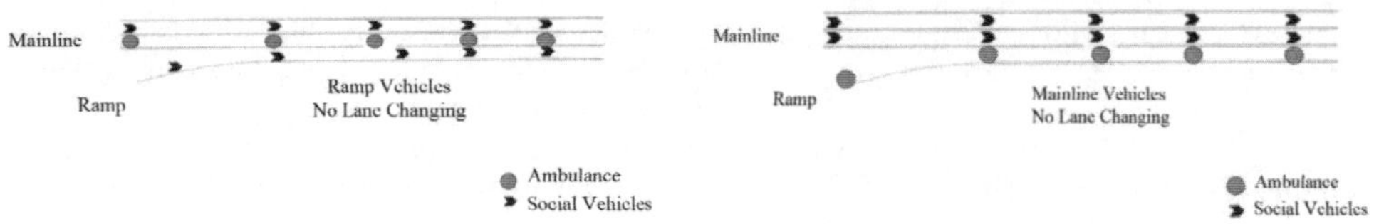

(a) The Control Strategy of Ambulances Entering the Tunnel Through the Mainline

(b) The Control Strategy of Ambulances Entering the Tunnel Through the Ramp

Figure 6. The Control Strategy of Ambulances Entering the Tunnel

3.4 Enabling Efficient Ambulance Emergency Response

By integrating an IoT platform at the end of the tunnel and leveraging a multi-phase process of federated data layering and real-time vehicle navigation, the Smart Tunnel Livelihood System has successfully achieved its sensing function and integrated data effectively. This section will primarily delve into the system's operations and illustrate the process in detail.

The Smart Tunnel Livelihood System comprises three key components: sensing, decision-making, and control. The sensing function mainly entails bridging the Smart Tunnel Livelihood System with the ambulance system and the information exchange system of various tunnel systems to receive emergency passage requests. The decision-making function focuses on conducting comprehensive analysis and judgment of the information gathered from the federated data module and activating the response plan to grant access to a green channel if the conditions are met. Lastly, the control function involves generating broadcasts, large-screen information, and advance warning alerts for emergency avoidance situations, opening up green channels for ambulances, and providing necessary navigation assistance, showed in Figure 7.

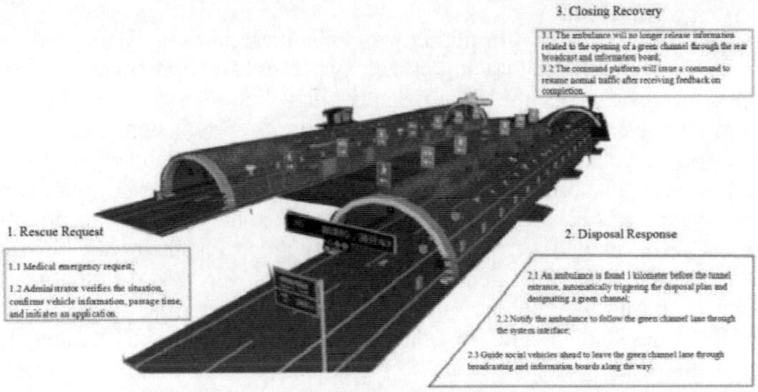

Figure 7. The Smart Tunnel Livelihood System

During an emergency, citizens call medical institutions for assistance, which evaluate the urgency and request green channel access for ambulances. Via the integration of real-time tunnel data and government data into the civil service system, relevant information such as vehicle details and passage times are fed into the system. Response plans are then distributed to the relevant parties.

During the response stage, following the instructions of the response plans, social vehicles are guided through the green channel using information displays and announcements. Required vehicles are discovered before entering the tunnel, and traffic control and green channel opening work is coordinated with tunnel management and transport departments as needed to facilitate ambulance passage.

During the recovery stage, after the ambulance has passed through the green channel, relevant broadcasts and large screen guides are turned off, and information is fed back to the command and dispatch platform. The platform confirms the recovery of normal traffic flow and cancels the green channel.

The overall smart tunnel livelihood system solution is achieved through three technological pathways. Firstly, an IoT platform based on edge computing is established within the tunnel, enabling efficient data transmission and interconnection between IoT devices. This eliminates data silos caused by independent communication and control systems within the tunnel and forms an efficient IoT platform. Secondly, a layered integration system based on federated data enables data sharing and collaborative analysis between different urban management and governing authorities. This enhances our understanding of the traffic situation and real-time emergency needs within the tunnel. It also solves issues relating to multi-location data sharing, privacy protection, and real-time data analysis. Thirdly, a real-time emergency vehicle perception and navigation system is established, which enables the timely opening of virtual green channels based on visual equipment at the tunnel entrance and various scenarios in response to emergency medical and civil needs. This helps to prevent urban tunnel congestion and traffic chaos, creating an efficient, collaborative, and safe intelligent vehicle navigation system. This realizes the closed-loop application of the smart tunnel livelihood system. This system represents a cutting-edge approach to emergency response, enabling efficient and real-time communication between emergency responders, government authorities, and medical institutions, streamlining traffic flow, and allowing emergency vehicles to reach those in need quickly and safely.

The Smart Tunnel Livelihood System's advanced technological pathways ensure its effectiveness, with the IoT platform based on edge computing creating an efficient data transmission system within the tunnel. The layered integration system based on federated data ensures seamless collaboration, comprehensive understanding, privacy protection, and real-time data analysis for the traffic situation and emergency needs within the tunnel. Lastly, the real-time emergency vehicle perception and navigation system enables the timely identification of required vehicles and dynamic opening of virtual green channels to prevent urban tunnel congestion and traffic chaos.

The Smart Tunnel Livelihood System represents a milestone in intelligent vehicle navigation systems and will serve as a model for other emergency response solutions worldwide. As cities continue to grow and demand for efficient emergency services increase, innovative solutions such as this will prove essential in ensuring the safety and wellbeing of citizens. It shows the achievable example of smart cities.

4. CONCLUSION

The creation of smart cities is faced with numerous challenges, particularly upgrading existing systems and devices for smart city construction and raising citizen awareness and appreciation of the significance of smart city development. This paper uses the Smart Tunnel Livelihood System as a case study and proposes innovative solutions for urban redevelopment and expanding its current status.

The Smart Tunnel Livelihood System constitutes a key component of smart city construction, encompassing both urban transportation and livelihood sectors. By creating urban smart tunnels connecting to healthcare centers and establishing an IoT end-system for integrating perception and control, real-time monitoring of rescue vehicles and barrier-free access to green channels can be achieved, improving the efficiency and quality of urban emergency response. Moreover, by formulating a multi-layered, data federated management and evaluation strategy for smart city governance, we can achieve transparent and seamless data availabilities and integration relationships between the city tunnel, hospitals, and medical emergency departments, leading to significant improvements in urban rescue operations and data security. The intelligent real-time ambulance green navigation system provides no-blockage passage solutions for the city tunnel using diverse scenario choices. The design and implementation of this entire solution represent a valuable attempt by the author's team in smart city construction, management, and operation, though challenges remain.

The potential of smart city construction and its innovations is far-reaching, offering a glimpse into the harmonious fusion of technology and society. The realization of these values requires a collaborative effort among all stakeholders, dedicated to the greater good of the community. To this end, we venture forth with boundless energy and enthusiasm, committed to shaping a world that transcends the limits of the present, thereby securing a brighter, smarter, and more dynamic future for all.

REFERENCES

Deakin, M., Al Waer H., 2011. From intelligent to smart cities. *Intelligent Buildings International*, 3(3), pp. 140-152.

Joss, S., et al, 2019. The Smart City as Global Discourse: Storylines and Critical Junctures across 27 Cities. *Journal of Urban Technology,* 26(1), pp. 3-34.

Kamtam, P., et al, 2022. The Role of Smart Cities in Building the Resilience of Vulnerable Communities: Three Case Studies from Europe, Asia, and Africa. *Resilient Smart Cities.* Springer, Heidelberg, Germany.

Kim, N. K., Jeon, M, Hong, K. K., 2019. Convolutional Recurrent Neural Network-Based Event Detection in Tunnels Using Multiple Microphones. *Sensors*, 19(12), pp. 2695.

Li, S., Xu, L. D., Zhao, S, 2014. The internet of things: A survey. *Information Systems Frontiers*, 17(2), pp. 243-259.

Li, S., Zhou, Q., Wang, K., 2019. Video-based tunnel luminance detection. *Automation in Construction*, 107(11), pp. 102900.1-102900.8.

Mohanty, R., Kumar, B. P., 2021. 7 - Urbanization and smart cities. *Solving Urban Infrastructure Problems Using Smart City Technologies.* Elsevier, Amsterdam, Netherlands.

Nagarajan, S. M., et al, 2022. Integration of IoT based routing process for food supply chain management in sustainable smart cities. *Sustainable Cities and Society*, 76, pp. 103448.

Nam, T., Pardo, T. A., 2011. Smart city as urban innovation: Focusing on management, policy, and context. *Proceedings of the 5th International Conference on Digital Government Research.* Tallinn, Estonia, pp. 282-291.

Ullah, Z, et al, 2023. Blockchain Applications in Sustainable Smart Cities. *Sustainable Cities and Society*, 97, pp. 104697.

Zanella, A., et al, 2014. Internet of things for smart cities. *IEEE Internet of Things Journal*, 1(1), pp. 22-32.

Zhong, Z., et al, 2023. Automatic tunnel lining crack detection via deep learning with generative adversarial network-based data augmentation. *Underground Space*, 09(04), pp. 140-154.

CITIZEN SCIENCE PLATFORM FOR RADON POTENTIAL ASSESSMENT AND RADON RISK AWARENESS

Paulo Meneses, Nuno Lopes, Patrícia Leite, Paulo Teixeira and Joaquim P. Silva
2Ai – School of Technology,
IPCA, Barcelos, Portugal
LASI – Associate Laboratory of Intelligent Systems,
Guimarães, Portugal

ABSTRACT

The WHO estimates that cancers caused by radon gas may account for 14% of all lung cancers. The concentration of indoor radon gas is a well-known public health threat and has been the subject of many scientific studies. The WHO has issued recommendations to encourage countries to act against radon exposure to reduce the associated cancer risks. However, despite the vast knowledge on this subject, the measures already taken and the information available, much of the population at risk remains unaware of what radon gas is and its impact on health. In this context, an approach has been designed to perform the pre-diagnosis of indoor radon potential based on the geographical location of the building, its construction characteristics, and its use profile. Recently, a model has been built that performs automatic classification of radon potential in indoor environments. This paper describes the development of a web platform integrating this model to enable citizens to obtain an assessment of the radon potential of the buildings where they work or live. The platform aims to promote radon measurements by the general population and raise awareness about radon risk. These measurements will generate new data that will be used to build more robust classification models.

KEYWORDS

Indoor Radon Potential, Citizen Science Platform

1. INTRODUCTION

Accumulation of radon gas inside buildings, often poorly ventilated, represents a known public health problem, widely reported by the World Health Organization (WHO). Indoor radon concentration is the second leading cause of lung cancer, after smoking. The WHO estimates that cancers caused by radon gas may account for between 3% to 14% of all lung cancers (WHO, 2009). Radon gas is a colorless, odorless, radioactive gas that results from the decay of uranium. It is most common in granite and shale soils and substrates, although it is also present in specific building materials such as concrete, brick and aggregates (WHO, 2009). Radon gas is approximately 8 times denser than air (Soltani-Nabipour, Khorshidi and Sadeghi, 2019), the reason for the highest concentrations being in basements and lower floors of buildings.

The concentration of indoor radon gas is a well-known concern for public health and has been the subject of many scientific studies. To minimize the health risks from indoor radon exposure, WHO recommends a reference level of 100 Bq/m3. Where this level cannot be achieved under country-specific conditions, the chosen reference level should not exceed 300 Bq/m3. In the European Community, the EURATOM Directive on Basic Safety Standards (BSS) states that reference levels for the annual average activity concentration in air should not exceed 300 Bq/m3 (European Union, 2013). In the United States, the Environmental Protection Agency (EPA) recommends remedial action in homes when radon concentrations are 4 pCi/L or higher, equivalent to 148 Bq/m3 (US EPA, no date).

Despite the WHO recommendations and the adopted radon legislation in many countries, the radon problem remains little known to the general population. In Portugal, no systematic measurements are performed in buildings to assess radon concentration levels. Only occasional studies are known, which in general include a few tens or hundreds of buildings. The most comprehensive recent study was conducted by the Portuguese Environment Agency (APA), in 2020, with the aim of obtaining data for the production of a radon exposure susceptibility map (APA, no date b). This was a nationwide radon gas monitoring campaign where passive radon sensors were used in 2224 buildings (APA, no date a). Following this study a radon susceptibility map was published by APA, where about one third of the territory of Portugal is considered to have a high radon potential.

The Portuguese government has recently approved the national radon plan which contains several measures to measure radon, prevent or remediate high concentrations in buildings and communicate the risks to the population (Presidência do Conselho de Ministros, 2022). The assessment of radon concentration inside buildings requires specific expertise involving the execution of a set of time-consuming tasks that may include radon potential assessment, short/long-term radon measurement, laboratory data analysis and technical reporting. Silva *et al.* (2022) proposed a radon potential assessment methodology, based on the analysis of historical data of a set of variables, to provide an estimate of the radon potential of a room of a given building, The radon potential classification model was based on the geographical location of the building, its construction characteristics and its use profile (Meneses *et al.*, 2023).

This paper presents the development of the web platform that will enable the use of the radon potential classification model and contribute to increase the awareness among the general population about the health risks caused by radon. This is a citizen science initiative which aims to involve citizens in combating the radon problem, motivating them, whenever appropriate, to carry out radon measurements and communicate the results on the web platform. This approach aims to engage the population to carry out radon measurements on a large scale and promote the implementation of prevention and remediation actions. These measurements will help improve the platform and could be a tool for indoor radon monitoring.

The next section discusses the concept of citizen-science platform and describes previous work where these platforms have been used in the context of radon gas. Section 3 describes the architecture of the web platform, its features, and the stages of its implementation. The results and presented and discussed in section 4. The last section consists of the conclusion.

2. RELATED WORKS

Citizen science platforms have been providing large amounts of data to researchers to aid in answering research questions. Citizen science concerns the active engagement of the public in scientific research. Citizens actively contribute to science in data collection and/or sampling for scientific research (Bonney *et al.*, 2009; Silvertown, 2009; Dickinson, Zuckerberg and Bonter, 2010). Silvertown (2009) highlights three factors factor driving the growth of citizen science: (a) the existence of easily available technical tools for disseminating information about projects and gathering data from the public; (b) the increasing realization among professional scientists that the public represent a free source of labor, skills, computational power and even finance; and (c) a condition imposed by research funders upon every grant holder to undertake project-related science outreach.

2.1 Citizen Science Platforms

There are large citizen science projects running today. In the Zooniverse platform, participants are invited to participate in data analysis projects at a large scale. The participants have to follow a simple guide on how to perform the analysis to the data that is displayed using images, video or audio and a. Galaxy Zoo, the first project created in Zooniverse, mobilized 165000 participants to classify images of galaxies (Simpson, Page and De Roure, 2014).

Citizen science has been largely adopted in biological science with the objective of collecting and sampling observations in the field (Kullenberg and Kasperowski, 2016). iNaturalist is a citizen science platform for citizens to observe and identify natural species. The citizens take pictures of natural organisms and upload the images to the platform and other participants help to identify the organism (Nugent, 2018). iNaturalist is a multi-projects platform that can be used for multiple purposes. It supported a biodiversity project to teach secondary school students biology (Echeverria *et al.*, 2021).

The eBird project collects information about distribution and profusion of birds through submissions of observations by users. Data quality is assured by an automated filter and a regional network of editors (Sullivan *et al.*, 2014). Despite the initial slow growth, eBird platform grew rapidly after the changes that provided new services to the participants and increased appeal (Wood *et al.*, 2011). New features allowed participants to keep track and view their observations and compare them to other participants (Wood *et al.*, 2011). The biodiversity data generated is being used for a large variety of uses, fueling scientific publications (Sullivan *et al.*, 2014).

Generally, these platforms require the user's registration to turn into active participant. The participants provide their own observations and cataloging. These observations can be viewed by users without the need to be registered, Unregistered users can also read news blogs about insights and data analysis. Participants can submit comments on observations gathered by other participants and share their opinions on discussion forums. Participants can view and edit the user profiles, track their own contributions, see interactive observation maps, and read observations made by other users. Some of the most successful citizen science platforms, such as Zooinverse and iNaturalist, are multi-projects.

The geographical scope for the three mentioned platforms is global, meaning anyone with internet access can use the platform. Zooniverse doesn't have a target scope. iNaturalist and eBird are targeted to the fields of natural science and ornithology, respectively. Zooniverse collects data about the classification for images, text, and graphs according to the project guidelines. eBird and iNaturalist gather images submitted by users with locations, date, and the identified bird species. These platforms have defined data gathering and accessing policies. To download data on eBird and iNaturalist platforms, the user needs to be registered. Registration encourages communication between users and eBird staff, provide metrics about data usage, improve data products and ensure proper use of the data according to the terms of use policy (Sullivan *et al.*, 2014). In Zooniverse, data generated by the users can only be accessed by the respective project collaborators.

These platforms adopted protocols for data quality assurance and monitoring. eBird employs a two-stage verification system. In the first stage, submissions are automatically evaluated based on species count limits for a given location and given date. In the second stage, submissions are reviewed manually by local editors (Sullivan *et al.*, 2014). iNaturalist assures the data quality by checking if the observation has the date and time, a location, a photography, a supported identification by the platform, etc. Some observation attributes must be approved by the community.

User engagement is essential for the success of the citizen science platforms. eBird built a useful tool for bird watchers to use while doing science and added features designed to improve user-reward. Each participant can define a specific list of features to manage his/her own observations. eBird incentivizes participants to insert current and historic data by providing recognition for individual achievements. It also integrates a wide range of visualization tools participants can use to view the data (Sullivan *et al.*, 2014).

2.2 Initiatives for Combating Radon

In a recent research work, Meritxell Martell et al (2021) found several citizen science initiatives related to radon and studied eight of them. In all the initiatives, participants are involved in radon data collection. The Evict Radon project, promoted by the University of Calgary, Canada, provides a radon measuring kit to allow citizens to measure indoor radon concentration in their home. The citizens must order the measuring kit and, after the data measurement, send the kit back to the project team. In Ireland, the Environmental Protection Agency (EPA) implemented the RadNorm project which allows citizens to borrow radon measurement devices in the Wexford libraries network to monitor their indoor radon levels during a period of 3 months. Two of the initiatives, one in Italy and the other in Israel, aim to educate students on radioactivity and to raise their awareness while the students help to gather measurements data. The other four initiatives were developed in the United States, three of them promoted by universities. Two initiatives describe pilot projects set up in Massachusetts and New Mexico, using community-based participatory research in marginalized communities. In the first case, citizens participate in the problem definition and are engaged in data collection. In New Mexico, short-term indoor radon kits were used to measure indoor radon levels in 51 homes. Another citizen science initiative in Tennessee offers several services and assistance such as test kits for homeowners, technical information for universities, and specific materials for targeted audiences, such as real estate professionals, home builders, home inspectors and school officials. Finally, a transdisciplinary community-academic project in Kentucky addresses community concerns about lung cancer by identifying geological and atmospheric conditions that increase radon intrusion into homes. This knowledge is translated into greater residential awareness of risk, enabling home radon testing and report back, and growing access to affordable and adequate radon mitigation. The project engages residents of rural Kentucky communities and trains them to rigorously collect data and interpret findings.

According to Muki Haklay (2013), there are four levels of participation and engagement in citizen science projects: (1) crowdsourcing, basic level where citizens are data collectors; (2) distributed intelligence uses the cognitive ability of the participants as a resource; (3) participatory science, when the participants define the problem and implement a data collection method, previously devised with the help of experts; and (4) extreme citizen science, where professional and non-professional scientists are involved in all kind of scientific activities while matching the motivations and interests of the participants. Most of the identified initiatives were classified as crowdsourcing by Meritxell Martell et al (2021) because citizens just participate in data collection and are not engaged in the production of scientific knowledge. However, the researchers pointed out that citizen science projects have potential to contribute to radon research and provide benefits for all participants and society in general. Citizen science initiatives can be used to raise the poor public awareness about radon and contribute to national radon action plans and implementation of legal requirements. The development of citizen science initiatives including testing and radon mitigation may contribute to a reduction in radon related lung cancer.

3. MATERIALS AND METHODS

From the literature review, we have concluded that citizen science is very helpful to scientific questions that have large spatial or temporal scopes, such as the case of radon research.

3.1 Guidelines for the Development

Yadav and Darlington (2016) recommend that citizen science platforms should be designed to be cost effective, and easy to maintain. The platform should allow good interaction between scientists and citizens, offer security and trust, and have a good usability for the participants. One key aspect for getting the best of volunteers is having an engaging user experience (Simpson, Page and De Roure, 2014).

The ten principles of citizen science (Robinson *et al.*, 2019) are a good tool to assess the extent to which a citizen science initiative can contribute to radon research. These principles correspond to the shared view of the international community of citizen science practitioners and researchers about the characteristics that support high quality citizen science. According to these principles, citizen science projects actively involve citizens and have a genuine science outcome. Both the professional scientists and the citizen scientists benefit from taking part. Citizen scientists receive feedback from the project and may, if they wish, participate in multiple stages of the scientific process. If there are no legal or mandatory constraints, citizen science project data and metadata are made publicly available. The development of the Zooniverse platform was an iterative process and using real users was vital for project development evolving to a flexible platform for web developers leading to a wide range of projects (Simpson, Page and De Roure, 2014).The user must understand the project and the goal in a few minutes (Jennett and Cox, 2014).

Some general design principles for citizen science are providing access to the data the users give, allow them to view it at any time, transparency with the data that it's collected, data sharing as much data as possible, modular development, each data point should have a unique identifier, design that appreciates time and lowers entry barriers to the users and secure data transmission and storage (Sturm *et al.*, 2018)

The collected data might have data quality issues and must be validated in some way. Data collection methods must be standardized, nevertheless very large sample sizes tend to lessen said issues (Silvertown, 2009; Dickinson, Zuckerberg and Bonter, 2010). Enhanced data quality can be achieved by providing a clear data collection protocol, simple data forms and support to participants (Bonney *et al.*, 2009).

3.2 Platform Development

A pre-diagnosis model was designed as a software tool for predictive diagnosis of indoor radon potential and radon awareness (Nunes *et al.*, 2022; Silva *et al.*, 2022). This paper describes the development of a web platform integrating this model to enable citizens to obtain an assessment of the radon potential of the buildings where they work or live. There are two main roles in accessing the platform: the user and the administrator. The user represents the citizen. The citizens will have access to the indoor radon potential estimation tool, and to the submission of radon measurements on the platform. The citizen is the one who provides the data to the platform. The administrator role will be assigned to researchers to give them exclusive access to user-entered data and model performance metrics. To bring some dynamic content to the web platform, the administrator will be able to publish news about radon-related issues.

The citizen science radon web platform will allow citizens to obtain an assessment of the radon potential of the buildings where they work or live. The platform aims to promote radon measurements by the general population and raise awareness of radon risks. These measurements will generate new data that will be used to build a more robust classification model. The functional requirements for the web platform were based on previous work that defined a pre-diagnostic model that allows a citizen to verify the need to perform an indoor radon assessment study based on a set of characteristics of the building and its location (Silva *et al.*, 2022). Thus, the platform should motivate citizens to obtain an estimate of radon potential by providing a form where the characteristics of the building and its geographical location are entered. At a later stage, the citizen should be able to consult the assessment and, if indoor radon measurements have already been performed at the site, insert the radon readings taken. The web platform shall provide clear and objective information on indoor radon. To avoid citizens feeling constrained in sharing sensitive data, radon potential estimates and subsequent submission of measurements can be performed without registration. Administrators will be able to view all assessments and citizen-entered measurements, check the model performance, and obtain metrics on the use of the platform.

REST architecture was used for the web platform, resulting in four main components, as depicted in Figure 1: Web server, API, classification model, and database. The database stores the radon measurements and the radon potential classifications. The API provides authenticated access to the data and execution of the indoor radon potential classification model based on location, building and room characteristics. The Web server runs the web application that provides citizen and administrator access to platform functionality in a graphical interface. This architecture implements separation of concerns and allows easy updating of application features.

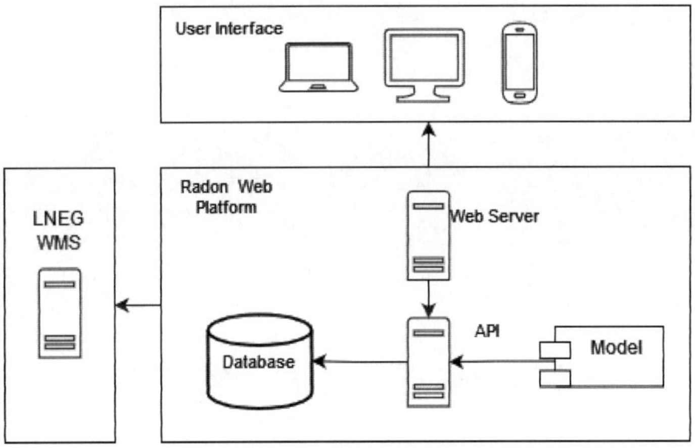

Figure 1. High-level architecture of the radon potential classification web platform

The platform must comply with the ISO/IEC 25010 standard. It should be easy to use and not require advanced computer skills. It should implement security measures to protect the data shared by citizens. It should be continuously tested for compatibility with the most used browsers, namely. Chrome, Firefox, Edge, and Safari, both on desktops and mobile devices.

The platform requires access to the LNEG map server to access the radiological map of Portugal and obtain the gamma value for the location selected by the citizen, which is an essential input for the model (Laboratório Nacional de Energia e Geologia, 2014). The radon potential estimation model runs through the RnHealth API. It will receive as input the location coordinates, the building construction year and the building number of floors, wall and floor materials of the room, the room floor and the number of floors below, thermal adaptation, and ventilation scheme. With these characteristics, the model estimates an indoor radon potential at the low, medium, or high level.

The platform was deployed into two containers: Mongo Container and RnHealth Platform Container. This design provides the isolation of the database from the web platform, so the web platform can be updated without compromising the database. The NoSQL database keeps flexibility to store arbitrary metadata as the platform evolves without having to be concerned with what exactly is stored (Simpson, Page and De Roure, 2014). The three services of the application used different technologies. The Mongo Container is a preloaded data image of a Mongo Database. The Rn Platform container supports the execution of two services: Rn API and RnHealth Web. The Rn API is running a Uvicorn server, a implementation of an asynchronous server gateway interface (ASGI) for python. The RnHealth Web service is running the Nginx open-source HTTP web server.

4. RESULTS AND DISCUSSION

The web platform is already online for functional testing by the researchers to ensure the functionality of the platform and to verify that the new radon readings data recorded through the platform are consistent with metadata from the dataset used to build the radon potential classification model. Figure 2 shows the form created to submit a radon potential estimation request on the web platform. The citizen can select the location of the building by clicking on the map and placing a marker on the building or in its vicinity, and all the fields required as input to the estimation model. Most of the fields in the form are drop-down lists with options that help the user experience.

Figure 2. Submission form for radon potential estimate

After the citizen fills the form and submitting it, it will receive in their email a link to access the estimate, for later access and feedback, and the page will show the result of the submission with the indoor radon potential estimate and recommended actions. The citizen can submit feedback by accessing the link sent to the email. This feedback should be an indoor radon assessment reading during a certain period to confirm or deny the estimate.

The web platform administrator has access to a dashboard that monitors the performance of the classification model and provides data about the radon potential assessments performed and the respective readings entered by citizens. The data provides detailed information about the geographic areas where the web platform is used and about the involvement of citizens through the ratio of the number of recorded radon readings to the total number of assessments performed. The indicators on the performance of the radon potential classification model allow researchers to decide whether the model needs to be rebuilt using our collected data.

This platform will need to be continuously updated, both in improving the performance of the radon potential classification model and the functionalities offered, to increase citizen participation and further their involvement. It is intended that this action will have a very positive impact on increasing public awareness of the health risks of radon inside buildings and significantly increase the number of buildings in Portugal where radon concentration measurements have been performed. The indoor radon concentration measurements that will result from this initiative will contribute to the increase of indoor radon data available for future scientific research.

5. CONCLUSION

The radon potential assessment web platform will allow citizens to obtain estimates from the radon potential classification model and provide new data for research. The development of the web platform is already online but restricted to the research team. After the testing and validation, the web platform will be opened to the citizens, and dissemination initiatives will be developed.

The performance of the current classification model needs to be improved. The dataset used in the construction was small and very unbalanced. Therefore, the estimates provided by the model should be carefully presented to citizens. New data collected by citizens will help to obtain much better performing models. The researchers are confident that the estimates of radon potential, combined with information on radon health risks, will increase citizens' awareness of radon health risks, trigger many measurements, and promote mitigation actions. This initiative will have a positive impact on the reduction of the radon health risks.

ACKNOWLEDGEMENT

This research was funded by the project TECH (Technology, Environment, Creativity and Health), Norte-01-0145-FEDER-000043, supported by Northern Portugal Regional Operational Programme (NORTE 2020), under the Portugal 2020 Partnership Agreement, through the European Regional Development Fund (FEDER).

REFERENCES

APA (no date a) *Campanha nacional de monitorização do gás radão 2020 | Agência Portuguesa do Ambiente (APA)*. Available at: https://apambiente.pt/prevencao-e-gestao-de-riscos/campanha-nacional-de-monitorizacao-do-gas-radao-2020 (Accessed: 29 May 2023).

APA (no date b) *Mapa de suscetibilidade ao radão | Agência Portuguesa do Ambiente (APA)*. Available at: https://apambiente.pt/prevencao-e-gestao-de-riscos/mapa-de-suscetibilidade-ao-radao (Accessed: 29 May 2023).

Bonney, R. *et al.* (2009) 'Citizen science: A developing tool for expanding science knowledge and scientific literacy', *BioScience*, pp. 977–984. Available at: https://doi.org/10.1525/bio.2009.59.11.9

Dickinson, J.L., Zuckerberg, B. and Bonter, D.N. (2010) 'Citizen science as an ecological research tool: Challenges and benefits', *Annual Review of Ecology, Evolution, and Systematics*, 41, pp. 149–172. Available at: https://doi.org/10.1146/annurev-ecolsys-102209-144636

Echeverria, A. *et al.* (2021) 'Learning Plant Biodiversity in Nature: The Use of the Citizen–Science Platform iNaturalist as a Collaborative Tool in Secondary Education', *Sustainability*, 13(2), p. 735. Available at: https://doi.org/10.3390/su13020735

European Union (2013) *COUNCIL DIRECTIVE 2013/59/EURATOM*. Available at: https://eur-lex.europa.eu/legal-content/EN/TXT/?uri=CELEX%3A32013L0059 (Accessed: 15 April 2023).

Haklay, M. (2013) 'Citizen science and volunteered geographic information: Overview and typology of participation', in *Crowdsourcing Geographic Knowledge: Volunteered Geographic Information (VGI) in Theory and Practice*. Springer Netherlands, pp. 105–122. Available at: https://doi.org/10.1007/978-94-007-4587-2_7

Jennett, C. and Cox, A.L. (2014) *Eight Guidelines for Designing Virtual Citizen Science Projects*. Available at: http://citizencyberlab.eu/

Kullenberg, C. and Kasperowski, D. (2016) 'What Is Citizen Science? – A Scientometric Meta-Analysis', *PLOS ONE*, 11(1), p. e0147152. Available at: https://doi.org/10.1371/JOURNAL.PONE.0147152

Laboratório Nacional de Energia e Geologia (2014) *Geoportal LNEG*. Available at: https://geoportal.lneg.pt/mapa/?mapa=CartaRadiometrica (Accessed: 15 April 2023).

Martell, M. *et al.* (2021) 'Evaluation of citizen science contributions to radon research', *Journal of Environmental Radioactivity*, 237. Available at: https://doi.org/10.1016/j.jenvrad.2021.106685

Meneses, P. *et al.* (2023) 'A Pre-Diagnosis Model for Indoor Radon Potential Assessment', *2023 11th International Symposium on Digital Forensics and Security (ISDFS)*, pp. 1–6. Available at: https://doi.org/10.1109/ISDFS58141.2023.10131887

Nugent, J. (2018) 'iNaturalist Citizen science for 21st-century naturalists'. Available at: https://www.proquest.com/docview/2009534937 (Accessed: 24 April 2023).

Nunes, L.J.R. *et al.* (2022) 'Designing a Multicriteria WebGIS-Based Pre-Diagnosis Tool for Indoor Radon Potential Assessment', *Applied Sciences*, 12(3), pp. 1–16. Available at: https://doi.org/https://doi.org/10.3390/app12031412

Presidência do Conselho de Ministros (2022) *Resolução do Conselho de Ministros n.º 150-A/2022, de 29 de dezembro, Diário da República n.º 250/2022, 1º Suplemento, Série I de 2022-12-29*.

Robinson, L.D. *et al.* (2019) 'Ten principles of citizen science', in *Citizen Science*. UCL Press, pp. 27–40. Available at: https://doi.org/10.2307/j.ctv550cf2.9

Silva, J. *et al.* (2022) 'A pre-diagnosis model for radon potential evaluation in buildings: A tool for balancing ventilation, indoor air quality and energy efficiency', *Energy Reports*, 8, pp. 539–546. Available at: https://doi.org/10.1016/j.egyr.2022.02.100

Silvertown, J. (2009) 'A new dawn for citizen science', *Trends in Ecology and Evolution*, pp. 467–471. Available at: https://doi.org/10.1016/j.tree.2009.03.017

Simpson, R., Page, K.R. and De Roure, D. (2014) 'Zooniverse: Observing the world's largest citizen science platform', in *WWW 2014 Companion - Proceedings of the 23rd International Conference on World Wide Web*. Association for Computing Machinery, Inc, pp. 1049–1054. Available at: https://doi.org/10.1145/2567948.2579215

Soltani-Nabipour, J., Khorshidi, A. and Sadeghi, F. (2019) 'Constructing Environmental Radon Gas Detector and Measuring Concentration in Residential Buildings', *Physics of Particles and Nuclei Letters*, 16(6), pp. 789–795. Available at: https://doi.org/10.1134/S154747711906030X

Sturm, U. *et al.* (2018) 'Defining principles for mobile apps and platforms development in citizen science', *Research Ideas and Outcomes 4: e23394*, 4, pp. e23394-. Available at: https://doi.org/10.3897/RIO.4.E23394

Sullivan, B.L. *et al.* (2014) 'The eBird enterprise: An integrated approach to development and application of citizen science', *Biological Conservation*. Elsevier Ltd, pp. 31–40. Available at: https://doi.org/10.1016/j.biocon.2013.11.003

US EPA (no date) *What is EPA's Action Level for Radon and What Does it Mean?, Radon*. Available at: https://www.epa.gov/radon/what-epas-action-level-radon-and-what-does-it-mean (Accessed: 15 April 2023).

WHO (2009) *WHO Handbook on Indoor Radon: A Public Health Perspective*. Edited by H. Zeeb and F. Shannoun. World Health Organization.

Wood, C. *et al.* (2011) 'eBird: Engaging birders in science and conservation', *PLoS Biology*, 9(12). Available at: https://doi.org/10.1371/journal.pbio.1001220

Yadav, P. and Darlington, J. (2016) 'Design Guidelines for the User-Centred Collaborative Citizen Science Platforms', *Human Computation*, 3(1), pp. 205–211. Available at: https://doi.org/10.15346/hc.v3i1.15

USING A SMART CITY DISTRICT SIMULATOR TO TEST NEW DIGITAL SERVICES

Philipp Neuschwander, Frank Elberzhager, Arian Ajdari and Phil Stüpfert
Fraunhofer Institute for Experimental Software Engineering IESE
Fraunhofer Platz 1, 67663 Kaiserslautern, Germany

ABSTRACT

Cities have become more and more digital in recent years. They have developed various digital services that they provide to their citizens. However, it is often difficult to test ideas for new services from both a technical and a user-centric perspective. Reasons for this are, among others, that existing technical infrastructures provide a complex environment where deployment is often time-consuming. Furthermore, certain qualities such as performance or data protection have to be ensured, which is something a prototypical solution might not be able to offer. We are developing digital solutions for a smart city district, and one of our main challenges is that the area is still under construction and no technical infrastructure exists yet. Therefore, we developed a mock environment for a digital ecosystem in which we can easily test new ideas and early service prototypes. In this publication, we report on how we used this environment in a hackathon event and provide experiences from the participants that used our smart city district simulator. We provide examples of the solutions and discuss how our technical solution supported the development of new digital solutions and how it might also be a solution for other cities.

KEYWORDS

Smart City, Digital Ecosystem, Simulation, Digital Services, Hackathon, Experiences

1. INTRODUCTION

The digital transformation is everywhere. It influences everyday life just as it does companies or municipalities. New business models and services are emerging. Cities have not been exempted from this development in recent years and are increasingly offering digital services [Neirotti et al., 2014].

In addition to the digital transformation, other topics are often also relevant for cities, such as sustainability or climate change. Specifically, cities have to deal with aspects such as mobility, energy, or heat, and develop new solutions [Lea, 2017]. Digital transformation can provide support here.

In the project EnStadt:Pfaff, we are developing a climate-neutral smart city district. Here, too, one question is how digitalization can help support climate-friendly behavior on the part of citizens and companies. For example, we developed a game that presents new mobility concepts in an interesting way and shows the player how the implementation of such solutions is received by citizens. Another example we developed is an app in the field of energy called Fish ´n Tipps, where a fish avatar gives the user energy-saving tips. The tips from the app can either be contributed by other users, or alternatively generated by the app itself by incorporating different data sources that are part of the smart city district ecosystem. For example, the user may be given a tip to use a rental bike for an upcoming appointment if the app has information about the availability of bikes at the nearby bike rental station, data about the current weather, and the user's appointment calendar [Elberzhager et al., 2021].

The last example already shows how important the networking of services and the use of data are in an ecosystem of this kind if real added value is to be gained for the user. A number of questions then immediately arise, for example regarding data protection, data usage control, technical feasibility, and business models.

In our project situation, we must deal with the following challenge: The district is currently under construction, i.e., there are no concrete apartments or stores yet, and the area is still largely unoccupied. This means that the digital ecosystem cannot be implemented yet. However, we certainly need to develop digital solutions and test prototypes and initial services. For this reason, we created a mock environment, or more precisely a mock platform, where we can get early feedback on digital services in a simple way [Elberzhager et al., 2021]. In doing so, we neglect requirements and qualities such as data protection or high performance, but in return we get a technically simple way to deploy services and evaluate them.

To analyze in more detail how such services behave, we recently developed a smart city district simulator [Ajdari et al., 2022]. With this, we are able to simulate different aspects and thus get more accurate feedback on the effectiveness and impact of the new digital services we want to test. Our aim is to present our technical environment in which early digital prototypes can be tested. This opens up a discussion on where such a solution is helpful and what kinds of solutions can be tested, i.e., how mature the environment is.

In this paper, we first present a compact overview of the basic operation of the simulator, and how it has evolved. Then we show how we used the simulator in the context of a hackathon event and give some examples of the different solutions developed in that environment and their added value. We discuss the results and the experiences made, and finally give an outlook on future work.

2. SMART CITY DISTRICT SIMULATOR

2.1 Concept and Implementation of the Smart City District Simulator

The smart city district simulator provides a virtual city district that can be used for simple simulations and quick prototyping of environmentally sensitive services in the field of smart cities [Ajdari et al., 2022]. The original concept and implementation comprise, among others, software-based sensors as well as controllable virtual actuators linked to a virtual world. Those components can serve as substitutes for their real-world counterparts and facilitate prototyping of smart city solutions (e.g., environmentally sensitive services) on a small scale. In combination with its visualization component, the smart city district simulator can support various stakeholders, such as the general public, future citizens, and developers, in exploring and empathizing with the presented scenarios and prototypical solutions. People can experience what it might be like to live in a smart city district and provide early feedback, which can be used to further improve the concepts and shape the future city district.

Figure 1 gives an overview of the realization of our smart city district simulator (a.k.a "Gameboard"). The idea is to substitute the whole real city district with a simplified virtual one and to provide an easy-to-use environment that supports exploring new ideas. The state of the virtual world (i.e., the world state) is managed in a central place called "Gameboard Backend". We differentiate between "fields" representing the map of the virtual world (i.e., the layout of the future city district) and "tokens" representing entities in the virtual world (e.g., pedestrians, vehicles, streetlights, and so on). It is possible to add scripted play to the virtual world by adding actor scripts to the "Gameboard Simulation" component, to add sensors to the virtual world by implementing software-based sensors (e.g., as part of the "Gameboard Sensor Server" component), and to add controllable virtual actuators by implementing corresponding control components that provide an API and introduce the related state changes to the world state. Environmentally sensitive services can be created that can consume the sensor data and reach out to the control components to manipulate elements in the world (e.g., turning a streetlight on or off). The "EcoLight" service is a simple example service that consumes pedestrian detection events from pedestrian detection sensors and controls streetlights in the city district in a demand-actuated way by using the corresponding example control component "Streetlight Control". Last but not least, there is the "Gameboard Frontend" component, which continuously renders and visualizes the current world state in a way that is suitable for human observers so they can actually observe what is happening in the virtual world.

Sensors provide their data as events using an event broker in order to have loose coupling with the services. For the same reason, control components should provide an API so that they can be accessed by the services without leaking too much implementation details or introducing tight coupling with the Gameboard Backend. By having such abstractions, the underlying world implementation (i.e., whether the environmentally sensitive service is connected to the virtual city district or the real city district) is hidden / encapsulated and ideally does not leak into the implementation of the prototypical environmentally sensitive services. This ensures that the prototyped innovations can be transferred to and implemented in the real world using real sensors and actuators.

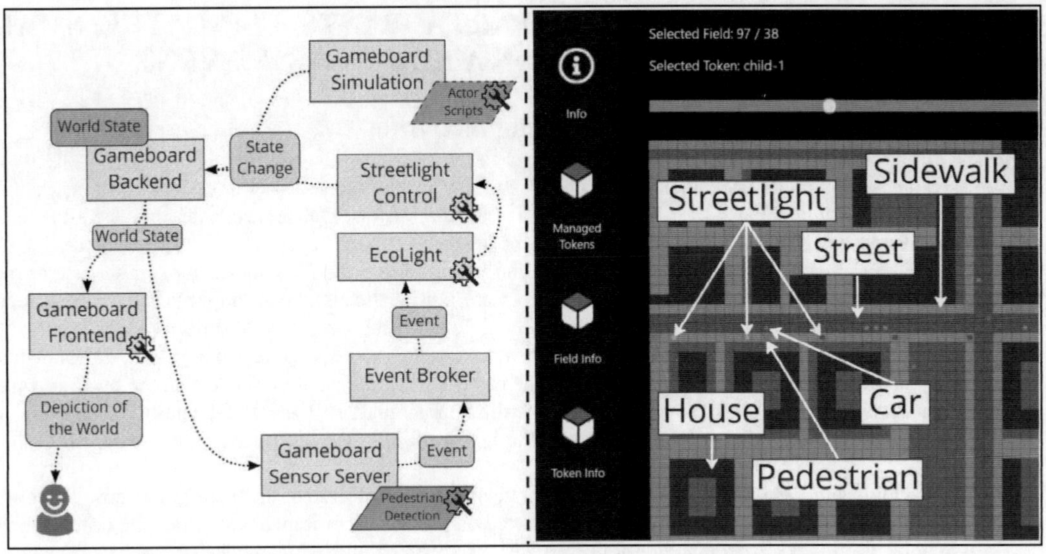

Figure 1. Smart city district simulator: implemented components, intended modification points marked using wrench icon (left) and gameboard frontend user interface (right)

2.2 Advancement of Smart City District Simulator Implementation

We further improved the implementation of the smart city district simulator and its integration with our mock platform environment. The goal was to make the simulator more inviting and easier to use, especially for people who are unfamiliar with it. This section provides an overview of the two main advancements.

2.2.1 Additional Actors Representing Example Inhabitants

In an early version of the simulation, there was only one single actor representing a pedestrian walking along a sidewalk. We added eleven additional actors representing example inhabitants of the future city district performing different activities (e.g., taking a walk in the park, driving a car, riding a bike). This is intended to make the virtual city district appear livelier and more natural. The goal is to help users empathize better with the shown scenarios and thus allow us to develop promising concept ideas. These ideas should not only improve the lives of people living in this urban district, but also lead to more climate-friendly and sustainable behavior.

2.2.2 Remote Control

The early version of the Gameboard Frontend was only able to display the current state of the world. To introduce changes to the world, one had to technically issue a request to a backend service. To facilitate getting to know the smart city district simulator and the testing of sensors and prototypes, we introduced remote-control functionality. Remote control is a special form of a control component that provides a user interface and allows creating and controlling tokens at runtime. In our implementation, we support moving a token one step in the direction of choice. We integrated the relevant user interface elements into the Gameboard Frontend for better usability. Using this feature, users can influence the virtual world more easily by introducing new entities and moving them around with just a few clicks. For example, they can create a token representing an inhabitant or a vehicle and navigate it through the virtual city district. Given there are sensors in place, the remote-controlled tokens will trigger them. The remote-control feature thus helps to stimulate the virtual world, possibly resulting in reactions that can then be observed. One obvious purpose for using remote control is stepping in and out of a sensor's range to test whether that specific sensor or an environmentally sensitive service behaves as expected. The feature can also be used for demonstration purposes, e.g., when presenting an idea or a prototype to a broader audience. Remote control complements the actor engine and its actor scripts.

3. EXPERIENCES FROM THE PARTICIPANTS USING THE SMART CITY DISTRICT SIMULATOR IN A HACKATHON EVENT

3.1 Hackathon Setting and Onboarding Activities

We used the simulator at a hackathon in November 2022. A hackathon is a 24-hour coding event in which the participants are assigned a task that they have to solve in small teams. During this time, they have to create a solution concept and present a programmed prototype at the end. At our event, 35 people participated in nine teams at one location. They were mainly students and young professionals with a computer science or other technical background. Their task was: "How can information from and about public spaces help citizens behave more sustainably?" It had to be implemented digitally. Of course, the main goal of the hackathon event was to get digital solution prototypes from the teams. However, we were also very interested in learning to what extent the smart city district simulator is able to support the participants; i.e., we used the hackathon event to gain as much feedback as possible from the participants regarding our simulator. We were especially interested in how quickly they would be able to use our simulator, and how well it would work from a technical and usability perspective.

As an additional incentive for the participants, prize money was offered to the three best teams, which were determined by a jury. The jury judged the idea as well as the technical implementation. The criteria for the assessment were: novelty, added value for the city district, technical maturity and design, and use of the mock platform environment.

To enable the participants to use the smart city district simulator as well as the rest of the mock platform environment, we relied on a multimodal learning concept. This consisted of a variety of different materials, such as documentation, source code examples, explanatory videos on YouTube, and a playful approach. To familiarize the hackathon participants with our mock platform, its services, and the smart city district simulator, we developed Pfaffhack Adventure. Pfaffhack Adventure is a web-based exit game based on the open-source solution “WorkAdventure” [WorkAdventure, 2023]. To solve the game, the players need to interact with the mock platform environment in a specific way to retrieve four-digit secret codes that can be used to unlock subsequent stages of the game towards the exit. The secret codes were integrated into the implementation of several mock platform components (similar to an Easter egg) and there are clues in Pfaffhack Adventure on how to trigger them. In this way, the participants were able to get to know and explore our mock platform environment, the services, and the smart city district simulator in a playful manner.

Regarding the smart city district simulator, we implemented the secret code in a modified pedestrian detection sensor that observes a specific spot in the virtual smart city district and publishes the code as part of the detection event once a token is detected entering that location. To solve the riddle, players have to subscribe to the pedestrian detection events at the event broker (to see the code) and must introduce or navigate a token to the right location in the virtual smart city district (to trigger the sensor, e.g., by using the remote-control feature in the Gameboard Frontend).

On the day of the event, we gave a talk to introduce and further explain our mock platform environment including the smart city district simulator. We explained the general concepts behind it and how the provided material, e.g., the source code of several of our software components, could be used by the teams.

In order to be able to gain experiences from the participants, all authors of this paper were deeply involved in the hackathon event. During the whole event, we continuously talked to the teams, helped them during the implementation, and listened carefully during the standup meetings. At the end of the event, we performed a feedback session where all participants were able to reflect the event itself, but also the technical environment and the simulator.

3.2 Hackathon Project Ideas and Prototypes Using the Smart City District Simulator

The participants developed several ideas and prototypes. Three out of nine teams used our smart city district simulator to realize their prototypes, as their ideas for digital prototypes were well supported by the simulator. This section presents the results of these three teams.

3.2.1 Team 1 (a.k.a. "bike punks")

The team "bike punks" brought up the issue that there are many cars on the road and there is not enough space for cyclists and pedestrians. In Germany, pedestrians are required to use the sidewalk and cyclists are required to use the road as long as there is no bicycle lane or signs allowing them to share the sidewalk with pedestrians. In everyday life, cyclists often have to share the road with motorists. One can observe that many car drivers do not show appropriate consideration for cyclists (i.e., tailgating and overtaking without a safety distance). Under these circumstances, riding a bike can become quite dangerous and fewer people are willing to use a bicycle to get around. This impedes certain forms of more sustainable and climate-friendly mobility.

The team's suggested approach is to adjust the usage of the roads to the needs of the people and to encourage sustainable and climate-friendly mobility. The road is not necessarily primarily for cars; there are other road users whose needs should be considered, too. For example, by introducing speed limits or other usage restrictions (e.g., assigning lanes to cyclists and/or pedestrians), one can address their needs, enable safe participation in traffic, and encourage climate-friendliness in everyday life.

The smart solution envisioned and prototyped by the team monitors the utilization of roads and dynamically adapts their usage restrictions. For example, given a multilane road and just a few cyclists and pedestrians on the road, it is likely that they can share a lane. If the number of cyclists increases, one of the car lanes might become a bicycle lane and a speed limit might be introduced for the other car lanes. In the case of roads with only two lanes, the whole road might temporarily become closed to cars. Once the number of cyclists decreases, the introduced restrictions could be reversed. Such a solution could provide priority to road users worth protecting, without banning cars in general. Having such a smart solution that provides space and safety for weaker road users will make the city district more inviting for them and might motivate even more people to use bicycles to get around. In more general terms, the idea of the team is about dynamically adjusting the use of public spaces using smart technologies to support the common good.

3.2.2 Team 2 (a.k.a. "GridScorer")

The team "GridScorer" brought up the issue of decision-making in the area of spatial planning. Decisions regarding spatial planning may sometimes appear to be arbitrary and hard to comprehend for the general public. According to the team, even the decision makers themselves (i.e., local representatives) sometimes do not feel well enough informed to decide and are therefore sometimes not confident in their own decisions. There is no adequate data basis to support informed decision-making in the field of spatial planning.

The team suggested creating such a data basis to support decision-making in spatial planning by gathering relevant information (e.g., about the utilization/visitor numbers) of numerous places in the city district. The collected information could be made publicly available and analyzed by the municipal administration to find out what parts of the city district need attention. Having such a data basis would provide decision makers with reliable data supporting informed and comprehensible argumentation and decision-making. By continually gathering and using this data to continuously improve the city district, city officials could ensure that it remains welcoming to all and supports them in living a more sustainable and climate-friendly lifestyle.

The smart solution envisioned and prototyped by the team monitors the spatially resolved utilization/visitor numbers of the public space over time. The data is recorded and stored to form a reliable data basis. Additionally, there is functionality to visualize aggregated data as a heat map.

3.2.3 Team 3 (a.k.a. "hacking.exe")

The team "hacking.exe" brought up the idea of providing the city district with a sustainable workout area, giving the community a place to exercise and improve their well-being. They got inspired by existing solutions in other areas and adapted the concept to the city district. Kinetic tiles mounted on the floor would generate electricity when stepping on them. The electricity generated in this way could be used to power nearby electronic devices (e.g., lighting). Every step would be a contribution supporting the environment.

The team suggested installing such kinetic tiles in the city district as part of a running or walking track to harvest and utilize the energy to power or charge nearby devices (e.g., lighting, sensors, e-bikes). According to the team, people from all over the city would come to the track and get stress relief while generating energy.

The smart solution envisioned and prototyped by the team captures/records the athletes' steps and the amount of energy produced. This information makes the individual contribution/impact perceivable for the individual athletes and the overall impact for the general public. Additionally, their solution contains elements of gamification; one can compete with friends, climb the leaderboard, complete challenges, reach daily goals (either personal or community goals), and/or earn rewards. This can support/motivate people to do more sports and lead to a healthier lifestyle. In addition, doing sports to generate electricity can teach 32Tmore conscious use of electricity as a resource and reduce energy waste. People will experience what it takes to produce a portion of electricity and may wonder how electricity is generated through means other than such climate-friendly ones.

3.3 Observations regarding the Use of the Smart City District Simulator

Team 1 ("bike punks") used the smart city district simulator to prototype their solution. They copied our Gameboard Simulation component and created their own actor scripts for a group of cyclists riding their bicycles in the virtual city district. Then they copied our Gameboard Sensor Server component and implemented sensors to observe several road sections, checking how they are currently used and capturing the number of pedestrians, cars, bicycles, and other types of road users currently using the road sections. Depending on how the observed road section is currently used and based on a set of example rules, the sensor modifies a newly introduced property of the gameboard fields that represent the road section to reflect the currently assigned usage restriction. Additionally, the sensed utilization information is published as an event to the event broker.

Our existing visualization component does not consider the newly introduced property when rendering the fields and therefore they are all rendered the same (i.e., the currently assigned usage restriction is not visualized). To display the usage restrictions assigned to the road, the team copied and modified the source code of our Gameboard Frontend component to consider the newly introduced property and draw the fields in different colors depending on the currently assigned usage restriction. Using their modified version of the Gameboard Frontend, it is possible to observe the change in road section usage restrictions each time cyclists enter or leave this road section. When pitching their solution to the audience, the team used a screenshot of their visualization component to explain their vision. Figure 2 shows an extract of that screenshot with additional explanatory annotations. As there are currently several cyclists on the road, the prototypical smart solution assigned individual lanes to pedestrians and cyclists. Furthermore, a speed limit is applied to the remaining car lanes. The entities using the road do not comply with the usage restrictions. For example, there are pedestrians and cyclists using the car lane.

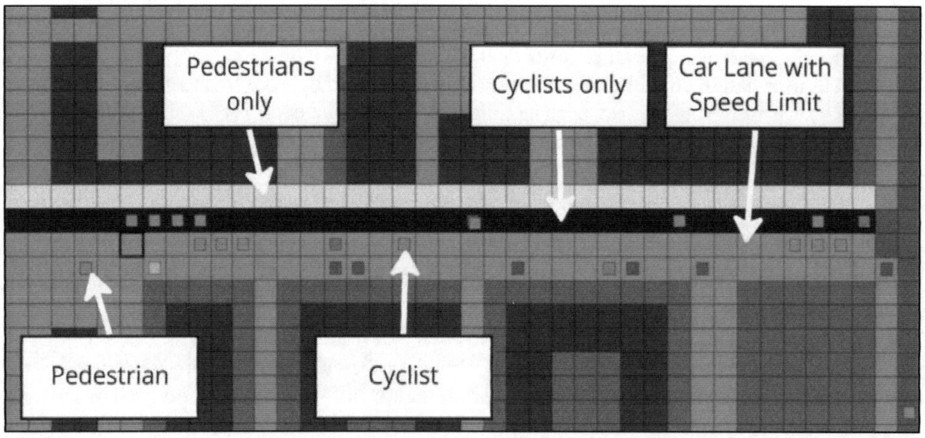

Figure 2. Visualization of the prototype of the team 'bike punks' with additional explanatory annotations

Team 2 ("GridScorer") also used the smart city district simulator to prototype their solution. They copied our Gameboard Sensor Server component and implemented a sensor that detects tokens representing persons within the sensor range. Then they divided the virtual city district into 110 cells (each consisting of 10x10 fields) to be monitored regarding their utilization. Each cell is observed by one sensor instance that publishes the sensor data to the event broker. Then they implemented a dedicated service to process the sensor data. There is a score for each cell that is incremented every time a detection event for that cell is received. The scores are persisted using the storage service provided by the mock platform environment. To visualize the scores, the team implemented a user interface. The user interface fetches the stored cell scores from the storage service and displays them in a grid. The cells are colored differently to make it easier for users to differentiate between cells with higher and lower scores (similar to a heat map). The team did not implement any actor scripts to introduce any additional entities or behavior to the virtual world; instead, they observed the already present actors (see section 2.2.1) and what the other teams added during the hackathon event. When pitching their solution to the audience, the team used a screenshot of the gameboard and a screenshot of their own user interface with the colored cells bearing the scores. They arranged the latter as a half-transparent layer on top of the first one so the audience could recognize the match.

Team 3 ("hacking.exe") implemented a mobile application for athletes with real-time tracking of their steps and the energy generated by them. There are also numerous mocked screens for the gamification aspects of their solution (e.g., leaderboard, challenges, daily goals). The team used the smart city district simulator to provide their application with step detection events from the virtual world. To do so, they copied our Gameboard Sensor Server component and implemented a sensor that reports a step detection event every time a token representing a person steps on the spot observed by the sensor. They instantiated several sensors that observe spots along a track in the virtual city district. The events are published to the event broker, which is part of our mock platform. They connected their mobile application to the mock platform to consume the detection events as they are generated. To generate the events, the team used the remote-control feature of our Gameboard Frontend component (see section 2.2.2). They navigated a token representing an athlete along the track, triggering the sensors to generate the step detection events that are consumed and further processed by their prototypical mobile application. In the application, they count the step detection events and display real-time statistics (e.g., total electricity produced).

3.4 Interpretation and Discussion of the Observations

Two of the teams used the smart city district simulator and its virtual world solely as a source for data to trigger their solution with input data from sensors (teams "GridScorer" and "hacking.exe"). They stimulated their prototype using the virtual world and the reaction is only observable within their prototypical application. However, the solution by the "bike punks" team additionally impacted the virtual world (i.e., the reaction of their smart service in the underlying world) and can therefore be easily observed and empathized with by human observers. The team was the only one that closed the loop by using the smart city district simulator as a source for sensor data as well as a sink for actuator instructions. Having the stimulus and the reaction visualized in the virtual world helps to empathize with the situation and the causalities of the presented solution. None of the teams used the smart city district simulator solely as a sink just to visualize the impact of their solution on the world (without having their solution triggered by the virtual city district). No team used the simulator solely to tell a story or present a certain scenario without implementing a prototype.

The "bike punks" team implemented multiple concerns in their sensor: the sensor logic to determine the utilization of the road section, the business logic to decide what usage restrictions to apply to the road section, and the logic to apply the new restrictions by updating properties in the Gameboard Backend. They could have separated these parts intro three building blocks: sensor, environmentally sensitive service, and control unit. Implementing everything as part of the sensor introduces high coupling and leaks implementation details of the smart city district simulator to the environmentally sensitive service. When we noticed this, we were unsure whether the team did not understand our concept regarding the different types of components and their separation or whether there were other reasons for this. We talked to the team, and it turned out that they were aware of our concepts and had tried to save effort by implementing it as a single software component. As one of the simulator's main purposes is to facilitate quick prototyping, we consider it valid to take shortcuts and simplify as much as possible. The teams created prototypes that should support presenting and testing their ideas to get early feedback; therefore, it should be totally fine to do such hacks and simplifications. This is more or less the same as we do with our mock platform; we intentionally neglect qualities that would be required for production-ready solutions in order to speed up and simplify test implementations.

We also observed that the "bike punks" team used four street lanes for their scenario, while there were actually only two street lanes available in the city district layout that we used for the simulator. In order to get two additional lanes, they repurposed the sidewalks next to the street and applied usage restrictions to them, too. In the real world, it would make no sense to assign a sidewalk for use by cars. Nevertheless, our city district layout was a limiting factor for them, and they made the best of it; and most importantly, people were still able to understand their general idea. Providing the participants with their own copies of the Gameboard Backend with an easy possibility to configure custom layouts might give them more freedom to choose a layout that supports their prototype ideally.

Another observation that is related to providing the teams with a single shared instance of the Gameboard Backend is that the scripted play provided by us and possibly by other teams will interfere with the solutions of the teams. As a result, we were able to observe that there is a pedestrian walking on a lane that is assigned to cars in the "bike punks" team's scenario, as the pedestrian's path is scripted using hardcoded coordinates that match the original sidewalk. On the one hand, it could be interesting to have all the teams share one world and explore how they can interact and influence each other. It might be possible to connect and combine different solutions that could complement each other. On the other hand, there could also be great potential in being able to provide each team with their own instance, if needed, in order to enable them to arrange their scenario without interference from elements already present or introduced by others without coordination.

None of the teams used the simulator to show a live demonstration of their prototype during the pitches. We do not know the exact reasons, but the short amount of time available during the final presentations might have prevented them from using the simulator.

4. CONCLUSION AND OUTLOOK

In this paper, we first presented our current smart city district simulator, which we developed to enable us to test new digital services for a smart city district. One main feature is that new digital prototypes can be evaluated quickly without worrying about certain qualities such as performance, security, or privacy, which are needed in digital ecosystems in actual operation.

In order to check what is useful and feasible and can be supported when developing digital solutions, we applied the smart city district simulator in a hackathon event. We provided a variety of background material for the participants so that they could prepare for the event. During the event, nine teams developed different digital solutions. Three teams used the smart city district simulator and were presented in this paper. It turned out that the teams were able to use the technical components of the simulator and could integrate their solutions into our mock environment. Based on our observations during the hackathon event and the participants' feedback, we can conclude that it was helpful for the teams that we provided them with the source code of our components. They could easily copy and adjust it to their needs to implement their own scenario.

While we applied the simulator environment in the context of a smart city district, which is reflected by a virtual map of the district in the simulator, we believe that this might not be the only context in the future. Some adaptations may make it possible to use the simulator in different environments to get early feedback during the development of digital prototypes.

ACKNOWLEDGMENT

The research described in this paper was performed in the EnStadt:Pfaff project (grant no. 03SBE112D and 03SBE112G) of the German Federal Ministry for Economic Affairs and Climate Action (BMWK) and the German Federal Ministry of Education and Research (BMBF). We thank Sonnhild Namingha for proofreading.

REFERENCES

Ajdari, A.; Neuschwander, P.; Elberzhager, F.; Mennig, P., Smart City District Simulator – How we Made a Virtual Smart City District Come Alive, 8th International Conference on Connected Smart Cities, CSC 2022, pp. 135-142

Elberzhager, F.; Mennig, P.; Polst, S.; Scherr, S.; Stüpfert, P. Towards a Digital Ecosystem for a Smart City District: Procedure, Results, and Lessons Learned. Smart Cities 2021, 4, 686-716. https://doi.org/10.3390/smartcities4020035

Lea, R.; Smart Cities: An Overview of the Technology Trends Driving Smart Cities, IEEE, 2017

Neirotti, P.; De Marco, A.; Cagliano, A.C.; Mangano, G.; Scorrano, F. Current trends in Smart City initiatives: Some stylised facts. Cities 2014, 38, 25–36, doi: 10.1016/j.cities.2013.12.010.

WorkAdventure, https://github.com/thecodingmachine/workadventure, last accessed 2023/01/26

BRIDGING THE DATA ANALYTICS – DATA EXCELLENCE GAP: REQUIREMENTS FOR A DATA CATALOGUE FOR EFFICIENT CROSS-ORGANIZATIONAL MOBILITY AND SMART CITY DATA ANALYTICS

Johannes Sautter[1], Jan Kräck[2], Rudolf Fischer[3], Nicolas Fähnrich[4], Kai Erlenhardt[5], Marco Soares[4] and Balthasar Weitzel[6]

[1]*Fraunhofer-Institute for Industrial Engineering IAO, Stuttgart, Germany*
[2]*ifeu – Institut für Energie- und Umweltforschung Heidelberg, Heidelberg, Germany*
[3]*University of Stuttgart, Institute of Human Factors and Technology Management IAT, Stuttgart, Germany*
[4]*Fraunhofer IAO, Stuttgart, Germany*
[5]*Municipality of Solingen, Solingen, Germany*
[6]*Fraunhofer Institute for Experimental Software Engineering IESE, Kaiserslautern, Germany*

ABSTRACT

Data Analytics projects often suffer from low data availability, inconsistent data formats and data models. In mobility data projects this especially means there are inconsistent attributes for specifying time and space dimensions. Also, non-data management and data analytics experts have so far had little opportunity to obtain an overview of data products, since data repositories and catalogs usually do not present the data in a self-describing manner. Data mostly exists in organizational silos as well as inadequate quality. In organizational departments data literacy is often focused on the own work area and does not include an overreaching view on data management. Arguing from a holistic view including data analytics as well as data management knowledge, here a "data analytics – data excellence gap" exists. This paper describes and motivates a smart city and mobility data ecosystem data catalog as a core component to bridge this gap.

KEYWORDS

Data Catalogue, Data Product, Data Analytics, Mobility, Real World Objects, Smart City

1. INTRODUCTION

Data is already used in various forms for everyday work in municipalities, research organizations and enterprises. In municipal as well as research and enterprise contexts the effort to collect and maintain data often outweighs the perceived added value. In general, in data analytics projects "data cleaning and preparation [even] takes approximately 80% of the total data engineering effort" (Zhang, S. et al 2003). From a data management point of view, there is a consensus in the scientific community that data quality (DQ) as "fit for purpose" (Wang and Strong 1996) for subject areas (DAMA 2017, p. 75) across departmental silos can only be achieved with data cleansing, master data management, and data quality management within organizations (DAMA 2017, Lis and Otto 2021, Sautter et al 2018). To deliver a solid and broad basis for decision, high quality data must be made available to others. In the discipline of research data management, and increasingly in other data management fields as well, it is becoming established that it should be kept FAIR, i.e., findable, accessible, interoperable, and reusable (European Commission 2021, Wilkinson, M. D. et al. 2016).

Coming from a citizen perspective, there are many digital services like social media platforms and online shops offering an excellent user experience. However, for beginner data scientists or civil servants, only a few possibilities exist to assess existing data products with a similar user experience.

Arguing from a holistic view, including data analytics as well as data management knowledge, a "data analytics – data excellence gap" exists: The enormous potential of FAIR as well as high quality data is far from being exhausted in data analytics projects, especially in mobility and geospatial analytics topics. However, as people in charge of data analysis tasks often are focused on their own domain of expertise only, they mostly have insufficient knowledge about broad and overarching data management. Thus, e.g., planning processes are often subject to the subjective experience of departments and resulting actions can be based on inadequate data. As a result, for instance in smart cities, the accurate estimation of future space requirements and necessary infrastructure becomes almost impossible and development stalls.

The MobiDataSol concept aims for designing a multiple organizations data trust model to make high quality data available using ecosystem data governance (Sautter, J. et al 2022). This paper serves as a subsequent step towards mobility data products provided by a Trusted Data Intermediary ecosystem that connects multiple organizations of municipal administration, research, and industry. The main paradigm, from a subject-matter data analytics point of view, therefore is: "Build and assess many small intermediate data products, document them, and hold them available also for others". A data catalog, motivated and proposed in this paper, serves as a basic tool towards those many "intermediate data products" as well as an "intelligent data product" for Greenhouse gas accounting use cases fulfilling the project's data analytics purpose. This paper first summarizes related work on data management and data analytics concepts, second describes the research method followed by section explaining the smart city mobility case study. A main section elaborates requirements for a data catalog in five subsections, before a short section summarizes the envisioned tool from a feature point of view.

2. RELATED WORK

Data Management Data products are the result of a value-added handling of data for data consumers (Wang and Strong 1996). From a technical and research point of view, two concepts for similar purposes aiming for storing and holding (and integrating) data can be distinguished: data lakes and data platforms (Schieferdecker, I. et al 2018) as well as data catalogues (Korte, T. et al 2019) and (meta-) data repositories (European Commission 2021). The former makes data directly accessible; the latter only provides metadata and links to the actual data sets hosted somewhere else. The boundaries between data lakes and metadata repositories can be blurred, depending on what is still interpreted as part of the actual data set and what is interpreted as metadata.

A data catalogue is understood as an integrated technical platform for data curation that aligns data supply and demand. It provides data inventory (data supply) and data discovery (data demand) functions. Further optional functions are data governance, data assessment and data analysis. Thus, a data catalogue is a critical tool for organizations seeking to optimize their data management practices.

When looking for data repositories and platforms with good user experience, the platform kaggle.com offers qualified metadata, a bronze, silver gold rating as well as user ratings for their open dataset allowing JSON, CSV SQLite and more datatypes[1]. Relevant platforms for mobility data in Germany are Mobilithek and research repositories situated at libraries like FID Move[2].

Metadata are information on data which can be subdivided in three subcategories: (1) descriptive metadata for identification purposes, (2) structural metadata on structure, attributes, and versioning as well as (3) administrative metadata for methodological and technical aspects related to data creation as well as access rights (Zeng 2004). An important initiative to enable better searchable and thus reusable public sector data is the DCAT-AP specification (European Commission 2022). It offers a unified data catalogue vocabulary that is the basis for a fine-grained search across national data portals within the EU. The actual search capability is realized by a hierarchical metadata harvesting mechanism, so that the European Data Portal[3] can be used to retrieve references to all Open Data published by public authorities of the EU.

[1] www.kaggle.com/datasets
[2] www.mobilithek.info, www.fid-move.de
[3] https://data.europa.eu

The processing of personal data requires compliance with data protection and privacy laws like the General Data Protection Regulation (GDPR) (EC 2016).

Data Analytics The CRoss Industry Standard Process for Data Mining (CRISP-DM) process model provides a step-by-step approach to data mining and data analytics processes, divided into the phases of business understanding, data understanding, data preparation, modelling, evaluation, and deployment (Wirth and Hipp 2000).

A major challenge in (big) data analytics is heterogeneity, "however, machine analysis algorithms expect homogeneous data, and are poor at understanding nuances. In consequence, data must be carefully structured as a first step in (or prior to) data analysis" (Jagadish et al 2014). Generating the right metadate to describe the recorded data is fundamental. E.g., in an experiment it is fundamental to offer details on the experiment conditions and procedures conducted. A special challenge in mobility data is representing the measures regarding space and "to acquire, store, analyze and visualize these datasets" (Demiryurek et al 2009). The data model shows the important key recorded variables such as sensor id, measure, space (i.e., coordinates) and time (timestamp), which allows further data aggregation such as hourly values, weekly values (cf. ibid.). To address these topics, among others in the domain of traffic information, the European Committee for Standardization developed DATEX II standard[4] for exchanging traffic information between traffic information centers, traffic operators and others, like media.

Regarding spatial information, the European union developed a nomenclature (NUT) to classify European regions into harmonized zones which can be easily referenced across datasets (Eurostat 2021). Other challenges mentioned are data inconsistency and incompleteness, scale (specific to big data), timeliness, and Privacy and Ownership (cf. ibid; Wang and Strong 1996).

Furthermore, a work conducted by Ardagna et al (2016) indicates Accuracy as a data quality issue: "To be aware of the accuracy level of a dataset before processing it. This problem may be even stronger with nonstationary data series.". It also refers to Usability and states that "On one side, data usability may increase with the availability of a more compact description of the dataset". Finally, it mentions challenges regarding Trust and Provenance "Accuracy and usefulness of the results in fact depend on the quality and origin of data, which in turn contribute to increase the trust of the final users in the process producing such results" (Ardagna et al 2016).

According to (Munappy et al 2020) data pipelines support solving data accessibility challenges by enabling data controllers to grant teams permissions to access data, originated from heterogeneous data sources at any point of the pipeline, which means at any level of (pre-)processing, granularity or enrichment as needed or possibly accessible (cf. data catalogue features). This eliminates the need for repeated collection and duplicated storage of the same data by multiple teams so sharing among them is facilitated. Data pipelines can be applied for data acquisition, data management, data quality monitoring, and machine learning applications (ibid.).

3. METHOD

First, we used own experience and consolidated related work on data management as well as data analytics requirements and solutions. Second, we identified data sources and gained subject-matter insights in semi-structured interviews with seven civil servants and executives as well as representatives from data-related enterprises in the city ecosystem of Solingen. The interviewees are experienced in the field of data analytics and/or manage relevant data regarding CHG. Roles of the civil executives and servants are Office of Geospatial Information, Statistical Offices, Mobility Planning, and Technical Operations. Beyond the identified locally available data sources, external information regarding transport data has been drawn together. Based on analytical requirements, a metadata scheme evolved (cf. 5.5). The data pipelining process that aims for elaborating a GHG accounting data product evaluates and determines the usability of the data catalog from a practitioner's point of view. Finally, both the final intelligent greenhouse gas data product as well as the data products may be (re-)used by various stakeholders in various use cases (cf. section 4).

[4] https://www.datex2.eu

Thirdly, we designed a data pipeline, which will later be technically implemented, in the form of a data flow diagram to define the necessary merging and calculation steps in the correct order. As a result of the processing steps leading to the final data product, a road-specific greenhouse gas accounting with the highest possible update frequency, intermediate data products are created, such as a traffic flow model, which can be relevant for further applications, for example in urban planning. Due to the relevance of these intermediate products for the quality of the final data product, additional requirements arise for the metadata of the data catalog. Fourth, we transferred the made conclusions in the case study to a general conclusion regarding data catalogs for data analytics.

4. CASE STUDY GREENHOUSE GAS DATA PRODUCT

In the information age, it is becoming increasingly challenging for municipalities to fulfil their tasks in an appropriate and future-oriented manner. Every municipal department has its own applications – be it the residents' registration office, the foreigners' registration office, the environmental department, the civil engineering office or the statistics department. These departments sometimes collect their data themselves or commission external service providers as e.g., research institutes, to create reports. However, it almost always leads to minimalistic data management and data use for a single purpose. In municipalities, Geoinformation Services typically are an exception as they offer services to other departments. Furthermore, initiatives driven by Geoinformation departments may exist that strive for holding and managing data for multiple-purposes for cross-department needs. Often, however, these fail due to a shortage of personnel. GIS systems and geodata departments fulfill the function of collecting data from different departments (e.g., using linked open data portals), because there is a legal obligation to publish and version some types of data. Geospatial data, like maps for example, are accessible and easy to understand. But by no means this represents the whole picture of what is realistically possible or already implemented by municipalities.

Previous approaches for traffic and Greenhouse gas emission estimations on a regional level (municipality, county, or administrative districts) are often based on a rather rough or outdated estimation of the traffic volume and sources on site, which also lead to an unprecise and not up to date estimation of traffic-induced emissions. Greenhouse gas accounting, deriving measures and monitoring could be carried out for different relation types like source-destination, domestic or commuter traffic more accurately and specific. During the development of a basic high-quality database, that provides all the available data of the and regarding the local transport system could be used by many actors involved in environmental monitoring and transport planning.

A Data Catalogue is the core component to bridge between the data-based ways of operating a smart city and everyday challenges of municipalities – or to put it differently: to bridge the Data Analytics – Data Excellence Gap in the public context.

5. REQUIREMENTS FOR A DATA CATALOG

In Data Governance projects aiming to consolidate an organization's data excellence the introduction of a data catalog (DC) to improve the transparency of data types, qualities and quantities is a typical early step. From an (expert- and non-expert) user point of view, both is needed: Ensuring data quality as well as offering subject matter quality e.g., in mobility. Also, the DC should comply with high user experience standards.
From a data analytics point of view, the systematic documentation and description of intermediate steps and even making data products available to other (external) stakeholder often comes up too short. Further, main data management paradigms, such as "Get usage right also for second usage before you acquire or assess data" is often not considered, especially in municipal and research contexts. Regarding personal data, the minimization of storage is an important issue to consider. In general, the use of anonymous data is preferable since in this case the data protection requirements do not apply.

In the following specification, we distinguish the data product, the physical object of the real world (e.g. a vehicle) and the data object (e.g. table row/line in the data model/stream of the data product). Requirements for a data catalog described in the following reach beyond classical metadata specifications in the eGovernment, research data and geospatial field[5] and data catalog or data repository features (cf. section 2).

5.1 Bibliographic and Identifying Metadata for the Data Product

Regarding metadata (data describing data), we distinguish between bibliographic and identifying metadata and content descriptive metadata. Data for the purpose of identification (e.g., title, abstract/summary, author and keywords) and methodological texts that are recorded at the level of the data set. Information on the origin, collection period and nature of the data.

Connecting applications to the data catalog can further enhance its value in two keyways. Firstly, metadata contained within the data catalog can be leveraged in business intelligence or advanced analytics applications, enhancing the accuracy and completeness of analytical results. Secondly, by integrating external systems or data sources, data catalogs can facilitate user collaboration and streamline data governance workflows. Bibliographic and identifying metadata requirements are:

- Persistent Identifier
- Curator/Reviewer
- Data Provider (organization/geodata service/integration level/data warehouse)
- Data Consumer (and known Quality requirements)
- Data Privacy Level
- Data Type/Format/Data Model (i.e. DATEX II)
- Availability/Licensing
- (Restricted/Unrestricted)
- Certificate terms/Maturities of certificates
- Dependencies to other data products/sources

Further requirements are given by the metadata standards DublinCore, DataCite (research data), INSPIRE, OCG (geodata) and DCAT-AP (eGovernment/public data).

5.2 Object Reference

Spatial reference: Ideal, for the case of the EU and UK, usage of NUTS (Eurostat 2021) nomenclature up to Local Administrative Units (LAUs are the building blocks of the NUTS which comprise municipalities and communes of the European Union) and further sub census division levels. Time reference: to which date/ period does the data refer to. Here, things become complex because observations should be allocated to a standardized datetime format, even though they might refer to a certain period. For this, ideally there is a timestamp which indicates the beginning of the period and a further data feature which indicates the duration of that observation e.g., if we are talking about a monthly value then we need to make some standardized way to set it; one way is to set the value to the first day of the corresponding month (if year, first day of the corresponding year) and add a variable "period" that indicates the duration.

Unit reference: Another important reference is the observed entity. In the smart city context, traffic can be viewed from different perspectives. On vehicle level or on street and district level, do we view traffic as a whole or do we distinguish between cars, buses, trucks or bikes. This point is related to both spatial reference and aggregation level. Nevertheless, it deserves mention in this section.

In these cases, it is also important to indicate if and how the observed value was aggregated (e.g., average of multiple observations, sum of multiple observations, etc.). Therefore, in general, units should be properly indicated and contextualized: we should follow the SI system (when available) and in case of aggregated data this information should follow (e.g. kg/vehicle). The system of units used, and the sizes must always be indicated and if necessary a conversion to SI system should be offered. When data is aggregated, the aggregation method should also be clearly described (in the variable metadata?)

[5] https://datacite.org/, https://www.dcat-ap.de/, https://www.dublincore.org/ , https://inspire.ec.europa.eu/

5.3 Object Measurement

Another important information is the source of the data: the data provision system (e.g., automated traffic acquisition system) and the data provider (e.g., municipal traffic office), as well as the available technical documentation for data acquisition and processing, should be fully available in their most current versions. It should be clear what devices/tools were used for the data collection (do they follow a standard as e.g., ISO) or at least their capabilities and restrictions. Techniques/methods for calculating the final value (e.g., if the specified value is the altitude but this is measured with a pressure sensor) must also be described, assumptions made must be mentioned. In addition, it is important to know in which intervals the data were collected. Is it a one-time survey or are the measurements performed in a regular cycle (daily, monthly, annually) and will they be continued to be recorded in the foreseen future? In order to be able to assess the accuracy of an analysis, the uncertainties and reliability (e.g., failure rates) are crucial..

5.4 Data Product Representativeness

Finally, it is important to understand the representativeness of the collected data: What was the sample of the data and what does it represent? (i.e., on surveys, what was the sample? Does it represent the spatial population composition?) Representativeness might not be mandatory but information whether it is, should be indicated particularly if there was a special scope of the collection as e.g., public transport users.

5.5 Data Product Content Describing Metadata and Status

Under content-describing metadata, we understand characteristics, structure and versioning of structured data (e.g., columns of a table), which are either described at file/data stream level (e.g., file type, raw date vs. processed data) or contained in files. In addition, information is required about the status of the data within the data pipeline (reference to data pipeline system cf. section 2). This means that, on the one hand, status messages must be forwarded from the processing chain that indicate errors. On the other hand, there are cases where an update of an intermediate product is not mandatory for the creation of a valid final product.

- Data product name
- Content summary and object description
- Data Type/Format/Data-Model (i.e., DATEX II)
- Geographical Scope (regional/national/global)
- Reference unit Scope (road/car/distance)
- Scope Means of transport (car/semi/bike)
- Boundary User group affiliation
- Data collection period
- Update frequency
- Foreseen availability (one-time vs. continuous survey or collection)
- Precision or uncertainties if available
- Data pipeline acquisition method (Push- vs. Pull-API)
- Aggregation of units/timeframes/spatial
- Suitability for analytical questions
- Analytical/Interpretation Restrictions
- Precise Description of key metrics and corresponding units
- Static or dynamic data (list of parking spaces vs. near real-time capacity utilization of parking spaces)
- Status of upstream data (products)
- Status of maintenance ("error has been recognized and will be fixed")

6. THE DATA CATALOGUE BUILDING BLOCK

Summing up the requirements, key features of an envisioned data catalog solution can be summarized as follows:

1. Data cataloging for metadata and corresponding full data of intelligent data products (that may refer to other data products)
2. Integration with data analytics processes (data pipelining) and support for "intermediate" data products

3. Manage the reference to the real-world objects standing behind data catalogue items (reference, measurement and representativeness)
4. Inherent features for cross-organizational access right management and data governance allowing compliance and data quality.
5. User Experience for public access by citizens (open data platform) as well as internal users
6. Fostering global identification (also for internal data products) and global findability

Beyond this basic concept for a data catalog, the following other enhancing aspects are relevant to tackle a sociotechnical solution for (ecosystem) data governance: (1) a data governance role model/curator and data steward per data product (who cares if the (dynamic) data product is broken?), (2) a data pipeline (tool/process) used by data analyst (validity/quality etc.), (3) interoperability for data products that allows an automatic creation of data pipelines based on bibliographic metadata (4) an automatic update of metadata in case of measurement parameters/assumptions/boundaries/tolerances change (5) An easy-to use automatic or half-automatic integration with public data platforms as well as research data repositories[6]. (6) a data space ecosystem connector as well as (7) an integration with municipality, enterprise and research institute structures and processes.

7. CONCLUSION AND OUTLOOK

This paper focused on deriving requirements for a data catalog based on expert's insights as well as own case study experiences. As a contribution to the state of the art, we define four main data requirements: Object reference (where does the data refers to?), object measurement (how is/was the measured quantity collected?), data product representativeness (what does the data under which assumptions represent?) and data product status how valid and of which operational value is the data?).

In data analysis, especially in the field of mobility, environmental and spatial data analysis, aspects of validity, accuracy, representativeness as well as the coverage quality of the spatial and temporal value progression are guiding questions to be answered in the data preparation phase. Coming from a data product-oriented approach of data analytics especially in the field of mobility and smart city data, using such a data catalog with an inherent management of metadata, and thus the knowledge about data, could significantly reduce the efforts for data analytics projects. To foster the adoption of the data catalog in public, private and research contexts, a mapping to respective vocabularies is required. Next steps should tackle the question of how these data can be put into value.

ACKNOWLEDGEMENT

We kindly thank the whole MobiDataSol consortium, especially Dominik Lis and Nicolas Ortiz, who had an essential contribution to the work of the paper. The MobiDataSol project, funded by the German Federal Ministry of Education and Research (funding code 16DTM103B) and supervised by VDI/VDE, aims at elaborating a concept for ecosystem data governance for mobility data products in the context of smart cities. Funded by the European Union – NextGenerationEU.

REFERENCES

Ardagna, C. A. ; Ceravolo, P., and Damiani, E.: "Big data analytics as-a-service: Issues and challenges," 2016 *IEEE International Conference on Big Data (Big Data)*, Washington, DC, USA, 2016, pp. 3638-3644, http://doi.org/10.1109/BigData.2016.7841029.

---

[6] E.g. http://www.mobilithek.info, http://www.fid-move.de etc.

DAMA (Ed.), 2017. *DAMA-DMBOK: Data management body of knowledge (second edition ed.)*. Technics Publications, Basking Ridge, New Jersey.

Demiryurek, Ugur & Banaei-Kashani, Farnoush & Shahabi, Cyrus, 2009. "TransDec: A Data-Driven Framework for Decision-Making in Transportation Systems," 50th Annual Transportation Research Forum, Portland, Oregon, March 16-18, 2009 207726, Transportation Research Forum.

Dobrokhotova, Ekaterina; Engelbach, Wolf; Sautter, Johannes (2015): *Marktstudie 2015 Multidomänen-Stammdatenmanagementsysteme.* Stuttgart: Fraunhofer-Institut für Arbeitswirtschaft und Organisation (IAO), https://doi.org/10.24406/publica-fhg-297449.

Eurostat, G., 2021. *Statistical Regions in the European Union and Partner Countries. NUTS and Statistical Regions* 2021. Eurostat, p.188.

European Comission 2016. General data protection regulation (GDPR): Regulation (EU) 2016/679 of the European Parliament and of the Council of 27th April 2016 on the protection of natural persons with regard to the processing of personal data and on the free movement of such data, and repealing directive 95/46/EC. http://eur-lex.europa.eu/legalcontent/EN/TXT/?uri=CELEX:32016R0679.

European Commission, 2021. *Recommendations on FAIR Metrics for EOSC.* Publications Office of the European Union, Luxembourg. https://doi.org/10.2777/70791

European Commission, 2022. *DCAT Application Profile for data portals in Europe.* Publications Office of the European Union, Luxembourg. https://joinup.ec.europa.eu/collection/semantic-interoperability-community-semic/solution/dcat-application-profile-data-portals-europe/release/211

Franklin, M. et al, 2005. From databases to dataspaces. *ACM SIGMOD Record* 34, 4 (2005), 27–33. https://doi.org/10.1145/1107499.1107502

Jagadish, H. V.; Gehrke, Johannes; Labrinidis, Alexandros; Papakonstantinou, Yannis; Patel, Jignesh M.; Ramakrishnan, Raghu; Shahabi, Cyrus (2014): Big data and its technical challenges. *In: Commun. ACM 57 (7)*, S. 86–94. https://doi.org/10.1145/2611567.

Korte, T. et al, 2019. *Data Catalogs - Integrated Platforms for Matching Data Supply and Demand: Reference Model and Market Analysis (Version 1.0)*. Fraunhofer Verlag, Stuttgart.

Kühling, J. 2021. Der datenschutzrechtliche Rahmen für Datentreuhänder. *Zeitschrift für Digitalisierung und Recht* 1 (2021*)*, 1–27.

Sautter, Johannes; Litauer, Rebecca; Fischer, Rudolf; Klages, Tina; Wuchner, Andrea; Müller, Elena et al. (2018): Beyond Data Quality: Data Excellence Challenges from an Enterprise, Research and City Perspective. *In: Proceedings of the 7th International Conference on Data Science, Technology and Applications.*, S. 245–252, DOI: 10.5220/0006912902450252, Available online at https://publica.fraunhofer.de/handle/publica/401256.

Sautter, Johannes; Lis, Dominik; Kräck, Jan; Helsper, Andreas; Erlenhardt, Kai; Schnieders, Fabienne; Lambrecht, Udo (2022): Mobility Data Products for Smart City Ecosystems – a Greenhouse Gas Balancing Case Study. In Yingcai Xiao (Ed.): International Conference on Connected Smart Cities (CSC2022), Lisbon, Portugal, 19-22 July 2022. Red Hook, NY: Curran Associates Inc. Available online at https://publica.fraunhofer.de/handle/publica/425148.

Schieferdecker, I. et al, 2018. *Urbane Datenräume - Möglichkeiten von Datenaustausch und Zusammenarbeit im urbanen Raum.(Urban data rooms – possibilities of data exchange and cooperation in the urban environment).* Fraunhofer FOKUS, Berlin. http://publica.fraunhofer.de/documents/N-500021.html

Wang, Richard Y. and Strong, Diane M. 1996. Beyond accuracy: What data quality means to data consumers. *Journal of management information systems* 12, 4 (1996), 5–33.

Weber K. and Klingenberg C. (Eds.). 2021. *Data Governance: Der Leitfaden für die Praxis.* Carl Hanser Verlag GmbH & Co. KG, München. https://doi.org/10.3139/9783446466746

Wilkinson, M. D. et al, 2016. The FAIR Guiding Principles for scientific data management and stewardship. *Sci. Data* 3:160018 doi: 10.1038/sdata.2016.18.

Wirth, Rüdiger; Hipp, Jochen (2000): CRISP-DM: Towards a standard process model for data mining. *In: Proceedings of the 4th international conference on the practical applications of knowledge discovery and data mining.* Citeseer, S. 29–39. Online available: https://citeseerx.ist.psu.edu/viewdoc/download?doi=10.1.1.198.5133&rep=rep1&type=pdf.

Lis, D. and Otto, B. 2021. Towards a Taxonomy of Ecosystem Data Governance. *In Proceedings of the 54th Hawaii International Conference on System Sciences*, Tung Bui (Ed.). , https://doi.org/10.24251/HICSS.2021.733.

Zeng, M. (2004). *Metadata types & functions*. http://marciazeng.slis.kent.edu/metadatabasics/types.htm.

Zhang, Shichao; Zhang, Chengqi; Yang, Qiang (2003): Data preparation for data mining. In: Applied Artificial Intelligence 17 (5-6), S. 375–381. https://doi.org/10.1080/713827180.

HOW DOES DIGITAL SOCIAL CREDIT SYSTEM PROMOTE BUSINESS INVESTMENT? AN EMPIRICAL STUDY OF 45 CITIES IN CHINA

Jinghua Yin*
School of International Business
Zhejiang International Studies University
Hangzhou, China

ABSTRACT

The Chinese government leads the digital credit system construction, and its strategic intention is to shorten the long process of the market's spontaneous credit system construction and catch up with high-income countries. Whether the digital credit system construction improves investment is crucial to the strategic intention realization. Thus, this study empirically examined the mechanism and effect of the digital SCS (social credit system) on enterprise investment in 49 first, quasi-first-, and second-tier Chinese cities. We used methods of fixed, ML (maximum likelihood method) fixed, and GLS (generalized least squares method) random effects based on relevant data from the China City Business Credit Environment Index, City Statistical Yearbook, Judicial Civilization Index Report, and Open Forest Index Network. The higher the degree of SCS digitization, the larger the scale of enterprise investment. A digitized SCS improves data resource recall efficiency, expands information application, and reduces enterprise financing difficulty, thereby increasing enterprise investment. The "cost reduction" and "policy uncertainty reduction" effects of the digital SCS are not significant. Thus, our study recommends further utilizing SCS's role in facilitating judicial civilization and data opening to enhance the main function of the government in SCS construction, and to improve government-enterprise interaction through digital channels.

KEYWORDS

Social Credit System, Enterprise Investment, Financing Constraints, Information Disclosure, Contract Breach Cost

1. INTRODUCTION

Social credit system (SCS) is the main component of China's socialist market economy and social management system. It is based on laws, regulations, standards, and contracts that improve the coverage of social members' credit histories and credit infrastructure networks. It is established on compliance with credit information applications and credit service system to support and foster an honest culture and promote good faith and traditional virtue. It uses trustworthiness incentives and dishonesty constraints as a reward and punishment mechanism. The purpose is to raise society's integrity, consciousness, and credit levels.

In 2014, the State Council officially issued the Outline of the Social Credit System Construction Plan (2014–2020). By 2019-end, China's public credit system had made major breakthroughs in several basic areas and key links, laying a solid foundation for building a moderately prosperous society by 2020. China has established a unified SCS and achieved comprehensive coverage of SCS. Moreover, the public credit system of Chinese government departments (e.g., credit information sharing platforms) has become saturated, and the degree of credit information sharing has greatly improved. Thus, the construction of the local credit platform has entered its third stage. Its functions are constantly enhanced by establishing and improving the red and blacklist of credit, joint credit incentives, and credit punishment systems. Furthermore, China has instituted a social and economic management system centered on credit information. It has completed the construction of various public credit platform projects led by governments at all levels, public credit data integration, and digital services development for public credit.

* Corresponding author email: *bike789@163.com*

The improvement of SCS represents overall progress in the legal, financial, market, and regulatory environments. China intends to promptly improve its SCS to avoid the middle-income trap and catch up with high-income countries. However, the process of building a credit system through the bottom-up approach of a modern market economy is lengthy. Therefore, China has adopted a state-led top-down credit system building process (Krause et al., 2022).

The issue with China's SCS construction is that it does not match, coordinate, and adapt to the country's economic development level and social development stage. Existing problems include: (1) lack of a credit investigation system covering the entire society, (2) large scale missing of credit records of social members, (3) imperfect incentive and punishment mechanism of keeping faith, (4) insufficient incentive of keeping faith, and (5) inadequate consequences for breaking faith. Additionally, the credit service market is underdeveloped, service system is not mature, service behavior is not standard, credibility of the service institutions is insufficient, and protection mechanism of the rights and interests of the credit information subject is missing. The inadequacy of SCS construction directly affects enterprises' operating costs and investment choices.

1.1 Concept of Digital SCS

SCS is an intersection of legal, financial, market, and regulatory environments. The original concept of credit in Europe and the US was strongly moralizing and educational, and was later reduced to a financial concept. The original narrow concept of credit in China, which focuses on financial aspects, has been expanded to a broader understanding of social behavior and moral ideals (Krause et al., 2022). In practice, China's digital credit system focuses on financial and commercial activities rather than political activities (Liang et al., 2018).

The credit system in China differs significantly from those in Europe and the US. The US and German governments focus on providing environments in which trust among business actors can develop from bottom-up, while the Chinese government intervenes in a top-down manner through its centralized information infrastructure and extended legal enforcement function of the credit system (Krause et al., 2022). Specifically, the state's influence on the Chinese credit system is significantly stronger than in the US and Germany; the Chinese government extends the legal enforcement function of the credit system to ex-post enforcement rather than only ex-ante approval, the information infrastructure in the Chinese credit system exhibits a higher degree of centralization, and private platforms in China show significantly higher levels of disclosure than their counterparts in the US and Germany (Krause et al., 2022).

Song (2022) proposed that the essence of the concept of credit is the reference to past behavior records of credit subjects related to various economic and social activities in the present and future. From the perspective of basic governance tools, SCS construction promotes a push toward the equilibrium state of high credit by improving the degree of information sharing and implementation effect, and makes the fulfillment of promises the rational choice of all kinds of social subjects. In economic and social governance, the basis to ensure a good social credit order is that the subject who holds information and the subject who makes decisions coincide as much as possible. Simultaneously, attention should be paid to the combination of large-scale and small-scale credit constraints.

Han (2020) believes that the market transaction mode in the digital economy era involves cash transactions between anonymous strangers and off-site spot credit transactions. However, in the industrial economy era, the trading modes of the market are spot transactions and off-site spot transactions between strangers. The credit discovery mechanism is a public credit investigation system and a private credit investigation agency, and credit punishment mechanisms include laws and regulations, administrative and judicial punishment mechanisms, as well as complex contracts and contracts signed between transaction subjects.

1.2 Function Mechanism of China's Digital Credit System

Some scholars view China's digital credit system as a bold and innovative governance experiment. Bach (2020) identifies China's SCS as a bold national effort to create a "trustworthy" society by developing a system for collecting, digitizing, and sharing the data of hundreds of millions of people and business entities to produce an automated state reward and punishment system. It is an emerging experimental governance system in China's quest for usable "credit" scores as a means of allocating financial resources and explicitly measuring citizens' trustworthiness.

Other scholars have analyzed the operation mechanism of China's SCS from a relatively neutral perspective. For instance, Liang and Chen (2022) studied citizens' scores in China's SCS focusing on 50 cities covering 210 million people. The results showed that SCS is conceived and designed as a social quantitative practice comprising two aspects: (1) a normative device to encourage "good" citizenship and social morality, and (2) a regulatory device to constrain "deviant" behavior and enforce social management. SCS illustrates that state actors are increasingly becoming data processors, while citizens are reconfigured as data points that can be measured, compared, and governed. SCS, as the infrastructure of social quantification, plays a role in social management, constructing differences, and guiding people toward desirable behaviors defined by the state.

Other studies argue that China's SCS raises privacy and security concerns. Lee (2019) proposed that China's new SCS, which converts online and offline behaviors into credit scores through smartphone applications, created a "new normal" lifestyle for Chinese citizens. Data-driven authoritarianism uses data and technology to monitor citizens. Aho and Duffield (2020) argue that the overall goal of China's digital credit system is to shift the power of data surveillance from private to public sector, repurposing the existing surveillance infrastructure and technology to advance the national agenda. Additionally, Devereaux and Peng (2020) argue that China's social credit rating, which goes beyond existing financial credit ratings, attempts to project less specific individual characteristics such as trustworthiness, criminal tendency, and group loyalty onto a single scale. The emergence of personal credit ratings has been enabled by big data, automated decision-making processes, machine learning, and facial recognition technology.

Related research theory is fruitful and has strong referential significance. However, some problems remain unresolved. First, SCS studies tend to focus on government governance; only a few empirical studies have been conducted on SCS to promote enterprise investment. Second, most existing studies are based on the macro or province level, and a few studies are based on the urban, individual level. Third, there is insufficient theoretical attention given to how the digital SCS has a practical impact on the investment of enterprises in existing literature. Therefore, this study conducted an in-depth research on the influence mechanisms and channels ofthe digital SCS on enterprise investment.

2. INFLUENCE MECHANISM OF THE DIGITAL CREDIT SYSTEM ON ENTERPRISE INVESTMENT

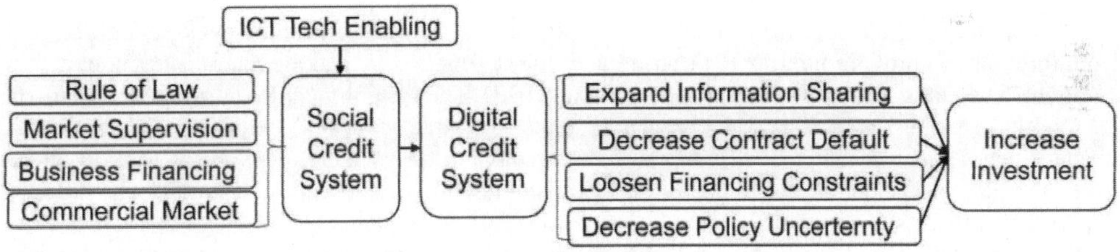

Figure 1. Theoretical framework

Figure 1 shows our theoretical Framework. First, the social credit system is a cross-cutting area of multiple dimensions of the business environment, including the rule of law, market regulation, financing environment and commercial market. Second, the digital credit system emerged after ICT empowered social credit system. Last, as a new digital governance tool, the digital credit system, expands information sharing, reduces the cost ofdefault, relaxes financing constraints, reduces policy uncertainty, and thus, promotes business investment.

2.1 Scope of Digital Information Applications

The greater the scope of digital information application of the credit system, the more efficient the information and the greater the business investment. Smart cities use big data to solve problems before they become prominent through knowledge discovery and reorganization, inter-organizational collaboration, and post-evaluation (Okwechime et al., 2018). The government's flexible use of information sharing, big data, and other emerging technologies to regulate corporate credit reduces the potential risk of corporate investment and increases the desire to invest. A more digitalized SCS implies that the way data resources are structured and information is applied becomes more sophisticated, thus reducing the difficulty of obtaining the required information before enterprises invest. The multi-subject, collaborative supervision of the big data credit system breaks the information asymmetry barrier, makes up for the deficiency of "hard information" of small, medium, and micro enterprises, promotes the development of private lending, and helps entrepreneurs cross the barrier of capital threshold.

2.2 Digital SCS, Business Investment, and Cost of Contract Default

Digital SCS influences business investments through the cost of contract default. The cost of a contract breach is mainly reflected as reputational damage. Intangible reputation cost refers to the cost incurred when the contract parties' reputation is damaged because of contract breach. However, considering the current situation in China, the credit system of enterprises and individuals has not yet been fully established. Therefore, when market players break their promises, some local governments and intermediaries often cannot immediately punish or expose them. Therefore, the reputation cost of a contract breach is almost zero. Pilot reforms of SCS were conducive to reducing agency costs and information asymmetry, thus promoting corporate social responsibility activities (Cao et al., 2022). When the formal system is not complete, pilot reforms of SCS play a more obvious role in promoting corporate social responsibility activities, and act as a substitute to the formal system. In non-state-owned enterprises and enterprises with weak corporate governance, pilot reforms of SCS play a more obvious role in promoting corporate social responsibility activities (Cao et al., 2022).

Digital credit systems narrow the treatment gap between firms and reduce transaction costs. Both state- and non-state-owned enterprises are required to comply more strictly with business regulations and better manage the relationships and company information with supply chain partners to build economic trust (Krause et al., 2022). China's previously rather opaque business environment required a significant amount of company information to reduce transaction costs. However, centralized public databases and new technologies can help improve transparency at a relatively low cost. More progress has been made in promoting credit systems among firms than among civil servants or the public (Tsai & Wang, 2021).

2.3 Digital SCS, Corporate Investment, and Corporate Financing Constraints

As an important digital governance tool, China's digital SCS contributes to the development of a high credit equilibrium by improving information sharing and implementation by making the fulfillment of commitments a rational choice for all types of social agents (Song, 2022). The system includes normative tools to encourage "good" citizenship, social ethics, and regulatory tools to restrain "deviant" behaviors and implement social management (Liang & Chen, 2022). As an infrastructure for social quantification, SCS plays an important role in social management, constructing differences, and guiding people toward desirable behaviors as defined by the state (Nicolas & Kim, 2020).

First, digital technology facilitated the formation of digital assets for enterprises, broadened the business boundaries of financial institutions, and reshaped the credit ecosystem. Digital SCS provides technical support for high-tech enterprises, as well as small- and medium-sized enterprises that lack collateral to transmit real information to form credit, effectively alleviating the credit rationing problem caused by information asymmetry in their loans (Zhao & Wang, 2022).

Second, managers control enterprise resources, make investment decisions, and may pursue in-service consumption and blind expansion of enterprise scale based on their interests. Creditors and investors, as outsiders, supervise enterprise managers to ensure their interests. This often manifests as an increase in the cost of financing capital or control over the amount of financing capital, thus affecting the scale and efficiency of investment spending. In contrast, smart governance can inhibit the opportunistic tendencies of management and shareholders, increase corporate earnings, and improve the external financing environment.

Third, digital credit systems have hitherto worked well. The ethical dimension of financial credit scoring in China has spread to non-financial areas, and the institutionalization of such ethical standards is viewed as an effective solution to various social problems that have long plagued China's economic development and social justice since the reform era (J. Wang et.al., 2022; L. Wang et.al. 2022). SCSs change resident behavior, and public SCSs have a greater impact on resident behavior than commercial SCSs (Kostka & Antoine, 2020). Further, SCSs work together with relational networks to promote a new stratified order driven by social consistency norms that reduce social mobility by designating some individuals as high risk, thus affecting their occupational achievement (Raghunath, 2020).

2.4 Digital SCS, Business Investment, and Government Information Disclosure

Digital SCS influences business investment through government information disclosure. Market players such as enterprises, face policy uncertainty from the government ex-ante before making business decisions. As an important player in China's economy, local governments often issue policies to regulate and develop the economy. Enterprises often cannot accurately anticipate the policy introduction timing, policy adjustment direction, and policy implementation intensity. An effective government information disclosure system will facilitate enterprises' access to government policy information so that they can form accurate expectations of macroeconomic trends, industrial development plans, supporting policies, financing environment, and the degree of market, and adjust their business strategies (Xie & Chen, 2021). An important goal of the current stage is to create a rule-of-law business environment that is open, transparent, fair, and just, and ensures that all types of market players participate fairly in market competition.

Government integrity promotes the investment efficiency of listed companies in China (Du et al., 2018). This implies that government initiatives that aim to promote information disclosure will create a stable and predictable policy environment for private and foreign-invested enterprises, reduce the cost of access to policy information, compensate for the "information disadvantage" in the interaction between government and enterprises, and help enterprises accurately assess the expected cash flow and potential returns of investment projects. Consequently, suitable investment opportunities will be identified to reduce the investment failure risk, enhance investment decisions' sensitivity to changes in the business environment, and avoid inefficient investments due to misguided policy expectations. This is of positive significance in creating a fair and competitive market environment and promoting the sustainable development of various market players.

In summary, establishing an SCS has brought many changes to government supervision and enterprise production and operation, impacting China's market economy and business environment as well as enterprise investment. First, business rivals must increase the cost of contract defaults. To a certain extent, a poor credit enterprise's default cost increase limits its investment. However, relatively low credit enterprise operating costs are conducive to enterprise investment. Second, the difficulty of financing for enterprises with poor credit increases, whereas the difficulty of financing for enterprises with good credit decreases, which promotes enterprise investment. Third, increasing the degree of government information disclosure enables enterprises to make timely adjustments and effectively reduces the adverse effects of policy uncertainty on enterprises. Therefore, the following hypotheses were proposed:

H1: The higher the degree of digitization of SCS, the larger is the investment of enterprises.

H2: Digital SCS reduces financing difficulty and increases enterprise investment.

H3: Digital SCS promotes smooth contract performance and increases investment in enterprises.

H4: Digital SCS increases the degree of government information disclosure and enterprise investment.

3. Methodology

The fixed effect model is adopted to be the baseline regression. The sample period is from 2004 to 2020. And the sample cities includes 45 Chines cities, including 4 tier 1 cities, 11 new tier 1 cities, and 30 tier 2 cities according to the "Top 100 Cities in China" list published by the Walden Institute of Economics in terms of GDP and hard and soft economic indicators.

Specifically, tier 1 cities includes Beijing, Shanghai, Guangzhou, and Shenzhen, Besindes, quasi-tier 1 cities contain Chengdu, Chongqing, Hangzhou, Wuhan, Xi'an, Zhengzhou, Qingdao, Changsha, Tianjin, Suzhou, Nanjing, Dongguan, Shenyang, Hefei, and Foshan. In addition, tier 2 cities involves Kunming, Fuzhou, Wuxi, Xiamen, Harbin, Changchun, Nanchang, Jinan, Ningbo, Dalian, Guiyang, Wenzhou,

Shijiazhuang, Quanzhou, Nanning, Jinhua, Changzhou, Zhuhai, Huizhou, Jiaxing, Nantong, Zhongshan, Baoding, Lanzhou, Taizhou, Xuzhou, Taiyuan, Shaoxing, Yantai, and Langfang.

Model 1 is the baseline model that aims to verify the overall impact of SCS on business investment. Model 2 examines the impact of the digitization of SCS on business investment. Model 3 examines whether digitized SCS affects business investments through financing constraints. Model 4 verifies whether digitized SCS affects corporate investment through contract performance costs. Finally, Model 5 examines whether digitized SCS affects corporate investment through government information disclosure. The model settings are shown in Equations (1) to (5), and the meanings of the variables are listed in Table 1.

$$\text{Logsum} = \lambda_0 + \lambda\ .CEI1 + \lambda_2\ .Sci + \lambda_3\ .\text{year} + \varepsilon \quad (1)$$

$$\text{Logsum} = C_0 + C_1\ .MCEI + C_2\ .Mobile + C_3\ .CEI + C_4\ .Sci + \mu \quad (2)$$

$$\log sum = \beta_0 + \beta_1\ .FCEI + \beta_2\ .Finance + \beta_3\ .CEI + \beta_4 .Sci + v \quad (3)$$

$$\log sum = Y_0 + Y_1\ .LCEI + Y_2\ .Law + Y_3\ .CEI + + Y_4\ .Sci + \text{负} \quad (4)$$

$$\log sum = \theta_0 + \theta_1\ .OCEI + \theta_2\ .open + \theta_3\ .CEI + \theta_4\ .Sci + z \quad (5)$$

Table 1. Variables

Variable	Explanation	Data resource
Sum	The total output value of domestic, foreign, Hong Kong, Macao, and Taiwan enterprises	China City Statistical Yearbook
Mobile	Number of cellphone subscribers at the end of the year	
CEI	City Credit Score	Blue Book of China's UrbanCommercial Credit Environment Index
Open	City and Provincial Open Forest Index	China Open Forest Index net
Law	Provincial Judicial Civilization Index	China Judicial Civilization Index report
Finance	City Digital Financial Inclusion Index	Digital Financial Inclusion Index of Beijing University
MCEI	Mobile*CEI	
OCEI	open*CEI	
LCEI	law*CEI	Blue Book of China's Urban Commercial Credit Environment Index
FCEI	finance*CEI	China City Statistical Yearbook
SCI	Expenditure on science and technology in regional public budgets	

4. RESULTS

4.1 Baseline Model

Table 2 shows that CEI corresponds to ap-value of $0 < 0.01$ (***), indicating that it is significant at the 1% significance level. Therefore, the degree of digitization of SCS is positively related to the value of new investments by enterprises, indicating that Hypothesis 1 holds. Moreover, this regression result is consistent with the expected result, indicating that the propensity of enterprises to invest is higher in regions with a high degree of digitization of SCS.

Table 2. Results of Model 1

logSum	Coef.	St. Err.	t-value	p-value
CEI	1236643.2	191279.91	6.47	.0000
Sci	-7.287	1.484	-4.91	.0000
year	-134132.06	246897.34	-0.54	0.588.
Constant	1.870	4.974	0.38	0.7070
Mean dependent var	5438496.786		SD dependent var	10080119.854
R-squared	0.183		Number of obs	198
F-test	14.529		Prob > F	0.000
Akaike crit. (AIC)	6914.693		Bayesian crit. (BIC)	6927.846

*p<.01, **p<.05, *p<.1*

4.2 Mechanism Analysis

Table 3 shows that the coefficients of SCS, digitization level, digital financial inclusion index, and judicial civilization index are significantly positive. This indicates that the higher the level of digitization, the more developed the SCS, the fewer the financing constraints, and the higher the quality of justice, the larger the investment of enterprises. However, the coefficients of the cross terms of Models 2, 3, and 4 are all significantly negative, possibly owing to multicollinearity. Additionally, the coefficients of the degree of government data openness and cross-term in Model 4 are insignificant.

Table 3. Results of Models 2, 3, and 4

Sum	Digital channel	Financing constraint channel	Contract execution channel	Information disclosure channel
Cross multiplication term	-0.0363***	-0.1611***	-0.700**	-0.0898
	(-5.53)	(-2.61)	(-2.2)	-0.29
Mobile	0.003***			
	(6.19)			
finance		0.014***		
		(3.04)		
Law			0.629***	
			(2.7)	
Open				0.002
				(0.09)
CEI	0.042***	0.034***	0.498**	0.022
	(6.44)	(2.95)	(2.3)	(1.53)
Sci	0	0***	0***	0***
	(1.06)	(2.72)	(6.84)	(6.22)
Constant	14.192***	15.263***	-28.092*	16.503***
	(27.83)	(17.28)	(-1.71)	(14.89)
Mean dependent var	18.064	18.134	18.064	18.162
Akaike crit. (AIC)	-226.114	129.921	94.639	58.26
R-squared	0.486	0.58	0.322	0.273

*p<.01, **p<.05, *p<.1*

Tables 4 and 5 confirm the validity of Hypotheses 1 and 2. The digital credit system promotes business investment through increased digitization and relaxation of financing constraints. However, the empirical results fail to strongly support Hypotheses 3 and 4.

Table 4. Summary of robustness test

	ML fixed-effect regression	GLS random-effect regression	Intro-group regress	Fixed-effect regression	Population mean regression	Test result
Digital channel	Positive	Positive	positive	insignificant	positive	robust
Financing constraint channel	Insignificant	insignificant	insignificant	negative	insignificant	not robust
Contract execution channel	Negative	insignificant	insignificant	positive	insignificant	not robust
Information disclosure channel	insignificant	positive	insignificant	positive	positive	robust

Table 5. Robustness test

sum	ML fixed-effect regression	Intro-group regress	Fixed-effect regression	ML random-effect regress	Population mean regression
MCEI1	0.007***	0.007***	0.001	0.002**	0.003***
	(5.8)	(3.03)	(0.79)	(2.32)	(2.64)
LCEI1	0.1	0.256	-0.563**	-0.127	-0.119
	(0.38)	(0.43)	(-2.31)	(-1.06)	(-0.89)
OCEI1	-0.093**	-0.102	0.741**	0.02	0.006
	(-2.18)	(-1.29)	(2.09)	(0.27)	(0.08)
FCEI1	0.018	-0.023	0.025***	0.023***	0.022***
	(1.32)	(-0.2)	(6.68)	(6.13)	(5.29)
Constant	17.183***	16.765***	18.581***	18.402***	18.371***
	(12.91)	(6.06)	(32.16)	(31.61)	(27.84)
Mean dependent var	18.224	18.224	18.224	18.224	18.224
Number of obs	115	115	115	115	115
R-squared		0.362	0.597		
Pseudo r-squared				0.965	

p<.01, ** p<.05, * p<.1

Digital credit systems promote business investments. The more digitalized the SCS, the more mature the data resources structuring and information application, and it reduces the difficulty in obtaining the required information before enterprises invest. The conclusions of Qian and Xu (2017) are consistent with those of this study. They proposed that to solve the difficulties of corporate credit regulation, the government needs to flexibly use emerging technologies such as information sharing and big data to regulate corporate credit, thus reducing the potential risks of corporate investment and increasing the desire for corporate investment, all of which can be achieved by the government's continuous improvement of digitization of SCS.

Digital credit systems promote business investments by reducing financing constraints. This is consistent with the findings of Han and Zheng (2021) and Zhao and Wang (2022). The former argued that firms face information asymmetry due to incomplete markets and are subject to financing constraints faced by bank credit rationing. In practice, the more information an enterprise has access to and the better the credit of the enterprise, the lower the financing difficulty faced by the enterprise. Under a good SCS, enterprises can obtain more comprehensive information, rely on their credit base, and obtain more financing, thereby improving the efficiency of enterprise investment. Zhao and Wang (2022) empirically showed that digital technology improves the risk preference and credit-granting efficiency of commercial banks for high-tech enterprises and medium-sized enterprises.

There is no strong evidence that digital credit system affects corporate investments through a contract enforcement channel. China's social credit system includes public credit system, enterprise credit system and individual credit system. The public credit system regulates the government's administrative and economic behavior, avoids the government's dishonest behavior such as changing orders and debt repayment, and improves the government's administrative and judicial credibility. Government data disclosure is an important measure to reduce policy uncertainty and improve government credibility, and is also an important step to improve the public credit system. Thus, the above results indicates that the rule of law in our credit system and the quality of the judicial environment for corporate investments must be improved. Moreover,

The more complete the legal system, the greater the sanctions for corporate default, which laterally reduces the instability of corporate investment, and thus increases companies' willingness to invest. SCS can operate more effectively and smoothly with good market regulations. Investors can rely on market regulations and reduce unnecessary expenses to improve their investment efficiency.

The current urban credit system has little to do with the availability of governmental data. Therefore, it fails to effectively reduce the policy uncertainty faced by enterprises, which in turn fails to promote corporate investment. This suggests that there is room for the government to play a leading role in constructing China's SCS. Additionally, it is a relatively long process for governments to build trust with their counterparts (Wang, 2021). This study supports the conclusion of Wang et al. (2020) that "when the degree of economic policy uncertainty is higher, firms stand to lower their investment and vice versa."

5. CONCLUSION

This study proposes that digital credit system promotes corporate investment by increasing contract default costs, reducing corporate financing constraints, and increasing information disclosure. We then conducted an empirical test for first-, quasi-first-, and second-tier cities in China. The following conclusions were drawn: (1) the higher the degree of digitalization of SCS, the larger the scale of enterprise investment; (2) digitalized SCS improves data resource invocation efficiency, expands information application fields, reduces enterprise financing difficulty, and thus, increases enterprise investment; (3) "cost reduction effect of contract performance" and "policy uncertainty reduction effect" of digital SCS are not significant; and (4) the relationship between the digital SCS and corporate investment is robust.

Therefore, we should further utilize SCS's role in promoting the construction of a judicial civilization, timely update the laws and regulations on enterprise credit supervision, and eliminate legislative conflicts and duplications. Additionally, it is imperative to strengthen the role of SCS construction in promoting data opening and disclose the credit of government agencies, including sensitive information such as letters and visits, job crimes, and disciplinary violations. Furthermore, the main function of the government in SCS construction must be enhanced, the government's network public service capacity must be strengthened, and public satisfaction and trust must be boosted. Finally, the level of government information disclosure and government-enterprise interaction through digital channels must be improved, and a standardized and transparent institutionalized government-enterprise relationship must be created.

ACKNOWLEDGMENT

We would like to thank Editage (www.editage.cn) for the English language editing.

FUNDING

This research was funded by Major Humanities and Social Sciences Research Projects in Zhejiang higher education institutions (grant number 2023QN111).

DATA AVAILABILITY

The data generated by this study is available from the corresponding author on reasonable request.

REFERENCES

Aho, B., & Duffield, R. (2020). Beyond surveillance capitalism: Privacy, regulation and big data in Europe and China. Economy and Society, 49(2), 187-212.

Bach, J. (2020). The red and the black: China's social credit experiment as a total test environment. The British Journal of Sociology, 71(3), 489-502.

Cao, Y., Kong, D., Tao, Y. (2021). Evaluation on the Effect of the Pilot Reform of the Social Credit System:From the Perspective of Corporate Social Responsibility..Journal of Finance and Economics,48(2), 16-25.

Devereaux, A., & Peng, L. (2020). Give us a little social credit: to design or to discover personal ratings in the era of Big Data. Journal of Institutional Economics, 16(3), 369-387.

Han, Y.,&Zheng, X.(2021). Research on the mechanism of the role of financing constraints in affecting the efficiency of corporate investment. Management Modernization, 41(01),4-7.

Kostka, G., & Antoine, L. (2020). Fostering model citizenship: Behavioral responses to China's emerging social credit systems. Policy & Internet, 12(3), 256-289.

Lee, C. S. (2019). Datafication, dataveillance, and the social credit system as China's new normal. Online Information Review, 43(6), 952-970.

Liang, F., & Chen, Y. (2022). The making of "good" citizens: China's Social Credit Systems and infrastructures of social quantification. Policy & Internet, 14(1), 114-135.

Okwechime, E., Duncan, P., & Edgar, D. (2018). Big data and smart cities: a public sector organizational learning perspective. Information Systems and e-Business Management, 16, 601-625.

Qian H., &Xu B.(2017). Research on rule of law indicators of administrative penalties in enterprise credit supervision. Journal of Zhejiang University (Humanities and Social Sciences Edition),47(4),32-47.

Raghunath, N. (2020). A sociological review of China's social credit systems and Guanxi opportunities for social mobility. Sociology Compass, 14(5), e12783.

Song L.,(2022). Discussion on the basic theory of social credit system. Macroeconomic Management, 5(7), 60-66.

Du, J., Li, W., Lin, B., & Wang, Y. (2018). Government integrity and corporate investment efficiency. China Journal of Accounting Research, 11(3), 213-232.

Nicolas, C., Kim, J., & Chi, S. (2020). Quantifying the dynamic effects of smart city development enablers using structural equation modeling. Sustainable Cities and Society, 53, 101916.

Krause, T., Chen, M., Wassermann, L., Fischer, D., & Grossklags, J. (2022). China's corporate credit reporting system: A comparison with the United States and Germany. Regulation & Governance.

Liang, F., Das, V., Kostyuk, N., & Hussain, M. M. (2018). Constructing a data-driven society: China's social credit system as a state surveillance infrastructure. Policy & Internet, 10(4), 415-453.

Tsai, W. H., Wang, H. H., & Lin, R. (2021). Hobbling big brother: Top-level design and local discretion in China's social credit system. The China Journal, 86(1), 1-20.

Wang, J., Li, H., Xu, W. W., & Xu, W. (2023). Envisioning a credit society: social credit systems and the institutionalization of moral standards in China. Media, Culture & Society, 45(3), 451-470.

Wang, L., Xie, Q., Xue, F., & Li, Z. (2022). Does smart city construction reduce haze pollution?. International journal of environmental research and public health, 19(24), 16421.

Wang, Y., Chen, C. R., & Huang, Y. S. (2014). Economic policy uncertainty and corporate investment: Evidence from China. Pacific-Basin Finance Journal, 26, 227-243.

Xie, G., Chen, J., Hao, Y., & Lu, J. (2021). Economic policy uncertainty and corporate investment behavior: Evidence from China's Five-Year Plan cycles. Emerging Markets Finance and Trade, 57(10), 2977-2994.

USING MULTI ADAPTIVE NEURO FUZZY INFERENCE SYSTEMS TO IDENTIFY FRAUD BANK CHECKS

Ying Bai[1] and Dali Wang[2]
[1]*Johnson C. Smith University*
100 Beatties Ford Rd., Charlotte NC 28216, USA
[2]*Christopher Newport University*
One Avenue Arts, Newport News, VA 23606, USA

ABSTRACT

This research is to use multiple Adaptive Neuro Fuzzy Inference Systems (ANFIS) to design and build an image processing algorithm to quickly and easily detect a fraud or altered bank check in real time. With the help of MATLAB Fuzzy Logic and Image Processing Toolboxes, this algorithm can be used to quickly detect most fraud or altered bank checks via an ATM scanner when a fraud or altered check is deposited to an ATM. The testing and validation processes have been performed to confirm the effectiveness and correctness of this detecting algorithm. The current correct detecting rate is about 80% for checks deposited via ATMs.

KEYWORDS

Fraud Bank Checks Detections, Altered Check Inspections, Multi Adaptive Neuro Fuzzy Inference Systems, Image Signal Processing and Detections

1. INTRODUCTION

Today check fraud is incredibly widespread at any corner in nationwide and worldwide. In fact, due to so many various types of check frauds, no one can have the exact numbers on how many people are affected and how much money has been lost each year.

Check frauds are also one of the largest challenges and lost to all financial and commercial related businesses and institutions today. With the fast developments of modern computer and micro-controller technology, it becomes easier for criminals, either in personal or in organized groups, to alter or modify checks in such a way as to deceive innocent victims expecting value in exchange for their money.

A great amount of check fraud is partly because of counterfeiting via advanced desktop printing and copying technologies to generate or duplicate some actual financial documents, or chemical fluid alterations, which consists of removing part or all of the information and modifying it to the benefit of the criminals. Most victims include financial related institutions and commercial related businesses that accept and issue personal or business checks, as well as the consumers. Generally, these crimes start with the theft of some financial documents or hand-writing customer bank checks. It can be perpetrated as easily as someone stealing a personal or a business blank check from victim's home or vehicle during a burglary, searching for a voided or old check in the garbage tanks, or stealing a check the victim has mailed to pay a bill from his/her home or office mailboxes.

Some typical fraud bank checks can be categorized into the following two categories based on their processes:

- Counterfeiting can be either wholly fabricating a check --using modern office desktop printing devices consisting of a personal computer (PC), scanner, professional software and high-level laser printer - or simply duplicating a bank check with advanced laser or color photocopiers.

- Alterations primarily refer to using chemical materials and solvents such as acetone, brake fluid, correction fluid to remove, or cover and modify the original handwriting and information on the check. When performed on specific locations on the check such as the payee's name or the amount area, it is called spot-alteration; when an attempt to erase information from the entire check is made, it is called check-washing.

It has been estimated that the annual losses could be around in the billions of dollars due to check fraud and this amount continues to grow steadily as criminals continue to seek ways to earn a living by defrauding others (Mary 2015). For the consumer, the amount of inconvenience and anxiety caused by resolving problems with the account, local merchants, as well as possible repercussions with credit bureaus can be considerable.

According to a global fraud study reported by the Association of Certified Fraud Examiners (ACFE) in 2022 (ACFE, 2022), a typical organization could lose 5% of its revenue every year due to fraud. The total loss caused by the cases in that study exceeded \$3.6 billion, with an average loss per case of \$1.7 million. That study was performed based on 133 countries around the world in 2022.

Although the check fraud case number is continuously increased around the world day by day, few related technologies and tools have been developed and built to protect victims from losing their funds. SQN Banking Systems reported their SENTRY Inspect™ software package called Check Image Analysis, which is used to detect fraud checks via an image processing tool (SQN 2014). SmartSearch reported their Anti-Money Laundering (AML) Check Monitoring and Alerts package to monitor and inspect the truthfulness of personal and business checks (SmartSearch, 2020). Kofax built and developed a FraudOne® fraud check image detection platform used to help customers to detect fraud checks in real time (KoFax, 2023). OrboGraph reported their Anywhere Fraud image analysis tool to perform check image detection for possible fraud checks (OrboGraph, 2022). ToolCASE released its Informant system that is an AI-Based solution for financial services industries (ToolCase, 2022). Parascript reported a machine learning software used to help commercial and retail banking organizations to combat fraud and support compliance such as Know Your Customer (KYC) and AML initiatives (Parascript, 2022). Experfy reported a fraud & risk detection system used to detect possible check tampering related issues (Experfy, 2022). CSI released fraud check detection software used to scan for check fraud and identify counterfeit, altered and unauthorized checks (CSI, 2022).

In addition to software packages and tools developed by various financial companies above, some researches and studies related to detect fraud checks are also reported in recent years. Lydia M. Rose, 2022 provided a review study for detecting methods used for inspecting the fraud bank checks and money orders with machine learning. Ayushi Maurya and Arun Kumar, 2022 reported a method to detect fraud credit card using machine learning technique. Badis Hammi, Sherali Zeadally and et al, 2022 built an algorithm to detect and prevent fake check scams by using a Blockchain-Based Solution.

Pradheepan Raghavan and Neamat El Gayar, 2019 reported a general way to perform the fraud detection using machine learning and deep learning. Saheb Chhabra et al., 2017 reported a method used to detect fraudulent bank checks. Abd-ElZaher, 2014 discussed certain inks' affection to physically erased handwriting in fraud checks. Kennard, Barrett and Sederberg, 2012, reported a strategy used to perform signature verification and forgery detection using a 2-D geometric warping approach. Kumar and Gupta, 2016, released a method used to identify and detect the authentication of bank checks. Lampert, Mei and Breuel, 2006, analyzed some popular printing technique classifications for document counterfeit detection. Patil and Takale, 2015, reported a distance matrix method to verify signatures on bank checks. Prakash and Sharma, 2014, built a detection method used to verify signatures on bank checks with computer vision and fuzzy logic offline strategy. Moises Diaz et al., 2019, developed technique to analyze handwritten signatures for most bank checks. Rajendar and Pal, 2014, developed a method to detect manipulated check images using mismatched pixels. Xie et al., 2009, reported a new method to identify the authenticity of banknotes based on texture roughness. S. Tayeb *et al.*, 2017, used a convolution neural network (CNN) to analyze pixels from a signature image to recognize abnormalities in bank checks.

Most of studies and researches mentioned above are concentrated on financial data analysis, check security data identifications, check signature authentications and verifications, manipulated check identifications and analysis. However, one of the most important and often-occurred check frauds is the modification or manipulation of the payee's names on bank checks, either business or personal, to steal funds by depositing them into the perpetrators' accounts via ATMs. Compared with all other check frauds, the check alteration on the payee's name area is a major crime and it is more crystal and serious since it totally steals other party's funds and put them into the perpetrator's pockets and makes financial institutions lose significant funds year by year in recent years. There are so many cases in which only the payee's names were altered but all other areas kept with no change at all in either business or personal checks. It would be rare and weird if all other areas were changed without payee's name altered for a bank check since that kind of crime is very easy to be tracked and captured.

Also most studies and researches listed above involved complicated and overlong algorithms and mathematical derivations, and therefore made the detection system more complicated and time consuming on fraud check detecting and identification processes.

In this study, we developed a multi ANFIS algorithm to quickly detect and identify fraud bank checks in which only the payee's name area is altered with some chemical or correction fluid and a perpetrator's name is rewritten to cover the original payee's name. The detecting correct rate is about 80%.

This study is divided into five sections; after this Introduction part, the detailed fraud check detection types and definitions are given in section 2. An introduction to ANFIS is discussed in section 3. The experimental results are provided in section 4, and the conclusion and future works are given in section 5.

2. THE FRAUD CHECK TYPES AND DEFINITIONS

2.1 Types of Fraud Checks

The major types of fraud checks can be categorized into the following sections:

1) The payment amount alterations or modifications
2) The payee's name alterations and modifications
3) The signature alterations or modifications
4) Both payee's name and payment amount alterations or modifications
5) The entire check making or alterations

Among all of these alterations or modifications, the payee's name alteration or modification is one of the most often-used methods by perpetrators due to its easy and simple in operations, and have been widely implemented by most criminals. The popular way to do this kind of crime is to cover the original payee's name on bank checks by using some chemical or correction fluid, and rewrite the perpetrator's name to replace the original payee's name, and deposit that altered checks to the perpetrator's account via ATMs. The possibility of success for this kind of crime is relatively higher than other fraud checks since any ATM did not have any ability to detect and identify this kind of alteration or modification when this check is scanned via a scanner in an ATM and deposited into a valid checking or saving account.

Some examples of altered checks are shown in Figure 1. Alterations include to replacing the entire original payee's name, or the partial payee's name by covering the original payee's name via correction fluid or a piece of cut-off paper, and rewriting the perpetrator's name on the payee's name areas. Regularly, the first case will be appeared in most alterations since it is not common to have both persons with the same first or the last names, of course, this case is really occurred in some situations.

To detect or identify these kinds of alterations or modifications, the ANFIS combined with MATLAB Image Processing Toolbox is a good tool or a candidate among available software tools. In fact, the core of this kind of image processing involved the image identifications via an ANFIS algorithm.

3. ADAPTIVE NEURO FUZZY INFERENCE SYSTEM (ANFIS)

3.1 The typical structure of an ANFIS

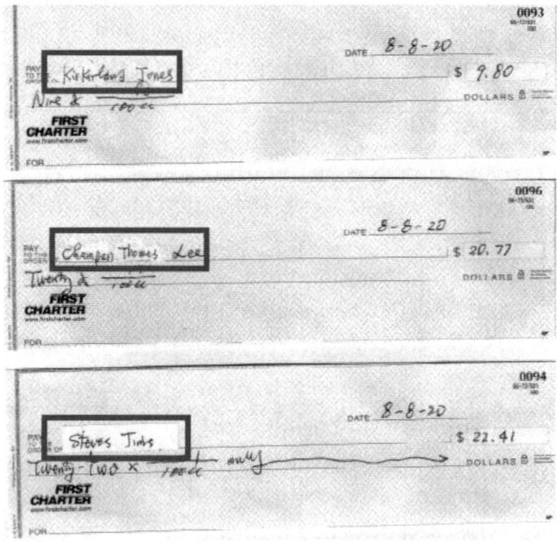

Figure 1. Examples of some altered checks

The so-called ANFIS is exactly a combination of two soft-computing techniques: Artificial Neural Network (ANN) and Fuzzy Inference System (FIS), which was first introduced by Jyh-Shing Roger Jang in 1992. The FIS used a Sugeno fuzzy inference system and its structure is similar to a multilayer feed forward neural network structure, but the difference is that the links between nodes in ANFIS define the signals' flow direction and there are no associated weight factors with the links. It consists of a network of neurons that communicate between the input and hidden layers and the hidden and output layers. Each layer consists of neurons constructed according to the principles of fuzzy control. Figure 2 shows a Sugeno fuzzy model with nine rules along with a corresponding ANFIS architecture. In our case, four rules in the method of "If-Then" for the Sugeno model are considered with x and y as inputs and f as output (Mohammed Imran, Sarah A. Alsuhaibani, 2019). Four rules are defined as below (*pixels in x direction* = *x*, *pixels in y direction* = *y*):

R_1: If x is H and y is L, then $f_{11} = p_{11}x + q_{11}y + c_{11}$

R_2: If x is H and y is H, then $f_{12} = p_{12}x + q_{12}y + c_{12}$

R_3: If x is L and y is L, then $f_{21} = p_{21}x + q_{21}y + c_{21}$

R_4: If x is L and y is H, then $f_{22} = p_{22}x + q_{22}y + c_{22}$

where H and L means reflection rates on the paper checks. The detection key for true and fake checks is based on the reflection rate on the bank checks. The checks covered with any correction fluid have different reflection intensity compared with the normal checks.

The input layer includes a check image taken by an ATM scanner or a camera, which is to be identified and detected. That image can be considered as a 2D matrix with different intensities. The first layer is the fuzzification layer to get desired membership functions in both x and y direction for the input image. The layers 2 and 3 are the fuzzy inference process combined with our rules. The fourth layer is the defuzzification layer to prepare to get the final or real outputs. Combined with layer 5, the actual output can be obtained.

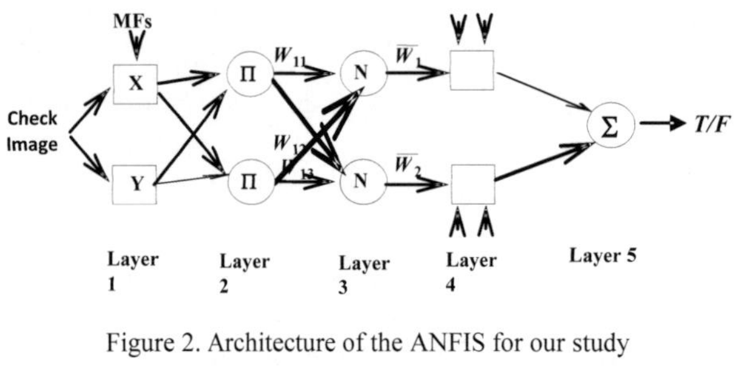

Figure 2. Architecture of the ANFIS for our study

3.2 The Fraud Checks Detection Implementation

Prior to applying the ANFIS algorithm to identify bank checks, some pre-processing jobs are necessary due to the requirements for inputs to the ANFIS. All check images are colored with relatively large size, and maybe inappropriate to be processed. Thus we must do some pre-processing for those images or 2D matrices.

Another important point is the limitation on the input size or the number of inputs to an ANFIS. The running time and the fuzzy rule numbers would be significantly increased if the number of the input variables is equal or more than to ten. In some worst cases, the ANFIS even cannot generate the outputs corresponding to the inputs. In order to solve those issues, we need to do some necessary preprocessing jobs for our input images. These pre-processing jobs include:

1) Change the colored images to the grayscale images since we do not need any color information for this detection or identification process.
2) Reduce the size of each detected check image and only concentrate on the payee's name area, as shown in Figure 1. In this way, the detecting time can be greatly shorter and the identification speed can be significantly faster.
3) Convert each image or each 2D matrix to a 1D array due to the input requirement of the ANFIS.
4) Configure or divide each 1D input image array ($1 \times N$) to M pieces of $1 \times (L - 1)$ array, and set the L column as the output value (Boolean value) with 1 as true and 0 as false. In this way, if the number of input images is K, all those input images can be made as M group of $K \times L$ sub-matrix. Each sub-matrix is a $K \times L$ matrix, which will be used as a training matrix to be applied as the input to an ANFIS later to train it.

As shown in Figure 1, for three different types of checks, the size of this area is identical. In fact, the actual size of this rectangular area should be finally determined based on the real inspection process. Regularly, this size should be big enough to cover the monitoring area as much as possible. However, the detecting algorithm would be too time-consuming with larger size of selected area if this size is too big. Therefore a trade-off should be taken between the large size of detecting area and quick detecting speed.

Figure 3 shows some example rectangular area samples for payee's name blocks on detected checks for both true and false check images. The detecting principle is based on the following facts.

It can be seen from both true (Figure 3a and c) and false check (Figure 3b and d) images, the refraction rate and roughness of the background for both checks are different. The reason for that is due to the alterations or modifications for the background on the altered checks, and these differences can be mapped to the intensities in the backgrounds on inspected checks.

1) By implementing the ANFIS algorithms on these detected checks, these differences can be effectively distinguished and identified.

2) By comparing these differences on detected intensities, the altered checks can be easily detected and identified by applying this ANFIS algorithm.

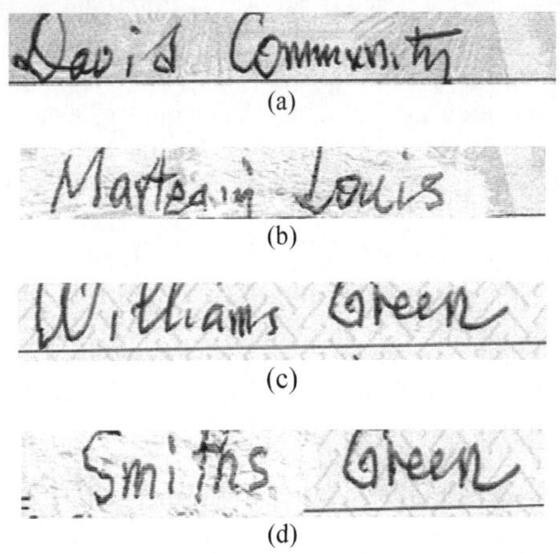

Figure 3. Testing samples for true and false checks

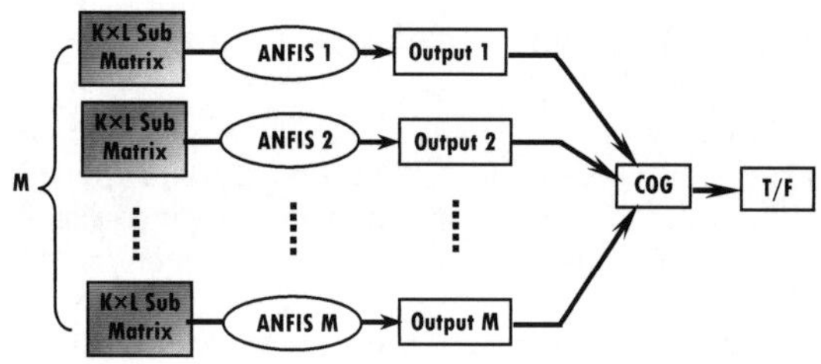

Figure 4. The structure of the multi ANFIS algorithm

One point to be noted is for the true (Figure 3c) and altered or false (Figure 3d) checks, the false check is only partially altered by modifying the first name without touching the last name. This is a coincidental case in which the perpetrator's last name is identical with that of the victim.

In order to effectively and accurately detect any altered or modified checks, an algorithm made by multi ANFIS with MATLAB Fuzzy Logic Toolbox is adopted (Benyamin Khoshnevisan, 2014). The inputs to this algorithm are a set of testing check images in grayscale format, and the outputs (**Y**) are a Boolean value, true for normal check and false for a fake check.

This detection algorithm must be trained first with a set of input arrays or input images represented by M group of K × L sub-matrix. Each row (1 × N) represented one image, which can be considered as an input row, and it can be decomposed into M pieces of 1 × L array and each last column (L – 1) is the output represented by a Boolean value, either a normal, true (1) or a fake (0) check.

The output values for all ANFIS algorithms can be fed into a Center of Gravity (COG) unit to derive the final Boolean output. The structure of this multi ANFIS algorithm is shown in Figure 4.

Determining or selecting an appropriate rectangular area for the payee's name block on the detected check needs some testing and checking processes to get a so-called optimal size for this area to make sure that a good trade-off between a big detecting size and a quick detecting speed can be achieved.

After the ANFIS algorithm has been trained and tested, it can be implemented to identify and detect real bank checks. A graphic user interface (GUI) maybe built to allow users to make different selections to do the detection to get real-time detection results.

4. DETECTING AND EXPERIMENTAL STUDY RESULTS

Due to difficulty to find a data set, in which both sufficient and good quality bank checks were involved, we collected forty (40) real checks coming from five (5) different US banks, including Bank of America, First Charter, ADVANTA Bank Corp., Ally Bank and Fifth Third Bank, and use them as samples to be tested and detected by this detecting system. These checks have different sizes, such as standard personal and business checks with a width × height of 6 × 2-3/4 inches, 7-1/2 × 3 inches and 8-1/4 × 3 inches, and different styles with various backgrounds (plain, color, palm trees, snow mountains and oaks tree).

All of these checks are transferred to 300 dpi color photos or images by using scanner and conversion software with their original sizes and styles, and stored in computers.

Figure 5 shows a sample testing result for this ANFIS algorithm. The training RMSE is 0.00094 and the checking RMSE is 0.528.

A GUI is developed with the MATLAB GUIDE tool, and the ANFIS algorithm built with MATLAB Fuzzy System Toolbox will be called by this GUI to perform related detecting and inspecting jobs when the detecting system works. Figure 6 shows the working status of this fraud check detecting system.

As the detecting system starts, a tested check should be selected from the ListBox, and the detecting process starts as the **Start** button is clicked. The testing result should be displayed with a textbox and the detected check image is also displayed, as shown in Figure 6.

With a total of 40 checks, 25 of them are used for the training and checking purpose and the other 15 are used to detecting purpose. The detecting results are: 3 of them are incorrect but 12 are correct, with a correct detection rate as 80%.

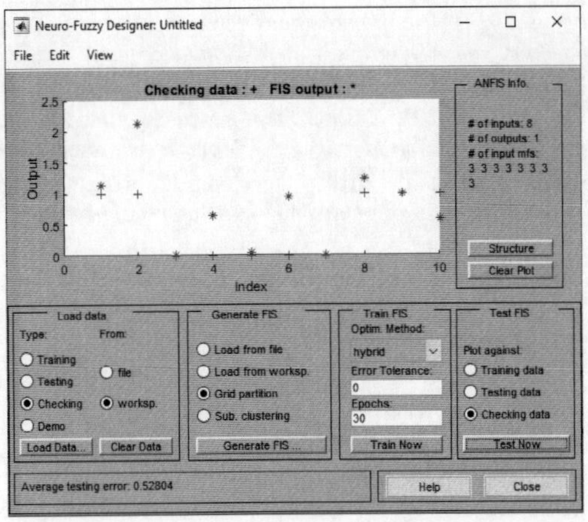

Figure 5. A sample testing results

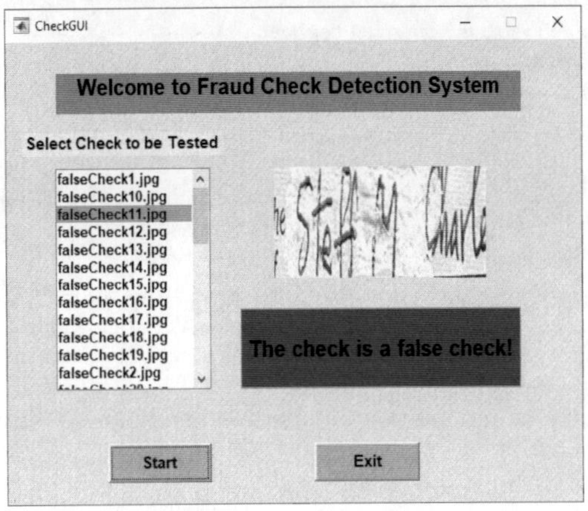

Figure 6. The running status of the fraud check detecting system

5. CONCLUSION AND FUTURE WORKS

With the help of an algorithm made by multi ANFIS built with MATLAB Fuzzy System and Image Processing Toolboxes, a fraud bank check detection system is developed and built in this study. Some real checks coming from five US banks with different sizes and styles are tested and detected via this detecting system. The current correct detecting rate is 80%.

For better detecting results, a similar system developed by using Deep Learning or Machine Learning techniques are needed, with which a better detecting result can be achieved. The correct detecting rate could be around 98% or better as that system is developed.

REFERENCES

Abd-ElZaher, M., 2014, Different types of inks having certain medicolegal importance: Deciphering faded and physically erased handwriting. Egyptian Journal of Forensic Sciences **4**(2), pp.39–44.

ACFE, https://acfepublic.s3.us-west-2.amazonaws.com/2022 +Report+to+the+Nations.pdf

Ayushi Maurya and Arun Kumar, 2022, "Credit card fraud detection system using machine learning technique", *2022 IEEE International Conference on Cybernetics and Computational Intelligence* (CyberneticsCom), Malang, Indonesia, June 16-18, pp 500-504.

Badis Hammi, Sherali Zeadally and et al, 2022," Blockchain-Based Solution for Detecting and Preventing Fake Check Scams", *IEEE Transactions on Engineering Management*, Volume: 69, Issue: 6, December, pp. 3710 – 3725.

Benyamin Khoshnevisan, ShahinRafiee et al. 2014, "Development of an intelligent system based on ANFIS for predicting wheat grain yield on the basis of energy inputs", *Information Processing in Agriculture*, Vol 1, Issue 1, pp. 14-22.

Chhabra S., Gupta G., Gupta M., Gupta G., 2017, "Detecting Fraudulent Bank Checks". Peterson G., Shenoi S. (eds), Advances in Digital Forensics XIII. Digital Forensics. *IFIP Advances in Information and Communication Technology*, Vol 511. Springer, pp. 245-266.

CSI, https://www.csiweb.com/how-we-help/platform-banking/core-bank-processing/fraud-risk-management/

Experfy, https://www.experfy.com/fraud-risk/banking-fraud-management

J. -. R. Jang, 1992. "Self-learning fuzzy controllers based on temporal back propagation," in *IEEE Transactions on Neural Networks*, vol. 3, no. 5, pp. 714-723, doi: 10.1109/72.159060.

Kennard, D., Barrett, W., Sederberg, T., 2012, "Offline signature verification and forgery detection using a 2-D geometric warping approach". *Proceedings of the Twenty-First International Conference on Pattern Recognition*, pp. 3733–3736.

KoFax, https://www.kofax.com/Products/fraudone/overview

Kumar, R., Gupta, G., 2016, "Forensic authentication of bank checks". Peterson, G., Shenoi, S. (eds.) Digital Forensics 2016. IFIPAICT, Vol. 484, Springer, pp. 311–322.

Lampert, C., Mei, L., Breuel, T., 2006, "Printing technique classification for document counterfeit detection". *Proceedings of the International Conference on Computational Intelligence and Security*, Vol. 1, pp. 639–644.

Lydia M. Rose, 2022. https://www.proquest.com/openview/5e7cd136814fe3b4c32ffa3384e6cf8c/1?pq-origsite=gscholar&cbl=18750&diss=y.

Mary, http://www.ckfraud.org/ckfraud.html

Mohammed Imran, Sarah A. Alsuhaibani, 2019, Chapter 7 - A Neuro-Fuzzy Inference Model for Diabetic Retinopathy Classification, Editor(s): D. Jude Hemanth, Deepak Gupta, Valentina Emilia Balas, In Intelligent Data-Centric Systems, Intelligent Data Analysis for Biomedical Applications, Academic Press, Pages 147-172, ISBN 9780128155530.

Moises Diaz, Miguel A. Ferrer, Donato Impedovo, Muhammad Imran Malik, Giuseppe Pirlo, and Réjean Plamondon, 2019, "A Perspective Analysis of Handwritten Signature Technology". *ACM Comput. Surv.* 51, 6, Article 117, p. 39.

OrboGraph, https://orbograph.com/anywhere-fraud/

Parascript, https://www.parascript.com/products-technology-solutions/fraud-prevention/

Patil, R., Takale, S, 2015, "Signature verification by distance matrix method for bank check process". *Proceedings of the International Conference on Electrical, Electronics, Signals, Communication and Optimization*, India, pp.1281-1285.

Pradheepan Raghavan and Neamat El Gayar, 2019. "Fraud Detection using Machine Learning and Deep Learning", 2019, International Conference on Computational Intelligence and Knowledge Economy (ICCIKE), Dubai, United Arab Emirates, pp 334-339.

Prakash, G., Sharma, S., 2014. "Computer vision and fuzzy logic based offline signature verification and forgery detection". *Proceedings of the IEEE International Conference on Computational Intelligence and Computing Research*, 2014, pp. 1-6, doi: 10.1109/ICCIC.2014.7238363.

Rajendar, M., Pal, R., 2014, "Detection of manipulated check images in a check truncation system using mismatch in pixels". *Proceedings of the Second International Conference on Business and Information Management*, Durgapur, India, pp. 28–33.

S. Tayeb *et al.*, 2017. "Toward data quality analytics in signature verification using a convolutional neural network," *2017 IEEE International Conference on Big Data (Big Data)*, Boston, MA, 2017, pp. 2644-2651, doi: 10.1109/BigData.2017.8258225.

SmartSearch, https://www.smartsearch.com/us/solutions/aml-monitoring-and-alerts

SQN, https://sqnbankingsystems.com/solutions/check-image-analysis/

ToolCase, https://www.toolcase.com/products/informant/index.html

Xie, J., Qin, C., Liu, T., He, Y., Xu, M., 2009, "A new method to identify the authenticity of banknotes based on texture roughness". *Proceedings of the IEEE International Conference on Robotics and Biomimetics*, Guilin, China, pp. 1268–1271.

POTENTIALS, LIMITS AND CHALLENGES OF USING DATA SCIENCE METHODS TO IMPROVE QUALITY PROCESSES IN MANUFACTURING INDUSTRY

Jens Kiefer[1], Lisa Ollinger[1], Melina Precht[2] and Kai Beer[2]
[1]University of Applied Sciences Ulm
Prittwitzstraße 10, 89075 Ulm, Germany
[2]Heidelberg Manufacturing Deutschland GmbH
Egelsee 10, 73340 Amstetten, Germany

ABSTRACT

This paper presents potentials, limits and challenges of using data science methods to improve quality processes in manufacturing industry. Apart from scientific discussions, this paper focuses on the specific application of data science methods based on real conditions and data from manufacturing industry. The application data, processes and all technical and organizational conditions originate from the foundry sector. As a first step, all necessary terms are explained and the most important processes and characteristics of the considered application scenario 'quality processes of a contract manufacturer for cast components' are illustrated. By means of two practical examples, potentials, limits and challenges of using data science methods in operational practice are critically evaluated. In terms of a socio-technical overall view, not only methodical issues but also information-technical and organizational aspects are discussed. The collected findings ultimately lead to recommendations for action, which are intended to give companies a kind of framework for a meaningful economic use of data science methods in operational practice.

KEYWORDS

Data Science, Artificial Intelligence (AI), Quality Processes, Manufacturing Industry

1. INTRODUCTION AND RESEARCH QUESTION

According to several studies (e.g. EIT, 2020), the profitable use of artificial intelligence (AI) in manufacturing industry is of outstanding strategic importance for most national economies. In a study on behalf of the German Ministry for Economic Affairs and Energy (BMWi) (PAiCE, 2018), it is even predicted that AI will be jointly responsible for around a third of the average economic growth in the future. In this context, quality management, data analytics and automation of processes in manufacturing industry are classified as having a particularly high potential (IFS, 2018).

Between these identified AI potentials and the current implementation status in manufacturing companies is (still) a large gap, which can be often referred to the following three reasons: inadequate application-oriented AI research activities, a reluctant technology transfer from science to companies as well as deficits in the development and implementation of AI technologies with a cross-sectional character such as machine learning, action planning and optimization (PAiCE, 2018).

In cooperation with the company Heidelberg Manufacturing Deutschland GmbH (HMD), the University of Applied Sciences Ulm (THU) wants to take on this identified topic field in the context of quality and process management in manufacturing industry. The superordinate research question is: What potentials do data science and particularly AI techniques offer in production-related engineering and manufacturing processes with regard to quality and process management, how can these potentials be most effectively developed and what are limits, challenges and recommendations for action of using AI concepts in this topic field?

Focus of the cooperation between the THU and HMD is the coherent processing of the following two subject areas:

- Development of a standards-based, semantic quality data and quality service model that transparently provides all relevant quality-related development, planning, production and customer data (and functions) to the identified applications within the quality process chain
- Development and use of effective and efficient AI algorithms for continuous planning and control of quality processes along the entire product life cycle

The scientific goals and the associated innovations are the development of a standards-based (and therefore system-independent) quality data and quality service model, the development and provision of modularly structured and independently executable services in the form of a standardized module library as well as in the implementation of a consistent overall solution. First, this paper focuses on the illustration of potentials, limits and challenges of using data science methods to improve quality processes in manufacturing industry.

2. DATA SCIENCE, AI AND QUALITY PROCESSES

2.1 Need for Data Science in Production

During the last decade, the massive expansion of digitization dominated the technical progress within the production industry. In Germany, these developments fall under the umbrella of the 'Industrie 4.0' initiative that internationally gained in great significance (Wahlster, 2021).

Digitization enables the rise of smart sensors that deliver information about production processes and products in real-time. They deliver information represented as data via communication standards that support both the horizontal and vertical integration, like OPC-UA and MQTT (Shi et al, 2020). Thus, potential data sources and data sinks can be seamlessly interconnected within and outside of the production system. This situation leads to drastically increasing amounts of data that provide substantial benefit for the processes at the company (PAiCE, 2018). However, the sheer volume of data makes it literally impossible for a human to parse it in a reasonable time frame (Ozdemir and Sunil, 2018). Furthermore, the value of the stored data is zero unless it is acted upon (Kotu and Bala, 2018). This where data science comes into play to make use of data as a resource in order to gain new knowledge about the production process.

2.2 Data Science and Related Disciplines

Data science is the art and science of acquiring knowledge and new insights through data (Ozdemir and Sunil, 2018). It requires the combination of three disciplines: computer programming, mathematics and statistics, and domain knowledge (Grus, 2019). The mathematical methods constitute the base for the algorithms that find patterns and relationships within data. These algorithms are translated into automatically executable programs by means of computer science. The domain knowledge depends on the respective application field. It is indispensable to provide the required data, to supervise the algorithms and to interpret the results correctly depending on the context of the application domain.

Artificial intelligence (AI), machine learning (ML) and data science are all related to each other and are often used interchangeably in popular media (Kotu and Bala, 2018). AI gives machines the capability of mimicking human behavior, particularly cognitive functions like facial recognition and automated driving. ML can be classified either as sub-field or as one of the tools of AI. It deals with gaining the capability of learning from experience whereas the experience for machines comes in the form of data (Kotu and Bala, 2018). Those methods are often used within data science projects to discover previously unknown patterns in data.

2.3 Usage of AI in Production and Quality Processes

This paper focuses on the application of data science for quality processes in the scope of industrial production. This area includes all activities that influence the quality of the product and services provided to the customer of a manufacturing company. A wide bandwidth of factors contribute to this effort; ranging from individual quality checks on the product itself to the optimization of business processes that determine the execution of the technical and commercial processing.

Today, applications with practical relevance can be particularly found in the fields of predictive maintenance, process analysis and control, and forecasting of quality-related properties (Zhang et al, 2019).

3. APPLICATION EXAMPLES OF DATA SCIENCE TO IMPROVE QUALITY PROCESSES IN FOUNDRY SECTOR

3.1 Application Scenario: Foundry Sector

The foundry sector supplies essential basic components for many branches of industry. The production portfolio of Heidelberg Manufacturing Deutschland GmbH (HMD) ranges from printing machine components for its parent company to large components for wind power systems and smaller components for truck steering and braking systems. Compared to other manufacturing processes, the casting process is fundamentally dependent on the interaction of different divisions. Due to many divisions and processes involved, the variables influencing the cast components quality are very diverse.

To be able to produce a cast component, a sand mold is made by taking an impression of a pattern. This sand mold represents the positive image of the cast component to be produced. In order to reproduce the inner contours, so-called cores are inserted, which are also made of sand. To produce these cores, a core box is required, which represents the positive image of the core to be produced. In the next step, the molded sand form (which consists of an upper and lower part) is filled with liquid iron. The liquid iron is previously produced from raw materials in another division and kept warm at a predefined temperature until casting. After casting, the filled sand mold must be cooled down. Then the finished component can be extracted from the sand mold and further processed (shot blasting, heat treatment, etc.).

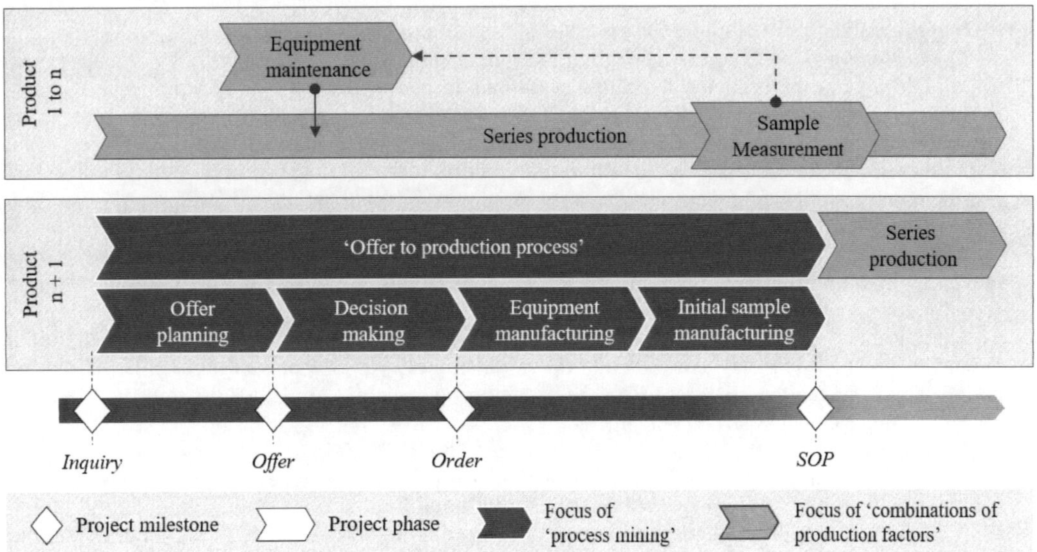

Figure 1. Considered quality processes

The interaction of all the processes is relevant for the quality of the cast component. In order to be able to produce new components, all process steps and manufacturing equipment (e.g. pattern, core box) must be planned. The so-called 'offer to production process' for new inquiries or change requests for cast components (cp. Figure 1) is accordingly dependent on the cooperation of various departments. In section 3.2 the optimization of this process with the help of the data science method process mining is discussed. At the end of a new inquiry with order, the component released by the series production release is produced in series (SOP). Series production usually refers to a multi-year cooperation in which agreed quantities are ordered on a regular basis. In general, minor changes are repeatedly requested. In some cases, products are also completely discontinued or replaced by substitute products. Each of these changes usually requires a new run through the 'offer to production process' in order to obtain the series release of the new or changed product.

Due to the constant contact of the manufacturing equipment with sand during series production, they wear out over time. Therefore, wear monitoring is required, in which the models and core boxes are regularly measured and, depending on the result of the measurement, revised so that their original condition is restored. Section 3.3 discusses which data science approach has been investigated for this wear monitoring.

3.2 Process Mining

The method process mining as one increasingly important subfield of data science has emerged strongly in recent years. The main concept of process mining is to gain information about real processes based on existing event data which is generated during the execution of processes (Berti et al, 2019). Whereas most methods in the field of data mining can be classified as data-centric, process mining is process-centric (Hand et al, 2001). During process mining, specific data mining algorithms are applied to event log data in order to identify trends, patterns or details contained in event logs recorded by an information system (Van der Aalst, 2016). One process mining method is the process discovery that aims to generate a process model exclusively based on log data.

Relevant business process:

The discovery technique is executed on existing event data from HMD. The goal of this implementation is to optimize the 'offer to production process'. The regarded process includes all process steps from a new inquiry for a product or product change request through planning, costing, offer negotiation and qualification up to the series release of the product. In the highly competitive international foundry industry, it is essential for a company to work in a customer-oriented manner. This also includes the fast and competent handling of new inquiries and changes. To improve in this field of customer interaction, the method process mining is applied.

Implementation:

The implementation of process mining requires preliminary work. As with most applications in the field of data science, data preparation and merging is essential for the application of process mining and represents the major part of the effort (Kiefer and Precht, 2019). The four major steps of the preparation process for the application of process mining are: selecting the right data (1), rating the individual data sources maturity level according to the Process Mining Manifesto (Van der Aalst, 2012) (2), generating a data model of the considered objects including their relations and the individual identification and attributes (3), merging the heterogeneous data sources (4).

The data, which has been produced through the preparation process, can then be played into the process mining tool. As a result, the process mining tool delivers the process model in a graphical manner and the according attributes e.g. the average duration of the process step.

The described application of process mining delivers a data based representation of the 'offer to production process'. This stated the basis for optimization measures to improve the customer satisfaction by optimizing the internal processes according to costs, time and quality. An extract of a process flow with the process mining method is shown in Figure 2.

Conclusion:

To sum up, the application of the data science method process mining has shown how processes can be optimized with the result of an improved customer satisfaction.

A prerequisite for the successful application of process mining is the quality of the data, which describe the process. In order to examine the data sources for suitability, a classification into maturity levels as stated in (Kiefer and Precht, 2019) is helpful. If a data source achieves a poor maturity level but is crucial for an analysis, the prerequisite for an analysis should be created in advance by revising/adapting the data source. Another prerequisite for the application of process mining is the underlying question. If the question is not aimed at an underlying target process, no useful result can be probably achieved.

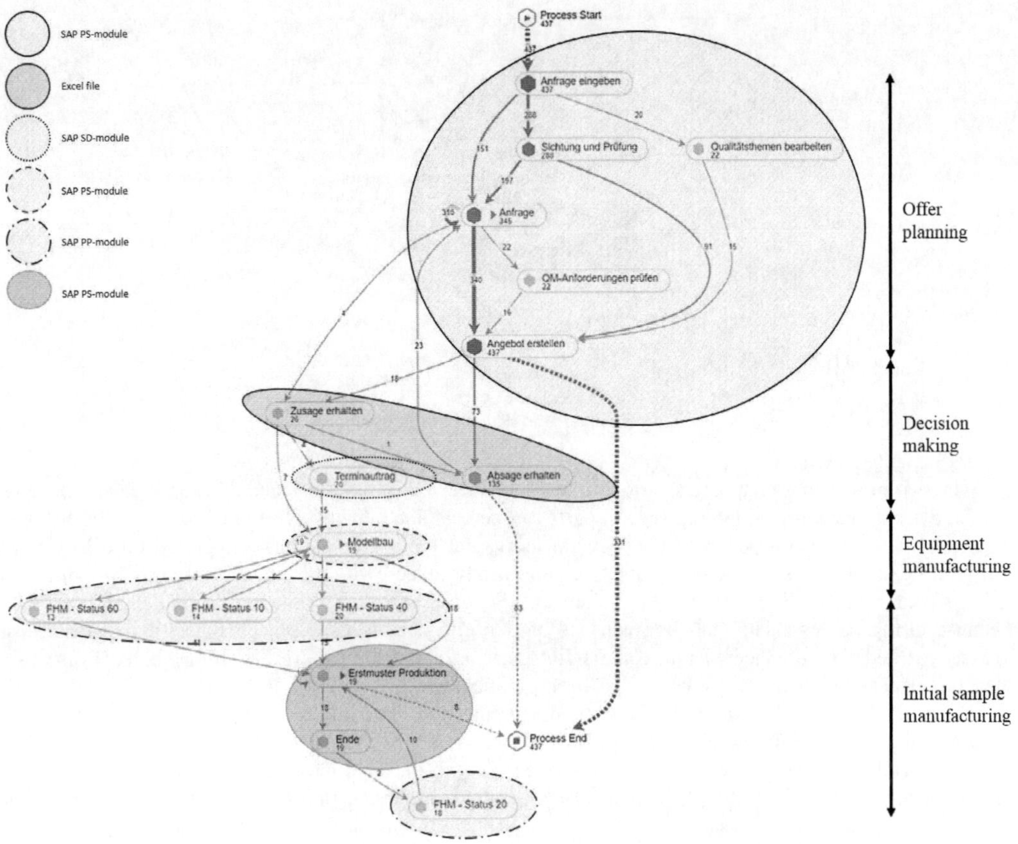

Figure 2. Schematic process flow generated by using process mining (Kiefer and Precht, 2019)

3.3 Combination of Production Factors

The goal of the second application of data science was the classification of different combinations of influencing parameters that lead to a finished casting. By classifying favorable and unfavorable pairings, favorable pairings should be systematically introduced to the production process and unfavorable pairings should systematically avoided.

Relevant business process:

In the foundry process flow, sand molds are produced with the help of patterns. Patterns are the positive of the finished component. In principle, a sand mold consists of a top box and a bottom box. At the same time, cores are made to reproduce internal contours or undercuts on the finished casting. The cores are inserted into the bottom box of the sand mold before the top box is placed above. Figure 3 shows the schematic structure of a molding box ready for casting. The molten iron is poured into the mold. Slow cooling of the liquid iron causes it to solidify. More information about the casting process can be find in (Beeley, 2001).

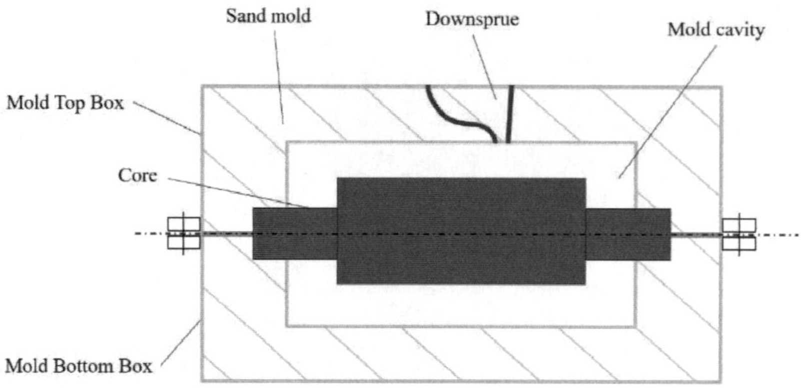

Figure 3. Schematic structure of a molding box

Influencing factors:

The two influencing parameters, core nest and model nest, are now considered in more detail. Depending on the dimensions and positioning, several parts can be cast simultaneously in one mold box. In order to be able to cast several parts per mold box, several individual models of the part are placed on a pattern plate (different pattern nests). At the same time, the cores are produced with the aid of a core box so that they can be inserted into the lower mold box later on. Several cores are produced simultaneously per one manufacturing process. The core box consists of multiple core nests. Each core nest and pattern nest is clearly marked. The finished casting contains information about the pattern nest it originated from (pattern nest number) as well as about the core it has been produced with.

Figure 4 visualizes the changes in sand mold and core over time due to wear on the model and core box. The dotted lines illustrate the development of the dimensions. These converge with increasing wear on the model and the core box. If the core has less and less play in the sand mold, it can move less and less in the mold during casting. This behavior should then become visible in the finished castings through a decreasing scatter of core-relevant dimensions.

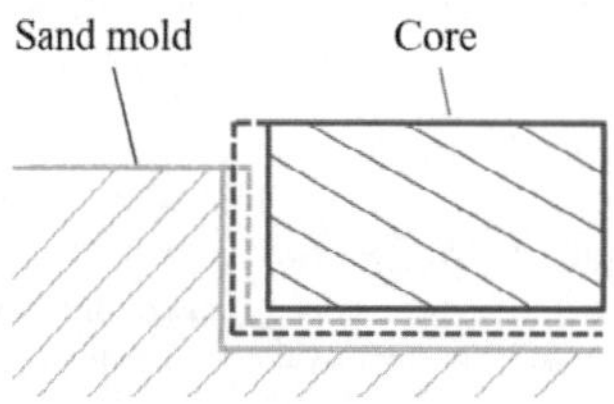

Figure 4. Wear effects on sand mold and core

The measurement data for determining this scatter comes from series production, where the finished castings are measured on a sample basis on a coordinate measuring machine. As portrayed in Figure 1, this measurement can result in the equipment (e.g. the pattern plate or the core box) having to be serviced. This happens, for example, when dimensional tolerance limits on the component are exceeded. The purpose of maintenance is to restore the equipment to its original or optimum condition.

Idea:

The dimensions of the sand mold and the core box converge. If there is no clearance between the sand mold and the core, stripping will occur. When the core is inserted, sand drifts off and lies at the bottom of the sand mold. During pouring, the loose sand is swept up and trapped in the finished casting.

By systematically identifying and classifying favorable (large clearance) and unfavorable (small clearance) pairings of pattern and core nests, with the help of data science methods, the service life of the pattern or core box can be extended by structurally inducing or preventing these pairings. This means that, in the optimum case, less maintenance of the equipment has to be carried out.

Implementation:

Measurement data of the finished castings serve as data basis. For each production batch, one casting per pattern nest is clamped onto the measuring machine. Defined points are then measured on this casting.

The scattering of core-relevant measurement data is then used to continuously classify the various combinations. The data analysis showed that it is not possible to make a statement or compare the different combinations on the basis of the scatter. Over a period of nine months, only a few measurements are available for the different combinations. An assessment of the current clearance between core and sand mold on the basis of the scatter of the measurements on the finished casting is not possible because the database is very limited. This also means that a comparison between the combinations is not possible.

Conclusion:

This use case cannot be used sensibly in the company. The database is very small compared to the number of different combinations. At the same time, data quality is limited by sometimes incomplete or ambiguous manual data entry, further limiting the usable database. No statement can be made about the quality of the different combinations, nor about whether there is a relevant difference between the combinations at all. What becomes clear here is that for the meaningful use of data science methods, the quality of the data basis is crucial.

This use case was initiated by a quality expert from the HMD in collaboration with an external IT expert with a scientific background. The IT side has to understand how the processes run in the company. The business experts from the company have to be able to objectively assess which IT methods and tools can do what, to what extent, and with what limits. For this, a deep mutual understanding is needed. At this point, it becomes clear that the application of data science in a company can only be used profitably if the competencies and knowledge from the various areas of data science (cp. Figure 1) are available and can be combined in a meaningful way.

4. RECOMMENDATIONS FOR ACTION

The experiences gathered by means of the presented use cases are used to formulate general recommendations for applying data science in manufacturing industry. Therefore, a separation into organizational aspects and technical ones is done in the following.

Today, many companies like HMD execute individual data science projects to gather experiences in this field. However, they often constitute isolated solutions that cannot prevail in the long term. To fully leverage the potentials of data science a fundamental data-driven culture needs to be anchored into the corporate strategy and all business processes. This requires the definition of business applications and their desired outcomes and the supply of required resources. Only then, comprehensive and long-term benefits can be derived that act self-reinforcing. Besides that, a successful implementation of the strategy is supported by interdisciplinary data science teams. They cover the required fields of expertise and collaborate with teams from the individual disciplines. This helps to overcome communication barriers between different subject matters and to disseminate basic data science knowledge within the company (Pricewaterhouse Coopers, 2020). The key task of this interdisciplinary team is to verify that a suitable data science method is picked regarding the desired outcome.

On the technical dimension the provision of data with the required degrees of quality and quantity is key. Both presented use cases at HMD have revealed major deficiencies in this regard. The current situation in industry is often characterized by insufficient amounts of data that would be required to extract the desired knowledge. Beyond that, existing datasets are usually spread over incoherent data sources that don't apply standardized data formats. Thus, an automatic processing isn't possible by default. The formatting, sorting, and merging of data as pre-task can require a disproportionate effort. Thus, a proper foundation has to be established first by providing the required amounts and maturity of the data and connecting the individual data sources (Kiefer and Precht, 2019). Furthermore, the datasets should include reference parameters so that its origin and semantic content is defined. For example, a raw dataset of temperature measurements is meaningless unless it is defined in which location the temperature has been measured and which measurement belongs to which product or batch. Each data science project should start with a review of these aspects that determines if the requirements are fulfilled or if substantial work has to be put in first.

5. SUMMARY AND OUTLOOK

This paper illustrates how data science methods can be usefully and profitably used in manufacturing industry taking the example of the foundry sector. Focus of these applications are production-related engineering and manufacturing processes with regard to the topics quality and process management.

The main research question of this paper is the investigation of potentials, limits and challenges of using data science methods to improve quality processes in operational practice. In terms of a socio-technical overall view, not only methodical issues but also information-technical and organizational aspects are discussed. The gained knowledge particularly results from studies of two processes: the 'offer to production process' and 'combination of production factors'. The collected findings ultimately lead to recommendations for action, which are intended to give companies a kind of framework for a meaningful economic use of data science methods in operational practice.

Although much progress has already been made in the field of data science and quality processes so far, the following topics remain to be addressed in further research activities:

- Development of a standards-based, semantic quality data and quality service model that transparently provides all relevant quality-related development, planning, production and customer data (and functions) to the identified applications within the quality process chain
- Development and use of effective and efficient AI algorithms for continuous planning and control of quality processes along the entire product life cycle

REFERENCES

Beeley, P., 2001. Foundry Technology. *Oxford*, Butterworth-Heinemann.

Berti, A. et al, 2019. Process Mining for Python (PM4Py): Bridging the Gap Between Process and Data Science. *In arXiv:1905.06169.*

EIT Digital, 2020. Overview on maturity of AI innovations in manufacturing. *Study.*

Grus, J., 2019. Data Science from Scratch. *O'Reilly Media, Inc.,* Second Edition.

Hand, D.J. et al, 2001. Principles of data mining. *In Cambridge, Mass: MIT Press.*

IFS, 2018. Künstliche Intelligenz und maschinelles Lernen. *Study.*

Kiefer, J. and Precht, M., 2019. Usage of process mining in the 'offer to production process' of a contract manufacturer for cast components. *Proceedings of the 4th International Conference on Big Data Analytics, Data Mining and Computational Intelligence – BIGDACI 2019.*

Kotu, V. and Bala, D., 2018. Data Science – Concepts and Practice. *Morgan Kaufmann,* Seond Edition.

Ozdemir, S. and Sunil, K., 2018. Principles of Data Science. *Packt Publishing*, Second Edition.

PAiCE Scientific Assistance, 2018. Potential of artificial intelligence in Germany's producing sector. *Study on behalf of the Federal Ministry for Economic Affairs and Energy (BMWi).*

Pricewaterhouse Coopers, 2020. An Introduction to Implementing AI in Manufacturing. *White Paper.*

Shi, H. et al, 2020. Construction of Industrial Internet of Things Based on MQTT and OPC UA Protocols. *In 2020 IEEE International Conference on Artificial Intelligence and Computer Applications (ICAICA).*

Van der Aalst, W., 2012. Process Mining Manifesto. *In Business Process Management Workshops*, Springer Berlin Heidelberg.

Van der Aalst, W., 2016. Process Mining: Data Science in Action. *In 2nd ed. Imprint: Springer*, doi: 10.1007/978-3-662-49851-4.

Wahlster, W., 2021. Ten Years of INDUSTRIE 4.0. *DFKI Interview with Prof. Dr. Wolfgang Wahlster, published in DFKI NEWS,* Issue 1/2021.

Zhang, W. et al, 2019. Data-Driven Methods for Predictive Maintenance of Industrial Equipment: A Survey. *IEEE Systems Journal 13,* pp. 2213-2227.

International Conferences ICT, Society, and Human Beings 2023;
and e-Health 2023; Connected Smart Cities 2023;
and Big Data Analytics, Data Mining and Computational Intelligence 2023

SOCIAL NETWORKS IN CREDIT SCORING: A MACHINE LEARNING APPROACH

Ahmad Abd Rabuh[1], Mark Xu[2] and Renatas Kizys[3]
[1]*Oxford Brookes Business School, Oxford Brookes University, Oxford, UK*
[2]*Faculty of Business and Law, University of Portsmouth, Portsmouth, UK*
[3]*Southampton Business School, University of Southampton, Southampton, UK*

ABSTRACT

This research examines if social network tie has an incremental predictive ability for borrower default in credit scoring and the precision of effect. With advanced digital technologies and increasing availability of non-financial behaviour big data, social network data has been explored to assess consumer credit scoring in research and practice. This research uses machine learning algorithms to analyse a large dataset (loan applications and defaults) obtained from a European lender. The results show that social network data, when working together with traditional financial data, improves predictive ability of borrowers' default. A Bayesian Analysis confirms the explanatory evidence of social ties of borrowers. This research generates insights of machine learning power in analyzing imbalanced large dataset using ten different classifiers, and contributes to the theoretical debate on social capital theory as well as practical guidance of using XGBoost algorithm for lenders.

KEYWORDS

Machine Learning, Predictive Analytics, Social Networks, Credit Scoring, Financial Inclusion

1. INTRODUCTION

Credit scoring has been evolving in terms of objectives, sources, modelling, and techniques. Particularly, with big data accessibility and the advances in digital technologies, e.g. intelligent systems like artificial intelligence (AI) and machine learning (ML) in data analysis. Objectives of lenders vary from market penetration, profit maximisation (Thomas, 2009) to more socially financial inclusion and equality (McKillop, et al. 2007; Wei, et al. 2015). Financial inclusion requires provision of financial access to high calibre with thin financial files such as students, refugees (Redrup, 2017), new entrants to the job market or homestayers coming back to workforce, immigrants with high skills, craftsmen working with cash-in-hand and many other talents (McEvoy and Chakraborty, 2014). For example, the Migration Observatory at the University of Oxford reported sharp increase in unemployment rates since 2020 among migrants in the UK (Fernandez-Reino & Rienzo, 2021). Finding ways to include those who cannot access credit will reduce social inequality (Wei et al., 2015). One of the challenges, for including more borrowers, is information asymmetry, which arises when one party has limited access to information that is kept by another contracted party (Yan, et al. 2015; Ackert and Deaves, 2009).

Traditional data represented 'hard' information (Lin et al., 2013) and was retrieved from borrowers' financial history at the time of application (Leow and Crook, 2016). Such data, like average balance, past credit performance in addition to demographics dominated credit scoring for quite some time, but alternative and non-traditional data including social data has been increasingly-used by innovative lenders (Lazarow, 2017). Meanwhile, models underpinning credit scoring vary from static and parametric to dynamic and non-parametric based (Chen et al., 2018). Machine learning models are capable of learning iteratively from training datasets to classify or predict risk level of prospectus borrowers, especially when a borrower has thin financial information and face unstable circumstances.

Research into the effectiveness of non-financial data modelling and the associated techniques (ML) for credit scoring is patchy. Particularly, when referring to social network data as one of the emerging non-financial data in credit, there is a lack of precision on how social network effects credit scoring.

This research aims to examine if social network data - specifically the default tie, has an incremental predictive ability for borrower default risk in credit scoring and the precision of such an effect. The remaining sections of this paper are structured as follows. Section2 reviews credit modelling, social networks and machine learning for credit scoring. The methodology is discussed in section 3 where the dataset preparation for ML and modelling are described. The results are presented with discussion in section 4. This is followed by a conclusion and proposed future research agenda.

2. LITERATURE REVIEW

2.1 Traditional Credit Scoring Models and Data

Credit scoring professionals traditionally relied on three categories of models to estimate targeted borrowers' abilities to repay their loans. Those are Probability of default (PD), Loss given default (LGD), and Exposure at default (EAD) (Tong, et al. 2016). According to Basel II committee's definition, a default happens when, at any point of time during the loan's tenure, 90 days' worth of instalments are in arrears within a 360-day period (Puri et al., 2017). The estimated PD, usually is set for a binary target - either default or repayment in which a reject or an accept decision to be made (Gordy, 2000). LGD estimates the amounts that the lender would lose in the event of default in an estimated point of time (Schuermann, 2004). EAD is to estimate the outstanding balance of an account at any point during the life of the loan, applying to revolving credit such as credit cards and overdraft facilities (Leow and Crook, 2016). In this research, PD is used as the target variable due to the nature of the available dataset.

A borrower's credit score is predicted by the lender and / or external credit referencing agencies (also refers to Credit rating authority/Bureaus) (Rajan, et al., 2010). Fair Isaac Corporation (FICO) is a typical credit score using financial data - with five components: (a) payment history (0.35), amounts owed (0.30), length of credit history (0.15), credit mix (0.10), and new credit (0.10). A FICO score ranges between 300 - 850 (Wei et al., 2015).

Data sources include banking transactions, shopping data from department stores, payments of energy bills to suppliers, county court judgements, and electoral registers (Banasik et al., 1999; Kshetri, 2016). However, data used for credit scoring has been dominantly financial data (Redrup, 2017). Financial data includes, but not limited to, the number of bank accounts and credit cards held, years with bank, employment category, residency type/ownership of home, credit history, debt-to-income ratio, average account balance, maximum credit payment, number of times over credit limit, average transactions value, number of cash withdrawals, credit limit changes (Leow and Crook, 2016; Freedman and Jin, 2017; Thomas, et al. 2017).

It is argued that, despite meeting the financial criteria, abrupt incidents such as divorce, job termination, diseases can lead to unexpected financial distresses, hence, undermines the trustworthy of the financial based credit scores (Lazarow, 2017). With the scalable powers of big data analytics, machine learning and data mining, lenders are able to incorporate various sources of data that are not bound by borrowers' accounts and financial records. The sources of behavioural data include e-commerce, psychometrics tests (Arráiz, et al. 2017), social media (Guo et al., 2016; Masyutin, 2015; Rusli, 2013; Weke and Ntwiga, 2016), web analytics, mobile phones' usage (Bjorkegren and Grissen, 2015), and social network circles (Lin et al., 2013). There is an increasing tendency of adopting such behaviour data by credit agencies, online lenders and P2P lending platforms.

2.2 Social Network Models and Theories

Social data has a relational aspect (Lin et al., 2013) where the borrower is affected and influenced by the connections within a network (Wei et al., 2015). Networks can be formed based on different social connections, hence come with different purposes, structures, strength and number of ties. When it comes to strength, the more two social networks overlap, the stronger the tie between the two individuals that the networks belong to (Granovetter, 1973). Lin, et al (2013) suggested types of social networks including alumni, geographical, military, medical, demographic, hobbies, business and religion social networks. Coleman (1988) suggests that, since persons are linked in more than one context (neighbour, fellow worker, fellow parent, coreligionist, etc.), the central property of a multiplex relation is that it allows the resources of one relationship to be appropriated for use in others. According to Currier (2020), persistent connections can shape up destinies of networked people. In social network science, an influential person has a high degree of centrality.

The assumption of using social network tie for credit scoring is that people tend to associate with others who share similar features when formulating social networks. The features include demographics such as gender, race, religion, nationality, or other discrete characteristics. On the other hand, the common features can be scalar characteristics such as age or income. The notion of 'homophily' indicates that people tend to form social ties with those that are like-minded (Newman, 2010). When it comes to credit, Wei et al. (2015) and

Freedman & Jin (2017) explained that individuals form social ties with others for a social utility and a posterior credit utility. Lin et al. (2013) tested data provided by prosper.com and confirmed that having friends makes it easier for borrowers to get funded in P2P lending while enjoying lower APRs. The scholars justified their findings with the term 'social stigma', which costs borrowers when their friends default.

From a theoretical perspective, the social capital theory provides a foundation to understand the value of social networks in credit. Coleman (1988) suggests that unlike other forms of capital, social capital inheres in the structure of relations between actors and among actors (i.e. the social network). Social capital impacts on human action in two ways - facilitate or constrain actions. For example, a group within which there is extensive trustworthiness and extensive trust is able to accomplish much more than a comparable group without that trustworthiness and trust. This explains why social networks is considered a source of credit worthiness. Key form of social capital includes obligations, expectations, and trustworthiness of structures, but it is the obligations that one person owes a second in relationship X, which the second person can use to constrain the actions of the first in relationship Y. Apparently, not all social relations bear financial obligations and not only dyadic relationships influence a behavior, but also the structure and the shape of a network. For example, 'herding' behavior was defined as the socio-economic decision-making process that is influenced by other's decisions and it can be seen particularly in ego-centric networks (Pokorna & Sponer, 2016)

How social network affects credit decision remains a practical and academic challenge. Among a few limited studies, Lin et al. (2013) suggest that having friends is considered a signal of good quality in borrowers. In practice, it is reported that online lender Lenddo considers the number of personal connections a borrower has, and ask those connections to endorse the borrower (Rusli, 2013; McEvoy and Chakraborty, 2014), but differentiates two social ties: friends vs. followers, which implies that not all network ties have relevance to credit decision. The lender also measures the time social media accounts have been active in order to estimate social ties strength. Nevertheless, how the social tie data is specifically affecting the Lenddo credit score, and if the social tie data is a complementarity (work with traditional financial data), or a sole determinant in credit decision remain unknown.

There are risks associated with using social networks for credit assessment, for example, social network members can deliberately falsify lenders (Wei, et al. 2015) by forming false association, because connections to individuals with high type signals (good borrowers) have an overall positive impact. Whereas individuals with low type signals (e.g. with default history) may be difficult to associate with or being endorsed by good borrowers, hence, exacerbate social exclusion problem. This is an important issue to pursue, however is beyond the scope of the current study.

In summary, despite a few limited studies on social networks' inference on credit scores, the results are largely inconclusive, and specifically lack of necessary precision as to the extent of social network tie impact. This gap is fulfilled by this study using a machine leaning approach on a large loan dataset that includes information about social ties of existing borrowers.

2.3 Machine Learning Models in Credit

The introduction of machine learning (ML) was prominent in last two decades. Dash et al. (2017) assert that machine learning and deep learning algorithms can make better and faster underwriting decisions in credit. Along similar lines, many researchers argued the superior suitability of ML models that would analyse and learn iteratively from both financial and non-financial data to predict credit scores (Turner & McBurnett, 2019; West, 2000).

The normal procedure in machine learning is to split the dataset into a large part (70-80%) of training set then test the model on the remaining (20-30%) holdout part (Han, Pei, & Kamber, 2011), using either supervised or unsupervised learning. A large literature review on ML models shows a tendency of using ML models, although traditional parametric models remain prevalent all the time including Discriminant Analysis (linear), Logistic Regression, Decision / Classification Trees, Linear Programming. On the non-parametric models, Neural Networks is dominating, this is followed (in popularity order) by K-Nearest Neighbors (West, 2000; Lee et al. 2002; Baesens et al. 2003; Chen and Li, 2010; Brown and Mues, 2012), Support Vector Machines (Bellotti and Crook, 2009; Huang, et al. 2007; Yang, 2007), Random Forests (Kruppa et al. 2013; Malekipirbazari and Aksakalli, 2015), Gradient Boosting (Fahner, 2018), Naïve Bayes (Kou and Wu, 2014), Genetic Algorithms (Huang et al. 2007).

The reported model accuracy from these studies has been moderately high in the range of 0.70-0.80 with some outliers. For probability default (PD) estimation, Yang (2007) suggests that logistic regression (LR) and linear discriminant analysis (LDA) have been the most commonly-used models, which supports the finding from our literature review. All the models mentioned above have been used to examine their suitability for the dataset, as well as their predictive ability as a PD ML module.

3. METHODOLOGY

In this research, we use a large dataset of loan by individual barrowers obtained from a European lender through an online repository. The dataset is large enough for the intent of this research. It contains a total of 307,511 borrowers data described in 120 variables representing the columns. In addition to that, the outcome of loans was available in binary setting – default or repaid. This enables supervised machine learning to be performed. The dataset covers dominantly financial and behavior data, with two columns of relevant social network data.

The process of the analysis is conducted in two stages - a) data preparation and b) fitting a machine learning model on the dataset for predictability.

3.1 The Dataset and Preparation

Within the total 307,511 distinct borrowers load data, 282,686 loans were repaid and the remaining 24,825 were not (default). The 120 columns cover wide range of attributes, which were classified into three main categories: financial, behavioural, and social attributes. For simplicity, behavioural and social data will be considered as non-traditional. Traditional attributes are those provided from the financial banking systems such as types of loans, which are classified into tenured loan with a maturity repayment date (e.g. car loan, mortgage, microloan loan, cash loan, etc.), and revolving loans indicating a credit limit that renews periodically whenever a borrower settles the outstanding (e.g. credit card).

In pre-processing, a number of steps were followed including data cleaning, data extraction, attribute construction, data aggregation, and loading (ETL). Missing values were treated by replacing with either mode, median or mean (based on variables' types and distributions) using the fillna() function in Python 3. Some observations with majority of their fields being missing were deleted. Also, attributes with many possible outcomes were aggregated in smaller number of categories. The dataset was loaded onto a virtual environment where a high processing unit that is set on a super computer node was used to run the models. Finally, the algorithm was run on a clean data made of 304,427 borrowers described in 48 variables. The new dataset included 279,767 repaid and 24,660 default cases. It is worthy to note the two social network attributes. One is a delinquent type – indicating friends of arears within 30 days and another defaulting type – network friends of actual default of the loan.

3.2 Machine Learning Models

Python 3 was used to perform machine learning on the full dataset. The functions that built our machine learning classifiers in the designed Python 3 algorithm are:

- *LabelEconder( )* to transform all categorical variables into numbers (encoding).
- *StandardScalar( )* to standardize all numeric variables within attributes.
- *fit( )* to initiate the model on the training sample set.
- *predict( )* to create a y_predict (y') object, which is compared to the corresponding y_test (actual loan outcome) to measure performance.
- *accuracy_score( )* to calculate each one of the classifiers' accuracy using the formula: (TP+TN)/(N_test)
- *log_loss()* is the function that computes the logarithmic loss which penalises false classifications during the iteration process of the machine learning classifier. The lower its value, the better. A log loss value of zero means that the model is perfectly classifying cases.
- *KNeighborsClassifier(), SVC(), NuSVC(), DecisionTreeClassifier(), RandomForestClassifier(), AdaBoostClassifier(), GradientBoostingClassifier(), GaussianNB(), LinearDiscriminantAnalysis(), QuadraticDiscriminantAnalysis()* are the machine learning classifiers. Algorithms trained on 70 per cent of the sample dataset and they were tested on the remaining 30 per cent.

To overcome the data imbalance class problem explained by Zięba et al. (2016), a balanced dataset is created to include all 24,660 default cases, and an equal sampled set from the class of repaying borrowers (24,660). This makes a total of 49,320 loans applications sample frame for the regress analysis. The target variable (loan outcome) was isolated in Y object. The X object contained different variations of attributes. The design is to examine the accuracy and precision of the bad social network tie impact on the default risks.

4. RESULTS AND DISCUSSION

The ML classification results are presented in the sub sections below.

4.1 Bayesian Analysis

Table 1 shows the default network ties and the associated metrics derived from posterior probability (Bayesian) analysis using proportionate figures.

Table 1. Profile of defaulting social tie from Bayesian analysis

Number of defaulting social ties	Repay	Default	Odds	Probability of repay	Probability of default	Information odds	Weight of evidence
0	248,523	20,989	11.84	0.92	0.08	1.04	0.04
1	25,276	2,862	8.83	0.90	0.10	0.78	- 0.25
2	4,667	616	7.58	0.88	0.12	0.67	- 0.40
3	1,032	150	6.88	0.87	0.13	0.61	- 0.50
4	214	34	6.29	0.86	0.14	0.55	- 0.59
5	48	8	6.00	0.86	0.14	0.53	- 0.64
6	7	1	7.00	0.87	0.13	0.62	- 0.48

The analysis indicates that a borrower, who has no defaulting ties (0) demonstrates 4% more evidence to being a good borrower. Conversely, a borrower with 1 - 6 defaulting social ties indicate a negative repayment effect.

4.2 Machine Learning Results

ML algorithms are run on both datasets (financial and full dataset). The results below (see Table 2) highlight the performance of the ten classifiers where those were measured by both accuracy and logarithmic (log) loss.

Table 2. Predictability of Machine Learning Classifier on PD

	Financial		Full dataset		Difference
ML Classifier	Accuracy	Log Loss	Accuracy	Log Loss	In accuracy
K-Nearest Neighbors	0.607	4.408	0.591	4.451	-0.016
Support Vector Machine	0.664	0.614	0.668	0.609	0.004
Nu-Support Vector Machine	0.613	0.673	0.636	0.643	0.023
Decision Trees	0.581	14.46	0.583	14.405	0.002
Random Forests	0.634	0.952	0.636	0.824	0.002
AdaBoost	0.668	0.689	0.673	0.689	0.005
Extreme Gradient (XGBoost)	0.676	0.604	0.680	0.599	0.004
Naïve Bayes (Gaussian)	0.604	0.914	0.635	1.316	**0.031**
Linear Discriminant Analysis	0.667	0.613	0.677	0.604	0.01
Quadratic Discriminant Analysis	0.583	0.907	0.535	10.159	-0.048

4.3 Discussion

The key points emerged from the findings are summarized below:

i. The results confirm that the ten ML classifiers yields average over 0.6 accuracy in predicting PD, with behavior and social network tie data, the accuracy improves particularly for Naïve Bayes, Nu-Support Vector Machine, AdaBoost, Extreme Gradient (XGBoost) models.

ii. The Naïve Bayesian machine learning algorithm improved significantly when adding social variables, which is consistent with the initial Bayesian analysis done in Table 1. That analysis demonstrated the weight of evidence added to the loan outcome when knowing in advanced the number of defaulting social ties a borrower has.

iii. The Extreme Gradient Boosting (XGBoost) algorithm achieved the highest accuracy while suffering the lowest log loss confirming that the said algorithm predicts the binary targets closest to their real values.

iv. It is believed that XGBoost algorithm outperformed other classifiers because, as heterogeneous behavioural data accumulate, a fast and parallel processing is needed. This is consistent with XGBoost with its ability to build decision trees (that are the basis for a simple credit scoring method) in a series of parallel analyses while minimizing its objective function (Dastile et al., 2020).

The findings of this study partially confirm the social capital theory that the default social tie does have a negative impact on other's credit scoring, other side of the social capital theory - the scalar homophily and its positive association with credit risks has not been examined in this study due to the aforementioned limitation of the dataset. We argue that the specific network ties pose impact on one's credit score. Whether or not the network size matter in effecting other's credit scoring deserve further investigation.

The machine learning approaches experienced in this study proves that the ML classification algorithms can uncover the potential trustworthiness of borrowers with thin financial files. Nevertheless, this research raises more questions than offering specific answer, for example:

1) Given the social networks data role in credit scoring, what types of social ties should be collected and how?
2) How such social network data shall be collected with necessary consideration of privacy, ethics and governance issues?
3) How to monitor and track the changing dynamics of such social network ties, so as to ensure a true dynamic credit assessment? i.e. moving from looking backwards of the history/current to forward prediction?
4) How to ensure healthy or defaulting network ties not exacerbate social exclusion and digital gaps in societies?
5) How to detect/ eliminate / counter fight social networks behaviour that intentionally falsify lenders?

5. CONCLUSION

It is concluded that default social tie has a negative influence on borrower's credit scoring, but the size seems not significant. Machine learning models overall are effective in analyzing and predicting probability default, but the accuracy and precision can vary from one model to another, this depends on the nature and classes of the dataset. The social capital theory on homophily and obligations within social relations need further examination and validation.

Analysing borrowers' social circles and network types may not be useful in classifying default risk, unless as revealed in this study, the very specific type of network tie – either with default association, or financial homophily can be identified. The main challenges, however, are to determine the relevant criteria of the social networks by the lenders and credit agencies, and overcome privacy and ethical concerns raised by regulators and data subjects. There are also needs to examine changes in social networks over time as well as the network attributes – the network nodes, edges, centrality, size, distance, layers, etc.. This can advance not only credit scoring research, but also the areas of customer relationship management and business intelligence. This study raised an important question addressing financial exclusion problem if social networks data are not fairly used by lenders and effectively governed for social equality and inclusion.

REFERENCES

Ackert, L., & Deaves, R. (2009). *Behavioral finance: Psychology, decision-making, and markets.* Cengage Learning.

Arráiz, I., Bruhn, M., & Stucchi, R. (2017). Psychometrics as a tool to improve credit information. *The World Bank Economic Review, 30* (Supplement_1), S67-S76.

Banasik, J., Crook, J. N., & Thomas, L. C. (1999). Not if but when will borrowers default. *Journal of the Operational Research Society*, 1185-1190.

Bjorkegren, D., & Grissen, D. (2015). Behavior revealed in mobile phone usage predicts loan repayment. *Social Science Research Network (SSRN), 2611775*, 1-10.

Chen, C., Lin, K., Rudin, C., Shaposhnik, Y., Wang, S., & Wang, T. (2018). An interpretable model with globally consistent explanations for credit risk. arXiv preprint arXiv:1811.12615.

Coleman, James S. (1988) Social Capital in the Creation of Human Capital. *The American Journal of Sociology*, Vol. 94, Supplement: Organizations and Institutions: Sociological and Economic Approaches to the Analysis of Social Structure (1988), pp. 95-120 URL: http://www.jstor.org/stable/2780243

Currier, J. (2020). Revealing the Unseen Forces That Guide Your Life. Your Life and Network Effects. Retrieved from https://www.nfx.com/post/your-life-network-effects/

Dastile, X., Celik, T., & Potsane, M. (2020). Statistical and machine learning models in credit scoring: A systematic literature survey. *Applied Soft Computing*, *91*, 106263.

Fernandez-Reino, M., & Rienzo, C. (2021). *Migrants in the UK labour market: An overview*. Migration Observatory-University of Oxford.

Freedman, S., & Jin, G. Z. (2017). The information value of online social networks: lessons from peer-to-peer lending. *International Journal of Industrial Organization*, 51, 185-222.

Gordy, M. B. (2000). A comparative anatomy of credit risk models. *Journal of Banking & Finance*, 24(1-2), 119-149.

Granovetter, M. S. (1973). The strength of weak ties. *American Journal of Sociology*, *78*(6), 1360-1380.

Guo, G., Zhu, F., Chen, E., Liu, Q., Wu, L., & Guan, C. (2016). From footprint to evidence: an exploratory study of mining social data for credit scoring. *ACM Transactions on the Web* (TWEB), 10(4), 1-38.

Han, J., Pei, J., & Kamber, M. (2011). *Data mining: concepts and techniques*: Elsevier.

Lazarow, A. (2017). What do emerging market consumers expect from InsureTech? Retrieved from https://medium.com/omidyar-network/what-do-emerging-market-consumers-expect-from-insuretech-e439b894b2c5

Leow, M., & Crook, J. (2016). A new Mixture model for the estimation of credit card Exposure at Default. *European Journal of Operational Research*, 249(2), 487-497.

Lin, M., Prabhala, N. R., & Viswanathan, S. (2013). Judging borrowers by the company they keep: Friendship networks and information asymmetry in online peer-to-peer lending. *Management Science*, 59(1), 17-35.

Masyutin, A. (2015). Credit scoring based on social network data. *Бизнес-информатика*(3 (33)).

McEvoy, M. J., & Chakraborty, T. (2014). Enabling financial inclusion through "alternative data". *Mastercard Advisors: Bentonville, AR, USA*.

McKillop, D. G., Ward, A.-M., & Wilson, J. O. (2007). The development of credit unions and their role in tackling financial exclusion. *Public money and management*, 27(1), 37-44.

Newman, M. (2010). *Networks: an introduction*: Oxford University Press, Oxford.

Pokorná, M., & Sponer, M. (2016). Social Lending and Its Risks. *Procedia - Social and Behavioral Sciences*, 220, 330-337. doi:10.1016/j.sbspro.2016.05.506

Puri, M., Rocholl, J., & Steffen, S. (2017). What do a million observations have to say about loan defaults? Opening the black box of relationships. *Journal of Financial Intermediation*, 31, 1-15.

Redrup, Y. (2017). How email and smartphone data could help you get a loan. Technology. Retrieved from http://www.afr.com/technology/how-email-and-smartphone-data-could-help-you-get-a-loan-20171212-h02zi0#ixzz534zFfQmg

Rusli, E. M. (2013). Bad credit? Start tweeting. *WALL ST. J.*, Apr, 1.

Schuermann, T. (2004). What do we know about loss given default?

Sullivan L. (2022) Nonparametric Tests. Retrieved 2022 from http://sphweb.bumc.bu.edu/otlt/MPH-Modules/BS/BS704_Nonparametric/BS704_Nonparametric_print.html

Thomas, L., Banasik, J., & Crook, J. (2001). Recalibrating scorecards. *Journal of the Operational Research Society*, 52(9), 981-988.

Thomas, L. C. (2009). *Consumer Credit Models: Pricing, Profit and Portfolios.* OUP Oxford.

Thomas, L. C., Edelman, D. B., & Crook, J. N. (2017). *Credit scoring and its applications (Second edition. ed.)*: Society for Industrial and Applied Mathematics.

Tong, E. N., Mues, C., Brown, I., & Thomas, L. C. (2016). Exposure at default models with and without the credit conversion factor. *European Journal of Operational Research*, 252(3), 910-920.

Turner, M., & McBurnett, M. (2019). Optimizing neural networks for risk assessment: Google Patents.

Wei, Y., Yildirim, P., Van den Bulte, C., & Dellarocas, C. (2015). Credit scoring with social network data. *Marketing Science*, 35(2), 234-258.

West, D. (2000). Neural network credit scoring models. *Computers & Operations Research*, 27(11), 1131-1152.

What's in my FICO scores. Retrieved from http://www.myfico.com/credit-education/whats-in-your-credit-score/

Yan, J., Yu, W., & Zhao, J. L. (2015). How signaling and search costs affect information asymmetry in P2P lending: the economics of big data. *Financial Innovation*, 1(1), 1.

SHORT PAPERS

EXPLORING THE POTENTIAL OF MUSICAL ACTIVITY SYSTEMS FOR PEOPLE WITH DEMENTIA AND THEIR CAREGIVERS

Kevin C. Tseng[1,2], Chieh-Yun Liu[2] and Tsai-Husan Tsai[3]
[1]*Department of Industrial Design, National Taipei University of Technology, Taipei, Taiwan*
[2]*Product Design and Development Laboratory, Taoyuan, Taiwan*
[3]*Master of Science Degree Program in Innovation for Smart Medicine, Chang Gung University, Taiwan*

ABSTRACT

As our population ages, more and more people develop dementia, which is a great concern for families and communities. Music has been found to be useful in relieving symptoms of dementia and in promoting interaction between patients, caregivers, and family members. However, music therapy requires a music therapist to plan and prepare equipment that is tailored to the preferences and needs of each patient. It can also be difficult to continue musical activities outside therapy sessions due to the lack of musical experience between patients and caregivers. To solve this problem, a study was conducted to investigate how caregivers can conduct musical activities using equipment commonly used in music therapy. A literature review was used to understand current issues affecting people with dementia and approaches to music therapy.

KEYWORDS

Dementia, Elderly, Human-Computer Interaction, Music Therapy

1. INTRODUCTION

The trend of an ageing society has caused a gradually rising number of dementia patients and has become a significant concern. Typically, the symptoms of dementia are divided into mild, moderate, and severe stages according to the degree of aggravation of the symptoms. The progression of dementia is irreversible. In other words, instead of fully recovering patients, treatment could only slow the aggravation of the symptoms. Patients may eventually need to depend entirely on caregivers to live. For dementia patients themselves, as well as their caregivers, this situation is undoubtedly a substantial burden (Fearnley, 2009).

Dementia is commonly treated through pharmacologic therapy and non-pharmacologic therapy. Music therapy, reminiscence therapy, and art therapy are standard non-pharmacologic therapies (Hulme et al, 2010). Through different interventions, non-pharmacologic therapy allows dementia patients to participate in activities to improve their symptoms, in which the music intervention significantly impacts patients with dementia (Riley et al, 2009). Furthermore, patients with dementia, even if their memory has been lost, can still remember the music to which they have listened (Cuddy & Duffin, 2005). Clinical practice has also proven that music intervention can effectively relieve emotional symptoms in patients with dementia (Hulme et al, 2010). Therefore, this study aims to explore the potential of music therapy activity systems for people with dementia and their caregivers.

2. CURRENT MUSIC THERAPY FOR DEMENTIA PATIENTS

Music therapy is a specialised approach that involves a trained music therapist working with people with dementia. The focus of therapy is to utilise music as the primary intervention to alleviate behavioural symptoms by fostering interaction between the therapist and the patient. The music therapist customises a series of musical activities to meet the unique needs of each patient (Park & Specht, 2009; Sung et al, 2010), which may include selecting appropriate songs and instruments. It is important to note that the goal of music therapy is not to improve the patient's musical abilities, but rather to enhance their emotional and cognitive well-being (Guetin et al, 2009; Hannemann, 2006). In addition to personalised therapy content, caregivers are encouraged to participate in musical activities, which can strengthen the caregiver-patient relationship and provide opportunities for increased interaction with the patient (Hanser et al, 2011). The music therapist serves as a planner and advisor during the activities, and conducts ongoing assessments to ensure that the therapy remains tailored to the patient's needs. Overall, music therapy is a valuable intervention that can improve the quality of life for individuals with dementia, and promote positive outcomes for both patients and caregivers (Cheston & Bender, 1999). This section presents an overview of music therapy for dementia, followed by a discussion of the related research on persuasive technology and musical activities for dementia.

2.1 Personalised Music Therapy

A music therapist utilises music as an intervention strategy by creating personalised therapy programs and activities tailored to each patient's specific clinical needs. The program's content may vary depending on the patient's unique case, and the therapist will assess the patient's condition before developing an individual or group therapy program. During the process, the therapist focuses on the interaction between themselves and the patient or between different patients. Since each patient has diverse backgrounds and musical preferences, the music therapist needs to learn about their symptoms, experiences, preferred music, and choice of instruments to design the most appropriate therapy program (Sung et al, 2010; Sung et al, 2012).

2.2 Selecting the Musical Activities Following the Symptoms of the Patients

Dementia is associated with many symptoms. Not every activity suits all patients, and not every activity can interest all patients. Thus, tailor-made music therapy is valued. The most common music therapy activity is conducted by "completing a song" with the therapist. The patients can participate by singing, playing a musical instrument, and other ways (Bruer et al, 2007; Lin et al, 2011; Raglio et al, 2008; Svansdottir & Snaedal, 2006; Takahashi & Matsushita, 2006). In addition, passive music therapy activities can effectively relieve the emotional and behavioural problems of the patients, and active music therapy can enhance the interaction between the patients and their caregivers.

2.3 Selecting the Songs Following the Experience and Preference of the Patients

Music therapy, like music for most individuals, must be tailored to the patient's interests. Due to varying backgrounds, patients may have different musical preferences. Furthermore, dementia patients' musical memory differs from their general memory; they may recall songs from their youth despite forgetting everything else (Cuddy & Duffin, 2005). As such, therapists choose music tracks based on the patients' backgrounds and preferences to elicit emotional responses and alleviate symptoms effectively (Sung et al, 2010). Additionally, therapists may select occasional tracks to create a celebratory atmosphere during treatment, potentially eliciting patient responses (Takahashi & Matsushita, 2006). If the music cannot induce responsive feelings and behavioural responses in patients, the music therapy activity loses its significance (Gerdner, 2010).

2.4 Selecting the Musical Instrument Following the Ability and Preference of the Patients

During the process of music therapy, active playing by the patients often requires the use of various musical instruments. For patients gradually losing abilities in everyday life, independently selecting a musical instrument can enhance their self-confidence (Hannemann, 2006). Additionally, by providing various musical instruments, the music therapist can instruct the patients to exchange in the group, allowing more social behaviour for the patients with a heavy sense of isolation and indifference and independent selection by the dementia patients (Gerdner, 2010). Preparing many musical instruments might be a good idea for the patients receiving group treatment or day care at a community centre. However, the music therapists or caregivers most likely need to provide all the musical instruments for patients receiving individual treatment. Considering the diversity of the musical instruments required for the music therapy activities, an inadequate budget and limited storage space for the musical instruments are other concerns for the therapists and caregivers.

2.5 Musical Activities for Dementia Patients led by Caregivers

Caregivers can assist in improving the quality of life for dementia patients by incorporating music into their daily routines, as recommended by music therapists (Götell et al, 2009). This can include participating in joint musical activities with patients or leading music-related activities themselves, such as passive listening to music using a music player (Clair & Memmott, 2008). However, to effectively engage individuals with dementia, it is recommended to involve loved ones in the musical activities. Although a book titled "Connect through Music with People with Dementia" provides recommendations for musical tracks and conducting sessions, caregivers may not have the necessary musical background and ability to plan appropriate activities (Rio, 2009). Therefore, it is recommended that only professional music therapists should be responsible for planning and designing musical activities for individual cases. Caregivers can still participate and select items to execute according to their ability levels, but leading patient musical activities may pose challenges for non-professionals (Hanser et al, 2011).

3. DEVELOPMENT OF TOOLS RELATED TO MUSIC THERAPY FOR PATIENTS WITH DEMENTIA

In the previous section, we learned about the potential problems that may arise during music therapy sessions. Therefore, some researchers have tried to improve the above issues through tool design. The design can be summarised through social interaction and collaborative music creation. Music-social interaction is a therapeutic tool developed for patients with dementia, such as the Computer Interactive Reminiscence and Conversation Aid (CIRCA), designed for two people (the patient with dementia and the caregiver) to improve the patient's difficulties in operating the device alone. While using the device, the patient must make choices, press the execute button, and communicate with the caregiver.

In contrast, the caregiver must prompt the patient's actions, continue to communicate with the patient, and provide the patient and caregiver with a common topic for discussion (Astell et al, 2010). Another example is the Intelligent Socially Assistive Robot, designed to stimulate, accompany, and increase the patient's attention and interaction during music therapy games (Tapus et al, 2009). Music collaborative creation is a therapeutic tool developed for patients with dementia; for example, the music creation tool is designed to compensate for patients with different musical backgrounds and levels, enabling them to achieve music creation and enjoy a sense of achievement during the process (Riley et al, 2009). Another example is the MINWii, designed as a musical game to increase the patient's sense of achievement by completing increasingly difficult levels (Benveniste et al, 2010). The above tools related to music therapy have their focus in the assessment. The assessment's method, process and results are shown in Table 1, which describes the contents of each evaluation.

Table 1. Music-related tools developed for patients with dementia

Name of the tool	Literature	Assessment method/process	Results
CIRCA	Astell et al. (2010)	Observing the verbal and nonverbal performance of the patients	The sociability and satisfaction of the users were significantly positive.
Music creation tool	Riley et al. (2009)	Questionnaires/interviews/observation	The satisfaction and ease of the test used by the users were significantly positive.
Socially assistive robot	Tapus et al. (2009)	Rate of error/level of difficulty	The result of the usability test was significantly positive.
MINWii	Benveniste et al. (2010)	Rate of error/speed	The result of the usability test was significantly positive.

In addition to researchers recording and interpreting participants' behaviour, questionnaires can be used to investigate their thoughts and suggestions (Riley et al, 2009). There has been a growing number of tools developed for music therapy. One such tool is the musical instrument designed for creating and listening to music during treatment, and another is the integrated assessment scale for music therapy assessment. Both aim to enhance the effectiveness of music therapy activities for dementia patients, enabling them to obtain maximum benefit without pressure.

4. CONCLUSION

This study explores how musical activity systems can benefit patients with dementia. By analysing existing literature on music therapy activities and assessing user needs, a systematic review was conducted to identify potential musical activities. Both passive and active musical activities were examined to provide a variety of multisensory stimulation options. The results of the study can be used to propose a new musical activity system that allows dementia patients and their caregivers to participate in music therapy activities together, following a plan created by a music therapist. This approach makes music therapy more accessible and reduces the preparation time and required musical ability.

ACKNOWLEDGEMENT

The author(s) disclosed receipt of the following financial support for the research, authorship, and/or publication of this article: This work was supported in part by the National Science and Technology Council under grant number NSTC 111-2410-H-027-021-MY3, by the Ministry of Science and Technology under grant numbers MOST 109-2410-H-027-003-MY2, MOST 108-2410-H-027-024-MY3, and MOST 106-2628-H-027-001-MY3, and by Chang Gung Memorial Hospital and National Taipei University of Technology joint research program under grant number NTUT-CGMH-110-05, NTUT-CGMH-112-02 and CORPG3L0141. The funders had no role in study design, data collection and analysis, the decision to publish, or the preparation of the manuscript.

REFERENCES

Astell, A. J., Ellis, M. P., Bernardi, L., Alm, N., Dye, R., Gowans, G. & Campbell, J. (2010) Using a touch screen computer to support relationships between people with dementia and caregivers. *Interacting with Computers*, 22(4), 267-275.

Benveniste, S., Jouvelot, P. & Péquignot, R. (2010) The MINWii project: renarcissization of patients suffering from alzheimer's disease through video game-based music therapy, in Yang, H., Malaka, R., Hoshino, J. & Han, J. (eds), *Entertainment Computing*. Lecture Notes in Computer ScienceSpringer Berlin / Heidelberg, 79-90.

Bruer, R. A., Spitznagel, E. & Cloninger, C. R. (2007) The temporal limits of cognitive change from music therapy in elderly persons with dementia or dementia-like cognitive impairment: a randomised controlled trial. *Journal of Music Therapy*, 44(4), 308-28.

Cheston, R. & Bender, M. (1999) *Understanding Dementia: The Man with the Worried Eyes*. Scotland: Jessica Kingsley Publishers.

Clair, A. A. & Memmott, J. (2008) *Therapeutic Uses of Music with Older Adults*. Maryland: American Music Therapy Association.

Cuddy, L. L. & Duffin, J. (2005) Music, memory, and Alzheimer's disease: is music recognition spared in dementia, and how can it be assessed? *Medical Hypotheses*, 64(2), 229-235.

Fearnley, K. (2009) *Understanding Dementia: A Guide for Young People*. Scotland: NHS Health Scotland.

Gerdner, L. A. (2010) Evidence-based guideline. Individualised music for elders with dementia. *Journal of Gerontological Nursing*, 36(6), 7-15.

Götell, E., Brown, S. & Ekman, S. L. (2009) The influence of caregiver singing and background music on vocally expressed emotions and moods in dementia care. *International Journal of Nursing Studies*, 46(4), 422-430.

Guetin, S., Portet, F., Picot, M. C., Pommie, C., Messaoudi, M., Djabelkir, L., Olsen, A. L., Cano, M. M., Lecourt, E. & Touchon, J. (2009) Effect of music therapy on anxiety and depression in patients with Alzheimer's type Dementia: randomised, controlled Sstudy. *Dementia and Geriatric Cognitive Disorders*, 28(1), 36-46.

Hannemann, B. T. (2006) Creativity with dementia patients. Can creativity and art stimulate dementia patients positively? *Gerontology*, 52(1), 59-65.

Hanser, S. B., Butterfield-Whitcomb, J., Kawata, M. & Collins, B. E. (2011) Home-based music strategies with individuals who have dementia and their family caregivers. *Journal of Music Therapy*, 48(1), 2-27.

Hulme, C., Wright, J., Crocker, T., Oluboyede, Y. & House, A. (2010) Non-pharmacological approaches for dementia that informal carers might try or access: a systematic review. *International Journal of Geriatric Psychiatry*, 25(7), 756-763.

Lin, Y., Chu, H., Yang, C. Y., Chen, C. H., Chen, S. G., Chang, H. J., Hsieh, C. J. & Chou, K. R. (2011) Effectiveness of group music intervention against agitated behavior in elderly persons with dementia. *International Journal of Geriatric Psychiatry*, 26(7), 670-678.

Park, H. & Specht, J. K. P. (2009) Effect of individualised music on agitation in individuals with dementia who live at home. *Journal of Gerontological Nursing*, 35(8), 47-55.

Raglio, A., Bellelli, G., Traficante, D., Gianotti, M., Ubezio, M. C., Villani, D. & Trabucchi, M. (2008) Efficacy of music therapy in the treatment of behavioral and psychiatric symptoms of dementia. *Alzheimer Disease and Associated Disorders*, 22(2), 158-162.

Riley, P., Alm, N. & Newell, A. (2009) An interactive tool to promote musical creativity in people with dementia. *Computers in Human Behavior*, 25(3), 599-608.

Rio, R. (2009) *Connecting through Music with People with Dementia: A Guide for Caregivers*Jessica Kingsley Pub.

Sung, H. C., Chang, A. M. & Lee, W. L. (2010) A preferred music listening intervention to reduce anxiety in older adults with dementia in nursing homes. *Journal of Clinical Nursing*, 19(7-8), 1056-1064.

Sung, H. C., Lee, W. l., Li, T. L. & Watson, R. (2012) A group music intervention using percussion instruments with familiar music to reduce anxiety and agitation of institutionalised older adults with dementia. *International Journal of Geriatric Psychiatry*, 27(6), 621-627.

Svansdottir, H. B. & Snaedal, J. J. (2006) Music therapy in moderate and severe dementia of Alzheimer's type: a case-control study. *International Psychogeriatrics*, 18(4), 613-621.

Takahashi, T. & Matsushita, H. (2006) Long-term effects of music therapy on elderly with moderate/severe dementia. *Journal of Music Therapy*, 43(4), 317-333.

Tapus, A., Tapus, C. & Mataric, M. (2009) The role of physical embodiment of a therapist robot for individuals with cognitive impairments, *The 18th IEEE International Symposium on Robot and Human Interactive Communication*. Toyama, September 27 - October 2. IEEE.

SUSTAINABLE TRANSPORT SYSTEM AS A BASIS IN THE FUNCTIONAL ASPECT OF SUSTAINABLE MOBILITY

Natalya Shramenko[1] and Christoph Hupfer[2]
[1]*Prof. Dr. Sci. (Engin.)*
[2]*Prof. Dr.-Ing.*
Baden-Württemberg Institute of Sustainable Mobility,
Hochschule Karlsruhe University of Applied Sciences, Germany

ABSTRACT

In the conditions of modern development of cities and urban agglomerations, there is an increase in the level of motorization, which leads to problems of congestion, noise, an increase in greenhouse gas emissions into the atmosphere, as well as the need for additional urban space and an increase in road capacity. The analysis of studies and official sources made it possible to conclude that the environment, social well-being and economic development of cities and urban agglomerations are closely related to the development of sustainable mobility. In sustainable mobility, a large role is given to the functioning of sustainable transport systems that are focused on the individual needs of people, guarantee comfort, accessibility, safety and operate on the principles of environmental friendliness, taking into account the interests of future generations. Thus, a socio-ecological transformation of transport systems is necessary, as well as changes in the planning, organization and management of transport processes that will determine the quality of life of citizens, their comfort and safety. A sustainable transport system is considered by us as the main component in the functional aspect of sustainable mobility. A comparative assessment of the development of transport systems with the traditional approach and the development of sustainable transport systems has been carried out. It has been determined that sustainable transport systems focus on a human-centered approach, are based on the priority development of public transport, and also involve the use of urban space focused on the preferences and needs of residents.

KEYWORDS

Sustainable Mobility, Sustainable Transport System, Climate Change, Greenhouse Gas, Urban Agglomeration, Transport

1. INTRODUCTION

The 2030 Agenda for Sustainable Development, which was adopted at the UN World Summit (United Nations, 2015), defines 17 sustainable development goals that require a rethinking of the worldview for the further development of society in the context of global problems and crises. To achieve Goal 11, it is necessary to ensure the access of the population to safe, inclusive and accessible, green and public spaces (paragraph 11.6), as well as to safe and sustainable transport systems (paragraph 11.2), in which priority development is focused on public transport, its suitability for people all categories.

The problem of climate change on Earth, which is associated with environmental pollution and the release of harmful substances into the atmosphere, is a global problem for mankind. The reduction of greenhouse gas emissions can contribute to the solution of the problem of climate change.

The Intergovernmental Panel on Climate Change (IPCC) report emphasized that «rethinking how cities and other urban areas function in future could help significantly in mitigating the worst effects of climate change. The right policies, infrastructure and technology…to enable changes to our lifestyles and behaviour, can result in a 40 to 70 per cent reduction in greenhouse gas emissions by 2050. "The evidence also shows that these lifestyle changes can improve our health and wellbeing." (IPCC, 2022).

In order to achieve the strategic goals of sustainable development and mitigate the consequences of global problems, it is necessary to rethink the traditional approaches to planning, organizing and developing cities and urban agglomerations, as well as the transition to the formation of sustainable systems and the development of sustainable mobility, taking into account the changing preferences of people in the context of the current level of digitalization and the development of information and communication technologies (Shramenko & Hupfer, 2023).

Urban mobility involves taking into account the individual needs of people in movement and is associated with the functioning of transport systems, so the article focuses on the development of sustainable transport systems and sustainable mobility.

The purpose of the study is to identify key aspects of the functioning of a sustainable transport system.

The theoretical and methodological basis of research are general scientific methods: a systematic approach, analogy, generalization, analysis and synthesis.

2. LITERATURE REVIEW

The Green Paper (1992) proposed the term "sustainable mobility" and assesses the overall impact of transport on the environment and presents a general strategy for "sustainable mobility" that should allow transport to fulfill its economic and social role while limiting its harmful impacts. on the environment (European Commission, 1992).

The UN report (April 2022) notes that «an increasing share of emissions can be attributed to towns and cities» (United Nations, 2022). At the same time, urban resilience focuses not only on the resilience of metropolitan areas, but also includes cities and small residential areas (Meyer & Auriacombe, 2019).

The transport policy of the European Union is focused on the development of sustainable mobility, which involves taking into account the individual needs of people in movement and taking into account the interests of future generations.

In the EU, transport accounts for almost a quarter of greenhouse gas emissions and is the main cause of urban air pollution. Achieving climate neutrality by 2050 will require a 90% reduction in total greenhouse gas emissions from transport compared to 1990.

The largest CO2 emissions in the world are in the energy sector 41.7% and transport 16.2% (Ritchie, n.d.). Among all modes of transport, the largest share of CO2 emissions in the world is accounted for by road transport (11.9%), which includes all types of road transport (cars, trucks, motorcycles and buses). Thus, the transport sector is the largest consumer of energy and plays a significant role in global greenhouse gas emissions. In this regard, in order to reduce harmful emissions, significant attention should be paid to the development of sustainable transport systems.

The main transport policies implemented by the European Union and the United States, as well as an analysis of the concept of sustainable mobility, are presented in the study (Gallo & Marinelli, 2020). It is emphasized that strategies aimed at achieving the goal of sustainable mobility through the production/use of less polluting transport systems are of fundamental importance, but this should not be limited. Consideration should be given to environmental aspects, which are of paramount importance, as well as social and economic implications.

Socio-ecological transformation, which is aimed at improving the level and quality of life of the population, as well as the implementation of innovative processes in the development of sustainable mobility, is possible with comprehensive political support and decisive actions from administrative structures (Shramenko & Hupfer, 2023).

The experience, creative ideas and desires of citizens, as well as their involvement and participation in decisions to develop livable cities, can serve as the basis for making progress in changing the behavior of citizens when choosing a mode of movement (Häußler et al., 2019).

The results of the study (Brůhová et al., 2020) show that there is still a big gap between the theory of sustainable mobility and its implementation in practice. However, there can be significant differences among stakeholders on how to achieve sustainable urban mobility.

Based on the analysis of the concepts of sustainable development, the definition of "a sustainable transport system is formulated as a system in which fuel consumption, vehicle emissions, safety, congestion and social and economic accessibility are at such a level that they can be maintained in the indefinite future without compromising the ability to future generations of people around the world to meet their own transportation needs" (Gordon, 1995).

In implementing the concept of sustainable mobility, the main priority is given to the development of public transport, which involves the collective use (bus, tram, trolley bus, metro) (Hermelin & Henriksson

2022), as well as expanding the opportunities for sharing individual vehicles (rental of cars, scooters, scooters, bicycles etc.).

According to the expert assessment of the International Union of Public Transport (UITP, n.d.), the total return on funds invested in public transport is five times higher than the volume of these investments.

In 2012, an analysis was made of the travel trends of young people from Germany aged 18 to 29, who represented one of the most car-oriented age groups (Kuhnimhof et al., 2012). It is shown that until the 1990s, car use increased for all age groups in Germany, including youth. At the turn of the millennium, youth car use declined. Young people are using alternative modes of transport, indicating an increase in multimodal travel behavior.

The development of cycling is crucial to reduce energy consumption and the negative impact on the environment. In this regard, scientists are actively conducting research aimed at improving the overall safety, efficiency of movement and comfort of cyclists (Oskarbski et al., 2021). Increasing attention is paid to smart cities, which are part of urban development and rely on information and communication technologies (Telang et al., 2021). Smart cities contribute to the improvement of urban planning, integrate the work of urban resources and infrastructure elements and are focused on improving the quality of services in order to ensure safety (Husain et al., 2020), comfort, high level of service and sustainability. One of the basic components of a "smart city" is an intelligent transport system (ITS) - a communication infrastructure that is based on the wireless interaction of transport infrastructure objects and vehicles with each other, and is focused on improving the overall characteristics of transport, such as reliability, accessibility, integrity, efficiency, improving road safety (Husain et al., 2020; Fitah et al., 2018). Many external factors influence the efficiency and development of ITS. The study (Cuong et al., 2022) identifies 28 factors, which are divided into 5 main groups: lack of close attention from the government, financial constraints, inadequate/incomplete transport infrastructure, excessive urbanization, readiness and integration for ITS. However, a wide range of studies show that transport systems in most cities and urban areas are unsustainable and very often unbalanced in the context of environmental, social and economic aspects, which poses a threat to future generations (Bamwesigye & Hlavackova, 2019). Under the current conditions, changes in the development trends of transport systems are necessary, which requires the cooperation of various stakeholders at the regional, national and international levels.

Thus, the analysis of studies and official sources led to the conclusion that the environment, social well-being and economic development of cities and urban agglomerations are closely related to the development of sustainable mobility. In turn, the development of sustainable mobility is based on the functioning of sustainable transport systems. However, the concept of "sustainable transport system" requires systematization, definition of characteristic features and functional role in the urban context.

3. KEY ASPECTS OF THE FUNCTIONING OF A SUSTAINABLE TRANSPORT SYSTEM

In sustainable mobility, a large role is given to the functioning of sustainable transport systems.

The transport systems of most modern cities traditionally develop based on the expansion of the road network and the use of additional urban space to increase road capacity in the face of an increase in the level of motorization. At the same time, social and environmental factors are often not taken into account, while these factors determine the quality of life of citizens, their comfort and safety.

A comparative assessment of the development of transport systems with the traditional approach and the development of sustainable transport systems was carried out (Table 1). So, a sustainable transport system should be focused on the individual needs of people, guarantee comfort, accessibility, and safety. At the same time, transport systems should function on the principles of environmental friendliness, taking into account the interests of future generations. The development of sustainable transport systems is based on the following main aspects: efficiency, safety, environmental friendliness, accessibility and comfort for all users. Essential in the functional aspect of sustainable mobility is the potential of sustainable transport systems in the urban context. With the development of a sustainable transport system, the main focus should shift to changing the consumption and mobility model, focused on a wide range of consumer preferences, objects and mobility services that are integrated into the transport and urban infrastructure based on information, communication, and innovative technologies.

Table 1. Comparative assessment of the development of transport systems in the traditional approach and sustainable transport systems. *Source: Created by the authors*

Transport systems under the traditional approach	***Sustainable transport systems***
Concentration of attention on vehicles when organizing traffic	Human-centered approach to the organization of movement
Urban space is focused on increasing the capacity of the city's road network	Urban space is focused on the preferences and needs of people in their pursuit of a comfortable life.
Inefficient use of urban space as a result of an increase in the level of motorization	Rational use of urban space and infrastructure through the priority use of public transport and public access to individual vehicles
Joint infrastructure for the movement of bicycles (scooters, scooters) when moving people and transporting small consignments	Separate infrastructure for the movement of bicycles (scooters, scooters) when moving people and transporting small consignments
The infrastructure of urban public electric transport is used primarily for the transport of passengers	Use of the infrastructure of urban public electric transport for the transportation of passengers and goods with a clear delimitation of time (to ensure safety)
Absence of massive restrictions on the movement of cars within the city	The presence of massive restrictions on the movement of cars within the city, which is supported by a list of motivational measures for the use of public transport
The transport system is considered as a set of modes of transport	The transport system is based on intermodality and interoperability of different modes of transport
Traffic regulation is carried out on the basis of preliminary monitoring of the traffic situation and analytical calculations	Adaptive traffic management based on continuous monitoring and expected forecasts using artificial intelligence
Information and telecommunication technologies are used as a basis for solving local problems of improving the efficiency of transport systems	Information and telecommunication technologies are used as the basis for the formation of an integrated platform for collecting, processing large amounts of data and making decisions on the sustainable functioning of the transport system as a whole
Separate payment systems for various types of movement	Development of a "smart payment system"; implementation of blockchain technologies; share economy development.

4. CONCLUSION

A sustainable transport system is considered as the main component of an integral system of sustainable mobility. The concept of "sustainable transport system", which is presented in official documents and scientific publications, does not fully reveal the key aspects of the functioning of such systems. In this regard, we have systematized and identified the characteristics of sustainable transport systems, as well as their functional role in the urban context.

A comparative assessment of the development of transport systems with the traditional approach and sustainable transport systems has been carried out. With the traditional approach, the focus is on the development of transport infrastructure and ways to increase the capacity of the city's road network. Sustainable transport systems involve the use of urban space based on the preferences and mobility needs of people of all categories, are based on the preferences and needs of residents, focus on the priority development of public transport and on public access to private transport, and also take into account the interests of future generations. Sustainable transport systems are based on a human-centered approach and are characterized by the following main aspects: efficiency, safety, environmental friendliness, accessibility and comfort for all users. Thus, a socio-ecological transformation of transport systems is necessary, as well as changes in the planning, organization and management of transport processes that will determine the quality of life of citizens, their comfort and safety.

Further research should be directed to the development and justification of technical and technological solutions for the development and management of sustainable mobility, and it is also necessary to focus on the methods and technologies used for decision-making.

REFERENCES

Asim Husain, A., Maity, T., Kumar Yadav, R. (2020). Vehicle detection in intelligent transport system under a hazy environment: a survey. *IET Image Processing*, *14* (1), 1-10. https://doi.org/10.1049/iet-ipr.2018.5351

Bamwesigye, D, Hlavackova, P. (2019). Analysis of Sustainable Transport for Smart Cities. *Sustainability*, *11*(7), 2140. https://doi.org/10.3390/su11072140

Brůhová Foltýnová, H., Vejchodská, E., Rybová, K., Květoň, V. (2020). Sustainable urban mobility: One definition, different stakeholders' opinions. *Transportation Research Part D: Transport and Environment. 87.* https://doi.org/10.1016/j.trd.2020.102465

Cuong, N. N. Tran, Thang Tran Huynh Tat, Vivian, W. Y. (2022). Tam and Duc Hoc Tran. Factors affecting intelligent transport systems towards a smart city: a critical review. *International Journal of Construction Management*, 1-17. https://doi.org/10.1080/15623599.2022.2029680

European Commission. (1992) *A Community Strategy for 'Sustainable Mobility', Green Paper on the Impact of Transport. on the Environment; COM (1992) 46 Final; Commission of the European Communities: Brussels, Belgium, 1992.* https://op.europa.eu/en/publication-detail/-/publication/98dc7e2c-6a66-483a-875e-87648c1d75c8/ (Accessed: 1 December 2022).

Fitah, A., Badri, A., Moughit, M., Sahel, A. (2018). Performance of DSRC and WIFI for Intelligent Transport Systems in VANET. *Procedia Computer Science*, *127*, 360-368. https://doi.org/10.1016/j.procs.2018.01.133

Gallo, M. and Marinelli, M. (2020). Sustainable Mobility: A Review of Possible Actions and Policies. *Sustainability.*, *12*(18), 7499. https://doi.org/10.3390/su12187499

Gordon, D., (1995). Sustainable transportation: what do we mean and how do we get there. *Transportation and energy: strategies for a sustainable transportation system*, 1-11.

Häußler, E., Blaszczyk, R., Hupfer, C. (2019). Den Fußverkehr in Städten fördern: Partizipative Forschung im Reallabor GO Karlsruhe. Gaia-ecological Perspectives for Science and Society. https://doi.org/10.14512/gaia.28.4.16Corpus

Hermelin, B., Henriksson, M. (2022). Transport and Mobility Planning for Sustainable Development, Planning Practice & Research, *37*(5), 527-531. https://doi.org/10.1080/02697459.2022.2082756

IPCC. (2022) *Climate Change 2022: Mitigation of Climate Change. April 2022.* Available at: https://www.ipcc.ch/report/sixth-assessment-report-working-group-3/ (Accessed: 1 December 2022).

Kuhnimhof, T., Buehler, R., Wirtz, M., Kalinowska, D. (2012). Travel trends among young adults in Germany: increasing multimodality and declining car use for men. *Journal of Transport Geography*, *24*, 443–450. https://doi.org/10.1016/j.jtrangeo.2012.04.018

Meyer, N, Auriacombe, C. (2019). Good Urban Governance and City Resilience: An Afrocentric Approach to Sustainable Development. *Sustainability. 11*(19):5514. https://doi.org/10.3390/su11195514

Oskarbski, J., Birr, K., Żarski, K., Bicycle (2021). Traffic Model for Sustainable Urban Mobility Planning. *Energies*, *14*, 5970. https://doi.org/10.3390/en14185970

Ritchie, H. (n.d.). *Sector by sector: where do global greenhouse gas emissions come from? Our World in Data.* Available at: https://ourworldindata.org/ghg-emissions-by-sector (Accessed: 12 December 2022).

Shramenko, N., Hupfer, C. (2023). Sustainable mobility: Changing mindsets and rethinking paradigms. *Proceedings of 9th International Conference, New technologies - development and application*. Bosnia and Herzegovina.

Shramenko, N., Hupfer, C. (2023). Sustainable mobility: Conceptual aspects of development and management. *Proceedings of 9th International Conference "New technologies - development and application "*. Sarajevo.

Telang, S., Chel, A., Nemade, A., Kaushik, G. (2021). Intelligent Transport System for a Smart City. In: Tamane, S.C., Dey, N., Hassanien, A.E. (eds) *Security and Privacy Applications for Smart City Development. Studies in Systems, Decision and Control*, *308*. Springer, Cham. https://doi.org/10.1007/978-3-030-53149-2_9

UITP. (n.d.) *Public transport moving Europe forward.* Available at: https://www.uitp.org/publications/public-transport-moving-europe-forward/ (Accessed: 17 November 2022).

United Nations. (2015) *Transforming our world: the 2030 Agenda for Sustainable Development.* Available at: https://sdgs.un.org/2030agenda (Accessed: 1 December 2022).

United Nations. (2022) *UN climate report: It's 'now or never' to limit global warming to 1.5 degrees. 4 April 2022.* Available at: https://news.un.org/en/story/2022/04/1115452 (Accessed: 11 December 2022).

ADOPTION ATTITUDE AND CONTINUANCE INTENTION TO ONLINE FINFLUENCER VIDEO USAGE: THE ROLE OF TRUST*

Emine Cigdem Canatan[1], Ali Coskun[2], and Aysegul Toker[3]
[1]*Institute for Graduate Studies in Social Sciences, Management Department, Bogaziçi University, Istanbul, Turkey*
[2]*Assoc. Dr., Institute for Graduate Studies in Social Sciences, Management Department, Bogaziçi University, Istanbul, Turkey*
[3]*Prof. Dr., Sabanci Business School, Sabanci University, Istanbul, Turkey*

ABSTRACT

The popularity of online financial videos created by finfluencers (finance influencers) has increased significantly in recent years, with a notable and enduring user preference for financial content encompassing general economic news, cryptocurrencies, stocks, and other investment instruments. This study investigated the antecedents of attitude towards watching online finfluencer videos and the factors influencing viewers' intention to continue watching them. In order to address this issue, a model is developed that examines trust-based determinants, as well as additional antecedents, including perceived usefulness (PU), perceived ease of use (PEU), perceived content quality (PCE), perceived behavioral control (PBC), and subjective norm (SN), perceived enjoyment (PE), and perceived behavioral control (PBC), as factors that impact users' attitude towards watching online finfluencer videos. We also explore financial literacy's role in moderating the relationship between trust-based determinants and other antecedents and attitude toward watching online financial videos created by influencers. This study is intended to provide insights into the antecedents of attitude towards watching online financial videos and continuance intention to use them. Online content creators can use the findings of this study to develop strategies to increase the popularity of their online financial videos.

KEYWORDS

Finfluencer, Online Finance Videos, ECM, Trust

1. INTRODUCTION

This study investigates the antecedents of attitude towards watching online finfluencer videos and the factors that influence viewers' intention to continue watching them. Scholars indicate that finfluencers are currently a niche group, but they are expected to proliferate in the coming years, especially as designated sections for them on social media platforms become more popular (e.g., #FinTok, #FinTwit, #StockTok, #Findependence) (de Regt, 2023) because there is a growing demand for financial videos especially by retail investors. Therefore, understanding the antecedents which create a habit in the user to watch a certain finfluencer will be beneficial to guide the finfluencers and other content creators to create more attractive finance videos.

As of present, the field of finfluencers remains largely under-researched. Notably, extensive research exists on influencers operating within other domains such as beauty and sports. The existing literature clearly favors fashion and beauty sectors according to the metaanalysis of Vrontis on social media influencer marketing (Vrontis et al., 2021). Numerous research on sports social media influencers such as studies of Silva et al. (2021), Peixoto (2022), Malenović (2016) illustrate the body of research on sports domain.

However, the investigation of the influencer phenomenon remains underexplored. Consequently, the purpose of this research is to alleviate the void in the academic literature by providing new insights. It is believed that this inquiry will hold some significant practical implications as well.

This paper is reflecting work-in-progress of the current research. We are reflecting the conceptual model in this theoretical paper. Therefore, this is still a concept paper and data will be fed to test in the future stages of the research.

*Research-in-Progress

2. LITERATURE REVIEW

2.1 The Rise of Social Media Influencers

Over half of the global population, equivalent to 4.76 billion individuals worldwide, utilize social media or social networking sites (SNS). The individuals consuming content in social media are also active in generating and disseminating it, resulting in assuming the role of content creators (Borchers, 2019). Due to the Internet's capacity for scalability and rapid diffusion, specific individuals are able to garner a significant audience and transform into social media influencers, SMI, providing guidance to their followers (Vrontis et al., 2021).

SMI is a "content generator who has a status of expertise in a specific area, who has cultivated a sizable number of captive followers by regularly producing valuable content via social media" (Lou and Yuan, 2019 & Farrell, 2022).

The outbreak of the pandemic has led to a significant expansion in the size of the influencer market. The global market for influencer marketing has exhibited a more than twofold increase in size since 2019 (Statista, 2023).

2.2 The Rise of Investing and Finfluencers

The pandemic has significantly impacted investment trends as well, with a rising prevalence of retail investors. In 2020, retail trading hit a 10-year high in the United States, representing 19.5% of all order flow in the equities markets, according to Bloomberg Intelligence (2021) figures, marking a rise of around 4.5% compared to the previous year and nearly twice the number of retail investor trades in 2010.

During the pandemic, more people became interested in investing due to increased savings gained with lockdowns and restrictions on travel. On the other hand, millions of people who faced increased uncertainty regarding their careers started to think more seriously about how to manage their financial resources (de Regt, 2023).

Low-interest rates in response to the pandemic made it less attractive to keep money in savings accounts. As a result, some people turned to investing to earn a higher return on their money (Coban, 2023). Also, the stock market experienced significant volatility during the pandemic, creating opportunities for investors. Furthermore, with more people working from home, retail investors had more time to research investment opportunities, increasing their interest in investing.

This environment created an increasing demand for accessible financial information, which has led to the proliferation of finance influencers, or "Finfluencers". Finfluencers provide advice on various financial topics through short, light-hearted videos posted on social media platforms such as Instagram, TikTok, and YouTube. This trend highlights a shift from traditional financial services towards informal, self-directed financial education, mostly free and easy to access (de Regt, 2023). The term "finfluencer" is also defined as a person or entity with an outsized impact on investor decisions through social media influence (Guan, 2022).

In addition to the increase in the volume of investment and the increasing number of new retail investors, all these factors also contributed to the diversification of types of investments. In the past, most people invested in traditional investments like stocks and bonds. However, in recent years, there has been a growing interest in alternative investments like cryptocurrency or equity crowdfunding (Alzahrani, S. et al, 2019) (Schwienbacher, A., 2019). Diversifying alternative investment types encouraged retail investors to learn and invest more. In parallel with the growing demand for finfluencer videos, there is increased content production by finfluencers on these diverse investment instruments.

3. THEORETICAL FRAMEWORK

To address the research question, a model was developed to examine the antecedents of finfluencer-watching continuance intention. This was done by exploring relevant theories and the constructs in each theory to determine the independent variables in the model. This study uses the theoretical foundations of the Expectation Confirmation Model (ECM), the Technology Acceptance Model (TAM), The Theory of Planned Behavior (TPB), and trust to investigate the factors that influence users' adoption attitude and continuance intention of watching finfluencer videos:

3.1 The Expectation Confirmation Model (ECM) of IT continuance

Bhattacherjee (2001) proposed the ECM of IT continuance based on the congruence between individuals' continued IT usage decisions and consumers' repeat purchase decisions. The ECM posits that an individual's intention to continue using an information technology (IT) system is dependent on three variables: user satisfaction, confirmation of expectations, and perceived usefulness. It applies to studying user behavior in finance video and finfluencer watching as it affects repeat use. The ECM has been used to understand e-learning continuance intention, as consumers watch finfluencers for financial information akin to e-learning via the Internet.

3.2 The Technology Acceptance Model (TAM)

TAM provides insights into how users will adopt and use a new technology. The model is based on two key constructs: perceived usefulness and perceived ease of use. TAM posits that perceived usefulness and ease of use directly impact users' attitudes toward using the system. Attitude, in turn, influences users' behavioral intention to use the system. Behavioral intention is the likelihood that users will actually use the system. (Davis, 1989). TAM can assist us in clarifying a crucial aspect of the relationship between the attitude to watch finfluencer videos and the intention to continue using them.

3.3 The Theory of Planned Behavior (TPB)

TPB is an extension of the Theory of Reasoned Action (TRA). It postulates that individuals' behavioral intentions are determinants of their actual behavior. Behavioral intention is a function of three determinants: attitude, subjective norm, and perceived behavioral control. It has been shown to be a reliable and valid predictor of behavior, also in the IT adoption context (Ajzen, I.,1991).

3.4 Trust

Trust is generally defined as the belief in the other party having a high level of ability, integrity and benevolence (Luhmann, 1979; Mayer et al., 1995). In situations where there is risk, uncertainty, and interdependence, trust plays a central role. By establishing trust, consumers can subjectively eliminate potentially undesirable actions by the trusted party, thereby reducing perceived risks to a manageable level (Mayer et al., 1995). Trusting beliefs in finfluencer videos is particularly important because this content consists of information about financial markets, news, dynamics, and diverse information about financial products. Furthermore, sometimes finfluencers imply or openly share investment advice which involves a great deal of risk, especially for novice and inexperienced investors.

In recent years, trust in technology has become a key factor in research on technology adoption in Information Systems research. Due to the interdependent relationship between trust in technology, video content, and content creators, our study incorporates trust models from both the Information Systems literature (McKnight et al., 2002, 2001, 2005) and the trust construct in Influencer Marketing and Celebrity literature (Vrontis et al., 2021) (Ohanian, R., 1990). Trust in technology use is important for initial and continued use (Thatcher et al., 2011, McKnight et al., 2011).

Scholars have concluded that social media influencers are more influential, credible, and relatable than traditional celebrity endorsers (Djafarova & Rushworth, 2017). Influencers are found to be trusted because they specialize in a category, like fashion or gaming. This trust is akin to the trust a person has in their friends (Swant, 2016, Campbell, 2022). Empirical evidence reveals that consumers are likely to invest in blind faith in the content shared on social media groups (Shareef et al., 2020; Vrontis et al., 2021). Studies have shown that trust in influencers leads to increased sales and stronger relationships (Vrontis et al., 2021).

In addition to trust, we incorporated two additional independent variables in the antecedents and explored their effect on attitude toward watching finfluencer videos. We hypothesize that perceived content quality and perceived enjoyment were significant predictors of attitude. These constructs are consistent with the influencer adoption literature (Vrontis et al., 2021).

Regarding perceived content quality, people are more likely to watch videos that they perceive to be high quality.

For the perceived enjoyment factor, people are more likely to watch videos that they find enjoyable. When individuals are in the playfulness state, they will find the interaction intrinsically interesting (Moon and Kim, 2001).

Lastly, we explore financial literacy's role in moderating the relationship between trust-based determinants and other antecedents and attitude towards watching online financial videos created by finfluencers. Financial literacy is defined as the ability to read, analyze, manage, and communicate about the personal financial conditions that affect material well-being (Anderson et al., 2000, Allgood et al, 2016).

4. RESEARCH MODEL AND HYPOTHESES

The research model in Figure 1 is based on the trust models, ECM, TAM, and TPB.

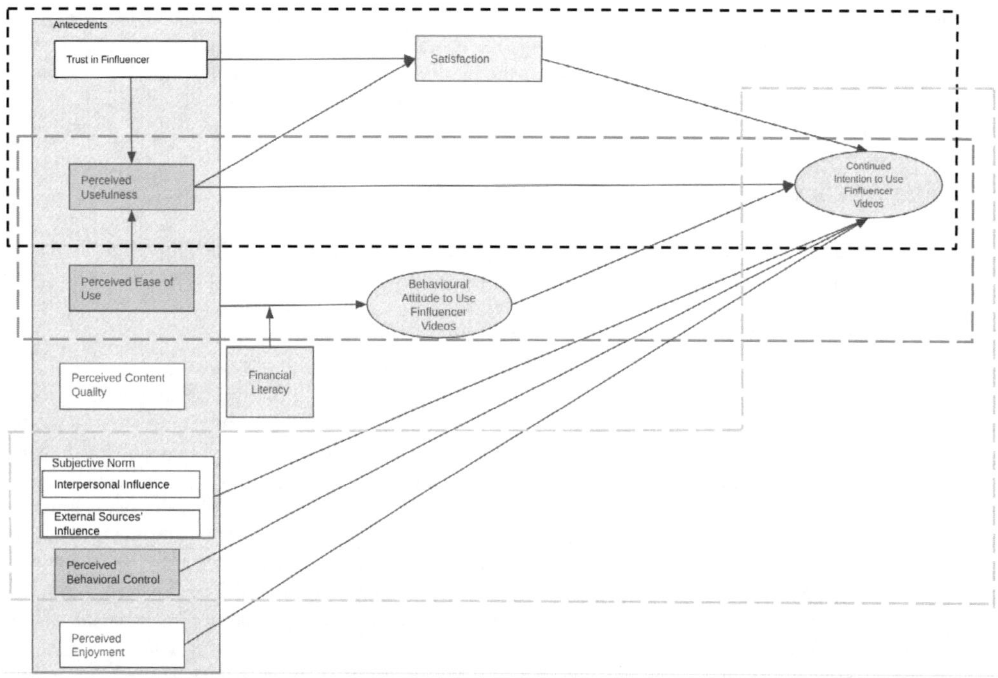

Figure 1. Research Model

H1. Users' satisfaction with Finfluencer Videos (FV) is positively related to their Continued FV Usage intention.
H2. Users' Trust is positively related to their Satisfaction with FV.
H3. Users' Perceived Usefulness of FV is positively related to their Satisfaction with FV. H4. Users' Perceived Usefulness of FV is positively related to their Continued FV Usage Intention. H5. Users' Trust is positively related to their Perceived Usefulness of FV.
H6. Perceived Usefulness is positively related to Behavioral Attitude toward FV.
H7. Perceived Ease of Use is positively related to Behavioral Attitude toward FV.
H8. Perceived Ease of Use is positively related to Perceived Usefulness of FV.
H9. Behavioral Attitude toward FV is positively related to the Continued Intention to use FV.
H10. Subjective Norm is positively related to the Continued Intention to use FV.
H11. Perceived Behavioral Control is positively related to the Continued Intention to use FV.
H12. Perceived Enjoyment is positively related to the Attitude toward FV.
H13. Perceived Enjoyment is positively related to the Continued Intention to use FV.
H14. Perceived Content Quality is positively related to the Behavioral Attitude toward FV.

H15. Perceived Financial Literacy is a moderator between the antecedents and the Behavioral Attitude toward FV; the higher the Perceived Financial Literacy, the stronger the relation between Antecedents of Attitude to Use FV and Attitude toward FV.
H16. Subjective Norm is positively related to the Attitude toward FV.

5. METHODOLOGY

The theoretical research model was constructed utilizing the previously mentioned Information Systems and Influencer Marketing constructs. This paper is yet still a concept paper and data will feed to test in the future stages of the research. In order to examine the proposed theoretical research model, a scale was developed and is intended to be distributed as an online survey. The objective is to gather feedback from a minimum sample size of 300 in order to evaluate the 16 hypotheses. We aim for 18-65 age and an equal representation of female and male. Data for this study will be collected from a convenience sampling and snowball sampling of internet users, who will participate in the research voluntarily via an online questionnaire

6. CONCLUSION

We think that trust plays a crucial role in watching a finfluencer and also in the continued watching behavior. Because relying on a finfluencer video involves uncertainty and risk, developing trusting beliefs in them is critical for the users. In the present study, we position trust in finfluencer videos as an immediate antecedent of perceived usefulness and satisfaction. Our trust conceptualizations are based on McKnight et al.'s trust models (2002) and Xiao's (2018) influencer marketing and Ohanian's (1990) celebrity trust aspect. Both trust and satisfaction have been shown to have a positive influence over the continuance intention to use an IT platform (Lee, 2010). Therefore, we propose that trust will positively moderate the relationship between users' satisfaction and continuance intention.

We hypothesize that high perceived financial literacy will strengthen the positive relationship between antecedents and attitude to watch finfluencer videos. This is because individuals with a higher level of financial literacy are more likely to recognize the value of financial information presented in these videos. Individuals with high perceived financial literacy may be more likely to be receptive to finfluencers' messages and engage with financial content.

By gaining insight into the antecedents of watching attitude, financial content creators on social media can enhance the quality and relevance of finance video content to cater better to viewers' requirements. The rise of individual investors has led to a correlation between finfluencer viewing habits and the demand for financial guidance. Investigating the factors that influence attitudes towards finfluencer videos can offer valuable insights to engage and educate individual investors on financial matters, leading to the development of more effective financial education programs and a more informed investor community.

REFERENCES

Allgood, S., & Walstad, W. B. (2016). The effects of perceived and actual financial literacy on financial behaviors. *Economic inquiry*, *54*(1), 675-697.

Alzahrani et al, 2019. Analysis of the cryptocurrency adoption decision: Literature review. In *2019 Portland International Conference on Management of Engineering and Technology (PICMET)* (pp. 1-11). IEEE.

Anderson, C. et al, 2000. Personal finance and the rush to competence: Financial literacy education in the US. *Journal of family and consumer sciences*, *107*(2), 1-18.

Ajzen, I.,1991. The theory of planned behavior. *Organizational behavior and human decision processes*, *50*(2), 179-211.
Bhattacherjee, A., 2001. Understanding information systems continuance: An expectation-confirmation model. *MIS quarterly*, 351-370.

Bloomberg (2021, January 28). *Trade Volume Explodes as Retail Traders Make Their Mark.* URL: https://www.bloomberg.com/press-releases/2021-01-28/trade-volume-explodes-as-retail-traders-make-their-mark (visited on 02/01/2023)

Borchers, N.S., 2019. Social media influencers in strategic communication. *International Journal of Strategic Communication*, *13*(4), pp.255-260.

Cheong, J.H. and Park, M.C., 2005. Mobile internet acceptance in Korea. *Internet research,* 15(2), pp.125-140. Coban, F., 2023. *Are finfluencers the new experts in the field of investment advice?* Master's thesis. Utrecht University. Davis, F.D. et al, 1989. User acceptance of computer technology: A comparison of two theoretical models. *Management science*, *35*(8), pp.982-1003.

de Regt, A. et al, 2023. Young People Under 'Finfluencer': The Rise of Financial Influencers on Instagram: An Abstract. In *Optimistic Marketing in Challenging Times: Serving Ever-Shifting Customer Needs: Proceedings of the 2022 AMS Annual Conference, May 25-27, Monterey, CA, USA* (pp. 271-272). Cham: Springer Nature Switzerland.

DeLone, W. H., & McLean, E. R., 1992. Information systems success: The quest for the dependent variable. *Information systems research*, *3*(1), 60-95.

Djafarova, E., & Rushworth, C., 2017. Exploring the credibility of online celebrities' Instagram profiles in influencing the purchase decisions of young female users. *Computers in Human Behavior*, 68, 1–7.

Djafarova, E., & Trofimenko, O., 2019. "Instafamous"–credibility and self-presentation of micro-celebrities on social media. *Information, Communication & Society,* 22(10), 1432–1446.

Farrell, J. R. et al, 2022. What drives consumers to engage with influencers?: Segmenting consumer response to influencers: Insights for managing social-media relationships. *Journal of Advertising Research*, *62*(1), 35-48. Forbes Agency Council. (2022). *The Rise of the Retail Investor.* Forbes. URL: https://www.forbes.com/sites/forbesagencycouncil/2022/11/04/the-rise-of-the-retail-investor/?sh=4e3fd26e1755 Guan, S.S., 2022. The Rise of the Finfluencer. *New York University Journal of Law and Business, Forthcoming, Santa Clara Univ. Legal Studies Research Paper Forthcoming.*

Lee, M.C., 2010. Explaining and predicting users' continuance intention toward e-learning: An extension of the expectation–confirmation model. *Computers & education*, *54*(2), pp.506-516.

Lou, C. and Yuan, S., 2019. Influencer marketing: How message value and credibility affect consumer trust of branded content on social media. *Journal of interactive advertising*, *19*(1), pp.58-73.

Luhmann, N., 1979. *Trust and power. John Willey & Sons*, pp.5-12.

Malenović, J., 2016. Social and digital mass Media impact on the perception of yoga. *International Scientific Yoga Journal Sense Sense*, *6*(6), 66-78.

Mayer, R. C. et al, 1995. An integrative model of organizational trust. *Academy of management review*, *20*(3), 709-734. McKnight, D. H. et al, 2002. The impact of initial consumer trust on intentions to transact with a web site: a trust building model. *The journal of strategic information systems*, 11(3-4), 297-323.

McKnight et al, 2001. What trust means in e-commerce customer relationships: An interdisciplinary conceptual typology. *International journal of electronic commerce*, 6(2), 35-59.

McKnight, D. H., 2005. Trust in information technology. *The Blackwell encyclopedia of management*, 7, 329-331. McKnight et al, 2011. Trust in a specific technology: An investigation of its components and measures. *ACM Transactions on management information systems (TMIS)*, 2(2), 1-25.

McKnight et al, 2009. Trust in technology: Development of a set of constructs and measures. Moon, J.W. and Kim, Y.G., 2001. Extending the TAM for a World-Wide-Web context. *Information & management*, 38(4), pp.217-230.

Ohanian, R. (1990). Construction and Validation of a Scale to Measure Celebrity Endorsers' Perceived Expertise, Trustworthiness and Attractiveness. *Journal of Advertising*, 19(3), 50.

Rodrigues Peixoto, A. J., 2022. *More than an exercise: the influence of social media on western yoga perception and practice motivation: a study on cultural appreciation and appropriation* (Bachelor's thesis, University of Twente)

Schwienbacher, 2019. Equity crowdfunding: anything to celebrate? *Venture Capital*, *21*(1), 65-74.

Shareef, M.A. et al, 2020. Group behavior in social media: Antecedents of initial trust formation. *Computers in Human Behavior*, *105*, p.106225.

Silva et al., 2021. The body as a brand in social media: analyzing digital fitness influencers as product endorsers. *Athenea digital*, 21(1), 1-34.

Statista, 2021. *Global influencer market size.* Statista. URL: https://www.statista.com/statistics/1092819/global-influencer-market-size/ (visited on 03/05/2023) Swant, M., 2016. "Twitter Says Users Now Trust Influencers Nearly as Much as Their Friends." URL: http://www.adweek.com/digital/twitter-saysusers-now-trust-influencers-nearly-muchtheir-friends171367/ (visited on 02/23/2023)

Thatcher, J. B. et al, 2011. The role of trust in postadoption IT exploration: An empirical examination of knowledge management systems. *Engineering Management, IEEE Transactions* on, 58(1), 56-70.

Vrontis, D. et al, 2021. Social media influencer marketing: A systematic review, integrative framework and future research agenda. *International Journal of Consumer Studies*, *45*(4), pp.617-644.

Xiao, M. et al, 2018. Factors affecting YouTube influencer marketing credibility: a heuristic-systematic model. *Journal of media business studies*, *15*(3), pp.188-213

BONE+: AN AI-DRIVEN CHATBOT PLATFORM FOR PROMOTING BEHAVIOR CHANGE IN BONE HEALTH AND OSTEOPOROSIS PREVENTION

Tsai-Hsuan Tsai[1], Yue-Shan Hsieh[2], Chan-Shien Ho[3], Yu-Cheng Pei[4] and Chien-Lung Hsu[5]
[1]*Master of Science Degree Program in Innovation for Smart Medicine, Chang Gung University, Taiwan*
[2]*Department of Industrial Design, Chang Gung University, Taiwan*
[3]*Department of Physical Medicine and Rehabilitation, New Taipei Municipal TuCheng Hospital, Taiwan*
[4]*Department of Physical Medicine and Rehabilitation, Chang Gung Memorial Hospital, Taiwan*
[5]*Graduate Institute of Management, Chang Gung University, Taoyuan, Taiwan*

ABSTRACT

Osteoporosis is a prevalent medical condition characterized by a reduction in bone mass and density, which significantly increases the risk of fractures. In order to promote optimal bone health, it is essential to acquire relevant information and adopt a lifestyle that safeguards bone integrity. In light of this, the present study employed the Health Belief Model as a framework to design and develop the Bone+ mobile health chatbot platform. Leveraging artificial intelligence technology, the platform utilizes image analysis of X-rays to assess bone density and predict the likelihood of bone loss. Moreover, it offers comprehensive information on osteoporosis prevention, empowering users to make well-informed decisions regarding their bone health. The results obtained from the heuristic evaluation questionnaire of the chatbot platform demonstrated satisfactory usability, thereby further supporting its efficacy as a valuable tool in osteoporosis prevention.

KEYWORDS

Osteoporosis, Bone Health, Mobile Health, AI Chatbot, Health Belief Model, Heuristic Evaluation

1. INTRODUCTION

Osteoporosis, a condition that primarily affects older individuals and postmenopausal women, is characterized by a loss of bone mass and density, leading to a higher risk of fractures (IOF, 2023). However, despite being a preventable condition, osteoporosis is often misconceived as an age-related disease (Stetzer, 2011). To effectively prevent osteoporosis, it is crucial to educate individuals about bone health and promote a lifestyle that safeguards bone health (Chan et al., 2018). Unfortunately, studies have shown insufficient public awareness of osteoporosis, resulting in a lack of preventive measures (D'Silva and Pinto, 2017). Although people may possess some knowledge and beliefs about osteoporosis, they often fail to take action to protect their bones due to their beliefs, diets, and lifestyles. Therefore, addressing the knowledge and practice gaps related to osteoporosis is essential for future prevention efforts.

Technology plays a vital role in identifying osteoporosis and facilitating long-term health tracking and educational interventions for patients (Chan et al., 2022). Mobile health (mHealth) can facilitate convenient monitoring of health status over time. mHealth is a public health and medical strategy that utilizes mobile technology, including cell phones, to provide assistance. It encompasses the development and utilization of more advanced functions and applications beyond simple mobile devices (WHO, 2011). With mHealth, users can easily monitor their health status and access remote medical advice regardless of time and location, enabling effective management of chronic conditions such as osteoporosis.

The Health Belief Model (HBM) is a widely used theoretical framework for developing and evaluating health interventions. It incorporates psychological and sociological concepts to examine individuals' attitudes and behaviors regarding health promotion, self-actualization, and fulfillment (Rosenstock, 2005). A study by Song et al. (2022) demonstrated that using the Health Belief Model (HBM) in a WeChat-based educational intervention improved disease knowledge, self-efficacy, and exercise adherence in patients with Ankylosing Spondylitis. This study highlighted the feasibility and effectiveness of WeChat-based patient education for individuals with chronic diseases. Alharbey and Chatterjee (2019) developed an innovative mobile device-based assistive technology grounded in the HBM to empower patients with Chronic Obstructive Pulmonary Disease, raising awareness and promoting engagement in self-care management activities. In the context of osteoporosis prevention, HBM-based health interventions can effectively promote the adoption of nutritionally healthy behaviors, enhance knowledge, and induce attitudinal changes, thereby mitigating the risk of osteoporosis (Panahi et al., 2021). Behavioral research often incorporates elements such as disease awareness and information access as conceptual frameworks to study people's health behaviors. Thus, the development of the Bone+ platform is guided by the usage of HBM as a theoretical framework in this study.

This study introduces Bone+, an mHealth service that utilizes AI and HBM to detect osteoporosis risk. The chatbot analyzes X-rays, predicts bone density, and educates users about their bone health. The community feature assesses risk and provides treatment information. By promoting healthy behaviors and preventive measures, the study aims to reduce fractures and raise awareness of osteoporosis.

2. MATERIALS AND METHODS

2.1 Enhancing Users' Perception of Susceptibility through Online Testing and Personalized Health Profiles

Perceived susceptibility plays a crucial role in motivating healthy behaviors, but misconceptions can hinder risk recognition. To address this, Bone+ employs personalized health profiles and an online tool to assess osteoporosis risk. Unlike conventional tools, Bone+ takes into account the user's specific body data and incorporates AI image interpretation of X-ray films to improve the accuracy of risk prediction for osteoporosis and fractures. Following the risk assessment, Bone+ displays the results, provides health advice, and offers feedback (refer to **Error! Reference source not found.**).

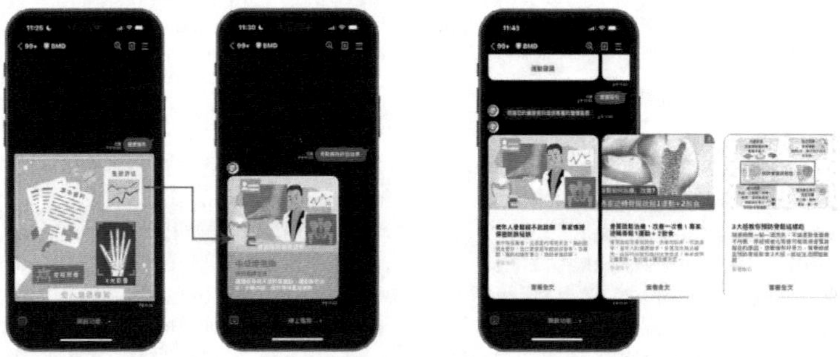

Figure 1. Personal health profile and online risk assessment for osteoporosis (left); personalized graded health education guidance for enhancing consumers' comprehension of osteoporosis risks (right)

2.2 Providing a Personalized Graded Health Information Guide to Enhance Consumers' Understanding of the Perceived Severity of Osteoporosis

Perceived seriousness refers to an individual's comprehension of the severity of their osteoporosis and the potential consequences of not receiving proper treatment. Osteoporosis can significantly impact an individual's quality of life, including daily activities and social interactions, and may even lead to depression, paralysis, and death (Panahi et al., 2021). In addition to offering a general introduction to osteoporosis, Bone+ evaluates and categorizes users' body data to deliver customized health education content tailored to their specific condition. The personalized graded health information guide, as depicted in **Error! Reference source not found.**, aims to enhance consumers' understanding of the perceived danger of osteoporosis and motivates them to take appropriate action.

2.3 Enhancing Users' Perceived Benefits through Medication Reminders and Personal Nutrition Records and Analysis

The adoption of healthy behaviors is influenced by individuals' perception of the benefits. After bone testing and health education, users gain increased awareness of osteoporosis, calcium intake, and the detrimental effects of caffeine consumption. To support medication adherence and motivate users, Bone+ incorporates feedback on medication usage and provides reminders. The nutritional record section offers users a seven-day analysis of their dietary intake, including statistics for each meal, enabling them to track their progress and identify any nutritional deficiencies. The images depicted in Figure 2 illustrate that users can access their medication reminders and associated information on the left-hand side of the interface. Meanwhile, the right-hand side of the image displays users' dietary intake record along with corresponding analysis.

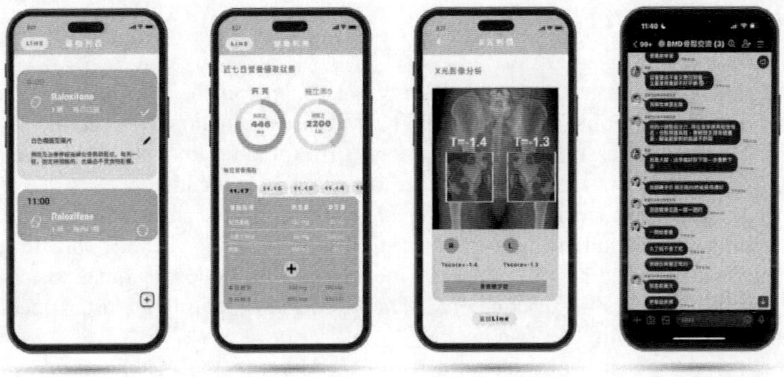

Figure 2. Medication reminders and information (left), diet records and analysis (middle-left), online AI X-Ray image analysis (middle-right), and the osteoporosis community (right)

2.4 Facilitating Access to Medical Services Recommendations to Reduce Perceived Barriers

Modifying health behaviors can be challenging due to perceived barriers. A study by Chan et al. (2022) revealed that only 41.86% of osteoporosis patients effectively changed their lifestyles. Regular educational reinforcement and visits were found to improve knowledge and habits. To address these perceived barriers, Bone+ has been designed to facilitate users' access to medical service recommendations through a systematic approach. The middle-right section of Figure 2 illustrates this feature, enabling users to conveniently take online tests and select the most suitable list of doctors on their mobile devices. This approach helps mitigate any conscious mobility limitations that users may have.

2.5 Facilitating the Adoption of Healthy Habits through the Osteoporosis Community as Action Cues

Action cues, such as public health campaigns, peer support, and educational resources, play a crucial role in motivating individuals to take action. These cues indirectly fulfill expectations of efficacy and efficiency. To promote the adoption of osteoporosis prevention habits, Bone+ provides users with a conversational platform where they can acquire health-related information and guidance from other users within the osteoporosis community. This feature is illustrated on the right-hand side of Figure 2.

2.6 Enhancing Users' Self-Efficacy through Personalized and Tailored Health Guidance

Self-efficacy, which refers to the belief in one's ability to overcome obstacles and adopt healthy behaviors, is vital for driving behavior change. Osteoporosis patients often work on increasing their self-assurance and control over their condition by learning about the illness, the risk of fracture, available treatments, self-management techniques, and engaging in informed decision-making with healthcare professionals (Ganda et al., 2013). Bone+ enhances self-efficacy by providing personalized health recommendations based on users' information (**Error! Reference source not found.**, right side). With practical guidance, Bone+ empowers users, fostering autonomy and self-efficacy.

3. SYSTEM VALIDATION

The system validation approach utilized in this study was based on the mobile device interface heuristic evaluation proposed by Yáñez Gómez et al. (2014). The usability evaluation involved the use of the remaining 12 mobile-specific heuristic assessment questions derived from the original set of 13 usability heuristics by Yáñez Gómez et al. (2014), with the exclusion of the "Help and Documentation" item. To assess the usability of the system, five designers and specialists in interactive interface design and mobile application development were recruited to conduct tests and evaluate the system's usability using a mobile-specific post-development checklist. Through discussions and the exchange of ideas, these professionals aimed to identify design flaws and describe any usability difficulties encountered. The findings of the usability assessment, based on the 12 heuristic evaluation guidelines, are presented below:

(1) Visibility of system status: The system effectively displays its status, making it easy for users to identify the current function and access information.

(2) Correspondence between system and the real world: Users can observe immediate changes in the interface as they interact with different functions, ensuring correspondence between the system and real-world actions.

(3) User freedom and control: The system provides clear execution of functions, offering users a sense of freedom and control.

(4) Standards and consistency: The system adheres to design principles by maintaining consistency in the use of fonts, colors, and icons.

(5) Error prevention: While there is no specific error handling mechanism in place, the system's ease of use contributes to minimizing the likelihood of errors.

(6) Recognition rather than recall: To enhance system recognition and reduce cognitive burden, the system provides text options alongside icons.

(7) Flexibility and efficiency of use: The system's functions are clearly visible and do not require specialized memory or extensive searching.

(8) Aesthetic and minimalist design: The interface is visually appealing, clear, and straightforward in its design.

(9) Help users recognize, diagnose, and recover from errors: It is recommended to incorporate a problem-solving feature to facilitate users' understanding of the system's functionality, particularly for new users.

(10) Compatibility with technical standards: Technical operations align with users' typical operational capabilities.

(11) Pleasant and courteous interaction: Users can easily obtain the required information and answers by selecting the desired mode through the dialogue.

(12) Privacy: Due to the sensitive nature of health analysis and the collection of personal information, concerns arise about ensuring the security and privacy of user data.

4. CONCLUSION

This study aimed to develop Bone+, an AI chatbot that incorporates the Health Belief Model (HBM) to raise awareness of osteoporosis. The chatbot provides remote medical advice and access to osteoporosis health information resources. To further enhance the Bone+ chatbot platform, comprehensive user testing will be conducted in authentic scenarios with osteoporosis patients and potential patients, including older individuals and postmenopausal women. This approach will gather realistic and reliable data on user engagement, satisfaction, and the effectiveness of the system. In addition, longer-term trials and randomized controlled trials in clinical settings will be conducted to optimize the system's effectiveness. These trials will provide valuable insights and improve the recommendations for Bone+, ensuring enhanced functionality and practicality.

REFERENCES

Alharbey, R. & Chatterjee, S. 2019. An mhealth assistive system "mylung" to empower patients with chronic obstructive pulmonary disease: Design science research. *JMIR formative research,* 3, e12489.

Chan, C. Y., Mohamed, N., Ima-Nirwana, S. & Chin, K.-Y. 2018. A review of knowledge, belief and practice regarding osteoporosis among adolescents and young adults. *International journal of environmental research and public health,* 15, 1727.

Chan, C. Y., Subramaniam, S., Chin, K.-Y., Ima-Nirwana, S., Muhammad, N., Fairus, A., Ng, P. Y., Aini, J. N., Aziz, N. A. & Mohamed, N. 2022. Effect of a screening and education Programme on knowledge, beliefs, and practices regarding osteoporosis among Malaysians. *International Journal of Environmental Research and Public Health,* 19, 6072.

D'silva, D. & Pinto, C. A. 2017. Knowledge level of pre-and post menopausal women on osteoporosis: A cross-sectional study. *IOSR J. Nurs. Health Sci,* 6, 70-75.

Ganda, K., Puech, M., Chen, J., Speerin, R., Bleasel, J., Center, J., Eisman, J., March, L. & Seibel, M. 2013. Models of care for the secondary prevention of osteoporotic fractures: a systematic review and meta-analysis. *Osteoporosis International,* 24, 393-406.

IOF. 2023. *About Osteoporosis: International Osteoporosis Foundation* [Online]. Available: https://www.osteoporosis.foundation/patients/about-osteoporosis [Accessed April 20].

Panahi, R., Siboni, F. S., Kheiri, M., Ghoozlu, K. J., Shafaei, M. & Dehghankar, L. 2021. Promoting the adoption of behaviors to prevent osteoporosis using the health belief model integrated with health literacy: quasi-experimental intervention study. *BMC Public Health,* 21, 1-11.

Rosenstock, I. M. 2005. Why people use health services. *The Milbank Quarterly,* 83.

Song, Y., Reifsnider, E., Chen, Y., Wang, Y. & Chen, H. 2022. The Impact of a Theory-Based mHealth Intervention on Disease Knowledge, Self-efficacy, and Exercise Adherence Among Ankylosing Spondylitis Patients: Randomized Controlled Trial. *Journal of medical Internet research,* 24, e38501.

Stetzer, E. S. 2011. Identifying risk factors for osteoporosis in young women. *Internet Journal of Allied Health Sciences and Practice,* 9, 6.

WHO 2011. *New horizons for health through mobile technologies.*

Yáñez Gómez, R., Cascado Caballero, D. & Sevillano, J.-L. 2014. Heuristic evaluation on mobile interfaces: A new checklist. *The Scientific World Journal,* 2014.

FROM DATA TO DECISION: EXAMINING REAL-WORLD DATA ANALYSIS TRENDS

Alfredo Tirado-Ramos[1], Xinwen Xu[1], Yuxuan Peng[1], Manuel Castañón-Puga[2], Luis Enrique Palafox-Maestre[2] and E. Dante Suarez[3]
[1]*Geisel School of Medicine at Dartmouth, Dartmouth College*
Lebanon, NH 03756, USA
[2]*Autonomous University of Baja California*
Tijuana, BC 22424, Mexico.
[3]*Trinity University*
One Trinity Place, Northrup Hall 210Q, San Antonio, TX 78212, USA

ABSTRACT

Although vast amounts of Real-World Evidence are available from Electronic Health Records, there is little information on current trends in Real-World Data analysis, thus hindering informed decision-making by healthcare professionals. Real World Evidence includes administrative claims, disease or product registries, patient-generated health data, and other digital health technologies. In this study, we review current and relevant developments, as well as guidelines across different countries, to summarize recent advancements and extract valuable insights from the existing literature on Real-World Evidence.

KEYWORDS

Real-World Evidence, Randomized Controlled Trials, Electronic Health Records

1. INTRODUCTION

The use of Real-World Evidence (RWE) in the initial evaluation of pharmaceuticals' benefits and the assessment of treatment effectiveness is becoming more prevalent. RWE refers to all data from any source that may contribute to more effective health care. As the limited eligibility criteria of traditional Randomized Controlled Trials (RCTs) underscores the importance of examining the translation of RCT findings to real-world settings, the exploration of Real-World Data (RWD) provides valuable context with more inclusive patient populations and representation of typically excluded patients in clinical trials. Even the U.S. Food and Drug Administration (FDA) defines RWE as the clinical evidence regarding the usage and potential benefits and/or risks of a medical product derived from analysis of RWD. RWE can be generated by different study designs or analyses, including but not limited to, randomized trials, large simple trials, pragmatic trials, and observational studies (prospective and/or retrospective). To date, RWE has been incorporated into a variety of regulatory approval procedures to assist with decisions regarding treatment safety, as well as to supplement randomized controlled trials in evaluating treatment effectiveness (Kirchgesner *et al.*, 2022).

2. PROBLEM STATEMENT

Over the past decade, the use of RWE has become increasingly prevalent in drug pre-approval and post-approval processes. However, there is a significant lack of comprehensive guidance on how to link RWE to routine clinical practices and research, posing challenges for healthcare professionals and researchers seeking to fully utilize its potential. Although vast amounts of RWD are available from sources such as Electronic Health Records (EHR), administrative claims, disease or product registries, patient-generated health data, and other digital health technologies, there is little information on current trends in RWD analysis, hindering informed decision-making by healthcare professionals (Derman *et al.*, 2022). In this study, we aim to review current developments and guidelines for RWE across different countries and summarize recent advancements in related fields to extract valuable insights from existing literature. Our goal is to help increase the implementation of real-world data in clinical practice and research.

3. METHODOLOGY

We conducted a targeted literature review, in PubMed using the following keywords: "real-world evidence", "real-world data", "effectiveness study", and "development and future directions". Such a strategy allows us to review key RWD/RWE studies, status in various countries, and regulatory guidelines. A total of 39 papers were examined and 6 of them were selected for a detailed evaluation of analysis-treatment effectiveness. In addition, the official websites of six different countries were examined to obtain the latest updates on regulations related to RWE.

3.1 Analysis: Treatment Effectiveness

Numerous studies globally employ real-world evidence to investigate the effectiveness of disease treatments, complementing the limited scope of randomized controlled trials. This approach offers several advantages, including evaluating the accuracy of clinical trial evidence, enhancing the generalizability of treatments used in clinical practice, investigating patient profiles of specific diseases, and analyzing treatment responses based on timeframes. In this review, we summarize treatment effectiveness for various diseases, outlining the methodology and main conclusions. In this way, we aim to understand and characterize the current use of real-world studies and to suggest novel approaches for leveraging real-life data effectively (Table 1).

Table 1. Real-World Evidence Usage in the Assessment of Treatment Effectiveness

First Author, Year	Agent	Study Focus	Disease Type	Study Result	Study insight
(Kuehne *et al.*, 2022)	US, Lilly Research Award Program	Methodological challenges and risk of bias when drawing casual conclusions using RWD, in the context of ovarian cancer	Ovarian Cancer	The "fully simulated trial" yielded a hazard ratio of 1.12 (95% CI: 0.96-1.28), which closely aligned with the reference randomized controlled trial. Immortal time bias is a significant source of bias that can impact the effect estimate in different directions.	This study offers guidelines for conducting causal analyses with real-world data and evaluates the effects of biases and confounding factors, such as time-independent and time-dependent confounding, selection bias, and immortal time bias. By employing causal frameworks, researchers can mitigate self-inflicted biases. Additionally, this study suggests that residual (unmeasured) confounding may have less impact on bias than previously thought in real-world data analyses.
(Kirchgesner *et al.*, 2022)	US, Clinical Pharmacology ε Therapeutics, Brigham and Women's Hospital and Harvi Medical School	Emulate the SONIC trial's findings on the efficacy of infliximab in Crohn's disease patients by utilizing healthcare claims data from the US and France.	Crohn's disease	Combination therapy reduced the risk of treatment failure in the overall cohort, consistent with the SONIC trial. Patients starting combination therapy had a 29% lower risk of treatment failure than those starting infliximab monotherapy. The study result was consistent across the three databases.	This study utilized real-world data to replicate the efficacy of combination therapy with two drugs in an RCT trial, providing strong evidence that real-world evidence analysis can complement the evidence obtained from RCTs by including patients who are generally excluded from RCTs and defining new outcome measures to improve treatment generalizability.
(Juarez-Garcia *et al.*, 2022)	Bristol Myers Squibb	Overview of the existing real-world evidence regarding the effectiveness of immunotherapy as a second or later treatment option for advanced non-small cell lung cancer in terms of patient survival.	Non-small cell lung cancer (NSCLC)	Of 66 RWE studies reviewed, 46 included a nivolumab-specific monotherapy arm. Nivolumab monotherapy showed higher 1-year and 2-year overall survival rates compared to mixed immune checkpoint inhibitors. Elderly patients and patients with poor performance status had	This study re-evaluated the overall survival rates of immunotherapy observed in clinical trials, using real-world evidence (RWE). The study highlights the advantages of utilizing RWE to analyze treatment effectiveness in vulnerable

				worse survival rates. Nivolumab had lower 2-year overall survival rates in patients with brain metastases than those without.	patient groups often excluded from trials.
(Peng *et al.*, 2022)	China, National Natural Science of China	Literature mining for studies on different clinical tumors and summarizing the antitumor effects of traditional Chinese medicine using real-world data	Common clinical tumors	The real-world clinical application of an integrated treatment approach using traditional Chinese and Western medicine has demonstrated anti-tumor effects regarding overall survival, progression-free survival, mortality hazard ratio, side effects, and medical adherence.	This study is a unique effort to evaluate the efficacy of traditional Chinese medicine by analyzing real-world data. The findings can establish evidence for clinical efficacy improvement and offer insights and cautions for other observational studies that utilize real-world data to determine treatment effectiveness.
(Zhao *et al.*, 2022)	UK, Cancer Research UK RadNet Manchester	Utilize real-world data to study radiation-induced heart disease in patients with lung cancer following radiotherapy and the subsequent change to cardiac dose constraints	Radiation-induced heart disease	An incidental dose to the base of the heart would increase the risk of early mortality in lung cancer patients. High-quality real-world data has the potential to provide robust evidence in the radiation-induced heart disease field by assessing the clinical impact of introducing a new cardiac avoidance region dose constraint as a substitute for traditional RCT.	This study considers the potential of using real-world data to assess dose-treatment effectiveness and corresponding patient survival in postoperative cancer where RCTs are generally impractical. The discussion of ways to achieve high-quality real-world data research in terms of developing auto-segmentation tools and cardiac calcification detection in different phases of the respiratory cycle is also valuable in the development of real-world studies to address cancer disparities.
(Escalera *et al.*, 2022)	Spain	A thorough scoping review of real-world data (RWD) that depicts the characteristics of patients with systemic lupus erythematosus (SLE) who received belimumab in addition to standard care in actual clinical settings in Spain.	Systemic lupus erythematosus	The inclusion of 19 retrospective studies explored the overall prevalence of SLE and the effects of treatment with belimumab on the use of corticoids in Spain. Adding belimumab to standard therapy reduced daily glucocorticoid intake by 1.4–11.1 mg at 6 months. Belimumab discontinuation was observed in 18.6% of patients.	This study demonstrates the potential of real-world data to analyze the clinical characteristics of patients receiving specific treatment. Incorporating long-term retrospective studies to examine treatment effectiveness, offers insights into usage shifts and aids in predicting future trends. This information can be used to identify patient populations that are most likely to benefit from the given drug.

3.2 Regulatory

We examine the regulatory and legislative developments regarding RWE and RWD by leveraging the main legislative frameworks published around 2018 and 2019, leading by FDA's publication of its RWE framework. Many additional countries, though, are still formalizing their drug approval process using RWE. Consequently, various guidelines regarding RWE and RWD for industries have been published. In Canada, for instance, several projects and initiatives have been hosted to examine the existing environment and infrastructure of RWD (Dai *et al.*, 2022; Government of Canada, 2018). The European Union has initiated the Data Analysis and Real-World Interrogation Network (DARWIN EU) project to collectively gather real-world evidence (European Medicines Agency EMA, 2021). China, on the other side, has also begun regional pilot applications of clinical RWD for drugs and devices in Hainan Province (CCFDIE, 2021)(Table 2).

Table 2. Regulatory Policies Updates

Country	Agency	Legislation	RWE Framework (Latest)	RWD Guidelines	Other Effort
US	FDA	21st Century Cures Act (2016)	FDA RWE framework (2018)	Considerations for the Design and Conduct of Externally Controlled Trials for Drug and Biological Products (draft guidance) (2023), Submitting Documents Using Real-World Data and Real-World Evidence to FDA for Drug and Biological Products (2022) (Guidance for Industry Final)	FDA grants for 4 projects exploring the utility of RWE (2020) Real-World Data: Assessing Registries to Support Regulatory Decision-Making for Drug and Biological Products Guidance for Industry (2021) (Draft) Considerations for the Use of Real-World Data and Real-World Evidence To Support Regulatory

					Decision-Making for Drug and Biological Products (2021) (Draft Guidance for Industry) Data Standards for Drug and Biological Product Submissions Containing Real-World Data (2021) (Draft Guidance for Industry)
Canada	Health Canada	NA	The regulatory review of drugs and devices (R2D2) project (2021), Optimizing the use of RWE to inform regulatory decision-making (2019)	Elements of Real-World Data/Evidence Quality throughout the Prescription Drug Product Life Cycle (2019)	Completed the project of Strengthening the use of real-world evidence and regulations for medical devices (2022)
China	National Medical Products Administration (NMPA)	NA	Guidelines for Real-World Evidence to Support Drug Development and Review (Interim) (2020)	Guidance for Real-World Data Used to Generate Real-World Evidence (Interim) (2021)	NMPA and Hainan Province jointly promoted the pilot application of clinical RWD of Drugs and Devices (2021) Technical Guidance for Real-World Study Supporting Research & Development and Evaluation of Pediatric Drugs (Interim)(2020)
Japan	Pharmaceutical and Medical Devices Agency	NA	Basic principles on the use of registries in approval applications (2021), Points to consider for ensuring reliability when registry data are used for approval applications (2021)	Points to consider for ensuring reliability when registry data are used for approval applications (2021) Utilization of Real-World Data - PMDA's approaches - (2021)	Established a RWD WG to deal with regulatory issues related to Real World Data such as using patient registry data and medical information databases. (2021) Posted a Q&A about Points to Consider for Ensuring the Reliability in Utilization of Data from Registry or Medical Information Database in Applications for Marketing Approval and Re-examination for Drugs (2022)
UK	Medicines and Healthcare Products Regulatory Agency (MHRA)	NA	MHRA guideline on randomized controlled trials using real-world data to support regulatory decisions (2021), MHRA guidance on the use of real-world data in clinical studies to support regulatory decisions (2021)	Points to consider for ensuring reliability when registry data are used for approval applications (2021) Utilization of Real-World Data - PMDA's approaches - (2021)	
EU	European Medicines Agency (EMA)	EU Pharmaceutical Strategy	EMA Regulatory Science to 2025 (2020), HMA/EMA EU Network Strategy to 2025 (2022)	EMA Patient Registry Initiative (2016–planned), and HMA/EMA Big Data Task Force data quality and representativeness framework (planned) Guideline on registry-based studies (2020)	Data Analysis and Real-World Interrogation Network (DARWIN EU) started working with hospital registries, and universities to collect real-world evidence. (2023)

4. ADDITIONAL CONSIDERATIONS

Another aspect of innovating in real-world evidence generation is by developing and employing digital health technologies. However, the use and interpretation of electronic health records (EHRs) data poses additional challenges that may arise out of potential biases imbedded in the original sources, records heterogeneity, and missing and/or non-standardized data formats. This notwithstanding, incorporating digital health data from wearable devices can provide more efficient access to care, and improve health literacy, adherence, and coordination. A recent advancement in this field comes from Germany's "Fast-Track" pathway, which established the market access for certain health applications known as DiGA, and passed through the regulatory process of the Federal Institute for Drugs and Medical Devices (BfArM) (Stern *et al.*, 2022). Currently, there are more than twenty approved digital applications under use for clinical practices, and BfArM promotes medical device development for more efficient patient care and clinical data collection (Stern *et al.*, 2022). Other countries, like the United States of America, are also making progress toward integrating regulated digital health applications. For example, "reSET", developed by Pear Therapeutics in Boston, MA (Stephanie Caccomo, 2017), received the first de-novo approval from the U.S. FDA as a digital therapeutic for examining product usage and efficacy through real-world data. Another example is "BlueStar", which focuses on diabetes and was developed by Welldoc in Columbia, MD, provides useful data collection with real-time feedback, drug usage guidance, treatment adherence promotion, and identification of clinical and non-clinical endpoints.

On the other hand, taking into consideration the recent and rapid progress of digital health initiatives, some concerns could be noted regarding the accuracy and reliability of collected data, along with the necessity of robust infrastructure and data management systems to handle the large amounts of data generated by digital devices. Further, there is still discussion about the optimal approach to managing and interpreting the implications of missing data (Stern *et al.*, 2022).

5. CONCLUSION

Upon analyzing targeted studies for specific diseases and worldwide regulatory developments, it is evident that all stakeholders encounter several common obstacles when utilizing RWE. These include challenges related to the quality and interoperability of RWD, as well as limited access to RWE (Schad and Thronicke, 2022). Methodologically, the main concerns center around the bias and confounding factors that may arise when employing RWE in experimental studies (Dai *et al.*, 2022). As such, randomized clinical trials (RCTs) are still the predominant type of study for generating reliable data used in drug approval processes.

Even though the current limitations of RWE outlined in this paper, there are multiple factors that point to its increased use. Aspects of the methodology that are driving RWE to become a major trend in the pharmaceutical industry include the growing digitalization of the healthcare industry, advancements in study designs, and the development of analytical methods that address statistical issues, alongside the growing support and promotion of the use of RWE by regulators worldwide. RWE has the potential to bridge the gap left by RCTs, particularly in addressing the limited number of patients in the rare disease market and promoting a fully representative patient population (Zura *et al.*, 2021).

REFERENCES

CCFDIE. (2021. NMPA and Hainan Province Jointly Promoted the Pilot Application of Clinical Real-World Data of Drugs and Devices. *The National Medical Products Administration (NMPA.* 28 December.

Dai, W. et al., 2022. Mapping Canadian Data Assets to Generate Real-World Evidence: Lessons Learned from Canadian Real-World Evidence for Value of Cancer Drugs (CanREValue) Collaboration's RWE Data Working Group. *Current Oncology*, Vol. 29 No. 3, pp. 2046–2063, doi: 10.3390/curroncol29030165.

Derman, B.A. et al., 2022. Reality check: Real-world evidence to support therapeutic development in hematologic malignancies. *Blood Reviews*, Vol. 53, p. 100913, doi: 10.1016/j.blre.2021.100913.

Escalera, C.R. et al., 2022. Use of belimumab in real-world in Spain: a scoping review about characteristics of SLE patients. *Clinical Rheumatology*, Vol. 41 No. 11, pp. 3373–3382, doi: 10.1007/s10067-022-06287-9.

European Medicines Agency EMA., 2021. Data Analysis and Real World Interrogation Network (DARWIN EU). *European Medicines Agency*, 26 March, available at: https://www.ema.europa.eu/en/about-us/how-we-work/big-data/data-analysis-real-world-interrogation-network-darwin-eu (accessed 12 June 2023).

Government of Canada., 2018. Strengthening the use of real world evidence for drugs. *Health Canada*, 22 August, available at: https://www.canada.ca/en/health-canada/corporate/transparency/regulatory-transparency-and-openness/improving-review-drugs-devices/strengthening-use-real-world-evidence-drugs.html (accessed 12 June 2023).

Juarez-Garcia, A. et al., 2022. Real-world effectiveness of immunotherapies in pre-treated, advanced non-small cell lung cancer Patients: A systematic literature review. *Lung Cancer*, Vol. 166, pp. 205–220, doi: 10.1016/j.lungcan.2022.03.008.

Kirchgesner, J. et al., 2022. Calibrating Real-World Evidence Studies Against Randomized Trials: Treatment Effectiveness of Infliximab in Crohn's Disease. *Clinical Pharmacology & Therapeutics*, Vol. 111 No. 1, pp. 179–186, doi: 10.1002/cpt.2304.

Kuehne, F. et al., 2022. Causal analyses with target trial emulation for real-world evidence removed large self-inflicted biases: systematic bias assessment of ovarian cancer treatment effectiveness. *Journal of Clinical Epidemiology*, Vol. 152, pp. 269–280, doi: 10.1016/j.jclinepi.2022.10.005.

Peng, L. et al., 2022. Real-World Evidence of Traditional Chinese Medicine (TCM) Treatment on Cancer: A Literature-Based Review. *Evidence-Based Complementary and Alternative Medicine*, Vol. 2022, pp. 1–10, doi: 10.1155/2022/7770380.

Schad, F. and Thronicke, A., 2022. Real-World Evidence—Current Developments and Perspectives. *International Journal of Environmental Research and Public Health*, Vol. 19 No. 16, p. 10159, doi: 10.3390/ijerph191610159.

Caccomo S., 2017. FDA permits marketing of mobile medical application for substance use disorder. *FDA News Release*, 14 September, available at: https://www.fda.gov/news-events/press-announcements/fda-permits-marketing-mobile-medical-application-substance-use-disorder (accessed 12 June 2023).

Stern, A.D. et al., 2022. Advancing digital health applications: priorities for innovation in real-world evidence generation. *The Lancet Digital Health*, Vol. 4 No. 3, pp. e200–e206, doi: 10.1016/S2589-7500(21)00292-2.

Zhao, X. et al., 2022. Integrating real-world data to accelerate and guide drug development: A clinical pharmacology perspective. *Clinical and Translational Science*, Vol. 15 No. 10, pp. 2293–2302, doi: 10.1111/cts.13379.

Zura, R. et al., 2021. Real-World Evidence: A Primer. *Journal of Orthopaedic Trauma*, Vol. 35 No. 1, pp. S1–S5, doi: 10.1097/BOT.0000000000002037.

AN IOT-BASED SYSTEM FOR MEASURING PENIBILITY AT WORK

Rémi Bastide
Champollion University
IRIT – University of Toulouse
France

ABSTRACT

The reporting of arduous work conditions in the workplace is a legal requirement in France. We present an IoT-based system that is designed to automatically collect most of the penibility factors as required by regulations, thus able to replace the declarative-only reporting that is currently in use. Another goal of our system is to investigate which of the hardship factors contribute most to the work penibility as perceived by the workers themselves.

KEYWORDS

Occupational Health, IoT

1. INTRODUCTION

Although the term "work penibility" is seldom found in the international scientific literature, it is of quite common use in France, both in the media and in the occupational medicine or in work sociology fields. It has been an important topic of discussion in the new retirement regulations that have recently been put in place. We have therefore chosen to use "work penibility" rather than "work hardship" or "word arduousness" that are more common in the English language.

France's workplace regulations make it an obligation for the employer to report on the arduous conditions endured by the workers, by maintaining a so-called "personal penibility account" (Compte Personnel de Pénibilité, CPP) for them, that may allow for compensations such as earlier retirement (up to 2 years), part-time work without wages reduction or benefit for professional education.

The identification of employees subject to a declaration is incumbent to the employer, who needs assessing the extent of each worker's exposure based on their typical working conditions, which are evaluated on an average basis over the course of a year.

We present an IoT-based system that is designed to automatically collect most of the penibility factors as required by regulations, thus able to replace the declarative-only reporting that is currently in use. Another goal of our system is to investigate which of the hardship factors contribute most to the work penibility as perceived by the workers themselves.

The paper first presents the various arduousness factors considered. It then presents the IoT-based architecture of the system we propose in order to monitor these factors automatically.

2. FACTORS OF WORK PENIBILITY

The arduous nature of work is distinguished by the exposure of employees to one or more occupational risk factors associated with significant physical constraints, a hostile physical environment, or specific work rhythms. The Labor Code enumerates ten such risk factors, that we classify into two categories: Ambient penibility factors, related to the environmental conditions at the workplace, and Task-related penibility factors, related to the physical aspects of the task performed.

- Ambient penibility factors
  - Mechanical vibrations: Refers to vibrations transmitted to the hands and arms, as well as those transmitted to the entire body, which may result in adverse effects.
  - Hazardous chemical agents: Includes substances such as dust and fumes that have the potential to cause harm or adverse health effects to individuals exposed to them.
  - Activities in hyperbaric environments: Refers to work performed in conditions of high pressure, such as underwater or in pressurized chambers.
  - Extreme temperatures: This refers to extreme temperature conditions encountered within the work environment, irrespective of external temperatures.
  - Noise: Denotes excessive or prolonged exposure to loud sounds or noise levels that may have detrimental effects on hearing and overall well-being.
- Task-related penibility factors
  - Marked physical constraints: This encompasses the manual handling of loads, which involves the transportation or support of objects through activities such as lifting, placing, pushing, pulling, carrying, or moving. These tasks require the physical exertion of one or more workers.
  - Awkward postures: Refers to forced positions of the joints that may be required during work, leading to discomfort or strain.
  - Work rhythms: Refers to the specific pace or pattern of work, including factors such as the duration of shifts and breaks, which can influence the physical and mental strain experienced by workers.
  - Night work under certain conditions: Relates to work performed during nighttime hours, subject to specific criteria or requirements that can impact the well-being and health of employees.
  - Work in successive alternating teams: Refers to shift work schedules, such as those involving three eight-hour shifts or two twelve-hour shifts, which can lead to challenges in maintaining regular sleep patterns and overall fatigue.
  - Repetitive work: Describes tasks that involve the execution of repeated movements, placing strain on some or all parts of the upper limb, typically performed at a high frequency and under constrained cadence.

The Labor Code mandates in great detail under which conditions the exposure to penibility factors should be taken into account in the personal penibility account (CPP), in terms of intensity and cumulative yearly duration. Table 1 gives an excerpt of these regulations for several of these ambient and task-related factors.

Table 1. Arduousness factors to be considered for inclusion in the CPP (excerpt)

Arduousness factor	Intensity	Minimal duration
Exposure to noise	8 hours above 81db	600 hours / year
	peak above 135 db	200 times / year
Repetitive movements	> 15 actions with a cycle time of less than 30s	900 hours / year
	> 30 technical actions / minute	
Extreme temperatures	outside of the 5°C-30°C range	900 hours / year

3. AN IOT SYSTEM TO MONITOR WORK PENIBILITY FACTORS

We have designed an IoT-based system that allows to automatically collect most of the penibility factors as presented above. The goals of this systems are twofold: It can be used to automatically collect the information needed for inclusion in the CPP, thus replacing the manual bookkeeping currently in place, that often requires approximative evaluations by the employer. Our second goal is to use the collected data to investigate which of the monitored factors contribute most to the penibility as perceived by the workers themselves.

3.1 Collecting Ambient Penibility Factors

The ambient penibility factors as defined in §2.1 are collected by an IoT node equipped with ad-hoc sensors. We have chosen to use Raspberry Pi™ as the hardware platform, due to its ease of use, low cost, and to the variety of off-the-shelf inexpensive sensors it can be equipped with. The device is equipped with temperature, humidity, noise, light, vibrations, and air quality sensors that collect all the measurements needed to assess ambient penibility. The measurements of these sensors are fed to a remote time-series database for real-time visualization and further data analysis.

3.2 Collecting Task-Related Penibility Factors

The penibility factors related to work schedules and shifts can in general be collected using the information system of the business under examination. However, we have chosen to collect more fine-grained data by using proximity sensors: the workers are equipped with a Bluetooth Low Energy (BLE) beacon, and the Raspberry Pi™ device tracks the proximity of the beacons, thus providing very detailed information on the actual time of presence of each worker in the workplace environment, allowing for an accurate calculation of the time of exposure to the ambient penibility factors stated in §3.1. The proximity sensors data is fed to the same time-series database used for sensors.

The repetitive movements are tracked by inertial sensors, in our case the sensors embedded in a commercial smartwatch (fitbit™) given to each user. Detailed information on the actual activity performed can be reconstructed from this inertial data (Lara and Labrador 2013)(Curone et al. 2010). The smartwatch moreover provides real-time information on heart rate, steps taken, calories burnt, skin temperature and oxygen saturation that give useful insights on the level of activity and stress of the worker. Although the data collected by the smartwatch is stored in a proprietary system, an API is provided to access it and store it at the same time-series database used for the other sensors.

3.3 Collecting Perceived Penibility

Beyond collecting the data mandated by the Labor Code for inclusion in the CPP, our goal is also to investigate which of the penibility factors contribute most to the penibility as perceived by the workers themselves.

- Instantaneous perceived penibility: By using the provided smartwatch, the workers can indicate in real time how they perceive the penibility of their work at any given moment, on a 0-9 scale, through a very simple touch-based used interface.
- Retrospective penibility: another simple smartwatch-based application allows the workers to retrospectively assess the penibility of their previous work shift, using the same 0-9 scale.

4. RELATED WORK

(Bavaresco et al. 2021) provide a detailed literature review of the use of IoT in occupational work, which has attracted attention in the recent years. (Palaniappan, Kok, and Kato 2021) systematically investigate the coupling of IoT and Artificial Intelligence in the field of occupational health and safety. Our proposal shares some concerns with the AAL-funded clockwork project (Pereira and Nunes 2018), which investigated the use of wearable sensors for monitoring physical activity for hospital workers.

5. CONCLUSION

This paper describes the goals and architecture of the system we propose for monitoring work penibility factors. The system will be deployed for validation in two different settings: we will investigate the work of pharmacy preparators in a hospital, representative of an office-like environment while still requiring significant posture changes and physical activity. The second experiment will address the training of firemen, a situation presenting very stressful and demanding physical activity in a hostile environment.

Our ongoing work concerns the development of the data analysis algorithms which will operate on the captured sensor data. These algorithms will address the identification of posture changes and the recognition of activities from the inertial data collected by the smartwatches, and the identification of the main factors contributing to the perceived penibility.

ACKNOWLEDGEMENT

This work is funded by the French Region Midi-Pyrénées under project PRESTON and is conducted by multi-disciplinary team of researchers in Sociology and in Computer Science.

REFERENCES

Bavaresco, Rodrigo, Helder Arruda, Eduarda Rocha, Jorge Barbosa, and Guann Pyng Li. 2021. "*Internet of Things and Occupational Well-Being in Industry 4.0: A Systematic Mapping Study and Taxonomy.*" Computers & Industrial Engineering 161 (November): 107670. https://doi.org/10.1016/J.CIE.2021.107670

Curone, Davide, Gian Mario Bertolotti, Andrea Cristiani, Emanuele Lindo Secco, and Giovanni Magenes. 2010. "*A Real-Time and Self-Calibrating Algorithm Based on Triaxial Accelerometer Signals for the Detection of Human Posture and Activity.*" IEEE Transactions on Information Technology in Biomedicine 14 (4): 1098–1105. https://doi.org/10.1109/TITB.2010.2050696

Lara, Óscar D., and Miguel A. Labrador. 2013. "*A Survey on Human Activity Recognition Using Wearable Sensors.*" IEEE Communications Surveys and Tutorials 15 (3): 1192–1209. https://doi.org/10.1109/SURV.2012.110112.00192

Palaniappan, Kavitha, Chiang Liang Kok, and Kenichi Kato. 2021. "*Artificial Intelligence (AI) Coupled with the Internet of Things (IoT) for the Enhancement of Occupational Health and Safety in the Construction Industry.*" Lecture Notes in Networks and Systems 271: 31–38. https://doi.org/10.1007/978-3-030-80624-8_4/COVER

Pereira, Ana, and Francisco Nunes. 2018. "*Physical Activity Intensity Monitoring of Hospital Workers Using a Wearable Sensor.*" In 12th EAI International Conference on Pervasive Computing Technologies for Healthcare. European Alliance for Innovation n.o. https://doi.org/10.4108/EAI.20-4-2018.2276323

COMPUTER MODEL FOR PREDICTING EFFECTS OF STRAIN-SPECIFIC IMMUNITY

Manuel Castañón–Puga[1], Camilo Khatchikian[2], E. Dante Suarez[3], Dustin Brisson[4] and Alfredo Tirado–Ramos[5]

[1]*Chemistry and Engineering School, Universidad Autónoma de Baja California*
Tijuana, BC 22424, Mexico
[2]*Department of Biomedical Data Science, Geisel School of Medicine at Dartmouth*
Lebanon, NH 03756, USA
[3]*Trinity University*
San Antonio, TX 78212, USA
[4]*Department of Biology, University of Pennsylvania*
Philadelphia, PA 19104
[5]*Department of Biomedical Data Science, Geisel School of Medicine at Dartmouth*
Lebanon, NH 03756, USA

ABSTRACT

Different outcomes may be expected when a particular host is exposed to an infectious disease, depending on its immunity status. For example, because of a previous encounter with the disease, the host may not become infected as consequence of the presence of antibodies. While sometimes it is difficult to establish when this disease avoidance occurred, the recognition of the existence and duration of such immunity is critical for the development of vaccines and accurate epidemiological models. In this paper, we propose a generalized computer model that can examine a wide range of epidemiological data and generate expected patterns of strain-specific infections and reinfection. These expected patterns can be compared and tested with empirical patterns existing in available databases.

KEYWORDS

Immunity, Reinfection, Model, Mining, eHealth

1. INTRODUCTION

Different outcomes can be expected when a particular host is exposed to a compatible infectious disease: a) the host become infected; b) the host does not become infected as, it is naturally immune to the disease; or c) the host does not become infected because it has developed resistance from a previous encounter with the disease. In the latter case—when immunity is acquired after a previous infection—the immunity could last for a finite timeframe or be permanent. The existence and duration of such immunity is a critical feature for the development of vaccines (for example, see Wormser 2021), as well as for the modeling of disease transmission, and the ultimate effects on entire populations (*i.e.*, Lee *et al.* 2021).

Identifying (and quantifying) the existence of acquired immunity from direct data is challenging, as the process involves having certainty of the existence of an "infectious event" (i.e., an exposure to the infectious agent) that did not result in an infection, where failure to produce a disease in the host could be due to innate or acquired immunity. It is feasible to distinguish between these two scenarios if there is a previous record of a successful infection, as such record would invalidate the innate immunity case.

The existence of records that confirm previous infections and exposure to an "infectious event" are difficult to find. Such difficulty is centered in the lack of evidence that a particular individual was exposed to an infectious agent, and that the reason of avoidance of the disease was the existence of acquired immunity. Previously, Khatchikian *et al.*, 2014, proposed that by examining and modeling the reinfection patterns of different Lyme disease strains (etiological agent *Borrelia burgdorferi*), it is possible to infer the presence and duration of strain-specific immunity. The model developed by Khatchikian and collaborators was limited to the outcomes of 17 clinical patients in the observed data set, which contained documented Lyme disease reinfections where the specific strain was identified, and correspondingly characterized according to a surface protein.

Current developments in electronic health records allow for the aggregation of systems that are publicly available, leveraging accessible research data. Furthermore, there has been recent and rapid growth of molecular sequences deposited in public repositories (such as Genbank, Benson *et al.* 2013), which allow the identification of the agents responsible for infections, and thus facilitate the accessibility and availability to such records (*i.e.*, Alex *et al.* 2017). Access to large amounts of epidemiologically relevant data highlights the potential utility of models that can take advantage of such data and produce pertinent inferences.

We propose a generalized computer model that can be used to examine a wide range of existing data. Such information may currently exist in clinical records, offering flexibility regarding the number of patients considered, infectious opportunities, number of strains involved, number of infections, and prevented infections. To this end, we provide examples and discuss the potential advantages and limitations of the proposed methodology.

2. METHODS

2.1 Model

The model allows the simulation of strain-specific immunity by generating expected patterns of infections and reinfections of different infectious agents' strains (*i.e.*, immunity-recognizable types), based on set model parameters. The model is intended to explore the parameter space, allowing the researcher to generate patterns that match patient-observed datasets. The model proposed here is the first attempt to expand and generalize the initial, infectious agent-specific (Lyme disease) model previously developed in Khatchikian *et al.*, 2014. The present model has greater flexibility and is intended to use data from multiple sources, including clinical records, online repositories, case reports, and simulated datasets.

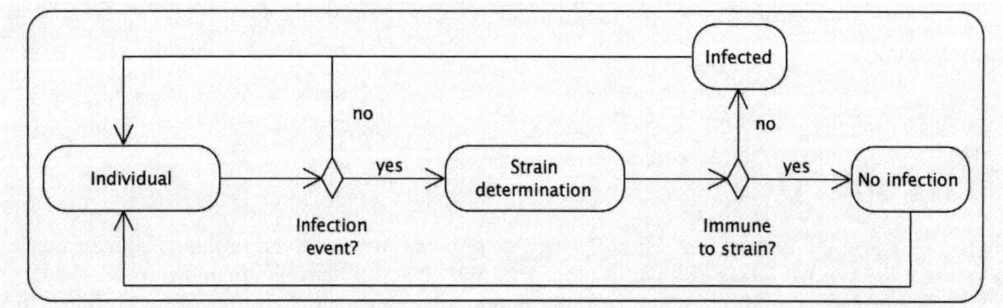

Figure 1. Schematic representation of the steps in the simulation model. Each individual's probability of exposure to an infection event are determined by the *probability of infection*, if it is exposed the model determines to which strain it is exposed according to the *strain distribution*. If the individual has been exposed to that same strain, the model considered that it is immune, and no new infection is recorded; otherwise infection occurs. Each individual goes thought this cycle n times according to the *time in simulation*. The total number of individuals is determined by the *number of individuals*

2.2 Script Coding

The model is coded in R (R Core Team 2017). We use the R implementation in RStudio (Rstudio Team 2015). The overall logic and data flow of the model is presented in Figure 1. To illustrate the model, we run it with simple settings (Table 1).

Table 1. General parameters of the model and specific values used in test simulation (right column)

Model parameter	Notes	Value in test run
Number of individuals	Number of potentially susceptible hosts	2,000
Time in simulation	Number of cycles in the simulation (*i.e.*, number of times that each host could be exposed to an infection)	20
Probability of infection	Chance of each host to be challenged by an infectious event	0.1
Strain distribution	Probability of infection with each specific strain	A=0.5 B=0.25 C=0.125 D=0.125

3. RESULTS

The model presented here allows to generate expected patterns of strain-specific infections and reinfection, which can in turn be compared and tested with observed or predicted patterns existing in databases.

Our test run of the script (see Table 1) allows for the comparison of the pattern (*i.e.*, distribution) of first strains vs. overall strain infections. It is therefore possible to infer the existence of strain-specific immunity by observing the changes in the strain prevalence (Figure 2).

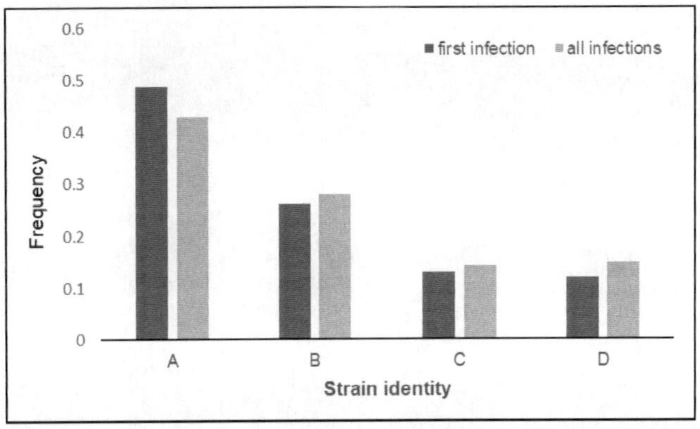

Figure 2. Frequency of strains (A, B, C, or D) in first strains vs. all infections. The strain present with high frequency (strain A) is less present in all infections, while less common strains are more frequent

4. CONCLUSIONS

Our proposed model opens the door to the exploration and simulation of different infections and reinfection patterns for diseases that have differentiable immune strains. The proposed methodology produces numerical expectations that can be statistically compared with observed datasets, generates inferences of an immune-specific event (as previously demonstrated by Khatchikian et al., 2014.), and thus allows the researcher to make inferences and predictions of the real epidemiological effect. The approach proposed here is limited to the existence of empirical records that can be compared with simulated values. Finally, this model has potential teaching applications, as it allows the easy manipulation of its parameters to generate a wide range of outcomes that demonstrate basic epidemiological concepts.

REFERENCES

Alex FB *et al.*, 2017. Exhaustively Characterizing a Patient Cohort by Prevalence of EMR Facts: a Generalized, Vendor-Agnostic Method for Quality Control and Research. *AMIA Annual Symposium Proceeding*, 458-464.

Benson DA *et al.*, 2013. GenBank. *Nucleic Acids Research,* 41, D36-D42.

Khatchikian CE *et al.*, 2014. Evidence for strain-specific immunity in patients treated for early Lyme disease. *Infection and immunity*, 82(4), 1408–1413. https://doi.org/10.1128/IAI.01451-13.

Lee TJ *et al.*, 2021. Network models in epidemiology. In: *Handbook of Statistics* (Eds. Srinivasa Rao ARS, Rao CR). Elsevier Vol 44. https://doi.org/10.1016/bs.host.2020.09.002.

R Core Team, 2017. R: A language and environment for statistical computing. R Foundation for Statistical Computing, Vienna, Austria. https://www.R-project.org/.

RStudio Team, 2015. RStudio: Integrated Development for R. RStudio, Inc., Boston, MA. http://www.rstudio.com/.

Wormser GP, 2021. A brief history of OspA vaccines including their impact on diagnostic testing for Lyme disease. *Diagnostic Microbiology and Infectious Disease*, 102(1), 115572. doi: 10.1016/j.diagmicrobio.2021.115572.

HOW TO INTEGRATE WIRELESS COMMUNICATION INTO THE HOSPITAL INFRASTRUCTURE TO ENSURE ITS SAFETY AND EFFECTIVENESS

Eisuke Hanada
Faculty of Science and Engineering, Saga University, Saga, Japan

ABSTRACT

Most Japanese hospitals have installed wireless LAN. It is currently not only used between PCs and Hospital Information System (HIS) servers, but increasingly for medical devices. Communication between medical devices and HIS servers enables monitoring of how well the devices are working, the exchange of data, and the distribution of instructions input by physicians. Wireless communication has been used in medical telemetry systems for decades in Japan, with the 400MHz frequency band used for signal transmission. Wireless voice communication by the in-house personal handy-phone system (PHS) has also long been used in Japanese nurse call systems. Due to the termination of the public communication service of PHS in 2023, such nurse call systems need to be replaced. Wireless LAN with Voice over IP (VoIP) technology is a strong candidate to be the next wireless communication system for use in nurse call systems. Wireless communication, if implemented properly, is effective for sharing patient information and raising patient safety. However, there are problems with the management of wireless communication systems, and more can be expected to arise in the future. When nurse call systems begin to use wireless LAN, the management of wireless communication may become much more problematic as the volume of data explosively increases. In this paper, we introduce two sets of Japanese guidelines that were promulgated to ensure the safe and effective use of wireless communication in Japanese hospitals.

KEYWORDS

Wireless Communication, Wireless LAN, Hospital Building, Medical Device

1. WIRELESS COMMUNICATION IN JAPANESE HOSPITALS

The results of a survey by the Japanese government showed that 90 % of Japanese hospitals have installed wireless LAN (EMCC, 2022). Its usage has mainly been for communicating between PCs and Hospital Information System (HIS) servers, but it is currently also being used for communication between medical devices and servers in Japan. This enables monitoring of how well the devices are working, the exchange of data, and the distribution of instructions input by physicians. Connecting medical devices to HIS servers is an effective way to share patient information and raise patient safety. Because of the widespread social use of wireless technologies, inpatients and visitors who wish to communicate with persons outside the hospital would like wireless LAN to be available to them.

Wireless medical telemetry systems (WMTS) have communicated wirelessly for decades. In Japan, they are analog systems that use the frequency band from 420 to 450 MHz for signal transmission (JEITA, 2020). The difference between Japanese WMTS and those used in other countries is only the frequency band. It is currently rare to see wireless LAN in Japanese nurse call systems.

The wireless communication systems that have historically been used in Japanese nurse call systems have mainly been based on the in-house Personal Handy phone System (PHS) (Hanada, 2000), which allows conversation between patients and nurses, even when they are moving around the hospital. PHS terminals emit a low output signal, 80mW maximum, so its use was safe from the viewpoint of electromagnetic compatibility (EMC). However, the public communication service of PHS was terminated in 2023, so other options must be explored. There are a number of candidates, including wireless LAN with Voice over IP network (VoIP) technology. If nurse call systems begin to use such systems, problems can be expected to arise in terms of their management. Also, to prevent the leakage of patient data, LANs used for hospital business, including wireless, must be separated from other LANs.

In this paper, we will introduce two sets of Japanese guidelines that were promulgated to ensure the safe and effective connectivity of Japanese medical devices and infrastructure.

2. PROBLEMS RELATED TO JAPANESE MEDICAL DEVICES

Here we illustrate from three viewpoints problems related to the connectivity of medical devices.

2.1 Technical Problems

The connection of Japanese medical devices to hospital networks has just begun. Each device currently establishes its own connectivity. For example, medical pumps tend to use the 2.4GHz band for wireless IEEE802.11 series LAN communication. In contrast, movable radiological devices tend to use the 5GHz band. Small medical devices, such as thermometers, pulse oximeters, and blood pressure meters use near field communication (NFC) with RF-ID or Bluetooth, including Bluetooth Low Energy (BLE). Any appropriate method can be used, and there is no need to unify the communication method. Unfortunately, no "medical band" has been assigned in Japan, which is also the case in many other countries.

Another problem is with data formats (expression), for which there are no worldwide standards. The Integrating the Healthcare Enterprise (IHE) association is working to improve connectivity. HL7 FHIR would be a viable candidate. Their target includes both medical devices and HIS, but their recommendations are not currently enforceable. Another problem is the amount of data processed in hospital LANs. It has been increasing and can be expected to continue to increase with the start of wireless communication by medical devices. If wireless voice communication were to be done through a LAN, the quality of voice communication would be degraded and there could be problematic congestion because of the huge volume of data. Standardization will be required not only for data expression but also for matters in the physical layer. Standardization of the physical layer would be effective for achieving EMC (Electro-Magnetic Compatibility), which involves operating in a way that does not interfere with other wireless communication.

2.2 Procedures for Medical Device Approval by the Japanese Government

Because Japan has a national medical insurance system, medical devices (and drugs) that can be used under the medical insurance system must be approved by the Ministry of Health, Labour, and Welfare (MHLW) under the Pharmaceutical Affairs Law. Applications for approval of a medical device by the Japanese government must follow strict procedures that take much time, one to two years on average. This is because high accuracy and safety are required for each device, and only if these are confirmed through extensive testing can devices be approved for inclusion in the Japanese medical insurance system. However, matters concerning connectivity are not included in this approval process. This often results in the wireless LAN standards used for active medical devices being outdated, as shown in Figure 1. Medical devices (covered by the insurance system) are required to be certified as meeting the technical standards for wireless communication before applying for pharmaceutical regulatory approval. This approval is by the Ministry of Internal Affairs and Communication (MIC). Thus, wireless communication devices that are certified and meet the newest standards cannot be used until they obtain pharmaceutical regulatory approval. Streamlining the certification and approval process would provide great benefits.

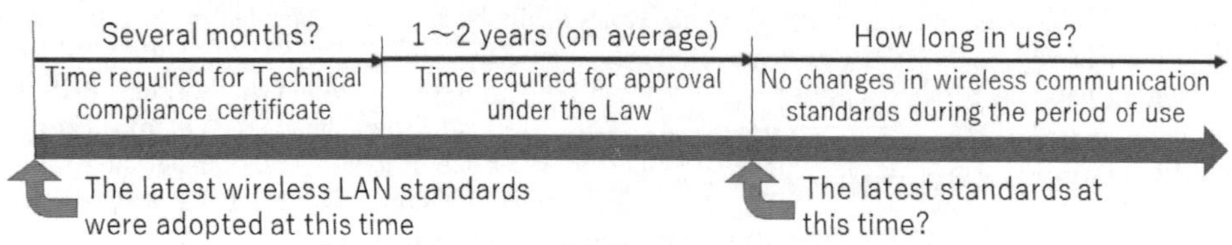

Figure 1. Timing of the adoption of standards for medical devices

Currently, if a manufacturer wants to change a communication function to one meeting the latest wireless LAN communication standards, it is treated as a "modification" of the medical device by the Pharmaceutical Affairs Law, which means the device must be re-approved.

2.3 Problems Related to The Electromagnetic Environment

The electromagnetic environment surrounding the wireless communication infrastructure is often not sufficiently considered when planning and constructing hospital buildings, especially concerning the following two aspects.

One is propagation problems, such as in the following examples;

- Many of the walls and doors in hospitals are made of or include metal
- Metal fixtures in rooms
- Metal ducts in ceilings

Electromagnetic propagation simulation is rarely done in the planning stage. Information on building component materials is rarely provided by the construction staff to the hospital staff members who will manage ICT / wireless communication. Also, metal fixtures that do not follow the original building plans are sometimes set without notice.

The other aspect is environmental problems such as electromagnetic noise. Even medical devices themselves can induce an electromagnetic field that acts as noise to a wireless communication system.

3. JAPANESE GUIDELINES FOR MANAGING WIRELESS COMMUNICATION IN HOSPITALS

In an attempt to solve these and other problems and to ensure the safe use of wireless communication in hospitals, two sets of guidelines were published in 2021.

The first is a “Guide to the Safe and Secure Use of Radio Waves in Medical Institutions”, first published in 2016 and revised in July 2021 (EMCC, 2021). This was developed by the Electro-Magnetic Compatibility Conference Japan (EMCC) in cooperation with MIC and MHLW. EMCC has also developed educational materials and videos for hospital workers.

The other is "Guidelines and descriptions for consideration when introducing radio wave-based equipment in medical institutions – Specific guidelines for wireless medical telemetry systems –", published in September 2021 by the Architectural Institute of Japan (AIJ) (AIJ, 2021).

EMCC guidelines have three main targets, WMTS, wireless LAN, and cellular phone use in hospitals. For each target, EMCC guidelines describe possible failures, countermeasures, precautions for safe implementation, explain methods for measuring the signal propagation / electromagnetic environment, and provide examples of countermeasures against tethering use by patients. Points to note in establishing a management system in medical institutions are shown in the guidelines. The EMCC guidelines recommend establishing a framework created by an interdisciplinary team for the management of an electromagnetic field that includes wireless communication.

When replacing PHS with another mobile communication system, some Japanese hospitals use smartphones. This will raise efficiency because the terminals can also be used for data communication (e.g., reference to patient information, input of vital signs, and patient identification using bar codes and RFID). For smartphones, the EMCC guidelines show four possible communication network construction methods: using VoIP to make calls through a wireless LAN, building a new self-owned LTE network (e.g. sXGP (ARIB, 2018)), building a self-owned 5G cellular phone system, or using a public mobile phone network. In all cases, data transmission speeds are faster than with PHS, but none of them is optimal in terms of cost, security, or precautions against electromagnetic interference due to the fact that some have a higher output than PHS terminals. This higher output may result in more problems related to EMC.

When using VoIP, existing hospital wireless LANs can be utilized, but the possibility of packet loss increases because the volume of information distributed suddenly increases. Packet loss in voice communication directly affects call quality. Voice communication is widely used in the medical field between patients, nurses, and doctors and, importantly, for emergency instructions. Packet loss can cause missed or misunderstood instructions, which can be life-threatening to patients. Therefore, communication networks should be constructed so that packet loss does not occur. Together with the increased use of wireless LANs for medical equipment described above, this is an issue that should be closely monitored in the future.

The introduction of self-owned 5G cellular phone systems into hospitals is also attracting attention. The 5G cellular phone system is a high-speed, low-latency, wireless communication system. In Japan, however, hospitals must be licensed to operate this kind of system. In addition, a full-time operation manager must be appointed and a fee must be paid for each terminal and base station. At present, self-owned 5G cellular phone systems in healthcare remain experimental.

The AIJ guidelines illustrate points to be considered for planning, installation, and evaluation of WMTSs in hospitals. The guidelines strongly recommend sharing information among the construction staff and the hospital staff who are charged with ICT management, especially for the materials of hospital walls, doors, and floors. This information is imperative to planning the area within which signals are reachable. Also, the guidelines strongly recommend the sharing of information between the hospital architect and the construction staff concerning the area within which patients who have a telemetry transmitter attached will move. This information is especially important when planning where antennas are to be set because Japanese wireless telemetry systems utilize two types of antenna systems. One is a "multi antenna system", and the other the is the so-called "leaky co-axial cable system". The former uses whip antennas, and cables between antennas and monitors will be radiant. In the latter, the cable itself acts as an antenna and the cable setting looks like a one-stroke drawing (Figure 2).

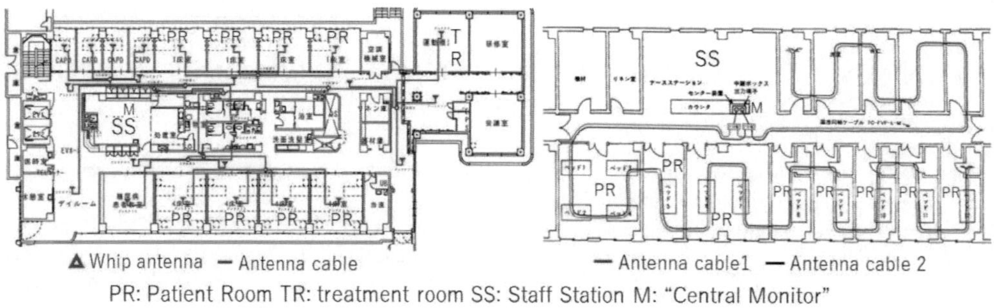

Figure 2. Antenna systems used in Japanese wireless telemetry systems (Left, Multi antenna system: Right, Leaky co-axial cable system)

The AIJ guidelines also show points to be considered when setting antenna cables, as shown in Figure 3. Interdisciplinary cooperation is necessary to make wireless communication in hospitals safe.

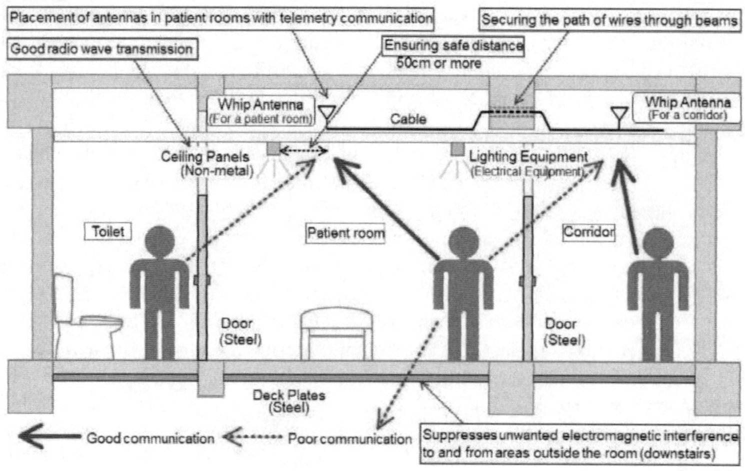

Figure 3. Points to be considered concerning where the cables and antennas are located (AIJ, 2021)

The AIJ guidelines also recommend conducting assessments of the electromagnetic environment and communication quality. WMTSs may identify transmitters using frequencies (channels) in the same frequency band as a wireless medical telemetry system, which can affect communication. This can result in electromagnetic noise that can affect machines monitoring the condition of patients in serious condition. Therefore, the AIJ guidelines require hospitals to evaluate wireless medical telemetry systems both before and after installation, and before operation. In addition, degradation of the antenna system can occur over time. The AIJ Guidelines require periodic communication tests to measure the strength of the received signal. As mentioned previously, the main difference between Japanese WMTS and those of other countries is only the frequency band assigned. Therefore, AIJ guidelines can be used worldwide. Only the settings of the apparatus at time of evaluation of signal acceptance will differ.

4. CONCLUSION

In this paper, we have identified a number of problems that affect the use of wireless communication in Japanese hospitals. We have also introduced two sets of Japanese guidelines that show how to adapt the infrastructure to ensure the safe and efficient use of wireless communication systems. ICT use in hospitals is expected to expand greatly in the future. To ensure its safe use, a well-designed infrastructure is required. Connection of Japanese medical devices to networks is just beginning, and each device currently establishes its own connectivity. Standardization is required not only for the application of systems but also for physical matters so they operate in a way that does not interfere with other communications. Assigning frequency bands dedicated to medical devices is desirable. We also recommend separating the frequency band for voice communication from that for data communication to prevent congestion of a wireless LAN, as shown in Figure 4; although we think only small amounts of data will be pass through voice communication systems.

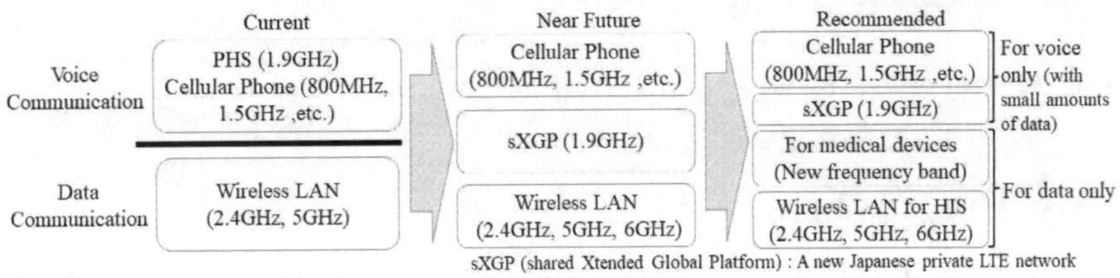

Figure 4. Wireless communication in Japanese hospitals, current, near future, and recommended (Hanada, 2022)

REFERENCES

Architectural Institute of Japan (AIJ), (2021). *Guidelines and descriptions for consideration when introducing radio wave based equipment in medical institutions – Specific guidelines for wireless medical telemetry systems –*. (in Japanese)

Association of Radio Industries and Businesses (ARIB), (2018). *ARIB STD-T118 OFDMA/TDMA TDD FOR DIGITAL CORDLESS TELEPHONE (sXGP) Ver. 1.1.* Association of Radio Industries and Businesses. https://www.arib.or.jp/english/html/overview/doc/5-STD-28v6_0-1p2-E.pdf

Electro-Magnetic Compatibility Conference Japan (EMCC), (2021) *Guide to the Safe and Secure Use of Radio Waves in Medical Institutions*. (in Japanese)

Electro-Magnetic Compatibility Conference Japan (EMCC), (2022). *Results of the 2021 Survey on the Promotion of Appropriate Radio Use in Medical Institutions (Hospitals)*. (in Japanese). https://www.emcc-info.net/medical_emc/pdf/22_R_R3_questionnaire_hsptl.pdf (Referred Feb. 24, 2023)

Hanada E., et al. (2000). Electromagnetic interference on medical equipment by low power mobile telecommunication system. *IEEE Transactions on Electromagnetic Compatibility* Vol.42, No.4, pp.470-476.

Hanada E., Kudou T. (2022). Future considerations for the management of hospital LANs. *In 2022 International Conference on Emerging Technologies for Communications (ICETC 2022)*, Tokyo, Japan, S10-1

Japan Electronics and Information Technology Industries Association (JEITA), (2020). *Low Power Medical Telemetry Device Usage Regulation*. (in Japanese). https://www.jeita.or.jp/japanese/standard/pdf/AE-5201B.pdf

DEVELOPMENT OF THE PROTOTYPE OF CLOUD-BASED FOOD SAFETY TRACEABILITY SYSTEM

Pradit Songsangyos[1], Arnut Ruttanatirakul[2] and Supakorn Iamamporn[2]

[1]*Faculty of Science and Technology, Rajamangala University of Technology Suvarnabhumi, Pranakhon Si Ayutthaya, Thailand*

[2]*Faculty of Business Administration and Information Technology, Rajamangala University of Technology Suvarnabhumi, Pranakhon Si Ayutthaya, Thailand*

ABSTRACT

Currently, concern about food safety is increasing dramatically. So food safety traceability system provides transparency and safety in consumer trust. This study aimed to develop a traceability system prototype which enables users to trace the food products from the manufacturer's production process back to raw materials from suppliers. Developing a prototype of the cloud-based food safety traceability system for food production management for chilled and frozen food and vegetable products is divided into two stages. Design a model of the cloud-based food safety traceability system for food production management. Besides, assessment of a prototype of the cloud-based food safety traceability system for food production management. Results of the evaluation of the system prototype by three experts, overall performance has the highest efficiency. Meanwhile, the users are satisfied with using the system prototype.

KEYWORDS

Cloud-Based, Food Safety, Food Production Management, Prototype, Traceability System

1. INTRODUCTION

Currently, the concern about food safety is increasing dramatically. So food safety traceability system provides transparency and safety in consumer trust. The traceability system lets final users verify feedstock quality in the food chain, from final users to production processes in the food industry (Foreign Agricultural Advisory Office for the European Union, 2022). Besides, traceability from the food industry to the supply chain.

Food traceability is essential in the industry, where consumers can trace each manufacturing process step. In addition, specialists can trace back when defects or problems occur in the production process. It is possible to ascertain the source of food quickly. Reduce the costs and quantity of recalled items because they can identify only the lot in question. They are increasing food safety requirements, besides complying with consumer demands that consumers want to know the origin of food. In addition, it prevents unsafe foods from reaching consumers (Foreign Agricultural Advisory Office for the European Union, 2022; Giuseppe T. et al., 2019).

In the study of Zheng M. et al. (2021), rice was used as an example and proposed food safety traceability systems using two-dimensional RFID code and big data technology with the Internet of Things. The article applies RFID technology to the system by analysing system requirements and designing tables and database systems. The process of information for food production is traceable through the design of dynamic query platforms and portable devices. In the study, a food-safety traceability system based on big data and the Internet of Things will guarantee traceable information's integrity, reliability and security. It is an effective solution that improves the reliability of traceable information. It ensures the integrity of information and has a high-performance data storage structure.

The main components in the application with RFID are (1) RFID Tag or Smart Label or Electronic Label. Microchips and antennas use for attaching electronic objects used to store information. There are several variations in their usage, (2) RFID readers or interrogators, devices used for the use and recording of data into the electronic label, which send radio signals to electronic devices for returning information and converting the received signals into digital formats for further processing. This device is both fixed and portable that can use as appropriate. RFID uses radio waves to read and record data, reducing barrier restrictions. It is possible to hide electronic labels in objects and read multiple tags simultaneously (Maimun, P., 2016; Kuawtaweekun, T., 2016).

Manufacturing is facing challenges in improving productivity. Modern factories will send several machines to the production line/assembly line. Each engine on the assembly line generates a large amount of detection and production data. The article aims to develop a distributed traceability system with Blockchain technology on IoT devices to improve the stability of factory production lines. To reduce the defect rate and raise the performance to a new level without relying on centralised data storage. It's a distributed and robust system. It can withstand node failures with better scalability and data interaction. The distributed network enables secure data collection and the potential of public traceability system protocols. In addition, it allows many factories to participate and build a shared traceability ecosystem (Shih, C.S. and Yang, K.W., 2019).

Improving services for pilgrims is one of the priorities of the Saudi Arabia government. Food supply is considered one of the primary services for pilgrims. In the current situation, meals are prepared under non-specific standards or regulations. Besides, the quality assurance process needs to work with multiple providers. The article proposed solutions for solving the problems by introducing the automation of all food supply services by establishing several centralised and specialised food factories operating under a single authority, widely distributed to cover a large area of the pilgrims. The factory will be responsible for food service to the performers on time without compromising the quality. The proposed system uses blockchain technology for quality control in the supply chain to record every transaction, verify the origin of every material used in food production, and track any problems at the source. Food will be cooked under specific standards and requirements, digitising the quality assurance process and making it more centralised (Radain, D. T. et al., 2021).

Tian, F. (2017) developed a food supply chain traceability system for real-time food tracking based on HACCP (Hazard Analysis and Critical Control Points), blockchain and the Internet of Things. Which can provide an information platform for all supply chain members with openness, transparency, impartiality, reliability and security. In addition, it introduced a new concept, BigchainDB, to fill the gaps in decentralised systems by scale. It concludes with a description of the use case and discusses the challenges of implementing blockchain technology in the future traceability system of the food supply chain.

In Taiwan, small-scale farming is common; most farmers grow crops based on their experiences. Therefore, excessive spraying of insecticides and fertilising often occur. In addition, raising livestock with management strategies in the enterprise can sometimes lead to antibiotics or other harmful chemical residues. The Taiwanese government has established a product traceability system with traceable markers. All agricultural products' information can be inquired through farmers' records on production and sales through Taiwan's agricultural product safety inspection information network. Qualified organisations will visit the production site to confirm whether the production record meets the requirements. By sampling the product, every batch inspected is recorded and traceable. Taiwan currently promotes a system of traceability of agricultural products and hopes that farmers will use it voluntarily. The traceability system can cause inconveniences that all farmers do not accept. Therefore, the study developed a traceability system of farm productivity using the Internet of Things technology, allowing the system to generate real-time production records (Jiang, J. A. et al., 2017)

The study of Tanthidolanet, N. and Boon Ying, S. (2020) has developed a model for the traceability of safe food crops in communities with RFID technology. The results showed that the community-based safety food crop traceability model with RFID technology consists of three modules: Module 1 for administrators, Module 2 for farmers, and Module 3 for customers.

Green Food operates a business related to processing agricultural products, vegetables and fruits by processing them into streaming, chilled and frozen products. The company uses raw materials ordered from various suppliers. In addition, there is a large number of documents related to the processes. Previously, Green Food implemented a manual archiving system that created problems in case of wanting to trace back which raw materials were received. When raw materials and processed goods are stored, such information is difficult to find and takes a long time. Sometimes a document is faulty, mismatched information cannot verify the correct information, or the document cannot be searched. The information is not up-to-date. At the same time, most studies of Thailand's agricultural food product traceability systems are aimed at traceback from manufacturers to raw material suppliers. This study contributes to developing a prototype of a cloud-based food safety traceability system, which enables final users and auditors to trace the food products from each production process of the manufacturer (Green Food) back to raw materials received from various suppliers.

Then conduct an implementation of a traceability system, a case study of Green Food as a pilot. The proposed system can manage the frozen fruit and vegetable production process semi-automatic, with RFID technology to meet the needs of Green Food.

2. METHODOLOGY

The research was conducted to develop a cloud-based food safety traceability system model. The study aimed at a food safety traceability system for the stream, chilled and frozen food and vegetable products industry, namely Green Food, as a pilot.

2.1 Data Collection and Analysis

The data collection using an assessment form for a prototype of the cloud-based food safety traceability system for food production management. The population in the study consisted of experts who have experience in the area of information and communication technologies. The sample group comprised of three experts who either held a doctoral degree or had at least five years' experience in information and communication technologies for evaluation and suggestions. A purposive selection method was selected. Jamil (as cited in Hamzah, F. et al., 2019) stated that a minimum of five years of working experience would ensure expertise in their respective fields. Analyse the results of the evaluation of the appropriateness of a system prototype using mean and standard deviation.

2.2 Research Processes

The research is divided into two stages.
Stage 1: Design a model of the cloud-based food safety traceability system for food production management.

1.1 Relevant articles and research papers were analysed and synthesised in the following topics: Cloud technology, traceability system, Internet of Things, information and communication technologies, food safety and food production.

1.2. Use the obtained information to synthesise and summarise as a model of the cloud-based food safety traceability system for food production management.

1.3 Research instrument was designed as an evaluation form. Five criteria for evaluation were considered using the 5-levels Likert scale anchored with the terms Strongly Agree, Agree, Undecided, Disagree, and Strongly Disagree (Samsudin, S. et al., 2019; Brown, S., 2010).

Stage 2: Assessment of a prototype of the cloud-based food safety traceability system for food production management.

2.1. A system prototype was proposed for evaluation by the three identified experts. There are five aspects of black box testing, such as (1) functional requirement test, (2) Functional Test, (3) Usability Test, (4) Performance Test, and (5) Security Test.

2.2 Analyse the system prototype's evaluation results using mean ($\overline{X}$) and standard deviation (S.D.).

2.3 Results of the Study

Stage 1: The model of the cloud-based food safety traceability system for food production management was designed as illustrated in figure 1.

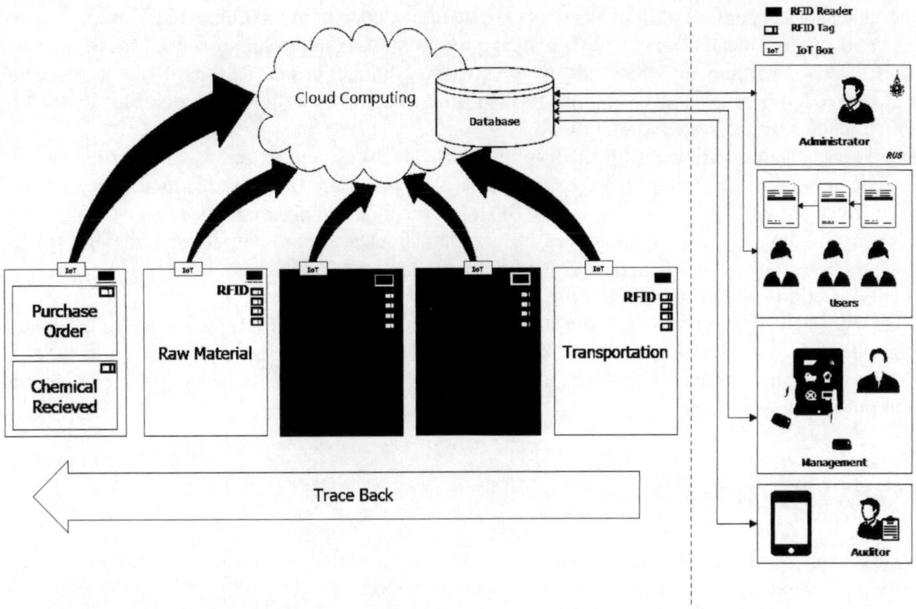

Figure 1. Model of the cloud-based food safety traceability system for food production management

Figure 1 shows a model of the cloud-based food safety traceability system for food production management consists of three parts as follows:

1. The online document recording includes (1) Purchase order, chemical received, (2) Raw material, (3) Production, (4) Packaging, and (5) Transportation.
2. The application which manages the daily documents.
3. RFID package for monitoring and tracking work processes.

Stage 2: Assessment of a prototype of the cloud-based food safety traceability system for food production management.

1. Conduct experiments using the system prototype to manage the production processes for the stream, chilled and frozen food and vegetable products industry, Green Food.
2. The system's users were divided into two groups: administrators and users, supervisor and operator.
3. Results of the assessment to determine the effectiveness of the system prototype by three experts, using five aspects of black box testing. The system's overall performance has the highest efficiency level, with an average of 4.57 and a standard deviation of 0.57. when considering each aspect, it found that the system's speed of operation has the highest efficiency, with an average of 4.73 and a standard deviation of 0.46.

The results of the system prototype evaluation in each aspect are as follows. (1) Functional Requirement Test has the highest efficiency, with an average of 4.67 and a standard deviation of 0.48. (2) Functional test has the highest efficiency with an average of 4.46 and a standard deviation of 0.59. (3) Usability Test has the highest efficiency, with an average of 4.54 and a standard deviation of 0.66. (4) Performance Test has the highest efficiency with an average of 4.73, and the standard deviation is 0.46. (5) Security Test has the highest efficiency, with an average of 4.42 and a standard deviation of 0.67.

4. It found that users are satisfied with using the system prototype. Whereas it is easy to use, reduces document processes and human errors, and enables to monitor of raw materials and finished products. Besides, enable to trace back sources of raw materials. So it increases overall productivity in the production processes.

3. CONCLUSION

The prototype of the cloud-based food safety traceability system for food production management for stream, chilled and frozen food and vegetable products of Green Food can work properly.

The systems model consists of three parts as (1) The online document recording, (2) The application which manages the daily documents, and (3) RFID package for monitoring and tracking work processes. It is easy to use and efficiently identifies raw materials' sources without human errors. Results of the assessment of the system prototype by three experts, overall performance has the highest efficiency level. Meanwhile, the users are satisfied with using the system prototype.

Presently the system prototype is under development with chilled and frozen fruits and vegetable products industry, Green Food. It can monitor and record various stages in managing production processes in chilled and frozen fruits and vegetables. There are a lot of audit documents to ensure the system works efficiently. In terms of changing the trajectory of employees from a manual inspection system to an automatic system using RFID technology. At the initial stage, it should be done as a parallel system. Both manual inspection and automated inspection systems with RFID technology.

Researchers will develop a system for streaming fruits and vegetable products for further study. Then should broaden the usage temperature range of RFID from -30 to 120 Celsius. In addition, comparisons are made to summarise the performance periodically by planning the work in a sequence of steps according to the document process flow system.

ACKNOWLEDGEMENT

Researchers express deep gratitude to the Talent Resource Management Project and the Rajamangala University of Technology Suvarnabhumi for the financial support. We are also grateful to the experts for their valuable evaluation and suggestions. Besides, we sincerely thank those who participated in testing the system prototype.

REFERENCES

Brown, S., 2010. Likert Scale Examples for Surveys. *In Iowa State University Extension*, pp. 1-4.

Foreign Agricultural Advisory Office for the European Union, 2022. The importance of the EU traceability system. [Online: 23 May 2022] https:// newsdatatoday.com/images/News/AO2019/102019/10474_3045_-_p. pdf

Giuseppe, T., et al., 2019. SISTABENE: an information system for the traceability of agricultural food production. *Proceedings of 2019 IEEE International Conference on Bioinformatics and Biomedicine*, pp. 2304-2309.

Hamzah, F. et al., 2019. Rasch Model Analysis for Evaluating Validity and Reliability of Authentic Learning Instruments in Malaysian Polytechnics. *In International Journal of Innovation, Creativity and Change.* Volume 9, Issue 7, pp. 51-62

Jiang, J. A. et al., 2017. "Integration of an automatic agricultural and livestock production management system and an agriculture and food traceability system based on the Internet of Things technology," *Proceedings of 2017 Eleventh International Conference on Sensing Technology*, pp. 1-7.

Kuawtaweekun, T., 2016. RFID innovations for applications in agriculture and food industry. *In Research Journal*, Faculty of Humanities and Social Sciences, pp. 119-131.

Maimun, P., 2016. RFID and traffic congestion management. Phetchabunsarn Rajabhat Journal, pp. 75-88.

Radain, D. T. et al., 2021. Towards Achieving the 2030 Vision, the Case study of Automating the Food Production Services during the Hajj Season and Quality Control Using the Blockchain Technology, pp. 144-154.

Samsudin, S. et al., 2019. Effectiveness of Outdoor Education Program on Physical Education Student Resilience. *In International Journal of Innovation, Creativity and Change*. Volume 9, Issue 12, pp. 218-229.

Shih, C.S. and Yang, K.W., 2019. Design and Implementation of Distributed Traceability System for Smart Factories based on Blockchain Technology, *Proceedings of the Conference on Research in Adaptive and Convergent Systems*. Chongqing, China, pp. 181-188.

Tanthidolanet, N. and Boon Ying, S., 2020. A prototype of the traceability of safe food crops in the community with RFID technology. *In Academic Journal of Information Technology Applications*, pp. 83-98.

Tian, F., 2017. "A supply chain traceability system for food safety based on HACCP, blockchain & Internet of Things," Proceedings of 2017 International Conference on Service Systems and Service Management, pp. 1-6.

Zheng, M. et al., 2021. Construct Food Safety Traceability System for People's Health Under the Internet of Things and Big Data. In IEEE Access, 2021 (9). pp. 70571-70583.

SITUATION-AWARENESS INTERACTIVE SELF-CHECK-IN SERVICES BY SMARTPHONES AND SMART LOCKS

John Yoon[1] and Jin Sup An[2]
[1]*Dept of Math & Computer Sciences, Cybersecurity Programs, Dobbs Ferry, New York, USA*
[2]*Dept of Information Technologies, SMA Group, New York, New York, USA*

ABSTRACT

One of the secure and intelligent functionalities in smart cities is a so-called smart lock and its operability. Smart door locks have been developed and deployed in home or hotel doors, building entrance doors, car doors, elevators, etc. It can be locked and unlocked by smartphone applications as well as it can be globally managed by a cloud server. However, the level of smartness in such door locks and its operability is not satisfied in part because the so-called smart lock is not inactive to the smartphone applications, and in part because the lock and the smartphone do not aware the situations surrounding one another. This paper surveys the current technologies and employ sensor and restful API server technologies to the smart lock development and the lock/unlock interoperability. The door locks will be smarter as the situations surrounding the smartphones is recognized or as the credentials of the human is verified. The most reliable requirement for smart self-check-in services includes 1) the appropriate services to guest's lock requests, and 2) the efficient and situation-awareness services in self-locking/-unlocking doors. To achieve the requirement, this paper proposes ways of resolving the errors by interoperating with sensors, and ways of improving the smartness of locks by interactive communications with a cloud server and smartphone applications. The contribution includes an improvement of the security and intelligence in smart self-check-in services.

KEYWORDS

WIFI, BLE, Smartphone Applications, PHP Servers, Smart Door Locks, Sensors and Actuators

1. INTRODUCTION

As sensors are widely used, the advances in the sensor technologies are profound and the power of microcontrollers gets stronger (J. Wilmsdorff, J. Kolf, A. Kuijper, 2022, T. Kalsoom, N. Ramzan, S. Ahmed, and M. Ur-Rehman, 2020, G. Baluyot, S. Chen, J. Villaverde, D. Chung, 2019, P. Poudel, B. Ray, A. Milenkovic, 2021), various devices can become connected one to another wirelessly and they behave intelligently in response to the change of (surrounding) data states (J. Xiao, H. Li, M. Wu, H. Jin, M. Deen, J. Cao, 2023, L. Xin, J. Becker, S. Gvozdenovic, D. Starobinski, 2019, T. Renzler, M. Spork, C. Boano, K. Romer, 2018). One of such devices is a smart lock. A smart lock is widely deployed in residential homes, hotels, and Airbnb (www.airbnb.com), automobiles, etc. (X. Zhang, M. Song, Y. Xu, Z. Dai, 2021). At a simple case, some of those smart locks are locked and unlocked by smart card, keypad, or smartphones. At another extreme end, as they are complicated, some of the smart locks are registered in a remote cloud system and a database. They are saved and managed in a server, and their profiles and accessibilities are controlled remotely from a cloud server. Lock access control is verified at the server side, and the decision is transferred to the smart lock and the guest's smartphones.

The simple smart locks are installed at residential homes and automobile cars, elevators, while in the complex case of multiple doors, those complicated smart locks are installed on hotel doors or Airbnb doors. The smart locks can be locked or unlocked by wireless communications with the web applications (or web App) from a server or the mobile applications (or mobile App) on a smartphone.

The smart door locks are equipped with a microcontroller which can run a program to switch the power supply in response to the request from a user (either through a web App or a mobile App). The smart door locks can communicate with other devices: in TCP protocol over WIFI with mobile applications (see 1 in Figure 1), in HTTP protocol with cloud server (see 1 and 2 in Figure1), and ad-hoc connection with mobile applications (see 3 in Figure 1). The ad-hoc connection can include a smart key card thru near-field communication (NFC), Bluetooth Low-Energy (BLE), Z-wave or Zigbee for home-devices for example.

There are two types of user interfaces (aka applications) to communicate to smart door locks: (1) Web App, where a web browser on a thin client enables a hotel guest to activate the door unlock request; (2) Mobile App, which runs on a smart phone to enable a hotel guest to click the door unlock button.

- A Mobile App presents an unlock request to the smart door lock. The request is transmitted to a TCP socket server in the smart door lock. The TCP socket server can receive lock requests from multiple socket clients (which means multiple guests or keys in the same room). This is illustrated in 1 in Figure 1.
- A Web App requests first unlock the lock to a cloud server (see 2 in Figure 1). Then, the server determines the door access to the smart door lock. The web server runs in a cloud server where a database contains the guests' profiles.

The former explains how the smart door lock functions directly by a Mobile App, and how it functions by a Web server which is requested by a Web App. Note that 3 in Figure 1 will be explained later.

This paper is motivated by the following motivating examples.

MOTIVATING **EXAMPLE 1** (*WIFI Shutdown*): Consider Figure 1, where a hotel guest tries to open the hotel door lock. When the guest approaches the cell phone to the door lock, if the cloud database does not respond (see 2 in Figure 1 has an issue), the request cannot be progressed, and the smart door will not respond (see 1 in Figure 1 where the router is down). In this case, rather than leaving the failure unfixed, the other route of data communication (possibly 3 in Figure 1). How can this be possible?

MOTIVATING **EXAMPLE 2** (*Intrusion Attack*): An attacker tries to open a hotel door lock as if the attacker presents as a legal hotel guest. The attacker requests to hotel door unlock through 1 to 1 in Figure 1 using a Web App or a Mobile App. If the attacker is not present in front of the hotel door, the server should not open the door for the attack guest. In this case, how can the server discern the attacker from the guest who must be in front of the hotel door?

The contribution described in this paper includes a TCP socket server that is equipped in a smart door lock as well as a cloud database server to cover the complexity of multiple doors, and an ad-hoc connections for emergency cases. We also employ sensor technologies and cloud database servers to improve the security and intelligence of self-check-in services.

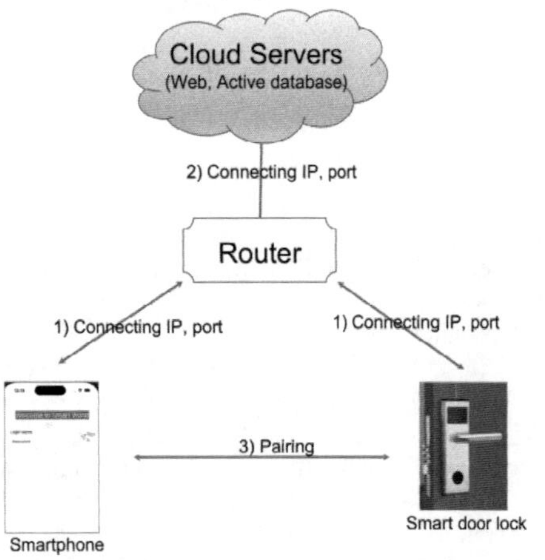

Figure 1. A self-check-in process in protocol view

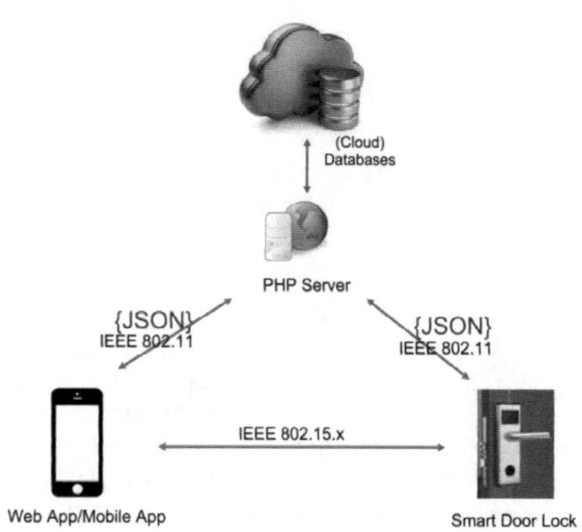

Figure 2. Global view of the architecture

2. PRELIMINARIES & RELATED WORKS

2.1 Smart Locks

Door locks become smart when they have both computing power and wireless connectivity. It is likely that a so-called smart door lock has a microcontroller embedded in a hardware lock fixture that has computing and communicating power as illustrated in Figure 1. An example of the microcontroller may be a Raspberry Pi (X. Zhang, M. Song, Y. Xu, Z. Dai, 2021) or a Pico W version, which can connect other devices thru WIFI.

A door lock will be smarter if the lock guides interactively the usage of devices which are held by the guest, or if the lock collaborates with one or more sensors (G. Bai, Y. He, C. Zhao, 2021) that can identify the ambient conditions to the door and the guest.

2.2 Wireless Communications

Theoretically, all types of radio waves in various frequency ranges can be used to communicate between the guest's smartphone and the door locks. IEEE 802.11-based WIFI is a typical wireless protocol to smart door locks (see 1 in Figure 1). While a network router assigns an IP address to each of the smart door locks and each of the Mobile Apps, TCP/IP network connections are possible. We can consider an ad-hoc network connection. IEEE 802.15-based wireless signals, e.g., BLE, ZigBee, etc., are also possible protocols (see 3 in Figure 1). A smart door lock and a Mobile App can be connected each other by pairing the key.

2.3 Web Apps and Mobile Apps

An interface to smart door locks can be available on a web browser and on a smartphone device, denoted as Web App and Mobile App, respectively.Web Apps are thin applications running on a web browser at a client, where the main logics are running at a server. The languages used to implement Web Apps are HTML, CSS, JavaScript including PHP. If user profiles are stored in a database, as a user requests to unlock a door lock, the user's access is verified based on the database. Then, the PHP server converts the verification result into a JSON data (D. Gope, D. Schlais, M. Lipasti, 2017). The JSON data will then be forwarded to a Web App and a smart door lock.

Mobile Apps are thick applications that run in their own implemented logics to request and respond to a door lock and a remote server. Mobile Apps are implemented in XCode Swift for iPhone devices or in Java for Android devices. They can communicate with a database through a PHP server by sharing data in a JSON format (I. Hrubaru, G. Talaba, M. Fotache, 2019) or communicate directly with a door lock through a socket coding. Of course, for the former communications, a PHP server at the cloud server should be implemented to receive the request from or send out the reply to a smartphone, while for the later communications, socket codes should be implemented on a smart door lock and open a port to receive TCP or UDP packets from or send encoded packets to a smartphone.

2.4 Sensors and Actuators

For security and intelligence improvement, sensors and actuators are employed (V. Modekurthy, M. Rahman, A. Saifullah, 2023, X. Cheng, M. Sha, 2023). The sensors that can be used for self-check-in services include thermal cameras, distance sensors, geo-code from GPS, motion detection sensor, etc. A thermal camera indicates the temperature of an object. Using it, the presence a human can be detected. A distance and motion sensor can detect the proximity of an object in front of the door lock. These sensors can be equipped and embedded in a smart door lock.

Since smartphones have a few of sensors available. The GPS sensor equipped in a smartphone can detect a geo-code. The geo-code can be used to determine the closeness of the smart door lock and a smartphone.

An actuator is a device that can react in response to the result of sensing. Monitoring the environment by one or more sensors may lead to take an action of an actuator. Sensing and actuating is coded at the smart door locks.

3. ARCHITECTURE

Our proposed approach consists of 4 major services: database service, web service, application services and smart lock service, which appear to be connected in Figure 2. The self-check-in services discussed in this paper relies on IEEE 802.11, WIFI signals to carry requests and responses among those four services. In addition to WIFI, cell signals are available to carry data for smartphones. Since smartphones have additional radio frequencies and sensors equipped, the ad-hoc signals in IEEE 802.15.x such as Bluetooth, GPS, NFC, etc. can be utilized.

The PHP services running on web servers 1) provide API's to relay the data between Web and Mobile Apps and Smart door locks, and 2) connect the backend databases to pose queries and receive answers to convert into JSON. The database queries are formed at the PHP services for the requests from Web/Mobile Apps.

In the normal situation, a smart door lock opens a port to receive a request packet from a guest. The Mobile App at the guest side has an IP assigned. In this situation, a guest activates the Mobile App to transmit the unlock request to the TCP socket server (to a specific IP and the port opened) at the smart door lock.

When it comes to multiple locks required, e.g., in a big hotel or in a large complex village, the security, safety and accuracy becomes the additional requirement for self-check-in services. Both the smart door locks and Mobile Apps connect to the PHP server to request additional data from the cloud database. Since the master data is available and maintained in the database, any services that require interrelationships among the guests and the smart locks need to communicate with the database through the PHP server. Note that the PHP server provides all Restful API's and connects to the database.

In an emergency, e.g., power outrage, network disconnection, etc., an ad-hoc network service, e.g., IEEE 802.15.x can be available. To cope with an emergent case, a smartphone needs to activate a special module to communicate with the smart door lock through Zwave, ZigBee, Bluetooth, etc. The communication in this protocol does not need a router.

4. SELF-CHECK-IN SERVICES

Our This section discusses three approaches of self-check-in services where smartphones and smart door locks are communicating through IEEE 802.11 WIFI signals or IEEE 802.15.x ad-hoc signals (IEEE 802.11 Standard, IEEE 802.15.x Standard). Depending on combination of services illustrated in Figure 1, we consider three approaches for smart door lock services: (1) Access control based on WIFI communications, (2) Centralized access control, and (3) Access control based on ad-hoc wireless communications.

4.1 Access Control Over WIFI – At Normal Situation

This approach does not refer to the authentication data in a remote database but makes it possible to communicate between the smart door lock and a Web or Mobile App. A network router enables them to IP addresses and the programs on the smart door lock and a Mobile App should open the IP with a port to communicate. This is illustrated in Figure 3(a).

4.2 Centralized Access Control – For Scaling Up

Consider a database that contains the information about all the guests and smart door locks. Data stored in a database can be accessed by Web and Mobile Apps and the smart door locks through a PHP server. The PHP server can function as a Restful server, which provides API's for the Web and Mobile Apps and for smart door locks.

In this approach, as a guest needs to open a door lock, a request for door opening is initiated on either Web or Mobile Apps. See Figure 3(b). The request can be transmitted to a web server through WIFI or a cell tower wirelessly. The web server includes the PHP server, which takes the request and relay it to the backend database. Since the database contains user profile table and authentication table, the answer to the request is returned to the PHP server. Then, the PHP server converts the result into the JSON data. The JSON data consists of the authentication for a given request as well as more data which will be used for self-check-in service improvement.

This approach is useful and necessary for a large scale of smart door locks, for example hotel room door management. Using this approach, a hotel with over hundred rooms can be managed automatically without having to much human interventions or manual services.

4.3 Access Control Over Ad-Hoc – At Emergency

Another approach does not use the network router WIFI protocol but uses ad-hoc communications. It means that this approach is appropriate for the case of emergency, e.g., power outrage or no WIFI available. Relying on IEEE 802.15.1, IEEE 802.15.2, or IEEE 802.15.4, ZigBee or BLE signal carries the lock or unlock request from a smartphone App. The program development and deployment in this approach is similar to the other approaches above. This approach is illustrated in Figure 3(c).

4.4 Sensor-Driven Improvement

Recall the sensors briefly introduced in Section 2.4. When a Web or Mobile App requests to unlock or lock the smart door lock, not only an authentication of the guest's credential but also one or more sensors can be installed at the smart door lock. For example, consider a distance sensor that measures the distance from an object. Only if the distance is in an expected range, additional authentication can be further processed.

The following is part of the TCP socket coding running on the smart door lock, which is the second approach as illustrated in Figure 3(d).

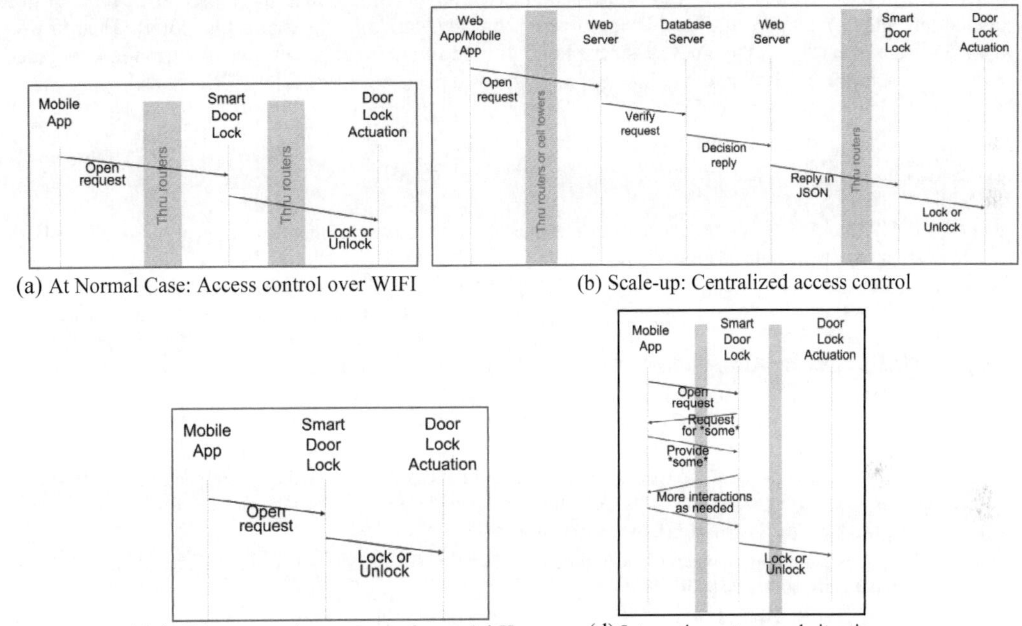

(a) At Normal Case: Access control over WIFI

(b) Scale-up: Centralized access control

(c) At Emergency: Access control over Ad-Hoc

(d) Interactions at normal situations

Figure 3. Approaches for self-check-in services

```
# some segments omitted
# the following computation depends on the specification of a distance sensor
pulse_duration = pulse_end_time - pulse_start_time
distance = round(pulse_duration * 17150, 2)

if distance < 50:    # if the distance is shorter than 50Cm
   good2To = False
   return True
else:
   reply5 = 'None appears in front of the door.'
   conn.send("From Lock: {}".format(reply5).encode())
   # interactively reply to guest's Web or Mobile App
   time.sleep(15)
   reply6 = 'Please approach to the hotel door lock'
      conn.send("From Lock: {}".format(reply6).encode())
   # this interaction can be iteratively provided for each different sensor data
```

As illustrated above, when the distance is beyond one half meter (or approximately 2 feet), the smart door lock can ask the guest to proceed closer to the door lock. This interaction can take place in response to various sensor's data.

5. CONCLUSION

Smart door locks presented in this paper become smarter and more intelligent than typical (commercial) locks. We have proposed to a WIFI-based TCP socket server that can be equipped in a door lock. For the management of multiple complicated door locks, a cloud server and PHP server can be involved. For an emergency case, e.g., WIFI is unavailable, or power supply is down, the ad-hoc network service can be activated to provide a seamless self-check-in service.

The functions that make the smart door locks smarter can be still improved by including sensors and machine learning technologies. Depending on sensors to monitor, interactions between the smart door locks and Mobile Apps can be made. Depending on the complexity of door locks, machine learning techniques can be applied to massive transactions of numerous guests' movements to determine the verification of guests' credentials and movements and then the result can be downloaded to the smart door locks. Then, the smart door locks interact with the guest's smartphone for additional guidance. The contribution includes an improvement of the security and intelligence in smart self-check-in services.

REFERENCES

Bai, G., He, Y., Zhao, C. (2021). "Research on task collaboration framework of mobile intelligent sensor cluster based on edge computing architecture," Proc. of the 1st Int'l Conf. on Control and Intelligent Robotics, pp. 86-91.

Baluyot, G., Chen, S., Villaverde, J., Chung, D. (2019). "Android application-based electrocardiogram design using microcontroller and bluetooth technology," Proc. of the 19th Int'l Conference on Biomedical Engineering and Technology, pp 27-30.

Cheng, X., Sha, M. (2023). "Autonomous traffic-aware scheduling for industrial wireless sensor-actuator networks," ACM Trans. On Sensor Networks (TOSN), vol. 19, pp. 1-25.

Gope, D., Schlais, D., Lipasti, M. (2017). "Architectural support for server-side PHP processing," Proc. of the 44th Annual Int'l Symposium on Computer Architecture, pp. 507-520.

Hrubaru, I., Talaba, G., Fotache, M. (2019). "A basic testbed for JSON data processing in SQL data servers," Proc. of the 20th Int'l Conf. on Computer Systems and Technologies, pp. 278-283.

IEEE 802.11 Standard, https://standards.ieee.org/ieee/802.11/7028/

IEEE 802.15.x Standard, https://standards.ieee.org/ieee/802.15.1/3513/, https://standards.ieee.org/ieee/802.15.3/6211/ and https://standards.ieee.org/ieee/802.15.4/7029/

Kalsoom, T., Ramzan, N., Ahmed, S. and Ur-Rehman, M. (2020). "Advances in Sensor Technologies in the Era of Smart Factory and Industry 4.0," Sensor, MDPI, vol. 20, pp. 1–22.

Modekurthy, V., Rahman, M., Saifullah, A. (2023). "Towards mixed criticality industrial wireless sensor-actuator network," Proc. of the 24th Int'l Conf. on Distributed Computing and Networking, pp. 425-430.

Poudel, P., Ray, B., Milenkovic, A. (2021). "Microcontroller fingerprinting using partially erased NOR flash memory cells," in ACM Transactions on Embedded Computing Systems (TECS), vol. 20, pp 1-23.

Renzler, T., Spork, M., Boano, C., Romer, K. (2018). "Improving the efficiency and responsiveness of smart objects using adaptive BLE device discovery," in Proc. of the 4th ACM MobiHoc Workshop on Experiences with the Design and Implementation of Smart Objects, pp 1-10.

Wilmsdorff, J., Kolf, J., Kuijper, A. (2022). "Reducing deployment cost for passive electric field sensors," Proc. of the 7th Int'l Workshop on Sensor-based Activity Recognition and Artificial Intelligence, pp 1-8.

Xiao, J., Li, H., Wu, M., Jin, H., Deen, M., Cao, J. (2023). "A survey on wireless device-free human sensing: application scenarios, current solutions and open issues," ACM Computing Surveys (CSUR), vol. 55, pp 1-35.

Xin, L., Becker, J., Gvozdenovic, S., Starobinski, D. (2019). "Benchmarking the physical layer of wireless cards using software-defined radios," in Proc. of the 22nd Int'l ACM Conf. on Modeling, Analysis and Simulation of Wireless and Mobile Systems, pp 271-278.

Zhang, X., Song, M., Xu, Y., Dai, Z. (2021). "Intelligent door lock system based on Raspberry Pi", Proc. of the 2nd Int'l Conf. on Artificial Intelligence and Information Systems, pp. 1-7.

HUMAN FACTORS ENGINEERING IN DESIGN OF SMART AUTONOMOUS TRAM SYSTEM

Jari Laarni[1] and Antti Väätänen[2]
[1] *VTT Technical Research Centre of Finland Ltd*
Tekniikantie 21, Espoo, 02044 VTT, Finland
[2] *VTT Technical Research Centre of Finland Ltd*
Visiokatu 4, 33101 Tampere, Finland

ABSTRACT

To promote the user-centered design of autonomous transportation systems a holistic integrated approach is required. We introduce a human factors engineering (HFE) program that provides guidance to the accomplishment of HF activities throughout the design process. The program consists of nine activities that can be performed in an iterative fashion. Here we present how main HFE activities have to be executed in order to promote the user-centered design of smart autonomous trams. Lessons learned from the ongoing SmartRail project and from some of our earlier projects have been identified and discussed.

KEYWORDS

Human Factors Engineering, Smart Autonomous Tram, User Centered Design, Concept of Operations

1. INTRODUCTION

Our aim is to develop a holistic framework for the user-centred design of smart autonomous tram systems. We will apply good practices of ecological design and Human Factors Engineering (HFE) approaches. Research is conducted under 'The SmartRail ecosystem', in which the aim is to foster the transition from traditional transportation systems towards an advanced service and business ecosystem consisting of a smart autonomous tram and integrated digital services (https://smartrailecosystem.com/). The SmartRail research project supports industry projects and promotes collaboration between different stakeholders and facilitates the building of an ecosystem around the smart tram concept. The research project aims to integrate knowledge areas of transportation service chain and users, connectivity and automation, and low carbon and energy efficient solutions (https://smartrailecosystem.com/).

There is a trend towards increased automation of transport systems and promotion of services that are based on high levels of automation, and there are previous and ongoing attempts to develop autonomous tram systems (e.g., the autonomous Siemens tram). The autonomous Siemens tram was launched in Potsdam during the InnoTrans 2018 Exhibition, and it performed a couple of demonstration drives with hundreds of passengers (Palmer et al., 2020). Furthermore, there are several ongoing attempts to develop autonomous public transportation systems, for example in Dubai (https://www.rta.ae/links/sdt/en/index.html) and in the Czech Republic (https://www.fdu.zcu.cz/en/Partnership/ForCompanies/AutonomousTram.html).

An autonomous tram is driverless, that is, the tram is circulating along its route without a driver sitting in a driver's seat. A fleet of trams is operated from a remote control center, where operators monitor the fleet or individual trams, and intervene in the operation of a tram in challenging situations.

The design task, i.e., the development of automatic tram, is challenging because the tram needs to operate in an open infrastructure. The automatic tram will also introduce new job roles, because tram drivers are replaced by fleet operators. Furthermore, there are several other worker groups such as technicians and traffic controllers whose work will change. Autonomous trams will also change the game from the perspective of passengers and other road users. Hence, operating a fleet of autonomous trams will introduce new human factors challenges that have to be taken into careful consideration (Laarni & Väätänen, 2023).

We introduce a HFE program that has been developed under the SmartTram project in order to better understand the human factors issues in the design of the automatic tram system. There are several reasons for taking a closer look at human factor issues in the design of autonomous tram systems. It has been found that complex engineering projects do not always proceed as has been originally planned. One of the main reasons for missed deadlines and budget overruns is the lack of socio-technical systemic approach in the design work, that is, the social and organizational complexity of the environment in which the systems are used is not considered to a sufficient degree (Mumford, 2006). Similarly, experiences from several design projects have shown that the demand to deliver systems within budget and on schedule has resulted in insufficient attention to HFE considerations in system engineering (Madni, 2010). It is also typical for the organization to drift away from the systems engineering and human factors standards it still believes it is following (Mumford, 2006).

HFE is a scientific approach to the application of knowledge about human factors and ergonomics to the design of complex technical systems. It can be considered as one engineering framework among several others under the systems engineering discipline. It can be divided into tasks and activities in terms of which stages of the design process they are related to. Typically, HFE activities are classified into four groups: analysis, design, assessment, and implementation/operation. In the analysis phase, all critical analyses providing background information and context for the design work are performed, in the design phase human-system interfaces, procedures and training activities are developed; in the evaluation phase, the design systems and activities are verified and validated; and finally, in the operation phase, in-service monitoring of the new systems is performed.

It is well known that the way the design work is carried out and how the human and organizational issues of the design are analysed during the design process has implications in how humans interact with the designed system in the future (e.g., Electric Power Research Institute, 2015). HFE activities are concurrent, iterative processes conducted to reduce project risks at different stages of the design work. HFE activities should both aid design work and progressively promote risk reduction during the lifecycle. HFE does not only enable to avoid accidents and to increase efficiency of technical systems, but it also enables to reduce project costs and to produce improvements in performance and productivity (Booher, 2003).

The following activities are typically included in the HFE approach for complex technical systems such as smart tram systems: management and planning including screening and grading of HFE activities, operating experience review, function allocation and analysis, task analysis, treatment of important human actions, manning, operator selection and qualifications, system design and development, design of maintenance activities, procedure design, training design, verification and validation, and commissioning and in-time monitoring. In the next section, the main HFE activities in design of smart autonomous tram systems are described in a more detailed fashion (see Figure 1).

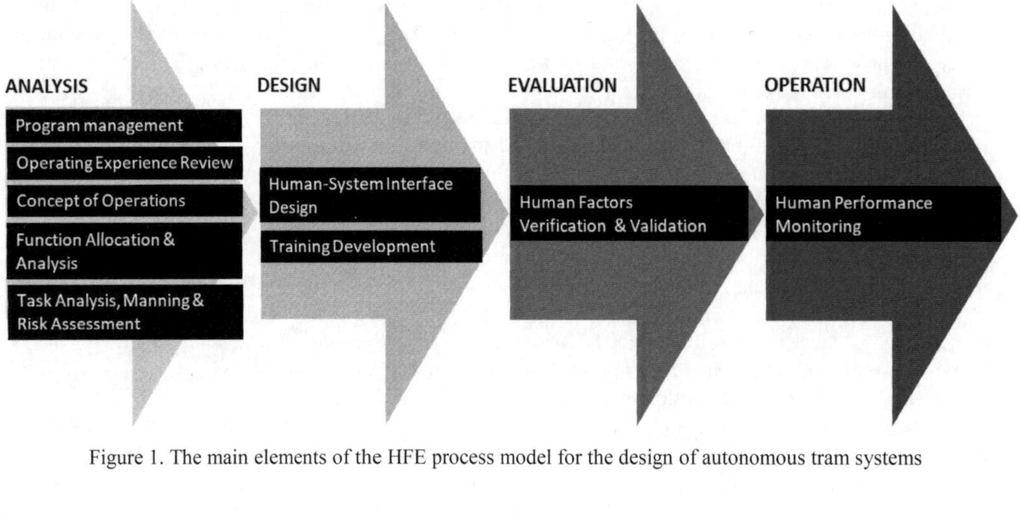

Figure 1. The main elements of the HFE process model for the design of autonomous tram systems

2. MAIN HFE ACTIVITIES IN DESIGN OF SMART TRAM SYSTEMS

2.1 Project Management

Project management is required at all the stages of the design process to efficiently integrate HFE activities into the design of an autonomous tram system (Bruseberg, 2008). In this first stage of the project, a detailed plan is needed for HFE, which is applied in parallel with the system development plan. The HFE program plan provides guidance to the accomplishment of individual HFE activities throughout the project.

Screening/grading, in which the impacts of the new design are determined in terms of complexity and safety impact in order to tailor HFE activities for design targets, is typically performed early in the project. An early-stage screening process is important in determining the emphasis and amount of effort allocated to HFE activities. In general, most of the efforts should be paid on the treatment of high-risk activities (Electric Power Research Institute, 2015).

2.2 Operating Experience Review

The aim of the operating experience review activity is to better understand current work of tram drivers, identify possible operational problems that can be addressed in smart autonomous trams and collect potential experiences with autonomous driving in other domains. Data is collected by interviewing tram drivers, trainers and traffic controllers. Tram drivers are experts in identification of potential hidden dependencies between activities that may at worst trigger an accident. It has also been valuable to discuss with experts working on research and development of autonomous road vehicles. For example, lessons learned accumulated over years in the development of autonomous cars are highly relevant in the design of autonomous tram systems.

2.3 Development of a Concept of Operations from the Perspective of Main Stakeholders

One of the first tasks in the analysis stage is to develop a Concept of Operations (ConOps) for the new system. The ConOps method was introduced by Fairley and Thayer (1997) as a bridge from operational requirements to technical specifications. ConOps can be considered as a transitional design artefact promoting the requirements specification activity during the early stages of the design (Laarni & Väätänen, 2023).

An overall Concept of Operations (ConOps) has to be defined for the smart autonomous tram system in the beginning stage of the design project. The objective is to specify the main aims of a smart rail system, potential challenges and open questions, identify key use cases and develop a Concept of Operations from the perspective of various stakeholders and user groups (Laarni & Väätänen, 2023; Figure 2). Recently, Laarni and Väätänen (2023) developed a ConOps for an autonomous smart tram system to demonstrate how smart trams can be applied in public transportation in the future. The ConOps for an autonomous tram system consists of, e.g., concept for human-autonomy teaming, identification of main operational states and of characteristics of user interfaces for supervising the tram fleet and occasionally controlling individual trams (Laarni & Väätänen, 2023).

2.4 Function Analysis and Allocation

Functions are activities that have to be performed in order to achieve a particular goal. Examples of main functions in the operation of an autonomous tram system are, e.g., approaching and passing a tram stop, approaching a pedestrian crossing, and crossing a street junction (Laarni & Väätänen, 2023). They can be presented in a hierarchical manner, from high-level functions to more detailed functions at the lower levels of the hierarchy. The primary aim in function analysis/allocation is to specify main functions and allocate functions of fleet operation in such a way that operator capabilities and capabilities of the autonomous tram system are used in an optimal manner. Furthermore, the aim is to identify the design consequences of allocation decisions.

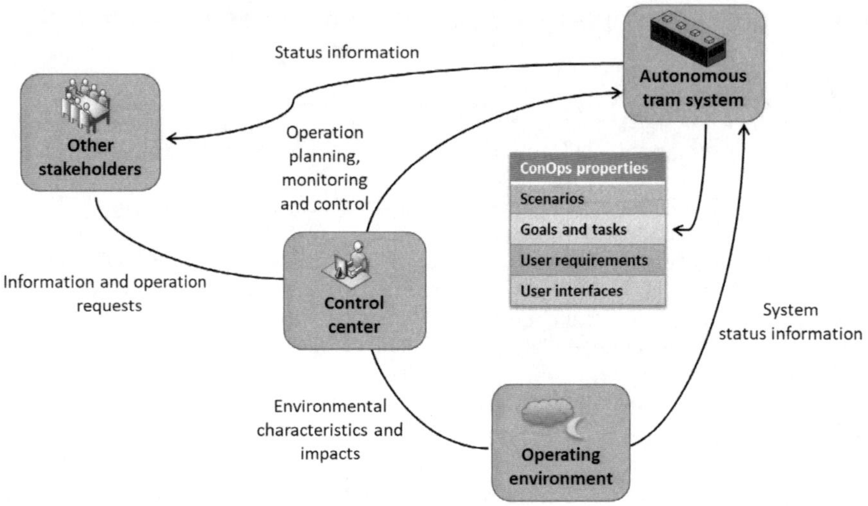

Figure 2. ConOps diagram for a smart autonomous tram system (modified from Laarni & Väätänen, 2023)

Function analysis determining what is required to perform a particular function provides input to function allocation (e.g., Electric Power Research Institute, 2015). Function analysis should be conducted to 1) determine the objectives, requirements, and limitations of the interface between the operator and the autonomous control unit; 2) specify the functions that must be accomplished to meet the objectives and required performance; 3) define the relationships between functions and the autonomous control unit responsible for performing the function; and 4) provide a framework for understanding the role of controllers (whether personnel or system) for controlling the tram (e.g., Electric Power Research Institute, 2015).

Function allocation describes the allocation of responsibility of the supervision and operation of a fleet either to an autonomous agent, a fleet operator or some combination of the two. Function allocation is optimized by selecting an optimal level of autonomy for each level of a function. Potential failures and malfunctions are also considered, including how they will be detected and what actions will be required to deal with the situation (e.g., Electric Power Research Institute, 2015). For each function, we have to determine whether an autonomous agent or human action is required and feasible and allocate the function either to an autonomous agent, to a human or share the allocation between both entities (e.g., Electric Power Research Institute, 2015).

2.5 Identification of Main Stakeholder Roles, Task Analyses and Risk Assessment

In the next activity, the aim is to first identify main worker groups and networks of groups (i.e., who participate in the fleet management); second, to identify and specify the main tasks included in the fleet management and analyze them (i.e., what kind of tasks are included in the use of tram systems, and what kind of subtasks can these tasks be divided into); and third, it is analyzed the dynamics of the activity and identified and specified the main risks and hazards associated with the activities.

Task analysis focuses on the analysis of functions that have been allocated to a fleet operator so that requirements can be specified for user interfaces that the operator needs in order to accomplish the human actions. Its main aim is thus to ensure that necessary fleet operator tasks can be successfully performed. Each task is described in sufficient detail so that requirements for task support can be specified. Typically, the task requirements include all the information, controls, alerts and other operator aids that are required for successful task performance (Electric Power Research Institute, 2015).

Based on our interviews we conducted with experts on public transportation, manning, operator selection and qualifications are important issues throughout the life-cycle of an autonomous tram system. The main goal of manning and qualification analysis is to show that the requirements for the number and qualifications of operators has been systematically analyzed and specified, and task requirements of manning and qualifications are well understood.

The objective of the assessment of risks and hazards is to identify and analyze critical operator activities and the potential of errors associated with the activities. The Functional Resonance Analysis Method (FRAM) was used in hazard analysis (Hollnagel, 2012). Several challenging activities were identified in tram operation related, e.g., to carelessness of pedestrians and car drivers, consequences of emergency braking and passengers' boarding on or off the tram. The FRAM method is used to establish a model of a certain activity representing the main features of how the activity is performed. The model specifies the potential variability of the system in terms of main functions (Hollnagel, 2012). FRAM is based on a couple of principles, for example, successes and failures are equally valued, fleet operators are considered as proficient in adjusting their behavior to the contingencies of an actual situation, and unanticipated events may be caused by unexpected interactions of multiple functions, e.g., in demanding weather conditions. FRAM has shown to be a valuable tool in understanding the reasons behind the gap between work-as-designed and work-as-done (Hollnagel, 2012).

2.6 User-Centered Design of Smart Tram Systems

Design of a control center for the management of autonomous tram systems is based on design dialogues between various stakeholders. Also, the design of training activities (that is, determination of what kind of training and education is needed) engages various stakeholder groups.

During the design process, function and task requirements are translated and developed into human-system interface characteristics and functions. The HFE activity aims to assess the process by which design requirements are identified, developed and refined. Typically, design guides are prepared in accordance with relevant human factors engineering principles.

Based on expert interviews, also HF aspects of the maintenance of the autonomous smart system has to be dealt with, including all activities that are conducted to prevent tram system failures and repair failed tram systems. The ultimate goal is to facilitate future maintenance activities by anticipating possible future malfunctions and to strive to make the systems more error-resistant.

Training must be performed to acquire the skills for using and maintaining the HSIs of the control center, and to adapt to new job practices. The aim is to develop training programs that teach skills fleet operators need in order to perform their critical duties in an adequate manner. Training design is based on systematic analysis of job and task requirements. The output of the design work is a concrete training program. Typically, the systematic approach to training design is based on four activities, analysis of training needs, design of training, implementation of training, and evaluation of training effectiveness (e.g., Laarni et al., 2011).

2.7 Verification and Validation, and Human Performance Monitoring

Verification and validation (V&V) activities are performed to ensure that the human-system interfaces of the control center are well designed and easy to use, and they fulfil main performance requirements (Koskinen, Laarni et al., 2021). A special attention has to be paid on the implementation of operator-autonomous system interaction. V&V is then planned to the evaluation and testing the operator's interaction with the autonomous system. It consists of the determination of acceptance criteria, execution of the test activities and the analysis of test results.

The verification and validation process must be iterative by nature, and there must be a systematic procedure for aggregating and systematizing V&V data. We have applied the Systems Usability Case (SUC) approach which enables a requirement-based human factors evaluation of complex systems (Koskinen, Laarni et al., 2021). SUC is based on Safety Case thinking and on the application of the Systems Usability construct. According to Koskinen, Laarni et al. (2021), SUC can be seen as a conceptual procedure within which a reasoning process takes place enabling a shift from abstract understanding of human performance and usability of complex tools to a concrete proof of an autonomous tram system, and finally to understanding of what is learned about the overall acceptability and level of Systems Usability of these systems.

In the implementation phase it is evaluated that all HFE issues have been adequately resolved and closed out. It is also designed the practices for the documentation and deployment of operating experiences throughout the system's life-cycle (i.e., human performance monitoring).

3. DISCUSSION

Many authors and organizations have called for a more integrated and holistic approach to HFE. According to Madni (2010), two kinds of integration of HFE activities with systems engineering are needed: Since the HFE area is quite fragmented itself, the HFE activities have to be first integrated with each other; after that they can be integrated with system engineering domains.

One of the key requirements for a unified HFE process for the design of autonomous tram systems is to start human factors activities as early as possible in the design process and continue them throughout the life-cycle of the autonomous tram system (cf., Koskinen, Laarni et al., 2021). When HF issues are considered late in the process the required changes become costly, and they are not necessarily any longer optimal to the problem (e.g., Bruseberg, 2008). In addition, it is also possible that HF issues will be treated in an inconsistent and fragmented manner, if the HFE process is started in the later stages of the design work.

There are clear benefits of HFE in autonomous tram system design: It is a support function through which HF problems and challenges can be avoided and mitigated throughout the project lifecycle (e.g., Bruseberg, 2008). HFE activities provide input to the design of autonomous tram systems, and they also function as a quality control at various stages of the design process. HFE experts play a mediating role between operating personnel and designers of an autonomous tram system and in establishing contacts between different parties. HFE also has a critical role in mitigating operational risks due to the fact that potential problems and risks can be better identified at early stages of the project when applying HFE.

We have focused here on the establishment of a HFE program for the design of a control center for fleet management. The focus has thus been quite narrow. To develop a similar integrative program for the SmartRail ecosystem, a broader Human Systems Integration (HSI) perspective may be feasible. According to Booher (2003), HSI takes an overarching holistic approach and integrates knowledge of various areas of human and organizational factors to promote systems engineering activities.

ACKNOWLEDGEMENT

The research was funded by Business Finland under the SmartRail 2 project.

REFERENCES

Booher, H.R., 2003. Introduction: Human Systems Integration. In Booher, H.R. (ed.), *Handbook of Human Systems Integration*. Hoboken, NJ, Wiley.

Bruseberg, A., 2008. Presenting the value of Human Factors Integration: guidance, arguments and evidence. *Cognition, Technology & Work*, Vol. 10, pp. 181-189.

Electric Power Research Institute (EPRI), 2015. *Human Factors Guidance for Control Room and Digital Human-System Interface Design and Modification (EPRI Report 3002004310)*. EPRI, Palo Alto, CA.

Fairley, R.E. and Thayer, R.H., 1997. The concept of operations: The bridge from operational requirements to technical specifications. *Annual Software Engineering*, Vol. 3, pp. 417-432.

Hollnagel, E., 2012. *FRAM, the functional resonance analysis method: Modelling complex socio-technical systems.* Farnham, Surrey, UK, Ashgate.

Koskinen, H. et al., 2021. Systems usability case in stepwise control room validation. *Safety Science*, Vol. 134, [105030].

Laarni, J. and Väätänen, A., 2023. Development of a Concept of Operations for an Autonomous Tram System. *Proceedings of the Intelligent Human Systems Integration (IHSI 2023)*, Vol. 69, pp. 541–549.

Laarni, J. et al., 2011. Development of training programs for nuclear power plant personnel – Lessons learned from Finnish development projects. *Proceedings of the Enlarged Halden Project Group Meeting*, Sandefjord, October 2nd-7th, 2011., Vol 1, p. C1.9.

Madni, A.M., 2010. Integrating humans with software and systems: Technical challenges and a research agenda. *Systems Engineering*, Vol. 13, pp. 232-245.

Mumford, E., 2006. The story of socio-technical design: reflections in its successes, failures and potential. *Information Systems Journal*, Vol. 16, pp. 317-342.

Palmer, A.W. et al., 2020. The Autonomous Siemens Tram. *2020 IEEE 23rd ITSC*, pp. 1-6.

UTILIZING BLOCKCHAIN TECHNOLOGY TO TACKLE FOOD WASTE IN SMART CITIES

George Plakas, Stavros T. Ponis, Dimitra Tzanetou, Eleni Aretoulaki and Antonios Kitsantas
School of Mechanical Engineering, National Technical University Athens
Heroon Polytechniou 9, Zografos 15780, Athens, Greece

ABSTRACT

Food waste is a major global issue and responsible for significant environmental and socio-economic repercussions, while millions of people worldwide, are affected by food insecurity. Food insecurity occurs even in large cities with high economic growth, where at the same time there is a large number of catering businesses and restaurants, which due to wrong forecasting and their effort to meet the predicted demand, prepare more fresh meals than necessary. As a result, quality food ends up in the garbage, in a daily basis. The research presented in this paper introduces a solution, based on Blockchain technology, that allows food-service establishments to offer the extra meals they have already prepared and would otherwise be disposed of, to food insecure people. The project's research focuses on the one hand on mitigating the food waste phenomenon in urban areas with high number of businesses providing fresh meals and on the other hand, on providing free, safe and nutritious meals to food insecure people. The integrated system ensures continuous communication between both parties and optimizes the allocation of food surplus and thus, minimizes food waste. The aim of the project is to develop a unique marketplace for food surplus, through a mobile platform and application, that can be utilized in the concept of smart cities, in order to mitigate the food waste phenomenon and actually contribute to the well-being of insecure members of our society, with the discretion and anonymity that is offered by the Blockchain characteristics.

KEYWORDS

Food Waste, Food Insecurity, Blockchain, Decentralized Applications, Tokens

1. INTRODUCTION

Food waste is a major global problem with multidimensional consequences. According to the Food and Agriculture Organization of the United Nations (FAO), one third of the global production, or else approximately 1.3 billion tons of edible food are lost or wasted every year (Ishangulyyev et al., 2019), creating a series of serious environmental and socio-economic impacts (Martin-Rios et al., 2018). Apart from the immediate consequences of the wasted resources, the problem acquires a moral component, due to the fact that 828 million people worldwide faced hunger in 2021 (FAO, 2022). Food waste occur mostly in developed countries (Buzby & Hyman, 2012), as the geographic profile of the phenomenon is heavily influenced by the level of income and the urbanization (Chalak et al., 2016). The above profile matches the description of smart cities, a term that represents a plethora of cities around the world with high levels of economic growth, trying to transform and upscale their services to their citizens, through the utilization of new technologies. In this context, such cities are capable of finding solutions to address the food waste problem, even through the use of technologies that they are not yet familiar with, such as Blockchain.

The alleviation of the food waste phenomenon in cities can be achieved through focusing on food waste taking place at the final stage of the food supply chain, the consumption stage. In particular, in food service establishments of a developed country, half a pound of food waste is created per meal served (Bloom, 2012), which indicates that for every 4 meals consumed, almost a kilogram of food is wasted. The waste is generated from food scraps either from customers, after completing their meal, or from the kitchen, due to ineffective demand forecasting. In the first case, awareness has to be raised against food waste among consumers that paid for it and they can easily avoid it, by packaging, freezing and consuming it, at some other time. However, the solution to the problem in the latter case is not simple, as demand cannot always be predicted with absolute precision. It is therefore imperative to create an integrated, clearly defined system through which food providing establishments, such as restaurants, cafes, pubs, etc., can find new ways to deal with food that is not consumed, such as the solution presented in this paper that offers them the capability to donate the generated surplus to food insecure people.

2. PROPOSED SOLUTION

Blockchain is a distributed ledger shared among the members of a network and is continually updated with the recording of all transactions transmitted by the network's nodes (Christidis & Devetsikiotis, 2016). The first Blockchain network (Bitcoin) was presented by Nakamoto (2008), but its operation is based on the combination of many pre-existing technologies and concepts, namely Peer-to-Peer networks, hash functions, public key cryptography, digital signatures and consensus mechanisms. This decentralized technology offers the opportunity to create self-executing contracts between parties (smart contracts) and create tokens (tokenization), which can function as the digital representation of real-world assets on the Blockchain network. The project presented in this paper takes advantage of the tokenization capabilities and develops a decentralized application (dApp), with a back-end based on an existing public Blockchain network.

The project's research focuses on the intersection of two rapidly growing areas of scientific interest i.e., Food Waste management and Blockchain technology, in order to simultaneously mitigate the food waste problem and provide safe and nutritious meals to food insecure citizens and their families. The decentralized ledger provided by the Blockchain technology is utilized to ensure the anonymous two-way communication between food establishments (suppliers) and citizens (food seekers), in order to optimize the allocation of the prepared meals surplus. In a high level, the solution in this paper can be described as a standard digital marketplace offering supply and covering demand. The sole product exchanging between the users are the meals prepared from food establishments that would be otherwise disposed of, in the end of the day. So, the suppliers connect to the project's database to upload data about the meals they would like to donate and the users pick up the food of their preference, anonymously, with the necessary discretion. The completed transactions are transmitted and verified transparently by the Blockchain network, and every transaction triggers the project's smart contract, that donates tokens to both parties rewarding them for their active participation in the effort to tackle the food waste problem. The food seekers can use the new tokens to purchase food again and the suppliers can donate their tokens to food insecure citizens or spend them in other businesses that also participate in the project's ecosystem against food waste, by offering discounts at their products and services, in exchange for the project's token.

The integrated system is composed of the three following subsystems:

- A data collection subsystem, gathering data streams such as photos and nutrition information, pertinent to the available portions of food, for visualization purposes, in the application's interface. The food establishments, upon their registration are connected to the project's network and upload this data about the meals to the subsystem, where they are processed using appropriate data analysis tools. This data, combined with data from external sources with an emphasis on geospatial computing, are analyzed for real-time decision-making and visualizing information.
- A blockchain subsystem, through which food donations are monetized, by providing tokens, as a reward, to the contracting parties, thereby creating a co-operative incentive. Every decision that saves a portion of food, by providing it to a user of the application, will be recorded in the blockchain network and will be translated into tokens.
- An application subsystem, enabling the visualization of the information gathered and offering value-added services to both food establishments and food insecure users of the application. The project's application is a dApp, which means that the front-end is designed like any other common app, but the back-end is running on a blockchain network, not a central server.

Any user of the application, either a citizen looking for meals, or a supplier offering the daily surplus, is able to proceed with the registration instantly through the mobile application. With the registration, a new anonymous account is created in the project's public Blockchain network, which means that the user is a new node of the network, with the same capabilities like any other node of the blockchain. If the user has already an account in the same public Blockchain, it is possible to use it as well, without a registration. With the blockchain account activated and registered in the mobile application, the user has to define the user profile, either as a food seeker or as a supplier, during the logging in process. The new food seeker accounts will receive tokens with their registration, in order to use them for their first meal. In the case of the suppliers, they have to provide basic information about their establishment (Name, Cuisine, Address, Work hours, E-mail, Telephone number etc.). Indicative screenshots are presented in Figure 1.

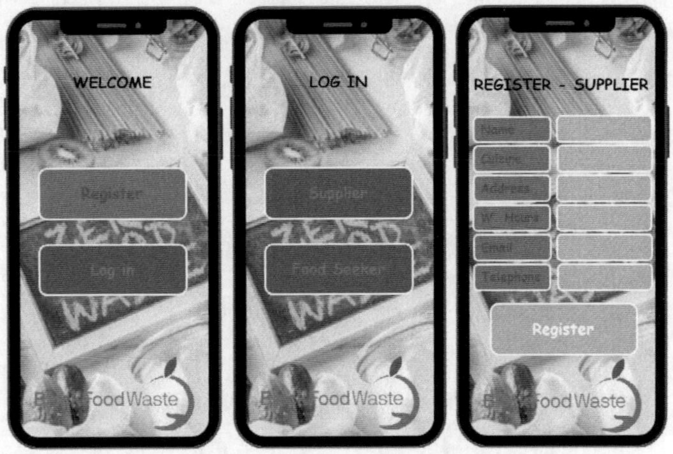

Figure 1. Application's interface (register and log in)

As a result, depending on the user profile, the application's interface is customized accordingly. For the food seekers, upon opening the application and activating geolocation using GPS, a map is loaded and displayed, highlighting the location of the user's closer suppliers, who offer at least one free meal available, in real-time. Through the mobile application the food seekers are easily informed about all the available choices from of all suppliers (with photos, food ingredients etc.) and then, they are able to select one of the available meals, taking into account both the distance to the establishment and the nutritional needs. The interface offers a booking option for the meals, as well as the option to navigate to the restaurant's location, in order to calculate the exact distance and expected time of arrival at the destination, based on the current traffic volume, using maps and routing services. The user selects the type of meal he/she wishes to consume, by making a reservation for it, with a validity period limited to one hour. In case the meal is not picked up after this time, it will be available again through the app, automatically. With the reservation, the user is provided with a QR code, which is required to be scanned by a food establishment's employee, in order for the transaction to be completed and verified in the Blockchain. Every QR is generated uniquely by the application and is related to one specific transaction. The QR code is the main oracle for the proper operation of the project's smart contract. Indicative screenshots for food seekers' interface are presented in Figure 2.

Figure 2. Application's interface (food seeker)

In the use case of suppliers, two options are presented in the application's interface: a) the uploading of new donations and b) the mechanism for the declaration of the completed pick up. The first option concerns fresh meals, which the supplier thinks should be distributed freely to consumers through the application, due to a poor forecast of daily demand. By selecting the listing of food for free distribution in the application interface, two columns are displayed, one for the type of food and one for the quantity of meals per type. Correspondingly, the meals become available and visible to application's users. When a food seeker arrives for the pickup, the supplier's employee uses the application's mechanism (scanner) to scan the QR code provided by the citizen and instantly the transaction is finalized and transmitted to the Blockchain network and is removed from the list of available products. Indicative screenshots of the application for the suppliers' interface are presented in Figure 3.

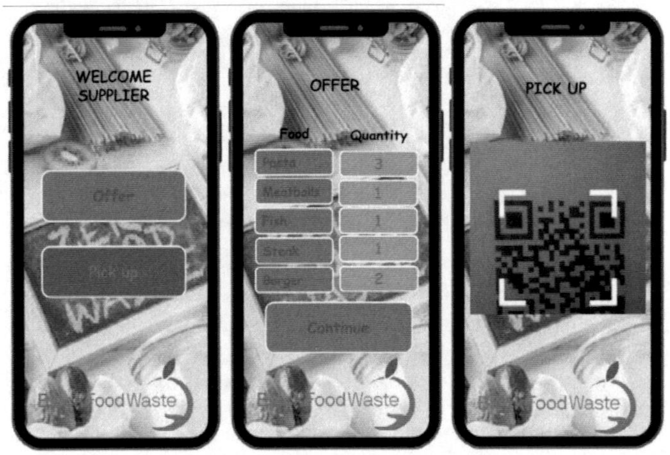

Figure 3. Application's interface (supplier)

The benefits of the integrated solution are not limited to just those involved in the marketplace through the dApp. Municipalities, NGOs, and other charitable organizations, which until now, have been exclusively responsible for providing food to people in need, through soup kitchens and donations in a centralized way, will be significantly relieved. Also, in addition to the decentralized operation of the application, the solution proposes also a decentralized last mile management, which is extremely important for the success of the project as a whole. Most of the restaurants have just a very small number of portions to offer every day and it would be too expensive logistics-wise to be transferred in centralized location to be distributed. With the project's solution restaurants can participate with even one meal, since the last mile is handled by the citizen that choses to reserve it.

The authors are currently in the phase of developing the final version of all the subsystems of the project and designing the specifications and scenario details for the project's large-scale pilot program. At the same time, the authors are already working to release another version of the dApp in order to create a complete ecosystem around the project's token targeting all citizens, by providing cheaper prices for those interested in paying less for food and helping to tackle food waste, at the same time. The project partners aspire to add businesses selling packaged food products close to their expiration date, which are also an important form of food waste.

3. CONCLUSION

The effort to reduce food waste is imperative, and the aim of the solution presented in this paper is a step towards that direction. Blockchain plays a cardinal role in the integrated system, due to the anonymity that provides to food insecure citizens and its tokenization capabilities. With the project's token, an ecosystem around food waste is created and also incentives are offered to food establishments, in order to participate in the program. The project, therefore, aims to contribute decisively to the fight against the phenomenon of food waste and to provide relief to a significant part of the food insecure population of the cities, with a simple mobile application.

ACKNOWLEDGEMENT

This research has been co-financed by the European Regional Development Fund of the European Union and Greek national funds through the Operational Program Competitiveness, Entrepreneurship and Innovation, under the call RESEARCH – CREATE – INNOVATE (project code: T2EDK-05051).

REFERENCES

Bloom, J. (2012). For Restaurants, Food Waste Is Seen As Low Priority. Retrieved from: https://www.npr.org/sections/thesalt/2012/11/27/165907972/for-restaurants-food-waste-is-seen-as-low-priority (Accessed: 10 April 2023)

Buzby, J. C., & Hyman, J. (2012). Total and per capita value of food loss in the United States. Food policy, 37(5), 561-570.

Chalak, A., Abou-Daher, C., Chaaban, J., & Abiad, M. G. (2016). The global economic and regulatory determinants of household food waste generation: A cross-country analysis. Waste management, 48, 418-422.

Christidis, K., & Devetsikiotis, M. (2016). Blockchains and smart contracts for the internet of things. Ieee Access, 4, 2292-2303.

FAO. (2022). Food and Agricultural Organization of the United Nations. Hunger and food insecurity. Retrieved from: https://www.fao.org/hunger/en/ (Accessed: March 30, 2023).

Ishangulyyev, R., Kim, S., & Lee, S. H. (2019). Understanding food loss and waste—why are we losing and wasting food?. Foods, 8(8), 297.

Martin-Rios, C., Demen-Meier, C., Gössling, S., & Cornuz, C. (2018). Food waste management innovations in the foodservice industry. Waste management, 79, 196-206.

Nakamoto, S. (2008). Bitcoin: A peer-to-peer electronic cash system. Decentralized business review, 21260.

MIXED APPROACH FOR BIKE AVAILABILITY FORECAST

Ignacio Sevillano Muñoz
Atos Research and Innovation
Madrid, Spain

ABSTRACT

We propose a predictive analysis model for bike-sharing systems based on a mixed model of statistical and machine-learning approaches. The model assumes a Poisson distribution on bike station arrival and departures. The model is validated by comparing the predicted arrival and departure rates to the actual rates. The case study results show that our model can outperform other naïve approaches using the mean on arrival and departure rates.

KEYWORDS

Bike Availability Forecast, Predictive Analysis, XGBoost

1. INTRODUCTION

The advancements in IoT and AI have brought about a significant transformation in our interaction with the physical world. It has enabled businesses to optimize their processes and make well-informed decisions. In this regard, the GreenMov initiative deserves accolades for its efforts toward developing eco-friendly transportation services that enhance the standard of living in urban areas. This article sheds light on the bike-station availability forecast (BAF for short) service that employs predictive analytics to ensure that bikes are readily available for users in Flanders (Belgium) and Nice (France).

Bike-sharing systems are an effective way to reduce traffic congestion and carbon emissions in urban areas. They work by providing a network of bike stations $\{s_i\}_{i=1}^{n}$ where users can rent and return bikes. Each station s_i has a capacity c_i, which is the maximum number of bikes that can be parked there. We represent the number of bikes at a certain station s_i at a given time t, or availability, by $a_i(t)$. Furthermore, the forecast we refer to in this context is an estimation of the station availability denoted as $\hat{a}_i(t)$.

In this article, we introduce a novel predictive availability service designed to tackle the issue of bike availability forecast in bike-sharing systems. Our proposed solution combines statistical foundations within a machine learning (ML) framework. By leveraging the inherent capabilities of Decision Trees algorithms, our approach enables us to provide accurate predictions while also accommodating the inclusion of additional regressors in the future. This article opens up new possibilities for the development of user-centric and data-driven services that can make our cities more livable and sustainable., and contribution of the paper.

This article begins (Section 2) with the introduction of a statistical model for bike availability in bike-sharing systems and its validation and forecast processes. The subsequent section (Section 3) presents and evaluates the case study of Blue-bike, in Belgium. Finally, the article concludes by proposing improvements and further applications of this service.

Gast, & Bierlaire, 2014, proposed a probabilistic forecast of bike-sharing station availability using a queuing-theoretic approach. Zhu, & Rasouli, 2016, also used a queuing-theoretic approach to design bike-sharing systems. Wang, W. & Madanat, S., 2017, developed a stochastic optimization model for bike-sharing system design.

Chen, & Guestrin, 2016, developed XGBoost, a scalable tree-boosting system. Guo, Chen, He, Ma, & Liu, 2017, studied the importance of tree diversity in boosting. Sun, & Srivastava, 2017, presented a comprehensive survey on Extreme Gradient Boosting.

For other approaches, (Lin, & Dou, 2016) proposed a research on urban public bicycle demand forecast based on a network perspective. Patil,, Musale, & Rao, 2015, used RandomForests to predict bike share demand. Duan, & Wang, 2017, proposed a moment-based availability prediction for bike-sharing systems.

2. STATIONS AS WAITING QUEUES

2.1 Regression Method

In this subsection, we present the statistical model for bike stations, which is modeled as a Markovian queuing system. Rather than focusing on the station's availability $a_i(t)$, we prioritize the arrival and departure rates, $\alpha_i(t)$ and $\delta_i(t)$ respectively, and define them as piecewise functions composed by 30-minute time intervals, i.e. $24\,[hours] \cdot 2\frac{[intervals]}{[hour]} = 48$ intervals. Thus, we need to determine $\alpha_{i,j}(t)$ and $\delta_{i,j}(t)$, with $i = 1, 2, 3, \ldots, n$ and $j = 1, 2, 3, \ldots, 48$. We assume that the arrival and departure rates of bikes at any station follow a Poisson process, which is characterized by the Poisson Probability Density Function:

$$f(k;\lambda) = \frac{e^{-\lambda}\lambda^k}{k!}$$

To verify if the arrival rate $\alpha_{i,j}(t)$ conforms to a Poisson distribution, we employ the Kolmogorov-Smirnov (K-S) test. Firstly, we generate an empirical probability function $g_{\alpha_{i,j}}(k)$ for each time interval. Then, using the previous equation for the Poisson pdf. with $\lambda_{\alpha_{i,j}}$ denoting the mean of arrivals at interval j in station s_i, we compute the cumulative distribution functions $G_{\alpha_{i,j}}(k)$ and $F\left(k;\lambda_{\alpha_{i,j}}\right)$. Subsequently, we calculate the K-S test statistic, defined by

$$D := \sup_k \left|F\left(k;\lambda_{\alpha_{i,j}}\right) - G_{\alpha_{i,j}}(k)\right|$$

We can then compare the test statistic D with the critical value $D_{0.95}$ at a chosen significance level 0.95. If D exceeds $D_{0.95}$, we reject the null hypothesis that the observed data follows a Poisson distribution. Same procedure applies for departures $\delta_{i,j}(t)$.

2.2 Regression Method

In Section 3, we will demonstrate that the majority of stations exhibit arrival and departure rates that can be approximated by a Poisson distribution during the daytime with fixed λ as the mean of arrivals or departures. However, our objective is to enhance this naive parameter by employing a machine learning model, thereby establishing a more flexible framework that incorporates various regressors. Specifically, our regression approach relies on the utilization of *XGBoost* decision trees, where lag values are chosen based on correlation analysis.

In our analysis, we discovered that in addition to the typical lag variables, such as the previous day, previous week, or previous hours, there exists a significant correlation between the departure rates $\delta_{i,j}(t)$ and the arrival rates $\alpha_{i,j}(t)$ in the time interval 6 to 8 hours later. This observation can be attributed to individuals using bicycles for their commute to work and relying on the bike-sharing service for their return journey home as well.

To estimate the expected station availability, we generate piecewise functions for the predicted arrival and departure rates, denoted as $\hat{\alpha}_i(t)$ and $\hat{\delta}_i(t)$ respectively, by utilizing the forecasted values from each time interval. Subsequently, we perform an integration of these functions over time to calculate the anticipated availability with the desired level of precision. Then, for each station s_i

$$\hat{a}_i(t) := a_i(t_0) + \int_{t_0}^{t} \hat{\alpha}_i(x)dx - \int_{t_0}^{t} \hat{\delta}_i(x)dx$$

The estimation for the availability $\hat{a}_i(t)$ at time t is defined for any $t \in R$, where t_0 represents the last available data time. This allows us to tailor the forecast resolution to specific needs. Furthermore, this flexibility is advantageous in cases where bike stations exhibit low activity levels, as it enables us to consider wider time intervals and assume a Poisson distribution underlying the data.

3. RESULTS AND PERFORMANCE

This section showcases the outcomes of our predictive analysis model for bike-sharing systems and offers insights into their implications for system design and operation. We evaluate the performance of our model through a case study: Blue-bike, utilizing data from Belgium.

We commence by validating the statistical model and subsequently present our results utilizing the coefficient of determination (R^2). This metric serves as an indicator of the accuracy of our forecast service, providing short-term precision assessments.

3.1 Blue-Bike

Blue-bike is a national bike share scheme in Belgium founded in 2011. It operates bike-sharing services through over 71 locations across the country. In 2019, Blue-bike recorded 276,000 rides with 20,000 members. Surveys indicate that 93% of members combine Blue-bike with train journeys, and 32% of rides were previously unmade or replaced car trips.

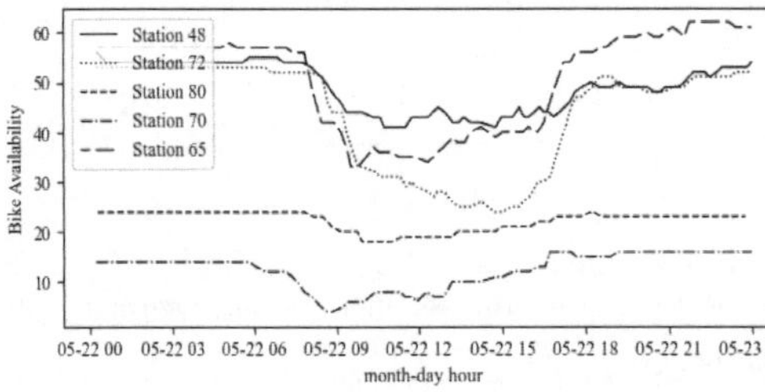

Figure 1. Bike availability at 22nd of May

For our analysis, we utilize one year of available data. Initially, we validate our model by distinguishing between workdays and non-working days. To provide clarity, we present our Kolmogorov-Smirnov (KS) tests in a 3D plot, where the x-axis represents the hour of the day, and the y-axis represents the p-value. We generate separate plots for arrival and departure rates, considering both workdays and non-working days, resulting in four distinct figures.

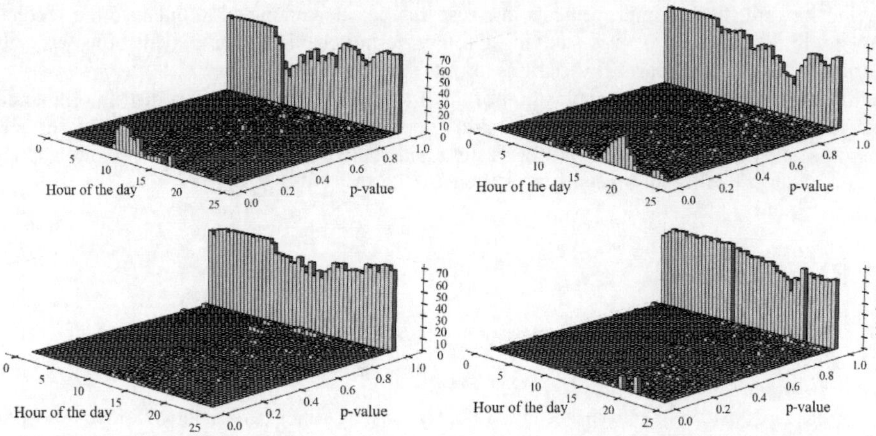

Figure 2. KS-tests for arrival and departure rates. From left to right, from top to bottom: Departure rates - Workdays, Arrival rates - Workdays, Departure rates - Non-working days, Arrival rates - Non-working days

In the given figures, we can observe that the arrival rates during the afternoon and the departure rates in the morning (Figures 2a} and 2b), which represent the busiest time periods overall (refer to Figure 1), do not seem to conform to a single Poisson distribution with a fixed mean parameter, as initially suggested in Section 2. To address this issue, employing Machine Learning models would offer a valuable opportunity to enhance the accuracy of these rates.

To conclude this section, we present the results obtained through training on multiple datasets (Figure 3). We compare two sets of results: the first one (left) represents the naive approximation, which yields an average R^2 value of 0.95, while the XGBoost approximation (the method showcased in this article, on the right) achieves an average R^2 value of 0.96. This improvement may seem subtle at first glance, but it is more significant than it appears. It should be noted that most of the test data consists of moments during the day when the arrival and departure ratios are trivial, such as nighttime periods. Hence, achieving higher accuracy in such cases demonstrates the robustness and effectiveness of the proposed XGBoost method.

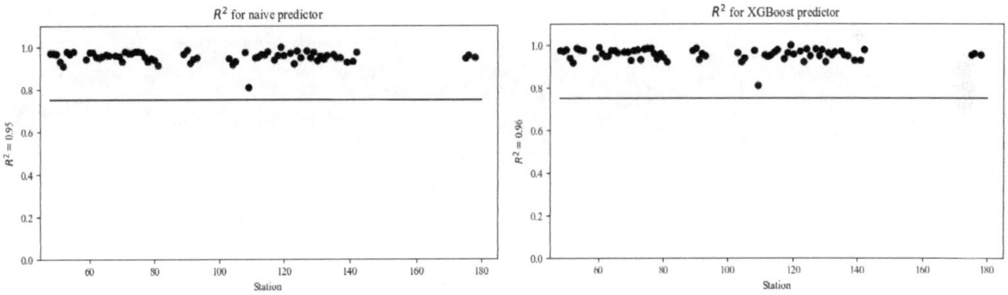

Figure 3. Results. Left: Naive approach. Right: XGBoost approach

4. CONCLUSION

This article introduces a novel predictive availability service for bike-sharing systems, employing statistical modeling and machine learning techniques. The proposed solution demonstrates high accuracy in predicting bike availability, with a strong correlation between the predicted and actual values. The approach presented here opens up possibilities for developing user-centric and data-driven services, contributing to the creation of more livable and sustainable cities.

Moving forward, further improvements and research can be conducted to enhance the predictive model. One potential area of exploration is studying the long-term behavior of bike availability predictions. Also, considering factors such as seasonal variations, events, and holidays.

Additionally, investigating the application of the predictive service in fleet redistribution strategies could optimize bike-sharing operations and improve user experience. Furthermore, extending the service to include other aspects, such as real-time demand forecasting and route optimization, could provide valuable insights and additional functionality for bike-sharing systems.

REFERENCES

Chen, T. & Guestrin, C. (2016). XGBoost: A scalable tree boosting system.

Duan, Z. & Wang, L. (2017). Moment-based availability prediction for bike-sharing systems. *IEEE Transactions on Intelligent Transportation Systems, 18(11)*, 2832-2841.

Gast, M. & Bierlaire, M. (2014). Probabilistic forecast of bike sharing stations availability. *Transportation Research Part B: Methodological, 68, 192-207.*

Guo, H., Chen, T., He, T., Ma, Y., & Liu, Y., (2017). On the importance of tree diversity in boosting.

Lin, Y. & Dou, W. (2016). Research on urban public bicycle demand forecast based on network perspective. *Journal of Intelligent Transportation Systems, 20(3)*, 234-245.

Patil, A., Musale, K. & Rao, B. (2015). Bike share demand prediction using RandomForests. *International Journal of Innovative Science, Engineering and Technology, 2(2)*, 83-88.

Sun, Y. & Srivastava, A. (2017). A comprehensive survey on Extreme Gradient Boosting.

Wang, W. & Madanat, S. (2017). A stochastic optimization model for bike sharing system design. *Transportation Research Part B: Methodological*, 105, 116-134.

Zhu, J. & Rasouli, A. (2016). A queuing-theoretic approach for bike sharing system design. *Transportation Research Part B: Methodological, 84*, 16-31.

A DATA LAKE PROTOTYPE SYSTEM FOR MULTIWAVELENGTH (MWA) AND MULTIMESSENGER ASTRONOMY (MMA)

Matheus Bento[1], Reinaldo Roberto Rosa[1,2] and Ulisses Barres de Almeida[3]
[1]*Applied Computing Graduate Program (CAP), INPE-MCTI*
Av. dos Astronautas, 1758, S. J. dos Campos, SP, Brazil, CEP 12227-010
[2]*Lab for Computing and Applied Mathematics, COPDT-INPE-MCTI*
Av. dos Astronautas, 1758, S. J. dos Campos, SP, Brazil, CEP 12227-010
[3]*Centro Brasileiro de Pesquisas Físicas (CBPF)*
R. Dr. Xavier Sigaud, 150 – Urca, Rio de Janeiro, RJ, Brazil, CEP 22290-180

ABSTRACT

At the beginning of the 21st century, large telescopes are coming into operation, making astronomy a Big Data science like few others. Recently, given this scenario, applied computing to astronomy and space sciences has pointed out major challenges involving data science and data engineering. Faced with such a challenge, this article presents a first prototype of a data lake as a strategic solution for heterogeneous data astronomy such as multi-wavelength astronomy (MWA), also including multi-messenger astronomy (MMA). The proposed prototype presents a three-layer architecture, composed of storage, processing and application layers. The storage layer incorporates S3 for raw data storage, MariaDB for simple data storage (crud) and TimescaleDB for complex data storage. The processing layer includes Laravel for CRUD operations and Python for handling large volumes of data. The application layer uses ReactJS with the Cubes.dev library for analytical data visualization. The solution proposes to deliver, when necessary, agility and concatenation between heterogeneous data (light curve, spectra and images) coming from different silos (instruments and telescopes). Thus, the solution, within the MWA and MMA paradigms, incorporates data cube technology and can be customized for applications other than astronomy, such as space weather monitoring and debriefing in the context of space medicine.

KEYWORDS

Data Lake, Multi-Wavelength Astronomy, Heterogeneous Data Science, Big Data, Agile Analytics, Data Cubes

1. INTRODUCTION

Modern astronomy faces a growing challenge to deal with the large quantity and diversity of data generated by observatories around the world (Ahmed et al., 2017). The exploration of the cosmos requires the analysis and integration of information from different kinds of sources, such as terrestrial telescopes, satellites, and space detectors. However, these data are often stored in a fragmented and isolated way in systems known as "silos" of row data which in practice can be strategically analyzed in a concatenated and agile way within the concept of a data lake (Rosa, R. R. 2020). This solution, compared to the traditional data warehouse approach, has been explored in several areas in which agile analysis of raw data, based on machine learning, is required (Che and Duan, 2020). Given the typical data warehouse limitations, there is a gap for a solution that allows efficient integration and analysis of multiple sources of astrophysical data, overcoming the barriers imposed by data silos. It is in this context that the Data Lake solution is useful also in space physics and astronomy. The Data Lake solution provides a centralized and flexible environment for the storage and processing of raw heterogeneous data. It allows the gathering of dispersed information in different silos, promoting the concatenation and integration of different kinds of data into a unified structure. Furthermore, the Data Lake can handle the heterogeneity of data formats, helping to analyze and obtain insights from large volumes of information.

Recently, the strategies in the analysis of large amounts of data for MMA have been established. Following these guidelines, we seek to develop a Data Lake Prototype that allows the agile and related concatenation of data from different observatories, respecting the heterogeneity of the data and handling the challenges imposed by the various silos of information present in the context of MMA and MWA, but that can be applied (and customized) to other areas that involve the same challenges that involve agile Big Data analysis. In this context, this paper presents a solution based on a Data Lake, with a focus on the application of MMA and MWA, which deals with the analysis of various heterogeneous astronomical data sources (silos). Furthermore, it stands out for the management of large volumes of data, seeking to simplify the management, security, sharing, and agile and concatenated big data analysis incorporating the data cube concept. Next, the details of the methodology used in the implementation of Data Lake will be presented in-depth, covering the layers of storage, processing, and application (Chen et al, 2012; Allen et al., 2019).

2. THE PROTOTYPE

The methodology adopted in this study involves the implementation of a data lake prototype (DLP) to be validated and assist the visual analysis of raw data in several projects that have the participation of Brazilian groups (e.g. LSST, SKA, CTA, among others) (see, Rosa 2018).

The DLP structure is composed of three main layers: storage, processing, and application as we can see in Figure 1. In the storage layer, we use different technologies to help us handle different types of data. Amazon S3 (Simple Storage Service) is employed as the primary storage service, enabling scalability and secure storage of astronomical data collected from concatenated observatories and other instruments. MariaDB is used as a relational database to store simple data such as Organization, Security data, etc., while TimescaleDB is used to store complex data in time series format, such as long-term astronomical event data (Zhang et al. 2018; Dean and Ghemawat. 2008).

In the processing layer, we adopted a hybrid approach to handle the manipulation and analysis of simple data/astronomical data. We use the Laravel framework, a PHP web development framework, to implement create, read, update, and delete (CRUD) operations on stored data, database as MariaDB. In addition, we use the Python programming language, along with libraries such as NumPy and Pandas, to perform statistical analysis, data transformations, and processing of large volumes of astronomical data, Laravel is able to orchestrate the connection with the databases and call Python API's to process data (Schreiber et al. 2018; Vouk 2008; Barchi et al., 2020; Jin & Finkel, 2018).

In the application layer, we are using the ReactJS framework to develop an interactive and friendly user interface. Additionally, we've incorporated the Cubes.dev library to make it easy to visualize astronomical data through dashboard analytics as charts, tables, and other visual analytics representations. This layer allows users to interact with the data stored in the data lake, performing custom queries, exploring relationships between data, gaining valuable insights, and also generating astrophysical reports that offer the user management the queries and data input following a given structure (Sidhu and Sehgal, 2015; Giommi et al., 2020).

This DLP is designed to allow a quick concatenation of information among counterpart observatories and heterogeneous astronomical data sources. The data lake and data cube concept offers flexibility and scalability to deal with the multitude of existing data silos in astrophysics, enabling integrated and multidimensional analysis, being possible to create different cubes to handle different data (Cuzzocrea, 2010; Ivezic et al., 2020; Buchschacher et al., 2019; Barres De Almeida et al., 2020).

Furthermore, as part of the methodology adopted, it will be necessary to establish a process for integrating and transforming data from different sources. This will involve using tools and techniques to deal with the heterogeneity of data formats, such as using Python libraries such as Pandas to handle structured data, Pickle Format, CSV, and other common formats. It will also be necessary to explore different technologies of each used storage system, such as Amazon S3, MariaDB, and TimescaleDB, to efficiently extract, transform, and load data. This approach will allow the creation of a flexible and adaptable environment, capable of handling the large volume of data characteristic of astronomy from multiple sources (Kelleher & Tierney, 2018; Allen et al., 2019; Carvalho et al., 2010; Bartos et al., 2017; Rosa, R. R. 2020).

The effectiveness of the proposed methodology is demonstrated through the implementation of a prototype, the DL-MMA (Data Lake for Multi-Messenger Astronomy), which is in testing and being validated in the context of the mockup data products which will come from the LSST ecosystem. However, we emphasize that the data lake with the data cube approach presented in this study can be applied in other areas of interest that face similar challenges related to the agile and concatenated analysis of data from different silos (Rosa, R. R. 2020). An analytical dashboard, developed in dash.py (there is also a version in react), is incorporated into the datalake front-end to quantify the percentage of data used in a given application (Figure 2).

A typical minimal data cube input organized within the data lake is shown in Table 1. The prototype is called ADALA (Analytical Data Lake for Big Data Astronomy).

2.1 Figures and Tables

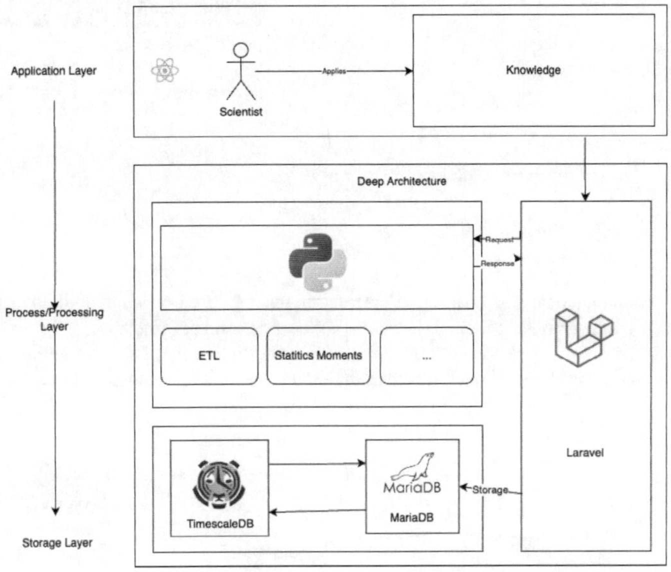

Figure 1. The DLP (also called ADALA in its preliminary version using Python)

Table 1. A hypothetical table based on five instruments that observe the same astronomical event within the MWA/MMA paradigm. The instruments are distributed as instances and the use of their respective dataframes incorporated into the Data Cube work as attributes. A Big Data usage analysis of this primary framework fed by the raw data from each silo is the input to the ADALA dashboard shown in Figure 2

INSTRUMENT	DF1(%) 5.1GB	DF2(%) 6.5GB	DF3(%) 31GB	DF4(%) 112GB
#001	98	95	89	71
#002	90	87	78	56
#003	67	46	45	43
#004	67	40	40	35
#005	59	42	35	30
#006	55	42	35	28
#007	45	38	27	17
#008	40	28	16	08

2.1.1 Results

The results obtained at this time are based on the implementation of a data lake for an active and flexible data system, following the MMA paradigm. The developed solution will offer access to a user-friendly interface, facilitating administration, security, sharing, and agile and concatenated analysis of different types of data.

Considering the deficiency of solutions for the concatenation of information between observatories and the heterogeneity of different kinds of data, the data lake makes it possible to integrate different sources of data in a single environment, overcoming a fragmented approach and data "silos". Through the layer of processing and application, the concatenation of information was implemented by means of ETL techniques (Extract, Transform, and Load) and analytical analysis (Allen et al., 2019; Rosa, R. R. 2020).

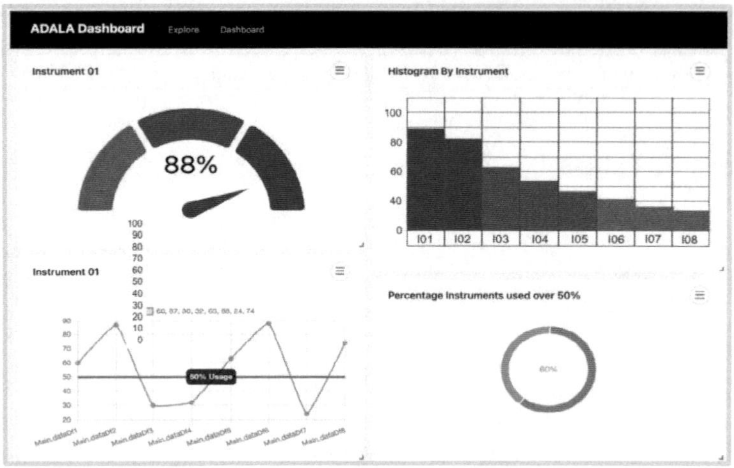

Figure 2. The 2nd order analytical dashboard, implemented in react on the front-end, applied to a hypothetical dataframe shown in Table 1. The info-graphs on the left measures, for each instrument (instances), the amount of data used (upper dash) in a given analysis of 1st order (ex. Figure 3) over time (bottom dash). The infographics on the right inform the percentage of data used by each instrument (upper dash) and the percentage of those whose data used were greater than 50% (lower dash)

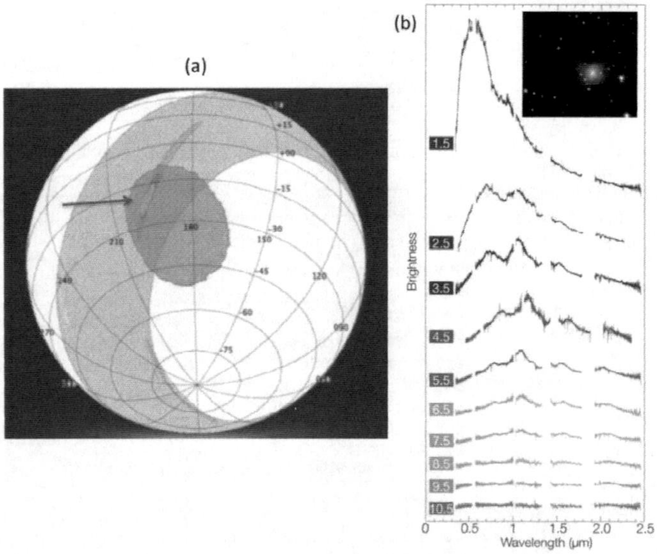

Figure 3. (a) Overlapping sky maps from the final localization of GW170817 via GWTC-1 (red arrow); (b) X-shooter composed spectra of kilonova NGC 4993 with the respective GROND image embedded. From ESCAPE Virtual Observatory. In the context of the developed prototype, the incorporation of the data cube concept allows the aggregation and multidimensional analysis of heterogeneous data. This approach facilitates the exploration and visualization of the two data more efficiently, allowing a deeper understanding of the astrophysical area

As a preliminary result of this application, MMA data were selected within the silo concept for future validation of the prototype, scheduled for the second half of 2023. The validation has been considering the following steps:

1. Loading sky maps with DSS2 in HEALPix FITS format;
2. Location of event GW170817 observed by LIGO and VIRGO;
3. Input of correlated data in other sources/frequencies (via ALADIN Desktop): PGC, GWGC, HyperLEDA, and 2MASS;
4. Generation of the Data Cube by Category (eg Sources and/or Telescopes);
5. Rendering of Location Maps by category (eg Telescopes) (Figure 3.1) and MWA (Figure 3.2).
6. Generation of the integrated MMA dashboard for the GW170817 event.

3. CONCLUDING REMARKS

The DLP presented in this paper is a comprehensive strategy for the development of a Data Lake capable of combining data from multiple sources in an agile way, overcoming the challenges imposed by the existence of different data silos. The implementation of the Data Cube concept embedded in a Data Lake, called DL-MMA (Rosa, 2018), shows promise for dealing with the analytical concatenation of redundant and valuable information in a continuous data cycle. At the conclusion of this test project, we recognize that the continuous evolution of two computational resources, such as programming languages, messaging services, and data banks for time series and other types of data, is essential to maintain the flexibility and portability of the developed solution. We believe that the IT concept based on the Data Cube immersed in the Data Lake has the potential to benefit the scientific community, allowing a more agile and concatenated approach to the analysis of information from different data silos (Beyer & Laney, 2012).

It is worth noting that the concept of a Data Lake with a Data Cube studied and developed in this work is not only limited to the DL-MMA prototype. What makes the concept specific are the attributes that make up the Dice Cube, allowing its application in other astronomical contexts, such as the demands of Virtual Observatories (VOs), as well as in other areas that require agility and analytical concatenation from silos of heterogeneous dice (Hajjat et al. 2011, Allen et al., 2019 and Rosa, R. R. 2020).

As a perspective for future work, it is intended to explore the application of the developed methodology in other astronomical Data Cubes and in areas of interest that involve agility and analytical concatenation. A promising example is the monitoring of Space Weather, which requires analysis of different types of solar, geomagnetic, and ionospheric data. Such data can be integrated with other data systems that involve visual analysis of raw data as is the case of NASA's Artemis mission, the initial step which will establish a sustainable exploration of the Moon to prepare for missions to Mars. In this context, digital platforms that can evaluate astronaut training remotely can be based on DLP like the one we present here. Therefore, the solution presented here has direct applications in other data paradigms, for example in environmental (Velho et al., 2001, Bolzan et al. 2005), and aerospace sciences (Lara et al. 2007, Mattedi et al., 2004), which also demand agility in analytics.

ACKNOWLEDGEMENT

The authors thank CNPq and CAPES for partial financial support. MB thanks for the data available from ALADIN and ESCAPE. RRR thanks FAPESP under Process No 2021/155114-8.

REFERENCES

Ahmed, E., Yaqoob, I., Hashem, I. A. T., Khan, I., Ahmed, A. I. A., Imran, M., & Vasilakos, A. V. (2017). The role of big data analytics in Internet of Things. Computers & Networks, 129(2), 459-471.

Allen, M. G., Dowler, P., Evans, J. D., Cui, C., & Jenness, T. (2019). The International Virtual Observatory Alliance. Proceedings of Astronomical Data Analysis Software and Systems XXVIII. Retrieved from arxiv.org/abs/1903.06636.

Barres de Almeida, U., Krone-Martins, Diaz, M. P., Do Nascimento, J. D., Leo, W. V., Rosa, R. R., & Saito, R. K. (2020). Information technology & astronomical data in Brazil: Perspectives and proposals. Boletim da Sociedade Astronômica Brasileira, 32(1), 142-146.

Barchi, P. H., Carvalho, R. R., Rosa, R. R., Sautter, R. A., & Soares-Santos, M. (2020). Machine and Deep Learning applied to galaxy morphology: A comparative study. Astronomy and Computing, 30, 100334.

Bartos, I., Kowalski, M., Multimessenger Asytonomy. (2017). IOP Publishing.

Beyer, M. A., & Laney, D. (2012). The importance of 'big data': A definition. Gartner, 6(13), 1-3.

Bolzan, M.A., Sahai, P.R., Fagundes, P.R., Rosa, R.R., Ramos, F.M. & Abalde, J.R. (2005). Intermittency analysis of geomagnetic storm time-series observed in Brazil. Journal of Atmospheric and Solar-Terrestrial Physics 67 (14), 1365-1372.

Buchschacher, N., Alesina, F., & Burnier, J. (2019). No-SQL Databases: An Efficient Way to Store and Query Heterogeneous Astronomical Data. In DACE 523, Astronomical Data Analysis Software and Systems XXVIII (pp. 405). ASP.

Carvalho, R. R. et al. (2010). The Brazilian Virtual Observatory - A New Paradigm for Astronomy. Journal of Computational Intelligence in Sciences, 1(3), 187-206.

Chen, H., Chiang, R. H., & Storey, V. C. (2012). Business intelligence and analytics: From big data to big impact. MIS Quarterly, 36(4), 1165-1188.

Cuzzocirea, A. (2010). OLAP Data Cube Compression Techniques: A Ten-Year-Long History. Lecture Notes in Computer Science, Springer.

Giommi, P., Arrigo, G., Barres de Almeida, U., De Angelis, M., Del Rio, J. V., Di Ciaccio, S., Di Pippo, S., Iacovoni, S., & Pollock, A. (2020). The Open Universe Initiative. In Space Capacity Building in the XXI Century (pp. 377-386). Springer.

Hajjat, M., Liu, X., Nguyen, T. N., Wang, C., & Zhang, Z. (2011). Cloudward bound: Planning for beneficial migration of enterprise applications to the cloud. IEEE Network, 25(4), 28-35.

Ivezic, Z., Connolly, A. J., Vanderplas, J. T., & Gray, A. (2020). Statistics, Data Mining, and Machine Learning in Astronomy: A Practical Python Guide for the Analysis of Survey Data. Princeton University Press.

Jin, Z., & Finkel, H. (2018). Power and performance tradeoff of a floating-point intensive kernel on OpenCL FPGA platform. IEEE Symposium on Parallel and Distributed Processing (pp. 716-720).

Kelleher, J. D., & Tierney, B. (2018). Data Science. MIT.

Lara, A., Borgazzi, A., Mendes Jr., O., Rosa, R.R. & Domingues, M.O. (2008). Solar Physics 248, 155-166.

Mattedi, A., Ramos, F.M., Rosa, R.R. & Mantegna, R.N. (2004), Value-at-risk and Tsallis statistics: risk analysis of the aerospace sector. Physica A: Statistical Mechanics and its Applications 344 (3-4), 554-561.

Rosa, R. R. (2020). Data Science Strategies for Multimessenger Astronomy. Anais da Academia Brasileira de Ciências, 93. DOI: 10.1590/0001-3765202020200861

Schreiber, R., Carpentier, M., Tchounikine, A., & Bresson, G. (2018). Towards a big data architecture for learning analytics. IEEE Transactions on Learning Technologies, 11(2), 198-208.

Sidhu, P., & Sehgal, R. (2015). A survey on data warehousing and big data analytics. International Journal of Computer Applications, 125(2), 28-33.

Velho, H.F.C., Rosa, R.R., Ramos, F.M., Pielke, R.A., Degrazia, G.A. & Neto, C.R. (2001). Multifractal model for eddy diffusivity and counter-gradient term in atmospheric turbulence. Physica A: Statistical Mechanics and its Applications 295 (1-2), 219-223.

Vouk, M. A. (2008). Cloud computing: Issues, research, and implementations. Journal of Computing and Information Technology, 16(4), 235-246.

Zhang, M., Zhang, Y., & Zhang, D. (2018). Big data storage systems: A survey. IEEE Access, 6, 18329-18352.

Zhang, Z., & Zhao, W. (2015). Astronomy in the Big Data Era. Data Science Journal, doi:10.5334/dsj-2015-011.

ANOMALY DETECTION IN CRYPTOCURRENCY TRANSACTIONS WITH ACTIVE LEARNING

Leandro L. Cunha and Miguel A. Brito
Centro Algoritmi
University of Minho, Portugal

ABSTRACT

Cryptocurrencies have gained tremendous popularity in recent years, with the rise of Bitcoin and other altcoins. However, this surge in popularity has also attracted fraudulent activities, such as scams, phishing, and money laundering. Particularly, machine learning (ML) algorithms have the potential to detect these fraudulent patterns. However, since in the fraud detection (FD) domain labels are scarce and most times very hard to get, traditional supervised ML models cannot be applied. Additionally, traditional unsupervised anomaly detection (AD) algorithms, generally, lead to high false positive rates. Therefore, this study is intended to explore the feasibility of using AD and active learning (AL) algorithms to uncover new fraudulent patterns in cryptocurrency transactions, assuming minimal access to labels.

KEYWORDS

Anomaly Detection, Active Learning, Fraud Detection, Unsupervised Learning, Cryptocurrencies, Machine Learning

1. INTRODUCTION

Computing technology has revolutionized ML, making it accessible for identifying opportunities and managing risks through data pattern recognition in various industries. The need for effective AD systems in the fraud detection domain remains a growing concern, particularly in the cryptocurrency market where digital currencies have increased the risk of fraudulent activities. The absence of regulation in this market makes it more vulnerable to fraud, resulting in significant losses for companies. Thus, companies must adopt efficient strategies for detecting and preventing fraud to avoid bearing the burden of substantial losses.

Training high-performance supervised classifiers for FD is crucial, but the lack of labeled data makes it challenging to train such ML models (Labanca *et al.*, 2022). The process of labeling an entire dataset requires a vast sample of manually reviewed transactions, which is often unpractical due to the high costs involved and the limited investigative time and budget available for companies (Barata *et al.*, 2021). Thus, an efficient mechanism is needed to help analysts focus their investigation on the most relevant transactions with a higher probability of being fraudulent and to efficiently uncover increasingly complex and sophisticated fraud patterns, given that fraudsters are constantly finding new ways to commit fraudulent activities.

The absence of labeled data to train supervised classifiers highlights the immense value of combining AD and AL techniques to detect anomalies in data, as potentially fraudulent transactions, and iteratively collect labels for training supervised ML classifiers. AD algorithms have the potential to alleviate the issue of labeled data by flagging unusual or suspicious behaviors and patterns within the data. However, being unsupervised, they may result in false positives, given the high-class imbalance and growing sophistication of fraudulent behaviors that are becoming more similar to normal patterns. AL has the potential to combine the benefits of unsupervised (i.e., exploration) and supervised (i.e., exploitation) methods by iteratively selecting the most valuable data to be reviewed and labeled by an oracle from an initial unlabeled pool. In other words, the most anomalous transactions are presented to an analyst for review according to an anomalous ranking provided by the anomaly detector and after the feedback, the labeled data can be used to train a supervised classifier that learns fraud. Thus, the usage of AD in conjunction with AL allows for more efficient utilization of the labeled data.

Is intended with this work to assess the feasibility of using AD and AL algorithms to uncover new fraudulent patterns in cryptocurrency transactions, assuming minimal access to labels. Thus, the main objective is to benchmark different AL setups that use AD algorithms and gain insights on how to improve the performance of an anomaly detector, incorporated in an AL process that iteratively gathers labels that would allow a supervised model to be trained with only the most informative transactions, thereby, reducing the necessary labeling effort. We organize the remainder of the paper as follows. Section 2, presents the main concepts related to our work, providing a clear understanding of the topic. Section 3 details the existing literature on this subject matter to gain a more comprehensive view of the problem. Finally, in Section 4, the main conclusions are discussed.

2. MAIN CONCEPTS

2.1 Anomaly Detection

ML algorithms play a fundamental role in FD by using transactional data to identify fraudulent patterns through high-level pattern recognition.

AD is the process of identifying patterns in data that deviate from the expected normal behavior (Injadat *et al.*, 2018). Anomalies are rare occurrences with little to no common characteristics and are not consistent with the majority of observations. The primary objective of AD is to create a profile of normal behavior and classify a data instance as anomalous if it deviates from this norm (Xuan *et al.*, 2018). Particularly, in transactional data, anomalies can detect fraudulent activity by identifying abnormal transactional patterns (Hilal *et al.*, 2022).

Essentially, anomalies can be categorized in three ways: point anomaly, contextual anomaly, and collective anomaly. A point anomaly is when a single data instance deviates from the normal pattern. A contextual anomaly is when data instances behave anomalously in a specific context. A collective anomaly is when a group of similar instances act anomalously when compared to the whole dataset (Ahmed *et al.*, 2016). Another issue to consider is how anomalies are represented in the output. This can be in the form of a score indicating the likelihood of a data instance being anomalous or a label assigning each data instance as either normal or anomalous (Ahmed *et al.*, 2016).

Unlike imbalanced classification, AD is an unsupervised learning technique that aims to identify outliers in a dataset. Generally, unsupervised AD methods can be divided into four different categories: nearest-neighbor, clustering, subspace, and tree-based techniques. Although the primary focus of this work is on unsupervised AD techniques, it should be noted that certain approaches incorporate semi-supervised discriminators trained to acquire knowledge about the normal instances' boundary. In such scenarios, instances that lie outside of the established boundary are considered outliers (Lorenz *et al.*, 2020).

2.2 Active Learning

AL is a method that improves ML performance by strategically selecting data for training. We may put queries, based on certain heuristics, in form of unlabeled data to be labeled by someone that understands the nature of the problem (Settles, 2012). This reduces the number of labels needed for training a supervised classifier by iteratively sampling the most informative samples for review from an initial unlabeled pool (Lorenz *et al.*, 2020). Thus, AL can be considered a specific instance of semi-supervised learning.

The AL process begins with an initial pool of unlabeled data or only a few labeled instances. The most informative samples, from the unlabeled pool, are then selected and labeled by an oracle based on a specific AL querying strategy. Further, the labeled samples are moved to the labeled pool and used to train and evaluate high-performance supervised ML models. This loop is repeated until a certain number of labels are collected (i.e., given a specific budget) or the performance of a supervised ML model trained on a fully labeled dataset is achieved. If the performance is still not satisfactory, the querying process continues to incrementally expand the labeled pool. It is important to note that the AL process is applied only to the training set of the data. The training set is divided into two pools: the unlabeled pool (containing the unlabeled transactions) and the labeled pool (containing instances that have already been queried and reviewed by an analyst). The supervised ML model is trained on the labeled pool and evaluated on the test set to assess its effectiveness. Therefore, the advantage of

AL is that we can have a human loop that reviews a set of data instances identified by the anomaly detector (e.g., the top 50 most anomalous transactions), and assigns them to the correct class label. So, with less effort, we can have a system with similar performance to a supervised baseline.

3. LITERATURE REVIEW

3.1 Traditional Approaches for Fraud Detection

Most of the cryptocurrency FD research often uses ML models applied to the Elliptic dataset, one of the largest labeled datasets for any cryptocurrency. The dataset is a graph network of Bitcoin transactions and is highly valuable for financial FD and research communities. Supervised machine learning models have proven effective in detecting fraudulent transactions. Weber *et al.* (2019) applied several binary classification methods to the Elliptic Dataset and found that Random Forest (RF) was the best-performing model, despite the graph structure information suggesting an advantage for graph-based approaches like Graph Convolutional Networks (GCN). Alarab *et al.* (2020) also achieved successful results using an ensemble of supervised learning techniques on the Elliptic dataset, outperforming Weber *et al.* (2019). Particularly, RF is a commonly used supervised ML model in cryptocurrency FD research. Studies have shown that RF performs well in classifying Ethereum transactions as fraudulent or not (Ostapowicz and Żbikowski, 2019; Ibrahim *et al.*, 2021). A recent study by (Melo-Acosta *et al.*, 2017) proposed a new approach using a BRF (Balanced Random Forest), which involves iteratively selecting samples from the minority class and constructing a tree-based classifier. This approach outperforms traditional supervised methods in both supervised and semi-supervised learning through co-training.

Additionally, a study conducted by Niu *et al.* (2019) compared supervised and unsupervised ML algorithms using an imbalanced credit card dataset. The results, as expected, showed that supervised models outperformed unsupervised ones.

Unsupervised techniques have shown promise in detecting fraudulent transactions, with AD methods yielding optimistic and encouraging results and achieving high detection rates. Several studies have highlighted the commendable processing speed and detection accuracy of these approaches in the field of financial fraud detection (Zengan, 2009; Monamo *et al.*, 2017). As with supervised classifiers, several authors focused on benchmarking the performance of several unsupervised AD techniques for FD (Domingues *et al.*, 2018; Ounacer *et al.*, 2018). The results revealed that the IF (Isolation Forest) algorithm outperforms all other techniques, proving to be an exceptional method for identifying outliers, even on large datasets. Additionally, some studies have even demonstrated the efficacy of AD techniques in uncovering fraudulent transaction patterns that were previously undetected or missed by conventional rule-based systems (Zengan, 2009).

While AD algorithms have shown promising results, their real-world effectiveness is still in question. Testing on a real-world Bitcoin dataset Lorenz *et al.* (2020) showed high false positive rates, highlighting the limitations of relying on synthetic anomalous data. Criminals can imitate normal behavior, making it harder to identify fraud. Therefore, solely relying on AD algorithms without human analyst review can lead to high false positive rates and may not be practical in real-world applications.

3.2 Anomaly Detection with Active Learning

Studies on financial systems are limited, but valuable insights have been provided. Particularly, combining unsupervised AD and supervised query strategies is advocated as a means of improving the AL process. AD algorithms have limitations in detecting fraudulent transactions, with slow identification and high false positive rates. Common supervised AL policies include uncertainty sampling, query-by-committee, and expected model change. Uncertainty sampling selects samples with the highest uncertainty based on the classifier's confidence in its prediction (i.e., the transaction with the predicted probability closest to 0.5) (Sharma and Bilgic, 2017). Query-by-committee employs a committee of multiple learners and leverages the disagreement among classifiers to make a decision (Kee *et al.*, 2018). Finally, expected model change chooses the instances that would result in the maximum change of the current model when queried (Cai *et al.*, 2017).

Labanca *et al.* (2022) proposed a framework that ranks anomalous data using unsupervised AD methods and trains supervised models on domain experts' feedback using two AL strategies. The authors emphasized the importance of aggregating transactions for capturing correlations and found that IF and RF were the best-performing ML models with high detection rates. Other studies have also contributed to the literature by designing multi-stage AL labeling policies (Lorenz *et al.*, 2020; Barata *et al.*, 2021), which initially apply unsupervised AD algorithms to rank the most anomalous transactions for review before switching to supervised AL policies. Barata *et al.* (2021) proposed an intermediate stage using ODAL as a warm-up learner, which efficiently alleviated the cold start scenario with high-class imbalance. Both studies found that switching to supervised learners improved the models' performance, with Lorenz *et al.* (2020) achieving promising results by matching the performance of a supervised baseline using just 5% of the labels. Lastly, Carcillo *et al.* (2017) developed strategies to handle high-class imbalance in streaming credit card FD scenarios. They evaluated different AL strategies and found that combining the baseline strategy of only querying high-risk transactions with Stochastic Semi-supervised learning resulted in higher fraud detection rates. This being said, an effective AL strategy for AD requires proper tuning.

4. DISCUSSION

Most literature shows that AD methods often underperform in the AL process, leading to suboptimal performance. The slow identification of fraudulent transactions in the unlabeled pool is the primary reason for this. One solution is to switch to a supervised hot learner after a threshold or combine unsupervised and supervised techniques to select the most informative transactions for querying. However, it's unclear if enhancing the anomaly detector's performance in an AL setup would lead to better results. If it can detect more fraudulent transactions in earlier iterations, it may improve the overall model performance.

We aim to improve the anomaly detectors' performance by benchmarking different AL setups on the Elliptic Dataset. Similarly, to Lorenz *et al.* (2020) we will use unsupervised and supervised AL strategies depending on the number of illicit transactions in the labeled pool (i.e., when a certain threshold occurs, we either switch from an unsupervised warmup learner to a supervised hot-learner or continue to use the unsupervised learner). The goal is to improve AD algorithm performance in the early stages by considering the intrinsic characteristics of the Bitcoin transaction network. Given the possibility that illicit transactions are often connected to other illicit transactions in the network, it is possible to leverage this information by combining it with the anomaly scores produced by the anomaly detector. We propose to investigate this, by benchmarking each AL setup against a supervised baseline.

5. CONCLUSION

The present article delves into the significant challenges posed by traditional ML approaches in detecting financial fraud using cryptocurrencies. To address these issues, a combination of AD and AL algorithms is proposed. The proposed approach aims to simulate a typical real-life scenario where a limited number of labels can be acquired through manual annotation by experts. By incorporating the intrinsic characteristics of the Bitcoin transaction network and graph information, the article proposes to investigate how the performance of the anomaly detector can be enhanced in an AL setup, leading to better results. The proposed approach has the potential to significantly improve the accuracy of the FD system while minimizing the need for labeled data. Furthermore, it can enable financial institutions to detect fraud in real-time, preventing monetary losses and ensuring customer safety.

ACKNOWLEDGEMENT

This work has been supported by FCT – Fundação para a Ciência e Tecnologia within the R&D Units Project Scope: UIDB/00319/2020.

REFERENCES

Ahmed, M., Naser Mahmood, A. and Hu, J. (2016) 'A survey of network anomaly detection techniques', *Journal of Network and Computer Applications, 60, pp. 19–31*. Available at: https://doi.org/10.1016/J.JNCA.2015.11.016.

Alarab, I., Prakoonwit, S. and Nacer, M.I. (2020) 'Comparative Analysis Using Supervised Learning Methods for Anti-Money Laundering in Bitcoin', *ACM International Conference Proceeding Series, pp. 11–17*. Available at: https://doi.org/10.1145/3409073.3409078.

Barata, R. *et al*. (2021) 'Active learning for imbalanced data under cold start', *ICAIF 2021 - 2nd ACM International Conference on AI in Finance* [Preprint]. Available at: https://doi.org/10.48550/arxiv.2107.07724.

Cai, W., Zhang, M. and Zhang, Y. (2017) 'Batch mode active learning for regression with expected model change', *IEEE Transactions on Neural Networks and Learning Systems, 28(7), pp. 1668–1681*. Available at: https://doi.org/10.1109/TNNLS.2016.2542184.

Carcillo, F. e*t al*. (2017) 'An assessment of streaming active learning strategies for real-Life credit card fraud detection', *Proceedings - 2017 International Conference on Data Science and Advanced Analytics, DSAA 2017, 2018-Janua, pp. 631–639*. Available at: https://doi.org/10.1109/DSAA.2017.10.

Domingues, R. *et al*. (2018) 'A comparative evaluation of outlier detection algorithms: Experiments and analyses', *Pattern Recognition, 74, pp. 406–421*. Available at: https://doi.org/10.1016/J.PATCOG.2017.09.037.

Hilal, W., Gadsden, S.A. and Yawney, J. (2022) 'Financial Fraud: A Review of Anomaly Detection Techniques and Recent Advances', *Expert Systems with Applications, 193, p. 116429*. Available at: https://doi.org/10.1016/J.ESWA.2021.116429.

Ibrahim, R.F., Elian, A.M. and Ababneh, M. (2021) 'Illicit Account Detection in the Ethereum Blockchain Using Machine Learning', *2021 International Conference on Information Technology (ICIT), pp. 488 493*. Available at: https://doi.org/10.1109/ICIT52682.2021.9491653.

Injadat, M. *et al*. (2018) 'Bayesian Optimization with Machine Learning Algorithms Towards Anomaly Detection', *2018 IEEE Global Communications Conference, GLOBECOM 2018 - Proceedings* [Preprint]. Available at: https://doi.org/10.1109/GLOCOM.2018.8647714.

Kee, S., del Castillo, E. and Runger, G. (2018) 'Query-by-committee improvement with diversity and density in batch active learning', *Information Sciences, 454–455, pp. 401–418*. Available at: https://doi.org/10.1016/J.INS.2018.05.014.

Labanca, D. *et al*. (2022) 'Amaretto: An Active Learning Framework for Money Laundering Detection', *IEEE Access, 10, pp. 41720–41739*. Available at: https://doi.org/10.1109/ACCESS.2022.3167699.

Lorenz, J. *et al*. (2020) 'Machine learning methods to detect money laundering in the bitcoin blockchain in the presence of label scarcity', *ICAIF 2020 - 1st ACM International Conference on AI in Finance* [Preprint]. Available at: https://doi.org/10.1145/3383455.3422549.

Melo-Acosta, G.E., Duitama-Munoz, F. and Arias-Londono, J.D. (2017) 'Fraud detection in big data using supervised and semi-supervised learning techniques', 2017 IEEE Colombian Conference on Communications and Computing, COLCOM 2017 - Proceedings [Preprint]. Available at: https://doi.org/10.1109/COLCOMCON.2017.8088206.

Monamo, P.M., Marivate, V. and Twala, B. (2017) 'A Multifaceted Approach to Bitcoin Fraud Detection: Global and Local Outliers', pp. 188–194. Available at: https://doi.org/10.1109/ICMLA.2016.0039.

Niu, X., Wang, L. and Yang, X. (2019) 'A Comparison Study of Credit Card Fraud Detection: Supervised versus Unsupervised'. Available at: https://doi.org/10.48550/arxiv.1904.10604.

Ostapowicz, M. and Żbikowski, K. (2019) 'Detecting Fraudulent Accounts on Blockchain: A Supervised Approach', ArXiv, *11881 LNCS, pp. 18–31*. Available at: https://doi.org/10.1007/978-3-030-34223-4_2.

Ounacer, S. *et al*. (2018) 'Using Isolation Forest in anomaly detection: The case of credit card transactions', *Periodicals of Engineering and Natural Sciences, 6(2), pp. 394–400*. Available at: https://doi.org/10.21533/PEN.V6I2.533.

Settles, B. (2012) 'Active Learning', *Synthesis Lectures on Artificial Intelligence and Machine Learning, 18, pp. 1–111*. Available at: https://doi.org/10.2200/S00429ED1V01Y201207AIM018.

Sharma, M. and Bilgic, M. (2017) 'Evidence-based uncertainty sampling for active learning', *Data Mining and Knowledge Discovery, 31(1), pp. 164–202*. Available at: https://doi.org/10.1007/S10618-016-0460-3/TABLES/9.

Weber, M. *et al*. (2019) 'Anti-Money Laundering in Bitcoin: Experimenting with Graph Convolutional Networks for Financial Forensics'. Available at: https://doi.org/10.48550/arxiv.1908.02591.

Xuan, S. *et al*. (2018) 'Random forest for credit card fraud detection', *ICNSC 2018 - 15th IEEE International Conference on Networking, Sensing and Control, pp. 1–6*. Available at: https://doi.org/10.1109/ICNSC.2018.8361343.

Zengan, G. (2009) 'Application of cluster-based local outlier factor algorithm in anti-money laundering', Proceedings - *International Conference on Management and Service Science, MASS 2009* [Preprint]. Available at: https://doi.org/10.1109/ICMSS.2009.5302396.

FRAUD DETECTION AND ANTI-MONEY LAUNDERING APPLYING MACHINE LEARNING TECHNIQUES IN CRYPTOCURRENCY TRANSACTIONAL GRAPHS

Ana P. Martins and Miguel A. Brito
Centro Algoritmi
University of Minho, Portugal

ABSTRACT

Cryptocurrencies have advantages such as lower costs, efficiency, and security, but are vulnerable to fraud due to a lack of controls and anonymity. Criminals use virtual currencies for quick, anonymous transactions. Robust measures are needed to prevent illegal activities like money laundering. Machine learning (ML) and graph analysis can help detect fraud in the cryptocurrency market, despite criminals mimicking normal transactions. This study aims to use cutting-edge technologies like ML and graph learning to find fraudulent patterns in cryptocurrency transactions.

KEYWORDS

AML, Cryptocurrencies, Graphs, Machine Learning, Fraud Detection

1. INTRODUCTION

Cryptocurrencies are digital currencies issued on blockchain technology with advantages over traditional monetary systems. However, cryptocurrency fraud is a risk due to the lack of central regulation and controls, and markets are susceptible to fraudulent schemes like Ponzi, phishing, and money laundering. Effective anomaly detection methods and anti-money laundering (AML) controls are needed to prevent illegal cryptocurrency activities. ML algorithms are a promising area of research in AML.

This article explores using blockchain data and deep learning, specifically neural networks, for detecting cryptocurrency fraud. Combining deep learning with graph analysis is essential for detecting fraudulent transactions and extracting hidden patterns. The study aims to prevent illegal transactions and comply with industry standards.

The remainder of the paper is organized as follows. Section 2 presents the theme of AML and the role of ML, providing a clear understanding of the topic. Section 3, it is introduced the concepts of graph learning and explained its importance in this domain. Section 4 incorporates a literature review of ML and graph learning applied to AML. Finally, the main conclusions are discussed in Section 5.

2. ANTI-MONEY LAUNDERING

Effective AML systems are necessary for organizations to reduce the risk of financial losses from money laundering. AML involves identifying suspicious activities through customer transaction analysis using software, databases, and analytical tools (Labanca *et al.,* 2022). Cryptocurrencies and blockchain technology present challenges in implementing traditional AML practices due to their decentralized and transparent nature.

Updated technologies and methods are necessary to comply with regulations and effectively detect new money laundering techniques. ML is crucial in AML to prevent and detect financial crimes.

Graph learning improves AML solutions by analyzing financial data, identifying patterns, and increasing accuracy and efficiency (Han *et al.*, 2020). This technology overcomes traditional AML system limitations and helps detect and prevent money laundering activities involving cryptocurrencies (Vassallo *et al.*, 2021).

3. TRANSACTIONAL NETWORKS

Analyzing cryptocurrency transactions using graph analysis and ML techniques can efficiently detect illegal financial activities. This method saves time by eliminating the need for extensive feature engineering and provides a better understanding of the interactions within the cryptocurrency ecosystem. Graph Anomaly Detection (GAD) is a graph learning technique that detects anomalies in high-dimensional graph data using deep learning methods. It involves using graph representations and algorithms to recognize patterns.

Graph-based methods, an extension of deep learning, can uncover hidden patterns in graph data that traditional ML cannot (Xia *et al.,* 2021). Transaction network graphs are frequently used, representing nodes and connections in a vectorial space. UTXO blockchain address, transaction, and user networks are analyzed for link prediction, node classification, and forensic investigation.

Graph embedding techniques such as graph2vec and node2vec are popular for learning latent representations of vertices on large-scale networks, effectively reducing the transaction network data's dimensionality (Lin *et al.,* 2020).

Graph neural networks (GNNs) process graph data and extract graph representations for classification, recommendation, and clustering. GCN is a commonly used GNN technique that outputs node embeddings using a learnable kernel on local graph neighborhoods (Alarab *et al*., 2020b).

4. LITERATURE REVIEW

This section reviews the use of ML algorithms for detecting fraud and money laundering and suggests the application of graph learning and analysis in this area, using concepts from the previous section for support and clarity.

4.1 Machine Learning for Anti-Money Laundering

Chen *et al.* (2018) studied how ML techniques are used for detecting suspicious transactions in banking. Effective data prep and ML models like SVM, neural networks, and Bayesian networks are used for accurate detection, scalability, and fast performance. Choithani *et al.* (2022) found that SVM, LSTM, and Artificial Neural Networks were the most effective AI and ML algorithms for predicting cryptocurrency behavior for risk management in banking. Kamišalić *et al.* (2021) reviewed studies on using blockchain and data mining to identify fraud and anomalies. They found that Gradient Boosting, SVM, and Random Forest (RF) were popular methods, but Deep Learning with Neural Networks and LSTM is becoming more prevalent, especially with their graph learning approach.

The Elliptic dataset, which was introduced by Weber *et al. (*2019), is a significant and extensively labeled dataset available for any cryptocurrency. It comprises a graph network of Bitcoin transactions and has gained widespread recognition as a valuable resource for research in AML. RF was found to be better than GCN by Weber *et al.* (2019), but graph-based methods had potential. Alarab *et al.* (2020a) used ensemble learning and showed better results than Weber *et al*.'s method.

Ostapowicz and Żbikowski (2019) and Fawaz Ibrahim *et al.* (2021) proposed ML models to identify fraudulent accounts on Ethereum, with RF having the best performance in recall and false-positive rate. However, Ostapowicz and Żbikowski (2019) cautioned that the recall of 85% obtained may not be adequate for a real-world anti-fraud system.

Supervised methods like RF show promising results but require large, labeled datasets, which can be costly and time-consuming to create (Labanca *et al.,* 2022; Song *et al.,* 2023). Labeling historical transactions to detect money laundering is particularly difficult due to the low percentage of labeled illicit transactions. Unsupervised methods and active learning can be effective alternatives when labeled transactions are scarce. Recent studies have shown that unsupervised and active learning can perform similarly to supervised methods with only a few hundred labeled transactions. An active learning system that includes an unsupervised model can identify both known and unknown anomalous patterns, making it effective in detecting money laundering in capital markets transactions. The Isolation Forest algorithm has been found to be the most effective for this task. (Lorenz *et al.,* 2020; Labanca *et al.,* 2022)

Zhang and Trubey (2019) used ML and sampling to detect money laundering in financial transactions, concluding that artificial neural networks were the most effective and addressing the problem of unbalanced data.

As blockchain data grows, finding information and detecting money laundering becomes challenging due to its rarity and resulting data imbalance (Chen *et al.,* 2018; Zhang and Trubey, 2019). Researchers suggest analyzing trade-offs between recall and False Negative Rate (FNR) to achieve high accuracy. Various methods have been proposed, such as clustering, semi-supervised techniques, and heuristic user address clustering. For example, Monamo (2017) used k-means clustering and supervised classification models to detect fraudulent activity, while Yuan *et al.* (2018) employed a heuristic user address clustering method followed by verification using the Gaussian Mixture Model. Chen *et al.* (2018) suggested clustering or semi-supervised techniques to address data imbalance.

The literature suggests that traditional ML algorithms such as RF have been utilized in detecting fraud in finance and crypto transactions, but they have drawbacks such as costly manual labeling, difficulty in detecting rare events, and challenges in balancing recall and FPR.

4.2 Graph Learning and Analysis for Detecting Money Laundering

Motamed and Bahrak (2019) analyzed five cryptocurrencies, focusing on Bitcoin and Ethereum. They devised algorithms to extract transaction features and used graphs to detect anomalous transactions. Their findings showed that the address and transaction graphs for Bitcoin and Ethereum's money flow transaction graphs were particularly helpful.

GCNs, DeepWalk, and node2vec encode graph structures into dense representations to group similar nodes based on their neighborhoods. However, they don't account for temporal information, which is crucial for some applications.

Alarab *et al.* (2020) enhanced node classification on transaction graphs by merging GCN with linear layers for superior performance compared to GCN alone. Alarab and Prakoonwit (2022) achieved 98% accuracy and 80% f1-score in detecting illicit transactions. They improved previous studies by combining LSTM and GCN models and using active learning to label unlabeled data, considering the temporal aspect of transactions.

Singh *et al.* (2021) used a GNN-based adversarial loss architecture to reduce temporal bias in the elliptic dataset, leading to better performance. Pocher *et al.* (2022) found that GCN outperformed baseline methods over GAT. Geng *et al.* (2022) proposed an enhanced version of GCN, a graph attention mechanism, with the potential for detecting illegal activities in blockchain networks, demonstrating the usefulness of graph learning in addressing blockchain system security.

Caglayan and Bahtiyar (2022) proposed a graph embedding algorithm called node2vec to detect money laundering in financial transactions. It proved to be superior to other methods for detecting money laundering in a banking transaction dataset. Lopes *et al.* (2022) also found node2vec effective in identifying criminal and non-criminal relationships in criminal networks when combined with predictive ML methods. The studies show that node2vec has the potential to be a valuable tool for analyzing financial and criminal transactions. Further, Zhou *et al.* (2021) used node2vec to classify and predict financial fraud by representing the topological features of a financial network graph as low-dimensional dense vectors. This approach improved the efficiency of financial fraud detection. These findings demonstrate the potential of using graph embedding and deep learning for Internet financial fraud detection.

Li *et al.* (2022) proposed TA-Struc2Vec, a novel method to identify fraudulent users in financial transaction networks. It uses ML to classify users by considering both structural and transaction amount homogeneity in the network. This approach was found to be more effective than previous methods and has the potential to improve fraud detection in financial transaction networks.

The literature review highlights the significance of graph representation techniques in detecting fraudulent transactions, including financial and criminal activities. Graph-based approaches such as GCN, DeepWalk, node2vec, and their variants (e.g., GNN-based adversarial loss architecture, GAT mechanism) have shown promise in improving fraud detection accuracy. Combining graph representation with deep learning and incorporating temporal and weighted information can further enhance the performance of fraud detection models. These findings have implications for future research and demonstrate the potential of graph representation in addressing security concerns in financial and criminal transactions.

5. DISCUSSION

ML algorithms struggle with complex and interconnected graph data. Representation learning automatically extracts features, and node embedding maps nodes to a vector, characterizing their features. Node2vec generates expressive node embeddings using biased random walks. GNNs are specialized deep-learning techniques that work on graph data, incorporating information from neighboring nodes and individual node features. GCNs are effective in identifying patterns of illicit activity, detecting suspicious transactions, and analyzing blockchain networks in cryptocurrency transactions.

Previous studies found that combining GCN with node embeddings yielded the best results for RF algorithm. However, recent research suggests that GCN alone may be sufficient. This raises the question of whether embeddings play a crucial role in achieving superior performance. We propose to investigate this by testing different scenarios on the Elliptic Dataset to establish a benchmark for evaluating results.

We aim to understand the impact of different features, embeddings, and graph information on model performance and establish the optimal way to integrate traditional supervised models with deep learning techniques for improved performance in analyzing complex graph data.

6. CONCLUSION

The article discusses the challenges of using ML algorithms for complex graph data and how representation learning and GNNs can address these issues. Specifically, the focus is on the use of GNNs, such as GCNs for identifying patterns of illicit activity, detecting suspicious transactions, and analyzing blockchain networks in cryptocurrency transactions. The article proposes to investigate the impact of different features, embeddings, and graph information on model performance and establish the optimal way to integrate traditional supervised models with deep learning techniques for improved performance in analyzing complex graph data.

ACKNOWLEDGEMENT

This work has been supported by FCT – Fundação para a Ciência e Tecnologia within the R&D Units Project Scope: UIDB/00319/2020.

REFERENCES

Alarab, I. and Prakoonwit, S. (2022) 'Graph-Based LSTM for Anti-money Laundering: Experimenting Temporal Graph Convolutional Network with Bitcoin Data', *Neural Processing Letters* [Preprint]. Available at: https://doi.org/10.1007/s11063-022-10904-8

Alarab, I., Prakoonwit, S. and Nacer, M.I. (2020a) 'Comparative Analysis Using Supervised Learning Methods for Anti-Money Laundering in Bitcoin', in *ACM International Conference Proceeding Series*. Association for Computing Machinery, pp. 11–17. Available at: https://doi.org/10.1145/3409073.3409078

Alarab, I., Prakoonwit, S. and Nacer, M.I. (2020b) 'Competence of Graph Convolutional Networks for Anti-Money Laundering in Bitcoin Blockchain'. Available at: https://doi.org/10.1145/3409073.3409080

Caglayan, M. and Bahtiyar, S. (2022) 'Money Laundering Detection with Node2Vec', *Gazi University Journal of Science*, 35(3), pp. 854–873. Available at: https://doi.org/10.35378/GUJS.854725

Chen, Z. *et al.* (2018) 'Machine learning techniques for anti-money laundering (AML) solutions in suspicious transaction detection: a review', *Knowl Inf Syst*, 57, pp. 245–285. Available at: https://doi.org/10.1007/s10115-017-1144-z

Choithani, T. *et al.* (2022) 'A Comprehensive Study of Artificial Intelligence and Cybersecurity on Bitcoin, Crypto Currency and Banking System', *Annals of Data Science*, pp. 1–33. Available at: https://doi.org/10.1007/S40745-022-00433-5/TABLES/2

Fawaz Ibrahim, R., Mohammad Elian, A. and Ababneh, M. (2021) 'Illicit Account Detection in the Ethereum Blockchain Using Machine Learning'. Available at: https://doi.org/10.1109/ICIT52682.2021.9491653

Geng, Z. *et al.* (2022) 'Novel blockchain transaction provenance model with graph attention mechanism', *Expert Systems with Applications*, 209. Available at: https://doi.org/10.1016/J.ESWA.2022.118411

Han, J. *et al.* (2020) 'Artificial intelligence for anti-money laundering: a review and extension', *Digital Finance*, 2(3–4), pp. 211–239. Available at: https://doi.org/10.1007/s42521-020-00023-1

Kamišalić, A., Kramberger, R. and Fister, I. (2021) 'Synergy of Blockchain Technology and Data Mining Techniques for Anomaly Detection', *Applied Sciences 2021, Vol. 11, Page 7987*, 11(17), p. 7987. Available at: https://doi.org/10.3390/APP11177987

Labanca, D. *et al.* (2022) 'Amaretto: An Active Learning Framework for Money Laundering Detection', *IEEE Access*, 10, pp. 41720–41739. Available at: https://doi.org/10.1109/ACCESS.2022.3167699

Li, R. *et al.* (2022) 'Internet Financial Fraud Detection Based on Graph Learning; Internet Financial Fraud Detection Based on Graph Learning', *IEEE Transactions on Computational Social Systems*, PP. Available at: https://doi.org/10.1109/TCSS.2022.3189368

Lin, D. *et al.* (2020) 'T-EDGE: Temporal WEighted MultiDiGraph Embedding for Ethereum Transaction Network Analysis', *Frontiers in Physics*, 8. Available at: https://doi.org/10.3389/FPHY.2020.00204

Lopes, D.D. *et al.* (2022) 'Machine learning partners in criminal networks', *Scientific Reports 2022 12:1*, 12(1), pp. 1–9. Available at: https://doi.org/10.1038/s41598-022-20025-w

Lorenz, J. *et al.* (2020) 'Machine learning methods to detect money laundering in the Bitcoin blockchain in the presence of label scarcity'. Available at: https://doi.org/10.1145/3383455

Monamo, P.M., Marivate, V. and Twala, B. (2017) 'A Multifaceted Approach to Bitcoin Fraud Detection: Global and Local Outliers', pp. 188–194. Available at: https://doi.org/10.1109/ICMLA.2016.0039

Motamed, A.P. and Bahrak, B. (2019) 'Quantitative analysis of cryptocurrencies transaction graph', *Applied Network Science*, 4(1). Available at: https://doi.org/10.1007/S41109-019-0249-6

Ostapowicz, M. and Żbikowski, K. (2019) 'Detecting Fraudulent Accounts on Blockchain: A Supervised Approach', *Lecture Notes in Computer Science (including subseries Lecture Notes in Artificial Intelligence and Lecture Notes in Bioinformatics)*, 11881 LNCS, pp. 18–31. Available at: https://doi.org/10.48550/arxiv.1908.07886

Pocher, N. *et al.* (2022) 'Detecting Anomalous Cryptocurrency Transactions: an AML/CFT Application of Machine Learning-based Forensics'.

Sabry, F. *et al.* (2020) 'Cryptocurrencies and artificial intelligence: Challenges and opportunities', *IEEE Access*, 8, pp. 175840–175858. Available at: https://doi.org/10.1109/ACCESS.2020.3025211

Singh, A. *et al.* (2021) 'Temporal Debiasing using Adversarial Loss based GNN architecture for Crypto Fraud Detection', in *Proceedings - 20th IEEE International Conference on Machine Learning and Applications, ICMLA 2021*. Institute of Electrical and Electronics Engineers Inc., pp. 391–396. Available at: https://doi.org/10.1109/ICMLA52953.2021.00067

Song, W. *et al.* (2023) *Blockchain Data Analysis from the Perspective of Complex Networks: Overview*. Available at: https://www.webofscience.com

Vassallo, D., Vella, V. and Ellul, J. (2021) 'Application of Gradient Boosting Algorithms for Anti-money Laundering in Cryptocurrencies', *SN Computer Science*, 2, p. 143. Available at: https://doi.org/10.1007/s42979-021-00558-z.

Weber, M. *et al.* (2019) *Anti-Money Laundering in Bitcoin: Experimenting with Graph Convolutional Networks for Financial Forensics*.

Xia, F. *et al.* (2021) 'Graph Learning: A Survey', *IEEE Transactions on Artificial Intelligence*, 2(02), pp. 109–127. Available at: https://doi.org/10.1109/TAI.2021.3076021

Yuan, Y., Member, S. and Wang, F.-Y. (2018) 'Blockchain and Cryptocurrencies: Model, Techniques, and Applications; Blockchain and Cryptocurrencies: Model, Techniques, and Applications', *SYSTEMS*, 48(9). Available at: https://doi.org/10.1109/TSMC.2018.2854904

Zhang, Y. and Trubey, P. (2019) 'Machine Learning and Sampling Scheme: An Empirical Study of Money Laundering Detection', *Computational Economics*, 54(3), pp. 1043–1063. Available at: https://doi.org/10.1007/S10614-018-9864-Z

Zhou, H. *et al.* (2021) 'Internet Financial Fraud Detection Based on a Distributed Big Data Approach with Node2vec', *IEEE Access*, 9, pp. 43378–43386. Available at: https://doi.org/10.1109/ACCESS.2021.3062467

REFLECTION

RESPONSIBLE TECHNOLOGY: THINKING ON THIS VITAL CONCEPT

Marcel J. Simonette and Edison Spina
Universidade de São Paulo, Brazil

ABSTRACT

Responsible use of technology is increasingly a business and social carefulness, motivated by diverse economics and people's living issues. Deploying responsible technology has become critical as the technology presence is unprecedented in our society and business operations. There is a global agreement that business needs to consider technology responsibly. There are businesses in which responsible technology is a core part of their mission, and some businesses use technology responsibly in explicit financial terms, such as returns on investment, talent acquisition, or improving attractiveness to investors. Technology may bring substantial cultural shifts in organisations and society. Therefore, responsible technology use is not merely to comply with regulations or manage risks. This work is a reflection paper about using technology ethically and sustainably, considering the potential impacts on society and the environment.

KEYWORDS

Responsible Technology, Ethics, Social Justice, Business Sustainability

1. INTRODUCTION

Technology has become an integral part of people's lives. Several "smart things" exist, from smartphones to smart homes and cities. In addition, technological advancements in artificial intelligence and virtual reality have brought people closer and helped people's lives to become more straightforward and efficient. Responsible technology has become critical with all companies' increasing reliance on technology, not just tech companies. Currently, every company use some technology.

Responsible technology is the development and use of technology ethically and sustainably, prioritising the well-being of individuals and society. Moreover, it considers the consequences of technological advancements and that technology is accessible to all (Davis, 2014; Jirotka & Stahl, 2020; Ocone, 2020).

Ethical technology is another term that also refers to technology development. This term considers the well-being of individuals. Ethical technology focuses on the moral principles that should guide the development and use of technology; it is grounded in humanistic principles and values (Buckley et al., 2017; Kiran et al., 2015; Palm & Hansson, 2006).

Both terms, responsible technology and ethical technology, can be appropriate for technology development, depending on the context and the specific aspects of technology being discussed. For example, it may be more appropriate to use responsible technology when discussing the practical implications of technology. In contrast, ethical technology may be more appropriate when discussing moral and philosophical considerations.

This reflection paper uses responsible technology to address companies' approach to deploying and driving business strategy and operations responsibly, integrating technology into people's everyday lives. Our reflection considers the technology's impact on individuals, society, and the environment, dealing with challenges such as privacy, environmental sustainability, social inequalities, and ethical implications. This paper deals with responsible technology as a socio-technical system, considering the complex interactions between social and technical factors.

The paper is structured as follows: this section, introduction, briefly present the responsible technology concept. Section 2 presents technology in the business context; section 3 is about our reflection on responsible technology; section 4 summarises the key point of our reflection, emphasising their significance, and the references follow it.

2. TECHNOLOGY IN THE BUSINESS CONTEXT

Technology in the business context is changing fundamentally. A few decades ago, technology for business was limited to supporting business operations with information systems. It was easier in many ways to know whether systems were working as intended. Responsible use of technology was related to software systems functioning properly, and Information Technology was concerned with issues such as business requirements, compliance, security, and user experience. Ethical concerns were related to the code of ethics as one of the IEEE Computer Society (1999).

Nowadays, businesses are supported by technology on dimensions beyond companies' offices, offering technological products and services embedded in people's lives. Artificial intelligence, machine learning, blockchain, and digital reality are technologies that are integrating into our lives. Moreover, the business landscape is changing at an unprecedented speed, and technology has become essential for companies to remain competitive and achieve their goals. However, using technology in business raises several ethical and social concerns, including privacy, security, and sustainability.

As technology becomes fundamental to business, the concept of responsible technology becomes critical to business, as there are consequences of technology misuse. It is necessary to ensure that technology is designed, developed, and used to benefit society as a whole, minimising its negative impacts. Unfortunately, it is not always clear how to apply technology responsibly. For many years tech people believed technology could solve the world's problems (Huesemann & Huesemann, 2011); it was usual to say that there is no bad technology; what exists is the bad use of technology.

Technology may be a business's invisible enemy. Even in systems constructed with the best intentions, there are ways in which technology may present unwanted consequences, as happens with artificial intelligence bias (Centre for Data Ethics and Innovation, 2020).

Responsible technology in a business context involves considering the ethical, social, and environmental implications of technology use and ensuring that these considerations are integrated into the business strategy and operations, which includes developing technologies that are accessible, inclusive, and transparent, as well as ensuring that technology is used in a way that respects privacy, promotes security, and supports sustainability.

Adopting a responsible technology approach can bring significant benefits for businesses, including enhanced reputation, increased customer loyalty, and improved employee satisfaction. It can also contribute to achieving the United Nations Sustainable Development Goals (United Nations, 2019, 2023), which provide a framework for sustainable development and address some of the world's social and environmental challenges.

It is essential for businesses to engage in dialogue with stakeholders, including customers, employees, civil society organisations, and policymakers, to ensure that technology is used in a way that meets the needs and expectations of all relevant actors. This engagement requires a collaborative and interdisciplinary approach that combines expertise from different fields, including science, engineering, social sciences, and humanities. Responsible technology is a critical aspect of modern business. Companies must adopt a responsible technology approach to succeed in a rapidly changing and socially and environmentally conscious world.

3. RESPONSIBLE TECHNOLOGY

Despite technology becoming increasingly central to what companies do, it is not always clear how to approach and apply technology with responsibility. Business leaders must acknowledge that they are responsible for offering technology to their customers. The assumption that technology can solve the world's problems and that there is no bad technology is utopic.

One of the primary concerns with technology is its impact on privacy and security. As personal data are being stored and processed by companies, there is the risk of data breaches, cyber-attacks, and the mismanagement of personal information. In order to ensure responsible technology, it is necessary to have adequate privacy and security measures in place, as well as regulations and policies that hold companies accountable for protecting personal data.

Another aspect of responsible technology is accessibility. Although technology has made life easier for many, there is still a significant part of the population that is unable to access and use technology; age, income, or location are some of the barriers to making digital inclusion (Chadwick et al., 2022; Pérez-Escobar & Canet, 2022; United Nations, n.d.). Companies need to ensure that their technological products and services are accessible to everyone, regardless of their background or circumstances, which will help to develop a more egalitarian society, as less than thirty per cent of the population uses the Internet, according to a 2021 work, developed by International Telecommunication Union (ITU, 2023).

Additionally, responsible technology also involves considering the environmental impact of technology (Cardinali & De Giovanni, 2022). There is a significant environmental impact, from the production of electronics to the disposal of outdated technology. Therefore, companies are responsible for reducing their carbon footprint, using sustainable materials in their products, and implementing environmentally friendly practices throughout their operations.

Responsible technology considerations must be present in business strategies and decisions, as financial considerations are present when companies make decisions about technology adoption. However, responsible technology investments are not tangible; their effect on business relates to brand reputation, customer and employee retention, and prevention of unintended negative consequences and associated brand risk.

Several responsible technology practices need to be focused on and prioritised, such as security, data privacy, inclusive design, data privacy, environmental impact, elimination of bias, and workforce diversification. Understanding the broader societal context of technology is becoming ever more critical as technology advances show no signs of slowing. Considering the complex problems involving people and society impacted by technology is necessary. Furthermore, business responsibility is as important as the technology used or developed by business demand. Businesses need to ask questions beyond the features of their products or services and understand that adoption of technology has consequences on their customers, society and even their ability to attract and retain future talent.

The business imperative of revenue as their measure of success needs to change. Currently, it is no longer possible for the business imperative to deploy technology in a not responsible way, focusing on revenue. Instead, companies need to measure value in other ways.

4. CONCLUSION

Technology has become necessary to people's lives and has changed how we live, work, and communicate. Therefore, business use of technology must go beyond the problem solutions. It is necessary to offer solutions with responsibility, ensuring that tech products and services are developed and used ethically and responsibly.

Getting something wrong, or even when there is a bad intention in using technology, is difficult to define. Likewise, some responsibilities are difficult to define, especially when it is necessary to define what is wrong. It is essential to understand that technology is not neutral and can have unintended consequences if not developed and used responsibly. For example, the rise of social media has led to increased cyberbullying, mental health issues, and misinformation.

Businesses must be aware that regulation of responsible technology is more than the existing laws, such as GDPR. On the other hand, there is a world movement about tech-responsible regulation, which also is a problem, as this scenario may be unstable. Moreover, it demands a consensus among the governments, which takes time and brings instability. Nevertheless, as a positive aspect, all this movement about regulation motivates the adoption of responsible tech practices until the industry develops accurate guidance.

In conclusion, responsible technology is an approach that acknowledges the challenges posed by technology and seeks to address them in a way that benefits individuals, society, and the environment. It involves being transparent about data collection and usage, promoting environmental sustainability, addressing social inequalities, and considering ethical implications. By adopting responsible technology practices, we can ensure that technology continues to enhance our lives in a safe, fair, and sustainable way.

ACKNOWLEDGEMENT

This work was partly supported by the Society and Technology Study Centre (or CEST – Centro de Estudos Sociedade e Tecnologia, in Portuguese) at the Universidade de São Paulo.

REFERENCES

Buckley, J. A., Thompson, P. B. & Whyte, K. P. (2017). Collingridge's dilemma and the early ethical assessment of emerging technology: The case of nanotechnology enabled biosensors. *Technology in Society*, 48, 54-63. https://doi.org/10.1016/j.techsoc.2016.12.003

Cardinali, P. G., & De Giovanni, P. (2022). Responsible digitalization through digital technologies and green practices. *Corporate Social Responsibility and Environmental Management*, *29*(4), 984-995. https://doi.org/10.1002/csr.2249

Centre for Data Ethics and Innovation (2020). *Review into bias in algorithmic decision-making*. Retrieved March, 21, 2023, from https://assets.publishing.service.gov.uk/government/uploads/system/uploads/attachment_data/file/957259/Review_into_bias_in_algorithmic_decision-making.pdf

Chadwick, D., Ågren, K. A., Caton, S., Chiner, E., Danker, J., Gómez-Puerta, M., ... & Wallén, E. F. (2022). Digital inclusion and participation of people with intellectual disabilities during COVID-19: A rapid review and international bricolage. *Journal of Policy and Practice in Intellectual Disabilities*, *19*(3), 242-256. https://doi.org/10.1111/jppi.12410

Davis, S. (2014). Responsible Technology. *The Serials Librarian*, 67:1, 12-20. http://dx.doi.org/10.1080/0361526X.2014.915607

Huesemann, M., & Huesemann, J. (2011). *Techno-Fix: Why Technology Won't Save Us or the Environment*. New Society Publishers. ISBN: 9780865717046

IEEE Computer Society (1999). *Code of Ethics*. https://www.computer.org/education/code-of-ethics

ITU. (2023). International Telecommunication Union, *Measuring digital development: Facts and figures*. Retrieved March 21, 2023, from: http://www.itu.int/en/ITU-D/Statistics/Pages/facts/default.aspx

Jirotka, M. and Stahl B.C. (2020). The need for responsible technology. *Journal of Responsible Technology,* vol. 1, oct., 10002. http://dx.doi.org/10.1016/j.jrt.2020.100002

Kiran, A. H., Oudshoorn, N. & Verbeek, P. (2015). Beyond checklists: toward an ethical-constructive technology assessment. *Journal of Responsible Innovation*, 2:1, 5-19. http://dx.doi.org/10.1080/23299460.2014.992769

Ocone, R. (2020). Ethics in engineering and the role of responsible technology. *Energy and AI,* vol. 2, Nov., 10019. http://dx.doi.org/10.1016/j.egyai.2020.100019

Palm, E. & Hansson, S. O. (2006). The case for ethical technology assessment (eTA). *Technological Forecasting and Social Change*, 73:5, 543-558. https://doi.org/10.1016/j.techfore.2005.06.002

Pérez-Escolar, M. & Canet, F. (2022). Research on vulnerable people and digital inclusion: toward a consolidated taxonomical framework. *Universal Access in the Information Society.* https://doi.org/10.1007/s10209-022-00867-x

United Nations (2019). Independent Group of Scientists appointed by the Secretary-General (2019). *Global Sustainable Development Report 2019: The Future is Now – Science for Achieving Sustainable Development.* https://sdgs.un.org/sites/default/files/2020-07/24797GSDR_report_2019.pdf

United Nations (n.d.). Round Table on Digital Inclusion. Retrieved March, 21, 2023, from: https://www.un.org/techenvoy/sites/www.un.org.techenvoy/files/general/Definition_Digital-Inclusion.pdf

United Nations (2023.). *The 17 goals.* Retrieved March 21, 2023, from: https://sdgs.un.org/goals

DEVELOPMENT OF THE ASSISTIVE TECHNOLOGY MODEL FOR THE ELDER CARE SYSTEM

Supakorn Iamamporn[1], Varit Kankaew[2], Pradit Songsangyos[2] and Jirawadee Yingkongdee[1]
[1]*Faculty of Business Administration and Information Technology, Rajamangala University of Technology Suvarnabhumi Pranakhon Si Ayutthaya, Thailand*
[2]*Faculty of Science and Technology, Rajamangala University of Technology Suvarnabhumi, Pranakhon Si Ayutthaya, Thailand*

ABSTRACT

Currently, the world demographic population, including Thailand, has changed. The number of senior citizens is increasing. Whereas their family members living in the same household is decreasing. Senior citizens may need the technology for a self-care assistant. This study develops the assistive technology model for the elder care system. The study aimed at medical assistants and social assistants. The research is divided into two stages. Stage 1: Design an assistive technology model for the elder care system. The assistive technology model for the elder care system consists of the following. 1. Senior citizen. 2. Family members or caregivers. 3. The Community of the senior citizens. 4. Assistive technologies. 5. Functions of the elder care system. 6. The outcome of the proposed model is medical assistance and social assistants. Stage 2: The experts evaluated the appropriateness of the proposed model. The appropriateness assessment of the proposed model by seven experts who have experience in the area of medicine and healthcare, besides information and communication technologies, overall is excellent.

KEYWORDS

Assistive Technology, Elder Care System, Senior Citizen, Telehealth, Activities of Daily Living

1. INTRODUCTION

Currently, the world demographic population, including Thailand, has changed. The number of senior citizens is increasing, whereas the birth rate is decreasing. Also, it drastically reduces family members living in the same household (Economic Research Institute for ASEAN and East Asia, 2021). The elder's health is genetic. Most are due to their homes, neighbourhoods, and communities, as well as their characteristics. Supportive physical and social environments enable senior citizens to do what is important to them, despite losses in capacity (World health organisation, 2022). A study by Sharif, M.S. (2018) found that the elder will need assistance with shopping, companionship, medicine, socialising, housework, exercise, meals, personal care and exercise. Whereas Martinez-Martin, E. and Costa, A. (2021) categorised technological assistants for the elderly into medical or rehabilitation assistants and social assistants.

This study was conducted to develop a model of assistive technology for the elder care system. The study aimed at medical assistance and social assistants.

Medical assistants

A cognitive assistant uses wearable devices as assistive technology to measure the elder's health status. So the elderly could be taken care of by family members or caregivers. In addition, they continuously monitor their health parameters at anytime and anywhere (Martinez-Martin, E. and Costa, A., 2021) (Alejandrino, M.A., 2018). According to Kemisetal (2012) (Cited in Martinez-Martin, E. and Costa, A., 2021), when the heart rate is greater than 110 bpm while the senior citizen is at rest, they confront abnormal heart rate patterns and require immediate care. So, the equipment should notify the condition or alert their family member or caregiver. Whereas Sumi, L.et al. (2019) studied IoT-based fall prevention and detection for senior citizens, physically and intellectually disabled. Fall detection triggers an alarm if a fall is detected, then notify family members or caregivers via emails and SMS. Bruun-Pedersen, J.R. et al. (2014) studied augmented exercise biking with virtual environments for elderly users: a preliminary study for retirement home physical therapy. The study had the exercise of a Chair-based exercise bike using a countryside base virtual environment (VE) as home physical therapy for senior citizens. The study found that a majority of senior citizens preferred the VE-based exercise to the conventional training.

Social assistants

Wungrath, J. (2021) studied the healthcare education process adopting the line application in conjunction with tele-counselling to improve knowledge, behavior and satisfaction among elderly with diabetes militias during the COVID-19 pandemic. The researcher sends video clips and infographics via the line group. Later, researchers have a discussion and explanation with group members. In addition, tele-counselling provided the senior citizen to call researchers if needed nutritional assistance. Researchers educated senior citizens by teaching and emphasising participation. Then described the essentials of health practices. An assessment was conducted and used to highlight the benefits of behavioural practices. In addition, speech to persuade the sample group to their ability to take appropriate care of themselves. As well as having an exchange of ways to deal with the problems and obstacles (Martinez-Martin, E. and Costa, A., 2021) (Supasri, S., 2021). Wungrath, J. et al., 2020 study Effect of Using Basic Elderly Care Handbook on Knowledge and Behavior Regarding Elderly Care Among Caregivers of the Elderly in Choeng Doi Subdistrict, Doi Saket District, Chiang Mai Province. The study found that the content in the manual includes the senior citizen's nature and dynamics, the senior citizen, mental health of the senior citizen, the role of family and community to the senior citizen, and nutrition for the senior citizen. In addition, exercise, an appropriate environment for the senior citizen and health problems the senior citizen. By the way, Chiangmai is a city with a high potential for retired people from many countries worldwide. Besides, the city has good infrastructure, including healthcare, with low living expenses.

2. METHODOLOGY

The data collection using an assessment form for the assistive technology model for elder care system. The population in the study consisted of experts who have experienced in the area of medical and healthcare, besides information and communication technologies. The samples group comprised of seven experts who either held a doctoral degree or had at least five years' experience in their field for evaluation and suggestions. A purposive selection method was selected. Jamil (as cited in Hamzah F. et al., 2019) stated that a minimum of five years of working experience would be enough to ensure expertise in their respective fields. Analyse the results of the evaluation of the appropriateness of a proposed model using mean and standard deviation.

The research is divided into two stages. Stage 1: Design the assistive technology model for the elder care system as follows. (1.1) Relevant articles and research papers were analysed and synthesised in the following topics: Assistive technology, Elder care system, Senior citizen, Telehealth, Activities of daily living (ADL), Smartwatch and smart band, Community of senior citizens, Elderly health care guide, Self-care, Internet of Things (IoT), Falling detection and augmented exercise biking. (1.2) Use the obtained information to synthesise and summarise as the assistive technology model for the elder care system. (1.3) Research instrument was designed as an appropriate evaluation form. Stage 2: The experts evaluated the appropriateness of the proposed model as follows. (2.1) Propose the concept idea and the assistive technology model for the elder care system to seven experts for evaluation and suggestions. (2.2) The proposed model was verified, and the seven identified experts filled in appropriate evaluation forms. (2.3) The proposed model was modified based on the advice of the experts. (2.4) Analyse the results of the evaluation of the appropriateness of the proposed model using mean ($\overline{\mathbf{X}}$) and standard deviation (S.D.). Five criteria for assessment were considered using Likert scales anchored with the terms Strongly Agree, Agree, Undecided, Disagree, and Strongly Disagree (Samsudin, S. et al., 2019; Brown, S., 2010).

Results of the study are as follows.

Stage 1: The assistive technology model for the elder care system was designed as illustrated in figure 1.

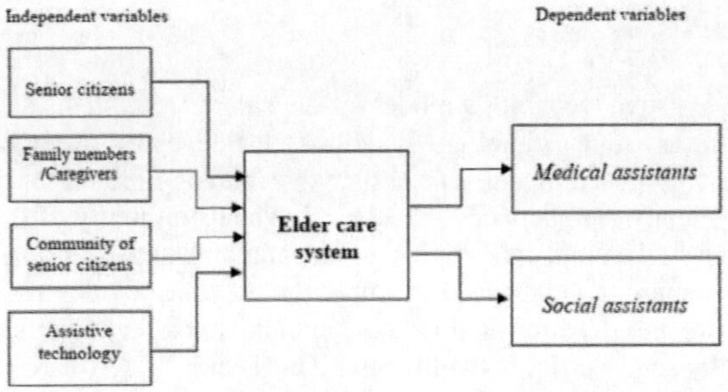

Figure 1. The assistive technology model for elder care system

The assistive technology model for the elder care system consists of the following. 1) Senior citizens aged 60 years and over. Older age is also characterised by the emergence of several complex health states commonly called geriatric syndromes (World health organization, 2022). 2) Family members or caregivers. Parents of a senior citizen or a person who is responsible for healthcare. 3) The Community of the senior citizens, as a group of elders or through online social. The community gathers to talk to each other, organise joint activities, exchange knowledge and experiences related to the direction, and deal with problems and obstacles. Then lead to a change in daily life to promote healthy behaviour. 4) Assistive technologies include the following: (4.1) Smartwatch and Smart band that has many functions to indicate the user's health at an affordable cost. (4.2) Exersight biking, present the user to view on-demand environment on the screen of smart Television, notebook, or smartphone. (4.3) Internet of Things (IoT) devices and various application software. Which support the senior citizens, including detection of falling and no movement, schedule for their exercise or appointment, and notifying them to take medication or an appointment. 5) The proposed model will apply hardware and software. The assistive technologies have the main functions: (5.1) Measuring heart rate and blood pressure. (5.2) Detection of falling and no movement. (5.3) Notify the time to take medication or have an appointment with a doctor (5.4) Exercise schedules, with Exersight biking. (5.5) Elderly health care guide. (5.6) Appointments Schedule for joint activities and conversations with friends or community of the senior citizens. 6) The outcome of the proposed model is classified into two areas. (6.1) Medical assistants (physician assistants) to reduce health problems, promote self-care ability, and facilitate family members or caregivers (6.2) Social assistants make senior citizens not feel isolated and proud in taking care of themselves. Then encourage them to follow the elderly healthy guide and collaborate to present the direction and how it works.

Stage 2: Assessment of the assistive technology model for the elder care system.

The experts evaluated the appropriateness of the assistive technology model for the elder care system, as shown in Table 1.

Table 1. Appropriateness of the assistive technology model for the elder care system

Descriptions	$\bar{X}$	S.D.	Level of appropriation
Senior citizens	4.71	.49	Strongly Agree
Family members / caregivers	4.57.	.53	Strongly Agree
Community of senior citizens	4.29	.96	Agree
Assistive technology	4.57	.68	Strongly Agree
Functions of Elder care system	4.71	.46	Strongly Agree
Medical assistant	4.71	.49	Strongly Agree
Social assistant	4.57	.79	Strongly Agree
Overall	***4.63***	***.56***	***Strongly Agree***

The experts evaluated the appropriateness of the assistive technology model for the elder care system overall as Strongly Agree, with a mean of 4.63, S.D. = .56. Senior citizens and family members or caregivers as Strongly Agree. Community of senior citizens as Agree. Assistive technology and functions of the elder care system as Strongly Agree. The outcome of the proposed model, medical assistant and social assistant, is Strongly Agree. Regarding the smartwatch / smart band, detection of falling and no movement and notification to take medication or an appointment had the highest score, as Strongly Agree, by a mean of 5.00, S.D. = 0.0. The exercise schedules had the lowest score, Agree, with a mean of 4.00, S.D. = 0.0.

3. CONCLUSION

This study develops the assistive technology model for the elder care system. The research is divided into two stages. Stage 1: Design the assistive technology model for the elder care system. Stage 2: The experts evaluated the appropriateness of the proposed model. The assistive technology model for the elder care system consists of 1. Senior citizens. 2. Family members or caregivers. 3. The Community of the senior citizens. 4. Assistive technologies. 5. Functions of the elder care system. 6. The outcome of the proposed model is medical assistants and social assistants. The appropriateness assessment of the proposed model by seven experts with experience in medicine and healthcare, besides information and communication technologies, is excellent. The study was designed based on healthy senior citizens maintaining their health by self-care or less assistance from family members or caregivers at affordable cost.

For further study, researchers initially developed a prototype of Exersight biking. Next, conduct the community of senior citizens as a pilot project. All components were then integrated to build a prototype of the assistive technology for the elder care system. Furthermore, a study of the critical success factor for Thailand's healthcare market, especially for retired foreign people, should be concerned.

REFERENCES

Alejandrino, M.A. et al., 2018. GiveCare: A monitoring and communication smart watch app with pulse sensor and analysis for caregivers assisting elderly and persons with speech impairment. *Proceedings of TENCON 2018-2018 IEEE region 10 conference*, p.p. 1136-1141.

Brown, S., 2010. Likert Scale Examples for Surveys. *In Iowa State University Extension*, pp. 1-4.

Bruun-Pedersen, J.R. et al., 2014. Augmented exercise biking with virtual environments for elderly users: A preliminary study for retirement home physical therapy. *Proceedings of 2014 2nd Workshop on virtual and augmented assistive technology, p.p.23-27.*

Economic Research Institute for ASEAN and East Asia, 2021. Long-Term Care Model: Review of Population Ageing Practices and Policies. *In Population Ageing in Thailand*, 2021, Volume 2, p.p. 21.

Hamzah, F. et al., 2019. Rasch Model Analysis for Evaluating Validity and Reliability of Authentic Learning Instruments in Malaysian Polytechnics. *In International Journal of Innovation, Creativity and Change*. Volume 9, Issue 7, pp. 51-62.

Joseph, S. et al., 2019. Wearable assistive device for human motion and health monitoring. *Proceedings of International conference on intelligent sustainable systems, p.p. 437-440.*

Martinez-Martin, E. and Costa, A., 2021. Assistive Technology for Elderly Care: An Overview. *In IEEE access*. Volume 9, pp.92420-92430.

Samsudin, S. et al., 2019. Effectiveness of Outdoor Education Program on Physical Education Student Resilience. *In International Journal of Innovation, Creativity and Change*. Volume 9, Issue 12, pp. 218-229.

Sharif, M.S. et al., 2018. An effective TeleHealth assistive system to support senior citizen at home or care-homes. 2018 *Proceedings of International Conference on Computing, Electronics & Communications Engineering, pp. 113-117, doi: 10.1109/iCCECOME.2018.8658877.*

Sumi, L. et al., 2019. IoT-based fall prevention and detection for senior citizens, physically and intellectually disable. *Proceedings of 2019 International Conference on Information Technology, p.p. 190-1141.*

Supasri, S., 2021. Effect of health promotion program on perceived self-efficacy and outcome expectations on self-care among elderly with hypertension, Chiang rai province, Thailand. *In Thai journal of public health*, 2021, Volume 51, Issue 1. p.p. 33-42.

World health organization, 2022. Ageing and health. [Online: 24 December 2022] https://www.who.int/news-room/fact-sheets/detail/ageing-and-health

Wungrath, J. et al., 2020. Effect of using basic elderly care handbook on knowledge and behavior regarding elderly care among caregivers of the elderly in Choeng Doi subdistrict, Doi Saket district, Chiang Mai province. *In Journal of Community Development and Life Quality*, Volume 8, Issue 1, p.p. 16-30.

Wungrath, J., 2021. Healthcare education process adopting the Line application in conjunction with tele-counseling to improve knowledge, behavior and satisfaction among elderly with diabetes mellitus during the COVID-19 pandemic. *In Psychology and education*, 2021, Volume 58, Issue 5, p.p. 6201- 6214.

OVERCOMING ZERO-TRUST ENVIRONMENTS FOR SMART CITY DATA AND DEVICES

Hebberly Ahatlan
Intertrust Technologies
www.intertrust.com
400 N McCarthy Blvd Suite 220, Milpitas, CA 95035, USA

ABSTRACT

Trusted interoperability for IoT devices is the backbone for today's smart cities. How can smart cities orchestrate disparate IoT networks into a unified secure working environment where data and devices can be fully trusted? We discuss the challenges and solutions to create or enhance trusted IoT and data fabrics in smart city ecosystems.

KEYWORDS

Device Interoperability, Data Security, IoT Orchestration

1. INTRODUCTION

Smart cities run on trusted data. Without it, many of the key advantages, fueled by a highly intricate web of interoperable systems, devices and data are either lost or can be compromised (Hamilton & D'Souza, 2022).

However, a significant number of IoT devices used in smart cities today are legacy devices. These devices were created before modern security practices and standards were widely adopted; hence, they often lack basic security features such as encryption and authentication (Gold, 2019). In some cases, these devices may no longer receive security updates or patches, making them vulnerable to attacks.

While modern IoT devices can be designed and manufactured with security in mind, they are only as strong as the weakest link in the data fabric they occupy. Traditional methods of protecting connected devices use connection-oriented protocols, like VPNs and TLS, and are all about securing the lines of transmission. In zero- or low-trust environments, this approach breaks down.

How can smart cities orchestrate disparate IoT networks into a unified secure working environment where data and devices can be fully trusted? In this paper, we discuss the challenges smart cities face while trying to implement infallible IoT security. We suggest a holistic IoT systems protection approach to securing IoT devices and the data they generate–from edge to cloud, in transit and at rest, and in any public, industrial, commercial, or private setting.

2. TRUST CHALLENGES IN IOT DEVICE INTEGRATION FOR SMART CITIES

Smart cities are complex systems of heterogeneous entities all of which need to interact and rely on one another (Ammara, et al., 2022). This vast, intertwined network of devices, applications, resources, and people are responsible for operating all manner of functions within the city.

Smart cities host heterogeneous IoT devices, such as traffic lights, parking meters, air quality sensors, and public safety cameras, which are supported by connectivity protocols, such as Wi-Fi, 5G, and low-power wireless networks like LoRa, ZigBee and Sigfox (Bauer et al., 2021). These entities represent a complex mix of manufacturers, platforms and formats optimized for fast data throughput, but not for protection of data in transit or at rest. Furthermore, legacy IoT fabrics go largely unauthenticated (Weinberg, 2021). Many data platforms are not set up to validate the integrity of the data they process–much less the authenticity of the devices the data came from.

Diverse IoT fabrics create hyperconnected environments where thousands of IoT devices send raw data where their trustworthiness cannot be assured because of the disparate technologies implemented by modern and legacy IoT devices (Thomas, et al., 2019). Furthermore, IoT meshes not only collect data, but bridge cyber-physical integrations where devices function as actuators to, for example, shut off grids, industrial robots, or cameras. Unprotected hyperconnectivity and vulnerable cyber-physical integrations are serious problems associated with poorly harmonized IoT networks composed of legacy and modern devices (Zetter, 2021).

2.1 Failure of Session-level Security

Session-level or endpoint security may be adequate to protect simple data and IoT fabrics, but what happens when data traverses multiple endpoints and is invoked by multiple sessions involving distinct communication protocols? Because there are no standard security protocols in place for IoT fabrics, session and endpoint security has a narrow protection scope. In addition, many IoT devices have limited computing power and memory, which makes it difficult to implement robust security mechanisms.

Vast numbers of legacy IoT devices - and even modern IoT devices, do not receive regular security updates, which leaves them vulnerable to new attacks. Moreover, some devices may not have the capability to receive updates at all. To make matters worse, in some cases, the data transmitted between IoT devices and servers may not be encrypted, making it vulnerable to interception and tampering. Data is not always protected in transit or at rest regardless of the number of networks or cloud ecosystems it traverses.

2.2 Lack of Data Interoperability

With a variety of IoT devices from different manufacturers, interoperability is a big challenge. Different devices may use different communication protocols, and it can be difficult to make them work together. Scaling cyber-physical networks is challenging, especially as different types of data or data commands may need to be processed in different ways.

2.3 Limited Network Connectivity

IoT devices can only connect to a network within a certain range, which can be a challenge when deploying devices in large or spread-out areas. Wireless signals used by IoT devices can be disrupted by physical objects or electromagnetic interference, leading to connectivity issues. Large volumes of IoT devices can put a strain on the network, leading to issues with connectivity and latency. IoT devices may use different communication protocols which can make it challenging to connect them to a common network.

This can require additional hardware or software solutions to enable interoperability. Stuxnet attacks on an IoT network demonstrate that isolating computers from using network security is totally inadequate, even if air gapped or intermittently connected through USB drives (Fruhlinger, 2022). NIST principles for a zero-trust architecture specify how this can be done (Rose, et al., 2020) but deploying NIST's specifications globally, with legacy devices in the mix is a real challenge.

2.4 Data Integrity Issues for Data Collection, Analysis and Sharing

Ensuring the integrity and accuracy of IoT data is essential. This is particularly challenging when attempting to assert data consistency, completeness and reliability in large networks. IoT devices generate data in different formats, making it challenging to integrate and process it into a unified data model that collects, stores and analyzes data in a scalable and flexible way.

3. SOLUTIONS FOR ESTABLISHING TRUSTED DEVICES AND DATA FOR SMART CITY INFRASTRUCTURES

Technical solutions for device deployment in the context of trusted smart city data and IoT networks can be broadly classified into three categories: hardware, software and cloud-based solutions.

- **Hardware solutions** include specialized security chips designed to be easily integrated into IoT devices.
- **Software solutions** include the development of network and device agnostic software applications and tools that simplify the process of deploying and managing IoT devices.
- **Cloud-based solutions** involve using cloud platforms and services to manage and harvest IoT data. These solutions typically provide a variety of services, including device management, data collection and analysis, and security features.

By applying standardized protocols to all three categories as well as testing devices thoroughly before deployment and developing clear documentation and training materials for device users, organizations can improve the efficiency and effectiveness of their data fabrics and IoT deployments.

3.1 Secure Systems Approach

Security protocols such as TLS often used to secure IoT devices and data in a smart city infrastructure are not able to completely secure complex networks of heterogeneous devices running on diverse platforms. A secure systems approach to IoT fabrics involves protecting availability of information, confidentiality, integrity, accountability and authorization through practices such as:

- *Authentication and Encryption*: Devices should be required to authenticate themselves before they are allowed to connect to the network. Data should be encrypted in transit and at rest to prevent unauthorized access.
- *Firmware and software updates:* Regular firmware and software updates can address vulnerabilities and improve security.
- *Security policies and procedures:* Organizations should establish security policies and procedures, and employees should be trained on how to follow them. This can help reduce the risk of human error and ensure that security best practices are followed.

Orchestrating trusted IoT networks within smart city infrastructures to mitigate issues arising from hyperconnectivity and cyber-physical integrations involves implementing robust security that not only protects the data channels, but the data packets themselves.

Technologies such as TLS can be greatly enhanced by protecting every packet of data entering or exiting an authenticated IoT system in addition to securing the data channel with canonical encryption schemes (Maher, 2022). It also involves enhancing existing technologies such as MQTT or MAC as well as tackling the challenges of comprehensive data security, data interoperability, device deployment, network connectivity, data collection and analysis.

3.2 Network Connectivity

There are several technical solutions to address the challenges of network connectivity in IoT-based smart city infrastructures, including:

- *Mesh networking*: Mesh networking is a topology where each device in the network is connected to several other devices. This helps in creating a self-healing network that is more reliable and can cover a larger area. However, meshes can be vulnerable unless they are protected with robust security hardware and software. It is in this context that trusted interoperability platforms become the kernel of well managed secure networks.
- *Broadband wired or wireless networks*: IoT devices rely on trusted interoperability platforms to integrate diverse connectivity modalities and protocols - each with its own set of benefits and vulnerabilities. It is critical to implement network and device agnostic trusted interoperability platforms for seamless device authentication and secure data sharing.

Overall, the choice of network connectivity solution depends on the specific requirements of the IoT deployment and the available resources. It is important to consider factors such as cost, coverage, bandwidth, power requirements and data security when selecting a network connectivity solution.

3.3 Data Collection, Analysis and Sharing

Smart city infrastructures depend not only on a mesh of IoT devices working within a secure VPN, but on packets of data moving to and from this VPN protected by a "digital bodyguard" that secures them no matter where they go (Kalima & Durand, 2023). Advanced data packet protection schemes are the new technologies that offer new methods of implementing a trusted multi-layered security approach that includes device authentication, enhanced data encryption, access control and regular security updates.

There are several approaches to address the challenges of data collection, analysis and sharing in IoT-based smart city infrastructures, including:

- *Edge Computing:* A distributed computing model that brings computation and data storage closer to the location where it is needed. In the context of smart city infrastructure, edge computing can help to reduce network traffic and latency by processing data locally on the devices, rather than sending it to a centralized server for processing. This can lead to faster response times, lower bandwidth requirements and more efficient use of resources. However, it is here where it is important to consider that IoT devices need to be properly authenticated to guarantee that their edge computing function is fully trusted.
- *Data Privacy and Security:* Ensuring the privacy and security of data collected from IoT devices and sensors is crucial in smart city infrastructure. Solutions such as encryption, authentication and access control can be used to protect the data from unauthorized access and to ensure the privacy of the citizens.
- *Data Visualization:* The use of graphical representations to make smart city data easier to understand and interpret. Visualizing and presenting complex data in an intuitive, trusted and accessible way enables city managers and other stakeholders to make informed decisions. Trusted interoperability platforms are the key to efficient data visualization across disparate data consumers.

4. CONCLUSION

A trusted interoperability infrastructure ensures that legacy and modern IoT devices can communicate and work together seamlessly. Interoperability can be achieved by using common communication protocols, such as enhanced MQTT or MAC that not only protect the data channels but also the data packets traversing through any networks or devices (Kalima & Durand, 2023). Furthermore, encryption and authentication mechanisms with digital signatures can be used to secure data transmissions and ensure the integrity of data and devices.

A holistic approach to these challenges comprises establishing integrated, trusted infrastructures, through synchronized collaborations among stakeholders and through well planned IoT fabric deployments. For efficient collaboration experiences across all data and device stakeholders, consider integrating a flexible interoperability trust stack that implements state of the art technologies that go beyond standard solutions, such as TLS or MAC where the trust and protection is agnostic to the network and device technologies and where data is protected beyond a session or endpoint, in zero-trust environments. Technologies such as XPN can enhance MQTT, MAC and TLS frameworks, thereby achieving a robust smart city zero-trust environment.

Addressing these key points through flexible interoperability trust stacks delivers environments where legacy and modern IoT devices can be integrated into trusted interoperability infrastructures for smart cities, enabling improved efficiency, enhanced safety and security, better citizen services, improved environmental sustainability and economic growth.

REFERENCES

Ammara, Umme, Rasheed, Khansa, Mansoor, Athar & Al-Fuqaha, Ala (2022, June) "Smart Cities from the Perspective of Systems." Researchgate.net. *https://www.researchgate.net/publication/361260619_Smart_Cities_from_the_Perspective_of_Systems*

Bauer, Martin, Sanchez, Luis & Song, JaeSeung (2021, June). "IoT-Enabled Smart Cities: Evolution and Outlook." National Library of Medicine. *https://www.ncbi.nlm.nih.gov/pmc/articles/PMC8271664/*

Fruhlinger, Josh (2022, August). "Stuxnet explained: The first known cyberweapon." csoonline.com *https://www.csoonline.com/article/3218104/stuxnet-explained-the-first-known-cyberweapon.html*

Gold, Jon (2019, July). "Report: Smart-city IoT isn't smart enough yet." Network World. *https://www.networkworld.com/article/3411561/report-smart-city-iot-isnt-smart-enough-yet.html*

Hamilton, Michael & D'Souza, Cedric B. (2022, February) "Technology of IoT Systems." UL Solutions. *https://collateral-library-production.s3.amazonaws.com/uploads/asset_file/attachment/44554/CS249800_-_Whitepaper_-_Technology_of_IoT_Systems_Final_Digital.pdf*

Kalima, Chris & Durand, Julian (2023, February). "How is XPN different from a VPN." Intertrust Technologies. *https://www.intertrust.com/platform/xpn-faq/#faq-item-6*

Maher, Dave (2022, June). "SPTIoTCoE fireside Chat." Intertrust Technologies. *https://www.youtube.com/watch?v=apwUjGm_okk*

Rose, Scott W., Borchert, Oliver, Mitchell, Stuart & Connelly, Sean (2020, August) "Zero Trust Architecture." Nist.gov. *https://www.nist.gov/publications/zero-trust-architecture*

Thomas, Ian, Kikuchi, Shinji, Baccelli, Emmanuel, Schleiser, Kaspar, Doerr, Joerg & Morgenstern, Andreas (2018, October). "Design and implementation of a platform for hyperconnected cyber physical systems." Sciencedirect.com. *https://www.sciencedirect.com/science/article/abs/pii/S2542660518300581*

Weinberg, Adam (2021, October). "Common IIoT & IoT Protocols and Their Security Flaws." First Point Magazine. *https://www.firstpoint-mg.com/blog/common-iiot-iot-protocolsand-their-security-flaws/*

Zetter, Kim (2022, April). "Israel May Have Destroyed Iranian Centrifuges Simply by Cutting Power." theintercept.com. https://theintercept.com/2021/04/13/iran-nuclear-natanz-israel/

TOWARDS EXPLAINABLE MACHINE LEARNING OPERATIONS (MLOPS)

Krzysztof Dzieciolowski and Ningsheng Zhao
Concordia University / Daesys. Montreal, Quebec, Canada

ABSTRACT

In the dynamically expanding AI applications, there is a need for an effective operational framework for managing ever increasing number of predictive models. A new field of Machine Learning Operations (MLOps) has recently emerged to address the issue of efficient operations and management of machine learning models. However, enterprises still struggle to develop effective MLOps systems due to lack of expertise and limited experience in AI operational needs. Attempts of an ad-hoc administration of ML algorithms have not been successful to deal with the challenges of fast proliferation of AI models. At the same time, existing concepts of MLOps frameworks don't adequately address an important question of accurate machine learning models diagnostics and explainability. In this reflection paper, we review commonly used MLOps approaches and explainability methods and suggest novel methods to address challenges of current model explainability methods. We discuss the need for university curriculum to address the issues facing management of AI models, their diagnostics and explainability.

KEYWORDS

Machine Learning Operations (MLOps), Model Explanations, Model Diagnostics, Importance Sampling

1. INTRODUCTION

Managing production of machine learning models has become challenging even for large organizations. To assist with this task a new field of Machine Learning Operations (MLOps) has emerged defined as a collaborative area of data scientists, data engineers and IT specialists. In most business and scientific applications, machine learning models are utilized repeatedly over time, to satisfy requirements for continuous usage of models' predictions. For example, the marketing department of a financial company may want to utilize machine learning models' predictions in its monthly campaigns to identify customers who are most likely to accept their credit card offer. These recurring predictions need to be updated, verified, and explained periodically with the ever-changing customer data. Given the need of recurring campaigns, there has been a requirement to maintain quality of models' predictions through automation of their updates, maintenance, diagnostics, and explanations.

A broad introduction to Machine Learning Engineering (MLE) of which MLOps is an integral part has been given by Burkov (2020). The author has distinguished nine principal stages of the machine learning project lifecycle: 1) goal definition, 2) data collection and preparation, 3) feature engineering, 4) model training, 5) model evaluation, 6) model deployment, 7) model serving, 8) model monitoring, and 9) model maintenance as shown in Figure 1. The project lifecycle stages 1) through 4) are concerned with data science model development process, while stages 6) through 9) are related to IT driven models' management and operations that can be construed as MLOps. However, such a definition of MLOps suffers from the narrow data engineering perspective and fails to recognize importance of data science needs for continuous machine learning models operations, diagnostics and explainability. Symeonidis et al (2022) provides a review of to-date, limited literature of the new MLOps field and suggests use of AutoML (Automated Machine Learning) retraining within MLOps framework so to obtain higher degree of models' accuracy. He also suggests that for high maturity MLOps systems, there is a need to complement models' monitoring with a capability for robust models explainability, though no further details are provided.

This paper has two main objectives: first, to provide suggestions for a modified design of a MLOps framework that includes capability to operate, accurately diagnose and explain ML models, and second, suggest new methods of explaining machine learning models that do not rely on an unrealistic assumption of feature independence. These two objectives are complementary and lead towards comprehensive and explainable MLOps architecture.

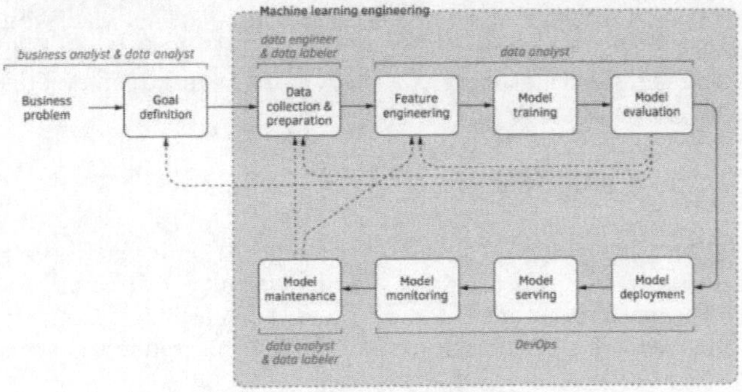

Figure 1. Machine Learning project life cycle

In section 2 challenges facing existing MLOps frameworks are reviewed, and enhancements are suggested. In section 3, a brief review of challenges of machine learning model explanations are provided and a novel approach to remedy untenable common assumption of uncorrelated features is provided. In section 4, we provide suggestions for further research in the area MLOps as well as model diagnostics and explainability.

2. IMPROVEMENTS TO MLOPS SYSTEMS

A growing need for MLOps systems can be illustrated by the fact that according to Forbes (2019) only 10% of machine learning models have been deployed into production. In addition, a single machine learning model deployment may take up to 3 months (VentureBeat, 2019). The growing use of AI is a challenge for most companies (Tech Republic, 2019). Traditionally MLOps systems have narrowly focused on solving the operational IT issues. On the other hand, we believe that the data science perspective of machine learning models diagnostics and explainability have not been so-far adequately addressed. We suggest that MLOps systems should contain an internal, robust capability for model quality monitoring and features explanations.

2.1 Classifier Rank

Typical machine learning model monitoring measures include metrics derived from the confusion matrix, such as F1, Receiver Operating Curve (ROC), Lift, Gain and Precision-Recall Curve (PRC) and others. However, these measures are sensitive to the imbalance rate, proportion of positive events. To remedy this problem, Zhao et al. (2022) suggested a new, robust measure of Classified Rank (CR) which provides the rank of the classifier in the space of all possible classifiers. Classifier Rank measure should be used for robust monitoring of models' quality in MLOps systems.

2.2 Model Explanations

Explaining machine learning models, often deemed as black box models, is an important practical problem.

Shapley values that have been derived from a game theoretic concept can be used to explain individual predictions and provide interpretation to a machine learning model. Aas et al. (2019) noted that Shapley value framework has desirable theoretical properties and can in principle handle any predictive model. Ribeiro et al. (2016) pointed out a question of trust (confidence) in a model. Therefore, monitoring explanations over time may reveal a potentially changing relationship between model predictions and features. MLOps systems that have capability to monitor model explanations may provide insights into their evolution. Both global and local model explanations should be used depending on the purpose of the model explanations. Fast developing applications of AI demand that universities adapt their curricula to prepare students for emerging technologies such MLOps systems and address interpretability of machine learning models to enable their more responsible applications.

3. A NEW APPROACH TO MODEL EXPLANATIONS

A popular way to explain machine learning models is understanding which features are important to the model output. Many possible approaches, such as L2X, LIME, RISE, etc., have been proposed to assign important scores to features, but Lundberg et al. (2017) shows that Shapley value attribution is the unique solution that satisfies certain desirable properties.

3.1 Shapley Value Explanation

Consider a machine learning model f that takes a set of features, denoted as $D = \{1,2,\dots,d\}$, as input and generates a prediction as output. We use the lowercase symbol $x = (x_1,\dots,x_d)$ to denote the values of the input vector and the uppercase symbol $X = (X_1,\dots,X_d)$ to denote the corresponding random variables measuring the input vector. Then, the Shapley value attribution of feature i can be calculated as

$$\phi_i(v) = \sum_{S\in D\setminus\{i\}} \frac{|S|!\,(|D|-|S|-1)!}{|D|!}\big(v(S\cup\{i\}) - v(S)\big),$$

where $v(S)\colon 2^d \to \mathbb{R}$ is a value function representing the model's output when only features in subset $S \subseteq D$ are considered. The Shapley value attribution captures the average marginal contribution, $v(S \cup \{i\}) - v(S)$, of feature i across all possible subsets of features that exclude i. To compute it, we must first specify the value function $v(S)$. However, the machine learning model f requires the input values $x = (x_1,\dots,x_d)$ of all features for prediction, so we cannot just remove features in $\bar{S} = D \setminus S$ and ask f to make prediction by only inputting a sub-vector x_S. There are two popular solutions: off-manifold and on-manifold value functions.

3.2 Proposed Importance Sampling Approach

To incorporate feature dependence into the Shapley value explanation, we propose the importance sampling method to approximate the on-manifold value function $v(S)$. Specifically, given $X_S = x_S$, we first draw a sample set $\{x_{\bar{S}}^{(1)}, x_{\bar{S}}^{(2)},\dots,x_{\bar{S}}^{(K)}\}$ from a known and easily sampled distribution $q(X_{\bar{S}}|X_S = x_S)$. Then, we can approximate $v(S)$ using the weighted average of $f\big(x_S, x_{\bar{S}}^{(k)}\big), k = 1,\dots,K$, over all samples, i.e.,

$$v(S) \approx \frac{1}{\sum_{k=1}^{K} w^{(k)}} \sum_{k=1}^{K} w^{(k)} f\big(x_S, x_{\bar{S}}^{(k)}\big), \text{ where } w^{(k)} = \frac{p\big(X=(x_S,x_{\bar{S}}^{(k)})\big)}{q\big(X_{\bar{S}} = x_{\bar{S}}\big|X_S = x_S^{(k)}\big)}.$$

We refer to $w^{(k)}$ as the *importance weight* of each sample $x_{\bar{S}}^{(k)}$, and trivial importance weights will be assigned to "impossible samples" with $p\big(X = (x_S, x_{\bar{S}}^{(k)})\big) \approx 0$, especially when it shows a high density value in distribution $q(X_{\bar{S}}|X_S = x_S)$. This helps prevent $v(S)$ from being estimated off the true data manifold, and thereby producing more informative model explanations. To compute $w^{(k)}$, we need to (1) train a density estimator $\hat{p}(X)$ to learn the true distribution $p(X)$, and (2) select the distribution $q(X_{\bar{S}}|X_S = x_S)$.

Training a density estimator $\hat{p}(X)$ is a common unsupervised learning problem. We need to choose and fit a state-of-the-art neural density estimator on the training or testing data. However, the selection of $q(X_{\bar{S}}|X_S = x_S)$ is a challenge, because it needs to be: (1) easy for sampling given $X_S = x_S$ for an arbitrary subset S; (2) fast to compute the conditional density value for samples; (3) as close to $p(X_{\bar{S}}|X_S = x_S)$ as possible to reduce the proportion of "impossible samples". In this work, we propose using a conditional mixture model for sampling. This approach assumes that the underlying distribution of interest contains J components, and within each component, all features are independent. In other words, for each component j, each feature i has its own marginal distribution $q_{ij}(X_i)$, such as Gaussian distribution. We further introduce a J-dimension multinomial random variable $Z \sim$ *multinomial(1, π)* as the indicator of component, where $\pi = (\pi_1,\dots,\pi_J)$ with each $\pi_j = P(Z_j = 1)$, which is referred to as the *component weights*. Then, the density function of the whole distribution can be written as $q(X) = \sum_{j=1}^{J} \pi_j \prod_{i=1}^{d} q_{ji}(X_i)$, and E-M algorithm can be used to estimate the unknown parameters π and each $q_{ij}(\cdot)$. Once the above mixture model is learned, given any subset of features $X_S = x_S$, we can easily compute the conditional distribution,

$$q(X_{\bar{S}}|X_S = x_S) = \sum_{j=1}^{J} P(Z_j = 1|X_S = x_S) \prod_{i \in \bar{S}} q_{ji}(X_i),$$

which can be referred to as the *conditional mixture model*, and where the posterior probability is given by,

$$q(Z_j = 1|X_S = x_S) = \frac{\pi_j \prod_{i \in S} q_{ji}(X_i = x_i)}{\sum_{j=1}^{J} \pi_j \prod_{i \in S} q_{ji}(X_i = x_i)}.$$

This conditional mixture model can be used for sampling as the following process: 1. Sample a $z \sim q(Z_j = 1|X_S = x_S)$, and return the corresponding j such that $z_j = 1$; 2. For each $i \in \bar{S}$, sample $x_i \sim q_{ij}(X_i)$.

The proposed importance sampling method has the following new contributions:

- It requires fitting two models $\hat{p}(X)$ and $q(X)$ only over the N samples of training data, rather than over $N \times 2^d$ samples, which are much easier to train and evaluate.
- It can well keep the estimation on the data manifold via a state-of-the-art neural density estimator $\hat{p}(X)$, such as Roundtrip, RealNVP and MAF.
- It can mitigate the data sparsity and the curse of dimensionality by re-sampling from a well-trained conditional mixture model.
- It is scalable to high dimension *d*, with computational complexity $O(KJd)$.

4. CONCLUSION

We have observed that current MLOps systems do not provide flexibility of managing, diagnosing, and explaining machine learning models. An expanded model diagnostics and explanation measures are needed to better understand model performance and its interpretability over time. A novel Classified Rank measure that considers data imbalance rate can be used to provide more robust and accurate assessment of model performance over time. We suggested a framework of importance sampling to overcome an unrealistic assumption of features' independence when explaining models. We point out the need to help university students understand the challenges facing organizations scaling adoption of machine learning models.

ACKNOWLEDGEMENT

The first author would like to acknowledge financial support from Concordia University Part-time Faculty Association CUPFA. The second author wants to acknowledge financial support from MITACS and Daesys.

REFERENCES

Burkov, Andriy., 2020. *Machine Learning Engineering.* True Positive. ISBN-13 978-1999579579.

Dinh, L., et al., 2016. Density estimation using real nvp. *ArXiv,* abs/1605.08803.

Lundberg, S. and Lee, S. 2017. A Unified Approach to Interpreting Model Predictions. *Proceedings of the 31st International Conference on Neural Information Processing Systems.* California, USA, pp. 4768–4777.

Masayoshi, M. et al., 2019. Explaining black box decisions by shapley cohort refinement. *ArXiv,* abs/1911.00467.

Papamakarios, G., et al., 2017. Masked autoregressive flow for density estimation. *ANIPS*, Vol. 30.

Ribeiro, M. T., et al. (2016). Why should I trust you? Explaining predictions of any classifer. *Proceeedings of te 22nd ACM SIGKDD International Conference on Knowledge Discovery and Data Mining (KDD). ACM.*

Sundararajan, M. and Najmi, A. 2020. The many Shapley values for model explanation. *Proceedings of the 37th International Conference on Machine Learning*. Virtual Event, pp. 9269–9278.

Symeonidis, A., et al. Definitions, Tools and Challenges. *IEEE Annual Computing and Communication Workshop and Conference (CCWC).*

Zhao, N., Yu, J.Y., Dzieciolowski, K. 2022. Classifier Rank – A new Classification Assesment Method. *7th International Conference on Big Data Analytics, Data Mining and Computational Intelligence (BIGDACI)*. Lisbon.

POSTERS

BLOOD PRESSURE PREDICTION SYSTEM BASED ARTIFICIAL INTELLIGENCE

Yonghee Lee and Hakjin Kim
Dept. of computer Engineering, Halla University
Wonju, South Korea

ABSTRACT

This paper is to implement the algorithm that estimates systolic and diastolic blood pressure through artificial intelligence learning by measuring optical blood flow signals. When the heart contracts, the blood that flows in through the pulmonary artery is supplied through the aorta to the peripheral blood vessels of organs and tissues that make up our body. Photo plethysmograph is obtained by measuring the amount of light absorbed according to the amount of blood supplied from the heart to ventricular contraction. It is a waveform that expresses the change in blood volume as a photoelectric signal and appears in proportion to the blood flow. PPG shows a close relationship with the activity of the heart. In proportion to the amount of oxygen in the blood, a photoelectric signal of red light according to the change in blood flow is detected. Based on the relationship between cardiac activity and PPG, blood pressure information can be obtained. The entire system consists of a PPG measurement module, signal processing, and artificial intelligence algorithm.

KEYWORDS

Blood Pressure, Artificial Intelligence, Photoplethysmograph, Prediction System, Continuous Monitoring

1. INTRODUCTION

In order to measure the PPG signal, it is measured through a sensor dedicated to photoplethysmography (Chandrasekhar, et al, 2020). Fluctuations in blood pressure are composed of the amount of blood flowing through the blood vessels and resistance components that impede the flow. The PPG signal shows similar changes in blood pressure, and shows periodic changes by repeating systole and diastole of blood vessels (Chandrasekhar, et al, 2020; Seung-Il Cho, et al., 2018). Figure 1 shows an example of PPG signal measurement.

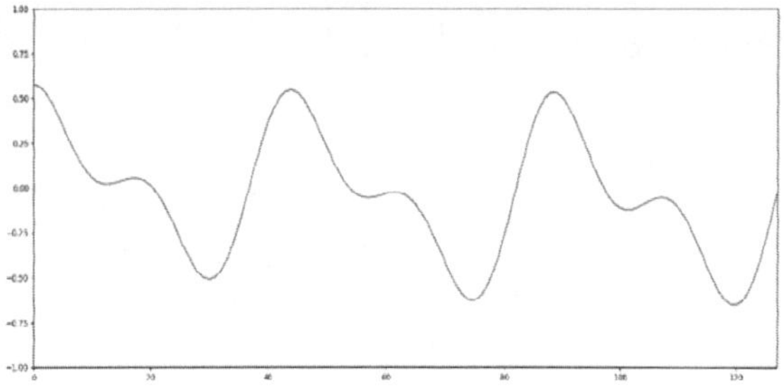

Figure. 1. PPG signal measurement example

The proposed method analyzes the correlation between PPG signals and blood pressure using deep learning artificial intelligence technology instead of statistical and mathematical methods, and predicts blood pressure by identifying meaningful patterns. In order to obtain a stable PPG signal, the signal from which breathing and motion artifacts are removed is used as an artificial intelligence learning and verification signal. In each frame, the average value within a certain level range was obtained, and frames showing rapid changes over a certain range were excluded from learning and recognition. This is because the statistical characteristics within the frame are not stable due to the DC component in the PPG.

2. ARTIFICIAL INTELLIGENCE MODELING

In this study, we develop an algorithm that estimates blood pressure using the PPG signal, which is closely related to the movement of the heart, and has systolic and diastolic blood pressure information, and weight and height signals. The anatomical structure of the heart and the contraction and relaxation activities of the heart appear in the PPG signal through blood flow and blood vessels. The PPG signal was directly used as an input variable in the artificial intelligence model, and weight and height, which are closely related to blood pressure, were added to the input variables. By configuring the system with the proposed method, the subject's height, weight, and PPG signals related to blood pressure were measured, and blood pressure was predicted in real time by learning using artificial intelligence. The artificial intelligence model consists of one input layer and one output layer, and three hidden layers.

3. CONCLUSION

The entire system allows you to check the most recently measured SpO2 average, the average diastolic and systolic blood pressure predicted by artificial intelligence, and the corresponding health status through the web. The accuracy of the proposed method is shown in Figure 2, and it showed 82% accuracy at convergence of 0.8254. It can be seen that the error rate of the data used for testing is slightly higher than that of the training data.

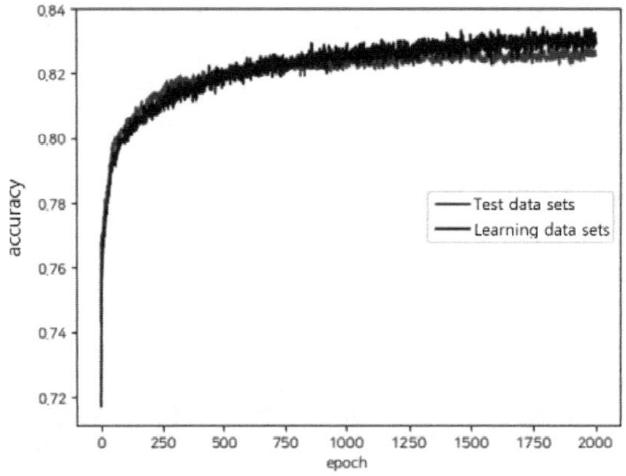

Figure 2. Accuracy comparison graph between test sets and learning sets

ACKNOWLEDGEMENT

This research was supported by the MSIT (Ministry of Science and ICT), Korea, under the ITRC (Information Technology Research Center) support program(IITP-2023-2018-0-01833) supervised by the IITP (Institute for Information & communications Technology Promotion).

REFERENCES

Chandrasekhar, et al., (2020). PPG Sensor Contact Pressure Should Be Taken Into Account for Cuff-Less Blood Pressure Measurement, *IEEE transactions on bio-medical engineering*, 67(11), 3134 – 3140.

Seung-Il Cho, et al., (2018). Estimation System of Blood Pressure Variation with Photoplethysmography Signals Using Multiple Regression Analysis and Neural Network, *International Journal of Fuzzy Logic and Intelligent Systems*, 18(4), 229-236.

CONSTRUCTING A STROKE DIAGNOSIS AND PROGNOSIS SYSTEM BASED ON THE BPN ALGORITHM USING TC-99M-ECD SPECT IMAGES

Jui-Jen Chen, Hung-Nien Chang Chien and Yen-Hsiang Chang
Department of Nuclear Medicine, Chang Gung Memorial Hospital, Kaohsiung Medical Center, Chang Gung University College of Medicine, 123, Ta Pei Road, Niao Sung Hsiang, Kaohsiung Hsien, Taiwan, Republic of China

ABSTRACT

Background: The technology of deep learning in artificial intelligence (AI) is increasingly applied to medical image recognition. These applications are focused mainly on the analysis of CT or MRI images and seldom intended to support research of nuclear medical images. SPECT is one of the few examination tools that can sensitively reflect abnormalities in the brain at an early stage. In addition to diagnosing cerebrovascular diseases, it can also be used as a tool for prognostic evaluation. Therefore, constructing a stroke image recognition system for diagnosis and prognosis is the goal of this study. Method: We collected all the Tc-99m ECD SPECT brain images from Kaohsiung Chang Gung Memorial Hospital over a period of five years from 2017 to 2021. A total of 144 medical records that met the ICD-10-Code I60-I69 cerebrovascular disease extraction rules were obtained. In the preprocessing of data, noise and defective images were removed. Data augmentation technology was exploited to avoid overfitting and underfitting due to the small amount of data and obtain more new images for higher generalization ability. The back-propagation neural network (BPN) algorithm was adopted to train stroke images and extract important features according to the distribution of blood flows in the brain. Result: The proposed model is compared with the VGG16 through transfer training. It delivers an accuracy of 94.4% (1.3% higher) and a sensitivity of 90.3% (5.4% higher). Its recall rate and F-score reach 90.3% and 94.2% respectively. The ROC curve and average AUC (0.89±0.08) indicate that this model has an excellent discrimination capability. Conclusion: Based on the strengths of the BPN algorithm, we construct a stroke image recognition system to support stroke image recognition, assessment of the possibility of a second stroke, and prognostic evaluation. This system can also assist physicians in performing a rapid diagnosis and reduce errors. It is hoped that the developed software can be ported to real-world medical systems for testing, so as to connect the theory with practical situations.

KEYWORDS

Stroke Diseases, Cerebral Perfusion Image, Transfer Learning, Back-Propagation Neural Network, Nuclear Medical

1. INTRODUCTION

Stroke is a common neurological disorder. It has been recognized as the leading cause of death and disability across the world. According to the 2017 and 2018 statistics on the top 10 causes of death in Taiwan released by Ministry of Health and Welfare (MOHW), cerebrovascular disease ranks fourth, claiming more than 10,000 lives each year on average [1], and the trend is rising year by year as shown in Figure 1. Stroke is scientifically referred to as a cerebrovascular disease. It is a condition in which sudden bleeding in the brain causes internal pressure, poor circulation or cell death due to ischemia, resulting in brain injury and neurological dysfunction, which can be severe and even fatal in the acute stage. Clinically, there are three types of stroke: cerebral infarction, cerebral hemorrhage and transient ischemic attack. Ischemic stroke induced by a blockage of the artery is more common, accounting for approximately three-fourth of all stroke cases, with bleeding-induced hemorrhagic stroke taking up the rest one-fourth. The most common diagnostic tools for stroke is single-photon emission computed tomography (SPECT) and drug used is Technetium-99m ethyl cysteinate dimer (ECD).

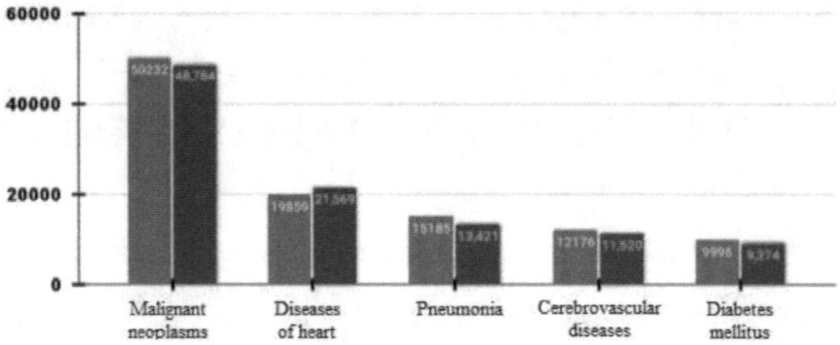

Figure 1. The study integration (data source：MOHW2018 ~2019year)

2. MATERIALS AND METHOD

In this study, the back-propagation neural network (BPN) algorithm is used. Feed-forward propagation is to transfer weighted data from the input layer to the hidden layer. Through the transfer function, the output value and error function corresponding to each neuron can be calculated. When the expected output cannot be obtained from the feed-forward output layer, the predicted output value and target output value are substituted into the error function. By means of the gradient descent method, the new weight can be obtained. This algorithm is known for backward propagation of errors (BP). Characterized by high accuracy and fast recall, BP is the most representative and widely used algorithm for training neural networks. The basic structure of a BPN is as illustrated in Figure 2 [2]. The data extraction procedure of the proposed system comprises three stages. In the first stage, it connects to the health information system (HIS) of the hospital to search for the medical records of the Department of Neurology and the Department of Neurosurgery. Based on the epidemiological distribution of the different types of strokes in Taiwan, the system collects the data of patients who underwent cerebral perfusion scan with SPECT (ECD examination item code N73-005) in the Department of Nuclear Medicine of our hospital. In the second stage, based on International Classification of Diseases (ICD-10-CM) and the criteria of the previous stage, patients whose disease meet any of the following code will be selected: I60 subarachnoid hemorrhage, I61 intracerebral hemorrhage, I63 cerebral infarction, I64, I66, I67, and I69 sequelae of cerebrovascular disease. I68 is cerebrovascular disorders in diseases classified elsewhere [3]. In the third stage, the system connects to the picture archiving and communication system (PACS) of the Department of Nuclear Medicine to collect the structured imaging reports for patients who meet the sampling criteria of the second stage. Finally, if the system's analysis result meet any of the structured extraction rules, the structured neurological medical reports and images will be stored in the examination report database for subsequent deep learning. More data is needed to ensure that training does not result in overfitting or underfitting. To make up for our lack of data, we exploit data augmentation technology within a reasonable range. The techniques of this technology include flipping, rotation, cropping, enlarge/lessen, horizontal flip, lean left/right.

3. RESULTS

In this study, we collect 144 sets of images, including (1) 72 sets of stroke images, and (2) 72 sets of normal images. Of these stroke patients, 45 are male (62.5%) and 27 are female (37.5%). Their average age is 65 years old. 48 patients have cerebral infarction or occlusion/stenosis of cerebral arteries, making up the largest proportion of the sample (66.7%). Cerebrovascular atherosclerosis or other cerebrovascular diseases account for 12.5% of the sample. Intracerebral hemorrhage and subarachnoid hemorrhage respectively constitute 6.9% and 1.4%. The images for machine learning are classified into two groups, image of stroke patient (labeled as 1) and image of non-stroke patient (labeled as 0). In machine learning, this is the method of supervised learning. Through use of labels in the analysis, the model can make predictions after training. The proposed system is designed to have two convolution layers and two pooling layers. Through binarization, all the images are converted from 3-dimensional color images into 2-dimensional grayscale images, with kernel size set as5x5 and filters respectively set as 32, 64, 128, and 256. To reduce generalization error, accelerate model convergence, and avoid internal covariate shift (ICS), we incorporate batch normalization into the operation. Excluding the dropout layer can reduce the training time by half. In nowadays, this has become a common approach. As to the activation function, ReLU is a common activation function for hidden layers. The pooling size is set as [2x2] or [3x3]. The parameters considered are very complicated.

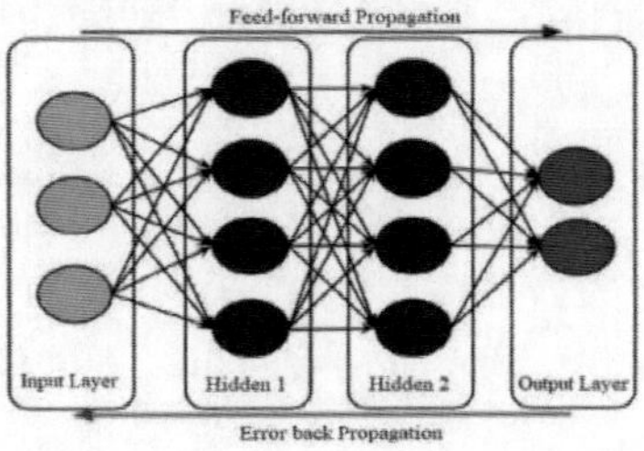

Figure 2. Back-propagation neural network

4. CONCLUSION

To address the limitation of the insufficient image data for research, we used the transfer learning technique to apply the pre-training weights of the VGG16 model to the Tc-99m ECD SPECT images in this study. VGG16 training has an accuracy rate of 93.1%, and the proposed BPN model achieves 94.4%. In comparison, the proposed model is not inferior at all. It delivers a 1.3% higher accuracy in image recognition, and 5.4% higher sensitivity. Under clinical requirements, the highest recall of 90.3% and the highest F-score of 94.2% are obtained. Although these statistics confirm the proposed model's differentiation ability, its learning curve is still not ideal, and its convergence is slower than VGG16. Unlike foreign studies that use huge data for training, this study relies on a small amount of data, which may be the cause of the high oscillation of the waveform. In our future studies, we will collect more data for training and consider the consumption of hardware resources required for processing huge data.

ACKNOWLEDGEMENT

Under Grant CMRPG8L1391 of Chang Gung Medical Research Program, this study was successfully conducted.

REFERENCES

Health Promotion Administration, Ministry of Health and Welf, 2021. Cerebrovascular disease, Available from: https://www.hpa.gov. tw/Pages/List.aspx?nodeid=213.

Jakub Nalepa, Michal Marcinkiewicz, and Michal Kawulok. 2019. Data Augmentation for Brain-Tumor Segmentation: A Review. Comput. Neurosci. Vol. 13, 1-18.

National Health Insurance Administration, Ministry of Health and Welfare, 2021. The Tenth Revision of the International Statistical Classification of Diseases and Related Health Problems, Ministry of Health and Welfare, Taiwan.

DOCTORAL

ADOPTION AND DIFFUSION PROCESSES OF TELEMEDICINE IN HEALTHCARE SYSTEMS

Yvonne Rauner and Harald Stummer
UMIT TIROL – Private University for Health Sciences and Health Technology Hall in Tirol, Austria

ABSTRACT

Background: Adoption and diffusion processes are integral parts of the dissemination of innovations; yet, they have been studied only fragmentarily in the healthcare systems field. This article outlines the goal of a PhD thesis in progress and describes how the dissemination of telemedicine in German health care will be analyze with a focus on innovation communication.
Methods: A three stage qualitative and quantitative research design consisting of expert interviews, focus groups and a partially standardized questionnaire consisting of demographic and sociocultural questions, the UTAUT2 on technology acceptance in the German translation by Harborth and Pape (2018) and questions on communication behavior will be used to investigate the research topic.
Results: Initial results of the exploratory literature review reveal that innovation communication between stakeholders regarding new technologies in healthcare systems has received little attention. Furthermore, the initial results indicate that adoption and diffusion theory can be used to analyze dissemination phenomena.
Conclusions: Based on the findings of the exploratory literature review, and considering Germany's struggle to implement telemedicine, an analysis of adoption and diffusion processes in the healthcare sector will be conducted.

KEYWORDS

Diffusion of Innovation (DOI) Theory, Telemedicine, eHealth, Innovation Communication, Public Health Information

1. INTRODUCTION

Despite the announcement of the Digital Healthcare Strategy (Bundesministerium für Digitales und Verkehr, 2022) and the creation of a legal framework for digital service provision, the German healthcare system lags behind many countries. According to the Digital Health Index, Germany's score of 30 is far behind Portugal's score of 67.2, Great Britain's score of 70 or Denmarks score of 72,5 (Thiel and Deimel, 2020). This index, which makes digitization strategies internationally comparable in healthcare sectors, indicates a need for improvement in Germany, despite many attempts at structural and system-related levels. The following article outlines the goal of a PhD thesis in progress.

As part of the *Digital Healthcare Strategy*, particular importance is attached to electronic patient records, electronic prescriptions, the expansion of the telematics infrastructure and the European Health Data Space (Bundesministerium für Digitales und Verkehr, 2022). In addition to these innovations, the expansion of telemedicine is intended to support and supplement the existing provision of services in the healthcare system.

The term telemedicine is defined by the National Library of Medicine as "the delivery of health services via remote telecommunications. This includes interactive consultative and diagnostic services" (National Library of Medicine, 1993). The terms *teleconsil*, *telediagnostics*, *teletherapy* and *telemonitoring* represent subcategories of telemedicine. In English-speaking countries, the terms *telemedicine*, *telehealth*, *telecare*, *eHealth*, *digital health*, *connected health*, *digital care* and *remote care* are used. Since 2014, alongside *telemedicine*, the term *mHealth* has been used. This refers to the concept of mobile health, which comprises the use of mobile telecommunication devices in the delivery of healthcare services (Hurst, 2016). In this paper the term *telemedicine* is used and includes the aspects of *mHealth.*

Telemedicine is a central characteristic of digitization in the healthcare sector. While particular attention is focused on the development and expansion of telematics infrastructure in Germany, other aspects, such as the (re)development of telemedical care structures, are also being targeted. Nevertheless, a sole focus on politics and legislative measures as the originator of implementation processes in society falls short. In implementation science, for example, a wide range of factors described in theoretical frameworks that influence implementation processes are relevant for innovations in the healthcare field.

According to Greenhalgh et al. (2004), an innovation in health care refers to a novel way of behaving and working that aims to improve the quality of care. Telemedicine can be understood as such an innovation. According to Gabriel, Stanley and Saunders (2017), an innovation can be divided into four main phases: (1) problem identification, (2) invention of new ideas, (3) adoption or implementation of new ideas into practice and (4) diffusion, (i.e., the spread and adoption of an idea more widely) (Cluley et al., 2022; Gabriel, Stanley and Saunders, 2017; Greenhalgh et al., 2004). Furthermore, an innovation can be categorized as one of three types: technological and clinical innovation, process and service innovation and system innovation (ibid.). In multidimensional health care, both the phased implementation process and the frequently overlapping types of innovation increase complexity. In the context of sector-based health systems, implementation is understood as the interplay between innovation and complex adaptive health systems (Greenhalgh and Abimbola, 2019; Plsek and Greenhalgh, 2001). This interplay, with its implicit determinants, will be examined from a sociocultural perspective based on adoption and diffusion theory, as this theory precedes the technical acceptance of innovations.

2. EXPLORATORY LITERATURE REVIEW

To determine the current use of theoretical concepts in the analysis of telemedical implementation projects, an explorative literature search was conducted. Due to the multidisciplinary research topic, databases from the social sciences, economics, health sciences and medicine were used for this purpose. Searches in Business Source, SSRN, CINAHL, CareLit, Wiso, Wiley Online Library and PubMed (Medline) were adapted to each database and included the following search terms: change management, innovation diffusion theory, innovation diffusion, implementation, barriers, eHealth, telemedicine, digital health and mHealth. The results of the database searches were checked for relevance using the abstracts, supplemented with the results of the freehand search.

2.1 Findings

In innovation and implementation science, there are three perspectives on the phenomenon of innovation implementation (Sepasgozar, Loosemore and Davis, 2016): a socioeconomic perspective based on Rogers' (1969) diffusion theory, a psychological perspective based on various frameworks of technology acceptance (Davis, 1989; Sepasgozar, Loosemore and Davis, 2016; Venkatesh et al., 2003) and an organizational-level perspective (Sepasgozar, Loosemore and Davis, 2016). However, these are not separable, as they can overlap depending on theory-based understandings. While adoption and diffusion theory, according to Rogers (2003), addresses individual (adoption-level) and societal (diffusion-level) implementation processes, acceptance research focuses on mental and behavioral determinants regarding individual products or technologies (Howard and Moore, 1982; Sepasgozar, Loosemore and Davis, 2016). Following Niknejad et al. (2021), who call for a precise distinction between the terms *adoption* and *acceptance*, acceptance research is assigned a secondary role in this context.

Another viewpoint on implementation research in health systems notes the distinction described by Scarbrough and Kyratsis (2022) between a diffusion and an implementation perspective. While the former examines the diffusion of an innovation in a social structure, the latter describes the implementation of an innovation in a particular context (ibid.). As understood by Scarbrough and Kyratsis (2022), the diffusion perspective illuminates the communication about an innovation in the social fabric while the innovation is still changeable (see also Rogers, 1969). The diffusion process is characterized by observation, imitation and influence among individual members of the social fabric and permeates a society when it reaches sufficient mass (Rogers, 2003; Scarbrough and Kyratsis, 2022). An analysis of innovation diffusion can include the types of information and communication channels used (Greenhalgh et al., 2004; Rogers, 2003; Scarbrough and Kyratsis, 2022).

To investigate the influence of awareness (knowledge) and demand (desire) concerning an innovation, Liang (2012) developed an awareness and demand matrix. This model demonstrated that the characteristics of an innovation and the desire for an innovation influence each other reciprocally (Liang, 2012). Thus, innovations are in tension between the technology push (introduction of innovative products to the market) and consumer pull (demand for the innovative products) (ibid.). The repetitive cycle of the technology push and consumer pull increases the awareness or knowledge of an innovation and thus the demand for it (Liang, 2012). This issue is relevant in regulated healthcare markets and requires further attention.

According to Jennett, Gagnon and Brandstadt (2005), the willingness to change and the perceived ability and necessity to change are significant factors in innovation implementation. These factors can be influenced by members of the social fabric, according to Armenakis, Harris and Mossholder (1993) and Romano (1995). As a subordinate factor of readiness to change, technology readiness should also be considered. According to Mauco, Scott and Mars (2018), technology readiness has been studied only to a limited extent in the context of health care. From a cultural perspective, it has not been studied at all (ibid.). However, the term *technology readiness* is not consistently used in the literature (Niknejad *et al.*, 2021; Vaittinen, Martinsuo and Ortt, 2017); therefore, this topic may have been studied using different terms.

Further research on user perception, perspective, experience, readiness and awareness in adoption was conducted in a literature review by Niknejad et al. (2021) and a cross-sectional study by Wei et al. (2020). Both papers highlight the relevance of perception, experience and readiness perspectives in the implementation of technological innovations in health care.

Moreover, a literature review by Dryden-Palmer, Parshuram and Berta (2020) underlines the centrality of context in innovation and knowledge implementation. Their study indicates that the complexity described by Rogers should be understood as a feature of the context, the implementation process and the innovation. Furthermore, this complexity represents a product of these factors (Dryden-Palmer, Parshuram and Berta, 2020). In addition to Dryden-Palmer, Parshuram and Berta (2020), Fitzgerald *et al.* (2002) also analyzed interactions in context.

2.2 Derived Research Topics

The literature review indicates that using established theories from implementation research can influence the diffusion of innovations and make implementation more effective (Wensing, Grol and Grimshaw, 2020). Despite these findings, previous research reveals few theoretical approaches to planning and designing implementation strategies (Lynch et al., 2018). Furthermore, according to Wensing and Grol (2019), the validity of these theories in the context of health care has been little researched and is thus limited.

In addition to general barriers and facilitating factors, the importance of perceptions or knowledge about an innovation, attitudes toward it and the prevalence of the combination of these elements in the social fabric have been highlighted (Jungwirth and Haluza, 2019; Liang, 2012; Lin and Bautista, 2017; Zolkepli and Kamarulzaman, 2015). The interaction between individuals' perceptions, attitudes and desire for innovation adoption in the social fabric has been identified as a significant diffusion factor (Jungwirth and Haluza, 2019; Wei et al., 2020). Despite the importance of interpersonal communication in the social fabric for innovation diffusion, as described in the literature (Fitzgerald et al., 2002; van de Ven et al., 1999), interprofessional, cross-sector communication about innovations (or aspects of innovation communication) in health systems has not yet been explored. Furthermore, the role of health professionals in the diffusion and implementation of innovations and their adoption is still unclear.

Another significant aspect of innovation diffusion and implementation represents the influence of sociocultural factors, which can operate differently at the organizational, professional and national levels (Heinzl and Leidner, 2012). The inclusion of cultural awareness in respective healthcare systems in analyzing the adoption and diffusion processes of innovations has not been the subject of research to date, nor has innovation communication.

Considering this research, the PhD project focuses on the descriptive analysis of telemedicine innovation as perceived by health professionals. The first study phase aims to describe possible sociocultural factors influencing the implementation of telemedicine from the perspective of experts in (tele)medicine (licensed physicians and/or telemedicine experts) in the German healthcare system. Based on the findings of the literature research, which indicate the relevance of sociocultural factors in the adoption and diffusion process of implementation, a restriction to the German healthcare system will be made.

The second study phase examines individual diffusion experiences of innovations in healthcare systems. For this purpose, licensed physicians with at least 1 year work experience concerning innovation communication in health systems will be studied. Here, the geographical limitation to Germany will allow the consideration of sociocultural and political characteristics.

Under the assumption that socio-cultural influencing factors and innovation communication in health care systems influence each other, a causal analysis will take place in the third phase within the framework of an explorative, partially standardized questionnaire survey. For this purpose, an analysis of the perceptions and attitudes of licensed physicians (like in phase 2) towards innovation communication is to be conducted. The question to be answered is whether sociocultural factors of health professionals contribute to innovation communication in health care systems and what effects this has on the perceived technology acceptance of innovations. The correlation hypotheses derived from this are:

ZH1: Sociocultural factors of health professionals contribute to innovation communication in health systems. (Undirected hypothesis)

ZH2: A high level of innovation communication among licensed physicians contributes to a higher perceived acceptance of innovations.

ZH3: A high level of perceived acceptance of innovations in health systems contributes to innovation communication among licensed physicians.

To answer the research question, a semi-standardized questionnaire consisting of demographic and sociocultural questions, the UTAUT2 on technology acceptance in the German translation by Harborth and Pape (2018) and questions on communication behavior will be used.

3. METHODS

The research questions will be answered using a two stage qualitative design with a subsequent quantitative design. For this purpose, guided expert interviews will be conducted in the first study phase. The number of expert interviews is based on the theoretical saturation to be achieved. The interview questions are based on the elements of adoption and diffusion theory, according to Rogers (1969). The interviews are conducted and evaluated according to the qualitative and quality criteria of Kuckartz (2018) in a systematic, rule- and theory-based manner.

Building on the findings from phase 1, moderated focus groups will be conducted in phase 2 on innovation communication in healthcare systems, using Germany as an example. The discussion topics are intended to revisit and deepen aspects of phase 1 and to extend the research topic to the aspect of innovation exchange in the social structure, according to Rogers (1969).

The medical and telemedical experts required to conduct the guideline-based interviews will be recruited from the professional environment and snowball sampling. Prerequisites for participation in the expert interview are a professional connection to medicine or telemedicine and professional activity in the last 12 months, the understanding of the German language in speech and writing as well as written consent to participate. At the beginning of the interview, an information sheet about the interview survey and a consent form for participation will be distributed. If desired, the interview can be conducted remotely via Zoom in compliance with applicable data protection conditions.

The focus groups will be conducted with the subjects of the expert interviews. An extension of the participant group is possible if inclusion criteria are fulfilled. Due to the pandemic, a remote execution is possible but not intended.

In the third study phase, a correlation analysis between socio-cultural factors, innovation communication and perceived acceptance will be carried out on the basis of the results from phases 1 and 2 and standardized questionnaires. The sample size is based on the total number of physicians practicing in Germany in 2021 (N=416100) (Bundesärztekammer, 2022). With a confidence interval of 95% and a standard deviation of 5%, the currently assumed sample size is 384.

4. CONCLUSION

Recent pandemic-related developments in health care have increased interest in implementing telemedicine care concepts. However, it is still unclear how telemedicine approaches can be implemented to improve the quality of care. Based on the findings of the literature review, it seems likely that the wishes, fears, attitudes and assumptions regarding an innovation, such as telemedicine, which have been only rudimentarily researched to date, can considerably influence the diffusion process. Likewise, the role health professionals play in the context of innovation communication in healthcare systems is still unclear. This dissertation project aims to contribute to answering these questions.

REFERENCES

Armenakis, A., Harris, S. and Mossholder, K. (1993) 'Creating Readiness for Organizational Change', Human Relations, 46, pp. 681–704. doi: 10.1177/001872679304600601

Bundesärztekammer (2022) Gesamtzahl der Ärzte in Deutschland im Zeitraum von 1990 bis 2021 (in 1.000). Available at: https://de.statista.com/statistik/daten/studie/158869/umfrage/anzahl-der-aerzte-in-deutschland-seit-1990/ (Accessed: 3 March 2023).

Bundesministerium für Digitales und Verkehr (2022) Digitalstrategie Deutschland: Gesundheit und Pflege. Available at: https://digitalstrategie-deutschland.de/gesundheit-und-pflege/ (Accessed: 1 January 2023).

Cluley, V. et al. (2022) 'Mapping the role of patient and public involvement during the different stages of healthcare innovation: A scoping review', Health Expectations, 25(3), pp. 840–855. doi: 10.1111/hex.13437

Davis, F. (1989) 'Perceived Usefulness, Perceived Ease of Use, and User Acceptance of Information Technology', MIS Quarterly, 13, p. 319. doi: 10.2307/249008

Dryden-Palmer, K.D., Parshuram, C.S. and Berta, W.B. (2020) 'Context, complexity and process in the implementation of evidence-based innovation: a realist informed review', BMC health services research, 20(1), pp. 1–15. doi: 10.1186/s12913-020-4935-y

Fitzgerald, L. et al. (2002) 'Interlocking interactions, the diffusion of innovations in health care', Human Relations, 55(12), pp. 1429–1449.

Gabriel, M., Stanley, I. and Saunders, T. (2017) Open innovation in health: A guide to transforming healthcare through collaboration. Available at: https://media.nesta.org.uk/documents/open_innovation_in_health_0.pdf (Accessed: 1 January 2023).

Greenhalgh, T. and Abimbola, S. (2019) 'The NASSS Framework – A Synthesis of Multiple Theories of Technology Implementation', Studies in health technology and informatics, 263, pp. 193–204. doi: 10.3233/SHTI190123

Greenhalgh, T. et al. (2004) 'Diffusion of Innovations in Service Organizations: Systematic Review and Recommendations', The Milbank Quarterly, 82(4), pp. 581–629. doi: 10.1111/j.0887-378X.2004.00325.x

Harborth, D. and Pape, S. (2018) 'German Translation of the Unified Theory of Acceptance and Use of Technology 2 (UTAUT2) Questionnaire', SSRN Electronic Journal. doi: 10.2139/ssrn.3147708

Heinzl, A. and Leidner, D.E. (2012) 'Informationssysteme und kulturelle Einflussfaktoren', WIRTSCHAFTSINFORMATIK, 54(3), pp. 103–104. doi: 10.1007/s11576-012-0319-1

Howard, J.A. and Moore, W.L. (eds.) (1982) Changes in consumer behavior over the product life cycle. Boston, Mass. (u.a.): Pitman.

Hurst, E.J. (2016) 'Evolutions in Telemedicine: From Smoke Signals to Mobile Health Solutions', Journal of Hospital Librarianship, 16(2), pp. 174–185. doi: 10.1080/15323269.2016.1150750

Jennett, P.A., Gagnon, M.P. and Brandstadt, H.K. (2005) 'Preparing for success: readiness models for rural telehealth', Journal of Postgraduate Medicine, 51(4), pp. 279–285.

Jungwirth, D. and Haluza, D. (2019) 'Information and communication technology and the future of healthcare: Results of a multi-scenario Delphi survey', Health Informatics Journal, 25(1), pp. 161–173. doi: 10.1177/1460458217704256

Kuckartz, U. (2018) Qualitative Inhaltsanalyse: Methoden, Praxis, Computerunterstützung. 4th edn. (Grundlagentexte Methoden). Weinheim: Beltz Verlagsgruppe; Ciando. Available at: https://ebooks.ciando.com/book/index.cfm/bok_id/2513416.

Liang, T.-H. (2012) 'Matrix analysis of the digital divide in eHealth services using awareness, want, and adoption gap', Journal of medical Internet research, 14(1), e11-e11. doi: 10.2196/jmir.1670

Lin, T.T.C. and Bautista, J.R. (2017) 'Understanding the Relationships between mHealth Apps' Characteristics, Trialability, and mHealth Literacy', Journal of Health Communication, 22(4), pp. 346–354. doi: 10.1080/10810730.2017.1296508

Lynch, E.A. et al. (2018) '"There is nothing so practical as a good theory": a pragmatic guide for selecting theoretical approaches for implementation projects', BMC Health Services Research, 18(1), p. 857. doi: 10.1186/s12913-018-3671-z

Mauco, K.L., Scott, R.E. and Mars, M. (2018) 'Critical analysis of e-health readiness assessment frameworks: suitability for application in developing countries', Journal of Telemedicine and Telecare, 24(2), pp. 110–117. doi: 10.1177/1357633X16686548

National Library of Medicine (1993) Telemedicine. Available at: https://www.ncbi.nlm.nih.gov/mesh/?term=telemedicine (Accessed: 1 January 2023).

Niknejad, N. et al. (2021) 'Understanding Telerehabilitation Technology to Evaluate Stakeholders' Adoption of Telerehabilitation Services: A Systematic Literature Review and Directions for Further Research', Archives of Physical Medicine and Rehabilitation, 102(7), pp. 1390–1403. doi: 10.1016/j.apmr.2020.12.014

Plsek, P.E. and Greenhalgh, T. (2001) 'Complexity science: The challenge of complexity in health care', BMJ (Clinical Research Ed.), 323(7313), pp. 625–628. doi: 10.1136/bmj.323.7313.625

Rogers, E.M. (1969) Diffusion of innovations. 7th edn. New York: Free Press.

Rogers, E.M. (2003) Diffusion of innovations. (Social science). New York: Free Press. Available at: http://www.loc.gov/catdir/bios/simon052/2003049022.html.

Romano, C.A. (1995) 'Predictors of nurse adoption of a computerized information system as an innovation', Medinfo. MEDINFO, 8 Pt 2, pp. 1335–1339.

Scarbrough, H. and Kyratsis, Y. (2022) 'From spreading to embedding innovation in health care: Implications for theory and practice', Health Care Management Review, 47(3), pp. 236–244. doi: 10.1097/HMR.0000000000000323

Sepasgozar, S.M.E., Loosemore, M. and Davis, S. (2016) 'Conceptualising information and equipment technology adoption in construction', Engineering, Construction and Architectural Management, 23, pp. 158–176.

Thiel, R. and Deimel, L. (2020) SmartHealthSystems: Digitalisierungsstrategien im internationalen Vergleich. online. Available at: https://www.bertelsmann-stiftung.de/de/publikationen/publikation/did/smarthealthsystems/ (Accessed: 23 September 2022).

Vaittinen, E., Martinsuo, M. and Ortt, R. (2017) 'Business customers' readiness to adopt manufacturer's new services', Journal of Service Theory and Practice, 28. doi: 10.1108/JSTP-03-2017-0053

van de Ven, A.H. et al. (1999) 'The innovation journey'.

Venkatesh, V. et al. (2003) 'User acceptance of information technology: Toward a unified view', MIS Quarterly, 27, pp. 425–478. doi: 10.2307/30036540

Wei, Y. et al. (2020) 'Physicians' perception toward non-invasive prenatal testing through the eye of the Rogers' diffusion of innovation theory in China', International journal of technology assessment in health care, 36(3), pp. 239–244. doi: 10.1017/S0266462320000136

Wensing, M. and Grol, R. (2019) 'Knowledge translation in health: how implementation science could contribute more', BMC Medicine, 17(1), p. 88. doi: 10.1186/s12916-019-1322-9.

Wensing, M., Grol, R. and Grimshaw, J. (eds.) (2020) Improving patient care: The implementation of change in health care. Hoboken, NJ: John Wiley & Sons.

Zolkepli, I.A. and Kamarulzaman, Y. (2015) 'Social media adoption: The role of media needs and innovation characteristics', Comput. Hum. Behav., 43, pp. 189–209.

AUTHOR INDEX